World Survey of Climatology Volume 3

GENERAL CLIMATOLOGY, 3

World Survey of Climatology

World Survey of Climatology Volume 3

General Climatology, 3

edited by

H. E. LANDSBERG

Division of Mathematical and
Physical Sciences and Engineering
University of Maryland
College Park, Md. (U.S.A.)

ELSEVIER SCIENTIFIC PUBLISHING COMPANY
Amsterdam-Oxford-New York 1981

ELSEVIER SCIENTIFIC PUBLISHING COMPANY
335 Jan van Galenstraat
P.O. Box 211, 1000 AE Amsterdam, The Netherlands

AMERICAN ELSEVIER PUBLISHING COMPANY, INC.
52 Vanderbilt Avenue
New York, New York 10017

Library of Congress Card Number 78-103353
ISBN 0-444-41776-1 (vol. 3)
ISBN 0-444-40734-0 (series)
With 190 illustrations and 101 tables

Printed in The Netherlands

World Survey of Climatology

Editor in Chief: H. E. LANDSBERG

Volume 1 General Climatology, 1
Editor: O. ESSENWANGER

Volume 2 General Climatology, 2
Editor: H. FLOHN

Volume 3 General Climatology, 3
Editor: H. E. LANDSBERG

Volume 4 Climate of the Free Atmosphere
Editor: D. F. REX

Volume 5 Climates of Northern and Western Europe
Editor: C. C. WALLÉN

Volume 6 Climates of Central and Southern Europe
Editor: C. C. WALLÉN

Volume 7 Climates of the Soviet Union
by P. E. LYDOLPH

Volume 8 Climates of Northern and Eastern Asia
Editor: H. ARAKAWA

Volume 9 Climates of Southern and Western Asia
Editors: H. ARAKAWA and K. TAKAHASHI

Volume 10 Climates of Africa
Editor: J. F. GRIFFITHS

Volume 11 Climates of North America
Editors: R. A. BRYSON and F. K. HARE

Volume 12 Climates of Central and South America
Editor: W. SCHWERDTFEGER

Volume 13 Climates of Australia and New Zealand
Editor: J. GENTILLI

Volume 14 Climates of the Polar Regions
Editor: S. ORVIG

Volume 15 Climates of the Oceans
Editor: H. VAN LOON

List of Contributors to this Volume

E. FLACH
Former Director of the
Physical-Meteorological Observatory
Davos (Switzerland)
(Now: P.O. Box 157, CH7270 Davos-Platz, Switzerland)

H. E. LANDSBERG
Institute for Physical Sciences and Technology
University of Maryland
College Park, MD 20742
(U.S.A.)

R. REIDAT
Etzelstrasse 22, D-2000 Hamburg 63
(Federal Republic of Germany)

A. Y. M. YAO
Environmental Data and Information Service
National Oceanographic and Atmospheric Administration
U.S. Department of Commerce
Washington, DC 20235
(U.S.A.)

Contents

Chapter 1. HUMAN BIOCLIMATOLOGY
by E. FLACH

Introduction 1
History and taxonomy of bioclimatology 2
Basic approach to bioclimatological aspects 5
Light and life 7
"Light" in geophysical and physiological phenomena 7
Extra-terrestrial solar radiation and illumination intensity, 7—The sense of light, 11
Climatology of solar and sky radiation, of global illumination intensity, and of circumglobal radiation 14
Global illumination intensity from a general climatological standpoint and in its relation to global radiation, 14—Climatology of global radiation T and its components S and D, 17—Bioclimatological significance of the individual components of solar radiation as well as of circumglobal radiation, 21—Changes of illumination intensity and their relation to reaction times for light impulses, 27
Taxonomy of energy fluxes active on the earth's surface and their bioclimatological significance 29
General taxonomy of radiation fluxes active on earth, 29—Structure of human skin in relation to influences of natural radiation, 30
Ultraviolet radiation 33
Ultraviolet radiation and its biological effects, 33—Climatology of ultraviolet radiation and climatic dosage, 35
Infrared radiation and radiation balance 39
Infrared radiation and its biological significance, 39—Atmospheric back radiation and heat radiation of the ground and human skin, 40—Radiation balance and its bioclimatological significance, 42
Further consequences of natural radiation fluxes for practical bioclimatology . . . 44
Radiation related to clothing, 44—Radiation on slanting surfaces and on surfaces facing different directions, 44—Special protective measures against heat and light radiation, 46—Special light effects of a physiological and psychological nature, 47
Radiation fluxes as aids in the study of atmospheric turbidity 49
Radiation measurement and air-hygienic orientation, 49—Application of measuring units of atmospheric turbidity and their significance for environmental protection, 51—Light scattering and water pollution, 54

Air and life 55
Basic aspects of atmospheric air 55
Gaseous composition of air near the ground. 57
Dependence upon altitude of atmospheric pressure and its climatic effects on man 58
Reduction of atmospheric pressure with altitude, oxygen partial pressure and its biological consequences, 58—Influence of altitude on sports, 64
Physiology of respiration 64
Atmospheric air as a colloid (aerosol). 66
Qualities of the atmospheric aerosol, 66—Aerosol assessment by nuclei and dust counts in different environments (large cities, medium and high mountains, forests, caves), 68—Results of pollen and spore counts, 71
Special features of atmospheric aerosol. 73
Size spectrum of atmospheric aerosol, 73—Meteorological influences of the constitution of atmospheric aerosol, 76—Electrical characteristics of natural aerosol, 78
Trace gases in the lower atmospheric layers (questions of environmental protection) 79
Origin of trace gases and their general air-hygienic significance, 79—Special meteorological and air-hygienic influences and analyses of aerosol foreign to the atmosphere, 83—Special trace substances in man's environment, 92—Radioactivity of air and its significance for the biosphere, 96
Temperature and life 101
Basic aspects of a synthesis of the complex terms temperature and life 101
General thermo-regulatory phenomena in man with regard to the terrestrial and thermal conditions 102
General aspects of heat production and mean tolerance limits 104
Processes of heat production and heat extraction in man. 106
General dependence and processes of heat production and heat emission, 106—Components of the inner and outer heat currents, 107—Evaluation of losses in the heat budget, and physical and chemical heat regulation, 111—Response of skin and issue temperatures and of skin humidity, 112—The process of perspiration in man under various climatic conditions, including its dependence upon work, 113—Heat and cold acclimatization, 114
Temperature sensations 117
General aspects of climatic measures, 117—Meteorological influences on skin temperatures, 118—Characterization of thermal environmental influences as well as special measuring techniques for determining "cooling intensity", or "cooling power" and "cooling temperature", 119—Sensation scales and difficulties in their interpretation, 127—Mathematical determination of cooling effects, 127—Complex explanation methods for the determination of the sensation climate of extreme climatic zones, 127—"Sultriness" and its common meteorological interpretation, 135
Climatological aspects of equivalent temperature in relation to "sultriness" and "comfort" 137
Equivalent temperature as a regular climato-physiological unit for the study of sultriness distributions (Central Europe) 146
Countryside-climatological behaviour of equivalent temperature and cooling power 148
Other methods for characterizing thermal sensations, in particular in connection

with physiological measures for physical performance and indoor climatological observations . . . 153
Protection against unfavourable climatic effects especially in purely thermal ranges 160
Bioclimatological evaluation systems . . . 163
Basic aspects of spatial evaluation and its scientific implication. . . . 163
Health resort climatology and its problems . . . 164
Methods of research in the thermo-hygric environment for bioclimatological requirements . . . 166
Seasonal differences in thermo-hygric behaviour, 166—Air mass climatology and its relevance to formation of water vapour aerosol, 168
Thermo-hygric environment as an aid for site evaluations based on aerosol, for the example of Alpine regions . . . 171
Further aspects on the application of aerosol climatology based on thermo-hygric factors . . . 175
Acknowledgement. . . . 177
References . . . 177

Chapter 2. AGRICULTURAL CLIMATOLOGY
by A. Y. M. YAO

Introduction . . . 189
Climatic factors affecting agriculture . . . 189
Radiation . . . 189
Incoming radiation, 190—Outgoing radiation, 190—The energy balance, 191—Spectral distribution of solar radiation, 192—Radiation and light distribution within the plant canopy, 194—Photosynthesis and climate, 197—Photoperiodism, 201
Temperature . . . 203
Air temperature, 203—Leaf and canopy temperature, 210
Moisture . . . 214
Precipitation, 214—Soil moisture, 214—Evapotranspiration, 217—Lysimetry, 224—Water requirement for plant growth, 225—Dew, 226—Fog, 227—Humidity, 227
Wind . . . 227
Wind effect on evapotranspiration, 227—Wind damage to plants, 228—Transportation of pollen, disease, and insects by wind, 228—Wind-profile near the ground, 228
Agricultural climate . . . 229
Crop climate . . . 230
Rice, 230—Wheat, 233—Corn, 238—Corron, 240—Soybeans, 243—Tomato, 245—Alfalfa, 248—Apple, 250 .
Forest climate . . . 251
Radiation, 252—Temperature and humidity, 253—Precipitation, 254—Dew and snow, 255—Wind, 255—Mathematical models, 256
Animal climate . . . 256
Temperature, 257—Radiation, 258—Air movement, 259—Humidity, 259—Animal breeding and housing, 260

Meteorological hazard and weather modification . . . 260
Radiation and heat . . . 260
Modification of radiation and heat near the ground, 260—Frost weather modification, 261
Moisture . . . 263
Cloud modification, 263—Hail suppression, 264—Drought alleviation, 264
Wind . . . 268
Windbreaks, 268
Weather and plant diseases . . . 270
Weather and pathogen . . . 271
The weather and the host . . . 272
Forecasting disease development . . . 272
Agricultural land utilization . . . 273
Climatic classification . . . 275
System analysis of land use in agriculture . . . 275
Remote sensing . . . 276
Remote sensing and agricultural climatology . . . 276
Remote sensing from satellites and agricultural meteorology . . . 277
Weather and crop yield and world distribution of productivity . . . 278
Yield model . . . 278
World distribution of productivity . . . 283
Acknowledgements . . . 283
References . . . 283

Chapter 3. CITY CLIMATE
by H. E. Landsberg

Introduction . . . 299
An urban model . . . 300
Thermal field of the city . . . 302
The wind field of the city . . . 313
Atmospheric pollution in cities . . . 317
Radiation, illumination, visibility . . . 321
Humidity, cloudiness, precipitation . . . 324
Conclusion . . . 329
References . . . 330

Chapter 4. TECHNICAL CLIMATOLOGY
by R. Reidat

Goal and scope of technical climatology . . . 335
The interrelation between technology and weather . . . 335
Engineering demands on climatologists . . . 336
Climatic aids for the engineers . . . 337

Fields of application for technical climatology 342
Indoor climate 342
Heating, 345—Cooling, 346
Application of climatology to building design 349
Wind, 350—The heat balance of buildings, 352—Moisture in building materials, 360—Rain, 363—Snow, 367—Fog deposit and glaze, 368—Weather construction, 369—Climatic aids for construction engineering, 370
Climatic conditions during transport 371
Climatic support for industry 373
Air pollution 377
Dust, 377—Waste gases, 378
References 381

REFERENCE INDEX 387

GEOGRAPHICAL INDEX 400

SUBJECT INDEX 404

Chapter 1

Human Bioclimatology

E. FLACH

Introduction

Bioclimatology is an interdisciplinary science with common boundaries to other sciences. The label, a synthesis of the terms "bios" and "clima", is fairly recent and dates from the first half of this century. This interdisciplinary field between geophysics and biology (in a general interpretation) is, however, one of the oldest of its kind. Since the beginning of history atmospheric influences on organic life have held a prominent place in the interests of mankind.

A purely historical review of the development of bioclimatology during the last millennia (questions of this nature are known to have been posed long before the time of Christ) would alone fill a voluminous monograph. Even more material could be published in a specific review of what has been accomplished so far and of what is yet to be achieved in such an extensive scientific field of research. In the present work only a short survey can be given of research results obtained by modern bioclimatology. Therefore, no claim is made for completeness.

Special attention must be drawn to the difficulties and limits in the solution of many individual bioclimatological problems. For centuries attempts have been made to explain all kinds of weather influences on biological activity which may result in meteorotropic reactions and illnesses. Among these aspects one finds, e.g., the far-reaching and economically not unimportant problem of *weather sensitivity*. Since time immemorial peoples and individuals have found it a fascinating subject. But despite laborious and expensive studies no valid interpretation has as yet been achieved. One must consider that within the range of meteorobiological activity intermediate and immediate as well as conscious and unconscious sensations and perceptions can be very closely related. Thus it is readily seen why so many questions remain unanswered.

Leaving purely meteorological discussions aside the present work deals with basic aspects of general *sensory physiology* which sets forth the mutual link between physical stimuli and physiological sensory perception. Furthermore, the senses possess—besides their objective qualities—secondary ones originating in the subjective world of the psychological experience. Hence sensory physiology must be considered on a dual psychophysical basis. This same dualism is revealed in bioclimatological studies on the various influences of weather, visibly expressed not only in relation to the body (discomfort, pain, etc.) but frequently also in the individual's psychic realm (depression, malaise, indecision, etc.).

In the explanation of purely meteorological causes difficulties are encountered when one tries to define and determine the various meteorological events which cause a gradation of stimuli in perception and experience. The meteorological approach to the problems in this specialized sector of bioclimatology does not simplify their solution, but tends towards the definition of meteorotropic reactions and illnesses by means of modern but to a certain extent *complex terms* (air masses, fronts, varied weather classifications). Therefore, the solution of bioclimatological problems was not supported by meteorology to the extent originally hoped for. The terminology of synoptic meteorology has been created from a large-scale view of weather and of its various development phases as a definition for atmospheric activity. Therefore, it does not accentuate details in the course of weather regarding either their temporal occurrence or their varying intensity. It thus replaces a fine analysis of the phenomena which should be the working method essential in the search for meteorobiological causal relations.

Methods promoting collaboration between biology (medicine) and meteorology (climatology) will be pointed out for cross-fertilization in the solution of bioclimatological questions. The preparation of basic meteorological material by easily comprehensible means will help anyone interested in understanding the objectives and activities of this interdisciplinary science.

In order not to extend unduly the range of purely factual statements only a very limited amount of instrumental data has been inserted here.

History and taxonomy of bioclimatology

Already in the pre-Christian era, for instance in the ancient Chinese empires, climatological characterizations had been established. They were even brought into connection with structural distinctions of the influence of climate and weather on organic life. During the Hellenistic period in ancient Greece Hippocrates (around 400 B.C.), the founder of scientific medicine, had already dealt with atmospheric influences on the human organism. These studies became guidelines for bioclimatological research. A few years later (around 350 B.C.) Aristoteles elaborated scientific interpretations of meteorological phenomena as such, which made him the founder of meteorology and climatology. During the Renaissance it was Leon Battista Alberti (Italy) who, at the end of the 16th century, in an extensive study on architecture discussed the influence of environment on construction. He included in his work informative bioclimatological methods regarding environmental hygiene, which have a very modern tone (Rodenwaldt, 1968). 300 years later the German naturalist Alexander von Humboldt (1769–1859), based on his vast treasure of experience, dealt with problems of climatic classification. A well-known example is his biologically orientated definition of the term "climate" (viz., "all changes in the atmosphere which perceptibly affect our senses").

The first half of the 20th century witnessed a world-wide intensification of climatological research as well as in biology and geography a rising interest for a more precise conception of atmospheric influences on environment. Subsequently increased endeavours were made to establish a practical *climate taxonomy*.

A large number of climatic classifications from the last hundred years (Knoch and Schulze, 1952) are based on large-scale subdivisions into different climatic zones. A distinguishing factor frequently employed was air temperature. Through knowledge gained on numerous fundamental climatic elements decisive progress was made in the already subtle subdivision of the earth's climates (KOEPPEN and GEIGER, 1936). The necessity, however, became more and more evident to meet the requirements of applied sciences by appropriate analytical methods. This was achieved for example by considering the earth's vegetable kingdom (forms of vegetation including ecological–physiological aspects), or by taking into account economic–geographic and hydrological requirements. In recent times *synoptic climatology* characterizing the succession and frequency of certain types of weather has gained in importance especially for bioclimatology (FLOHN, 1954). The same goes for the efforts made to create special climatological world maps showing elements whose grasping at a global level was up till now practically impossible, e.g., duration of sunshine, sun and sky radiation, etc. (LANDSBERG et al., 1963). Suggestions for a climatic taxonomy distinctly considering interests of human medicine are numerous (BERG, 1961). Their original intention was the demarkation of heat-physiological perception ranges that also include the determination of the earth's sultry zones (SCHARLAU, 1952). In order to achieve yet deeper knowledge efforts have been made in more recent times to establish detailed accounts also from bioclimatological standpoints on medical facts from various countries (JUSATZ, 1967).

Finally, experiments have been conducted which try to explain levels of human energy development, of civilization, of life spans, etc. by means of certain climatic characteristics (e.g., temperature threshold values) (HUNTINGTON, 1945; MISSENARD, 1949; LEE, 1958). In surveying works pertinent to this subject, DAMMANN (1951) points out the publications of MARKHAM (1947) and BROOKS (1950). MARKHAM delineates the zones of highest productivity and energy development by basing himself on a *comfort range* between 16° and 24°C and a relative humidity between 40 and 70% (whose maximum corresponds with a range between 30° and 55°C of equivalent temperature, t_e). Geo-medical efforts against the spreading of epidemics in the world are well-known and form an essential part of world-wide bioclimatological research (RODENWALDT and JUSATZ, 1952; MARTINI, 1955; DE RUDDER, 1960). In this context FLOHN (1961) pointed out that due to climatic changes of a secular nature as well as by man's encroachments (e.g., by further intensification of air pollution) which under circumstances may be irreversible, harmful divergences of a basically changed imprint are thrust upon human environments and living habits. In order to realize the recommendations made by the 1972 Stockholm U.N. Conference on the Environment, the 1973 Geneva Environmental Protection Conference in its "Programme des Nations Unies pour l'Environnement (PNUE)" demanded preventive measures to be taken also against damaging effects of climatic changes caused by man.

In order to clarify technical bioclimatological terms as to their objectives and goals, renowned representatives of this interdisciplinary science worked on their definition almost four decades ago: "Bioclimatology is an interdisciplinary science between physics and its applications on the one hand, and medicine and biology on the other. It deals with such conditions and forces of air and earth as influence living organisms." (LINKE, 1935) And: "Bioclimatology is the science of the influences of processes and conditions in the atmosphere on all forms of life." (DE RUDDER, 1938)

Both formulations include atmospheric occurrences in the form of various weather courses as well as stable climatic conditions. The above-mentioned definitions are therefore not limited merely to the influences of average atmospheric processes of the climate of one particular region or area. Furthermore, weather activity is consciously included in the basic definition as a quasi momentary condition over a determined area of the earth's surface. For this reason as well as due to the facilitated accessibility of many bioclimatic problems definitions have been set up which place the term "biometeorology" (occasionally "medico-meteorology" in particular contexts) as an overall definition over all types of atmospheric influences on organic life. One of these definitions goes as follows: "Biometeorology deals with the disclosure of meteorological fields in their relations to human, animal and plant life" (LEE, 1969). TROMP (1969) presents similar arguments. From a biological–medical point of view the tendency to use *exclusively* the term "biometeorology" instead of "bioclimatology" is a simplifying conception of this interdisciplinary science's *objectives*. Certainly the fact has to be considered that all biological systems (human, animal or plant), based on their individual constitutional receiving capacity and reactions, do not represent invariables. This is especially true under temporally conditioned meteorological influences. The long-term bioclimatological effects, however, demand first-rate attention in the scientific treatment of the ensuing problems. The important fact is the disclosure of average climate-conditioned behaviour patterns of organisms under a climate-conditioned control of physiological, pathological and psychic reactions.

Therefore, bioclimatology represents that field of research in an interdisciplinary science between biology and climatology (meteorology), which considers weather (as structural part of many climates) as well as the climate of a certain area, as the basic characteristics of an average atmospheric condition and its average biological influence (FLACH, 1957). Doubtlessly biometeorology forms an important part in the development and research of bioclimatology. In this context established definitions from climatology and human medicine deserve close attention. In the latter case special fields such as *acclimatization* and *climate therapy* demand interest. In a purely climatological relation, on the other hand, characterizations such as *local climate*, *health resort climate* (KNOCH, 1962) as well as *topo-climate* form parts of bioclimatology of a general nature and are not to be overlooked. From a climate–therapeutical aspect a certain significance is attributed less to momentary influences of a meteorological nature than to the very question of stress and adaptation during extended sojourns in foreign climates, e.g., long-term reactions to influences of unaccustomed climates. This acclimatization process, i.e., the ability of body and psyche for organic changes in a climatically alien environment, play a very important part. The objectives of *touristic medicine* (MOHRING, 1971) are most closely related to this subject. In world-wide travel and tourism as well as in steadily expanding development services all over the world there is a need for bioclimatological conditions which assure as little stress as possible on health and thus a salutary utilization of present possibilities and intentions. In this respect *preventive medicine* shows a particular interest in bioclimatological research and elimination of all possible climatic dangers. Most infectious diseases of tropical countries for example show accompanying idiotypical climatic symptoms. Many questions posed by environmental protection require not only meteorological but also climatological analyses.

Bioclimatology also has problems to solve regarding the distinctions between regional and district climates by local and topo-climatic influences (e.g., valley and slope climates, topoclimates in wooded areas and sea-coast zones).

This all leads to the taxonomy of bioclimatology given in Table I.

TABLE I

BIOCLIMATOLOGY (HUMAN BIOCLIMATOLOGY)

Meteorobiology (Biometeorology) Study of influences of weather on the organic world	*Climatobiology* Study of influences of different climates of particular basic character, including large-scale, regional and topo-climates
Main scientific and practical fields of application Human and veterinary medicine, botany, zoology, economy, technology	Agriculture, forestry, geography, ethnography, politics, traffic, environmental protection

Special fields of interest in human medicine (DE RUDDER, 1938)

(*1*) *Meteorological and climatological physiology*, especially in connection with the study of weather and climate sensations (cold, coolness, heat, sultriness, etc.)

(*2*) *Meteorological and climatological pathology* in connection with the *meteorotropic reaction* phenomena (weather feeling and weather sensitivity) and of *meteorotropic illnesses* due to weather occurrences and climatic conditions (e.g., such as accumulation of cases of influenza, apoplexies, heart attacks, laryngal group, meningitis, etc.)

(*3*) *Meteorological and climatological psychology* in connection with the study of psychic impressions and consequent influencing by weather and climate (landscape, light and olfactory sensations)

Specialized medical fields particularly interested in bioclimatology

Therapy with various objectives
Hygiene and sports medicine
Touristic medicine
Epidemiology
Balneology and health resort sciences

Basic approach to bioclimatological aspects

The principles outlined above must be observed in bioclimatological research programs. They require definition of elements of the surrounding atmosphere not only in terms of individual factors (such as temperature, oxygen contents of the air, etc.) but also in terms of *impact complexes* which are always at work.

The quality of atmospheric air as a mixture of permanent gaseous basic substances and as a colloidal system (aerosol) may almost be compared with an "accord group", for which reason it has been described as an "air-chemical impact complex". Similarly a "thermal impact complex" has been defined by influencing factors such as temperature, humidity, air displacement and heat radiation; furthermore, the "photoactinic impact complex" by the behaviour of visible and invisible radiation; the "meteorotropic impact complex" describing the influences of weather and climate on humans with consequent meteorotropic reactions or illnesses. Doubtlessly this procedure leads to a more precise definition of the complex world of activities in the atmosphere (FLACH, 1957).

On the other hand, a working system is conceivable which permits the consideration of bioclimatic influence forms with meteorological and geophysical causes under the primary aspect of their *typically specific effects*. Thus it appears as an advantage if the environmental systems to be investigated consist of ordered thoughts which present also the person less familiar with bioclimatic problems with a perfectly clear picture of the objectives and activities of this interdisciplinary science. Well-aimed key words from the

general wake of life and from the bioclimatological sphere as well are therefore practical expressive indications. The following bioclimatological environmental systems are explained on this basis and shall not be subjected to rigorous priorities (FLACH, 1968).
(*1*) *Light and life* (pp.7–55). In this section radiation fluxes active within the earth's atmosphere, in form of "visible light", of purely caloric character, and of the photochemical realm, are discussed. "Light and life" describes large-scale relations which connect these terms (PORTMANN, 1963).
(*2*) *Air and life* (pp.55–101). Included in this characterization is the entire quality behaviour of atmospheric air. Besides weather and orographically determined air pressures, effective factors of atmospheric aerosol including foreign constituents and the resulting variety of air hygienic situations play an outstanding role. Atmospheric water, in relation to a changeable atmospheric aerosol, takes its logical place as a quantitatively and qualitatively highly variable constituent of air, especially in liquid form.
(*3*) *Temperature and life* (pp.101–163). Air temperature plays an important part, but at the same time its limited range of significance and influence must be pointed out. It is indisputable that for the characterization of heat-physiological environmental conditions this element alone is not sufficient. The water vapour content of the air, air movements (wind) with their partial characteristics of turbulence and vertical exchange and, last but not least, heat radiation (sun, sky, reflecting radiation from ground and other surroundings) constitute factors of momentous importance beside air temperature. According to their interaction they regulate all kinds of heat sensations in a varying measure. Nevertheless, it is sensible to keep the term "air temperature" in the foreground of bioclimatological environmental characterizations, where its importance is often explained by its causal meteorological participation in partial phenomena of atmospheric humidity and wind conditions.
(*4*) *Biological phenomena influenced by weather and climate.* This system deals with the influence of atmospheric activity on biological occurrences in general. "Meteorobiology" as a research branch of bioclimatology forms the center of this field of interest. It comprises the positive and negative effects of weather on the entire biological experience complex of human beings. Closely related are bioclimatological considerations of the weather and climate-controlled daily and annual periodical phenomena (biorhythms). This special aspect has been subject to much investigation in recent years but will not be reviewed in detail here but will be dealt with in another monographic study elsewhere.
(*5*) *Bioclimatological evaluation systems* (pp.163–177). Difficulties and gaps in national evaluation systems are discussed following international efforts made within environmental research programmes. The climatology of control factors for atmospheric aerosols (temperature stratification, humidity and wind) is considered essential for the promotion of bioclimatological regional evaluation.

Light and life

"Light" in geophysical and physiological phenomena

Extra-terrestrial solar radiation and illumination intensity

Light as perceived by the human eye in its range of light sensitivity (Fig.1) physically signifies the radiation energy of the spectrum of electromagnetic oscillations in the range between $4 \cdot 10^{14}$ and $7.7 \cdot 10^{14}$ Hz. (For an explanation of terms and units see the Notation.)

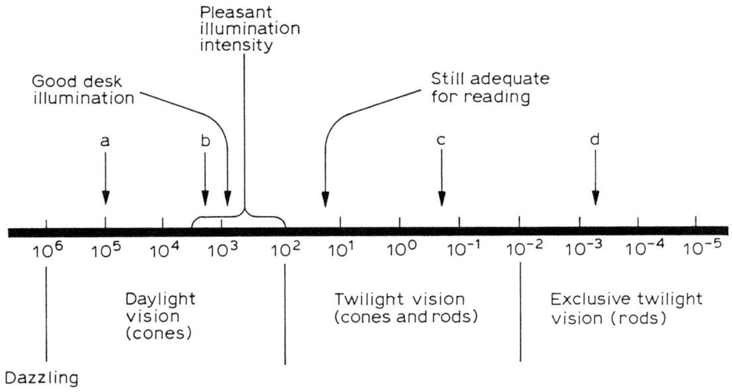

Fig.1. Illumination intensity range of the human eye (schematically, diagram by KEIDEL, 1971). Scale: lighting intensity in lux (illumination of a horizontal surface); a = by midday sun in summer; b = under clear sky at sunset; c = by full moon; d = in a clear starlit night.

Fig.2 demonstrates that light forms only a very small frequency or wavelength range in the known electromagnetic spectrum. Of the whole wavelength scale the narrow "optic" range amounts to less than the 10^{-20}th part. Judged by his innate electromagnetic antennas—the eyes—man is thus practically blind. This narrow band visible to the human eye is flanked by two areas of particular interest to bioclimatology. On the side towards the smaller frequencies the comparatively wide band of "infrared" radiation reaches from 10^{11} to $4 \cdot 10^{14}$ Hz. The smaller section immediately adjoining on the other side towards the larger frequencies in the electromagnetic spectrum—"ultraviolet" radiation—lies between $7.7 \cdot 10^{14}$ and $2 \cdot 10^{16}$ Hz. These two wavelength ranges are not perceivable by the human eye (cf. also p.11).

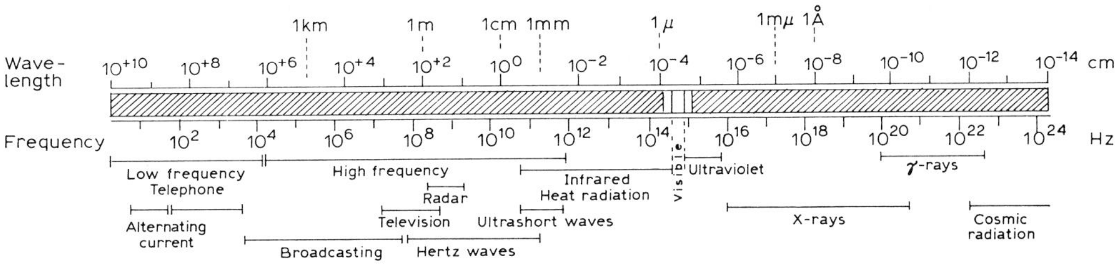

Fig.2. Spectrum of electromagnetic oscillations, indicated in wave length and frequency (TRENDELENBURG, 1961).

NOTATION

GLOSSARY OF GEOPHYSICAL AND LIGHT-TECHNOLOGICAL TERMS FOR THE MEASUREMENT OF RADIATION AND LIGHT

Radiation signifies the emission or transmission of energy in form of electromagnetic waves or corpuscles

Radiation energy I is measured in cal. cm^{-2} min^{-1} or in ly min^{-1} or in Joule (1 cal. = 4.186 J) cm^{-2} min^{-1}

1 cal. cm^{-2}min^{-1} = 73 klux	1 kWh m^{-2} = 105 klux
1 cal. cm^{-2}h^{-1} = 1.22 kluxh h^{-1}	1 kWh m^{-2}h^{-1} = 105 kluxh h^{-1} (105 klux)
1 cal. cm^{-2}d^{-1} = 1.22 kluxh d^{-1}	1 kWh m^{-2}d^{-1} = 105 kluxh d^{-1}

Solar constant I_0 (in the mean sun–earth distance)

I_0 = 2 cal. cm^{-2}min^{-1}
= 1.39 kW m^{-2} = 0.139 W cm^{-2}
= 1,200 kcal. m^{-2}h^{-1} = 0.033 cal. cm^{-2}sec^{-1}

I = radiation intensity on a receiving surface vertical to solar radiation, in or below the earth's atmosphere

S = solar radiation on a flat horizontal receiving surface in or below the earth's atmosphere = $I \sin h$ (h = solar altitude) = $I \cos z$ (z = zenith distance)

D = intensity of the diffusely scattered solar radiation on a horizontal flat receiving surface = *sky radiation*

T = intensity of direct and scattered solar radiation on a horizontal and flat receiving surface = $S + D$ (= *global radiation*)

CGR = intensity of radiation fluxes of S, D and the radiation reflected diffusely by the ground ($S + D$) = R_E reaching a spherical receiving surface = *circumglobal radiation*

G = atmospheric radiation = intensity of the diffuse radiation emitted by the atmosphere to the earth's surface (measured with a flat horizontal receiving surface)

E = *long-wave radiation of the ground* (not to be separated from reflected atmospheric radiation)

Light = characteristic of all perceptions and sensations pertinent to the organ of vision and transmitted by the same. Conversion of caloric units into light-technological units (cf. *Radiation energy*)

Spectral brightness perception degree = quotient from a radiation flux at wavelength λ_m and a radiation flux at wavelength λ (brightness-adapted eye)

$$V\lambda = \frac{\Phi\lambda_m}{\Phi\lambda}$$

Radiation flux = output emitted, transmitted or received in form of radiation (unit = 1 lumen (1 lm) = Φ = *light flux*)

Light intensity (= I_L) of a light source in a given direction is calculated from the quotient of the light flux and the space angle (ω) into which it radiates ($I_L = \Phi:\omega$; unit = candela (cd); formerly Hefner candle: 1 HK = 0.92 cd)

Illumination intensity (B) = quotient from light flux (Φ) and surface F at vertical incidence of rays (unit = lux = 1 lm m^{-2}; $B = \Phi:F$)

Exposure = product of illumination intensity (B) and duration of its influence (lux × sec)

Brightness density (L) is the physiological basic unit = quotient from light intensity I_L and the radiating surface F (unit = Nit (nt), formerly Stilb (1 Stilb = 10^4 Nit); 1 Stilb (sb) = 10'000 cd m^{-2} = 31'416 Apostilb (asb))

Data on illumination by natural light sources (according to TRENDELENBURG, 1961)
A horizontal surface is illuminated with:

approx. 10^5 lux	by noon-time sun in summer
approx. 2·10^3 lux	at sunset under clear sky
approx. 0.2 lux	at full moon
approx. 0.0005 lux	in a clear starlit night

Sufficient illumination requires approx. a 20 lux on upwards

Good desk lighting requires approx. 800 lux

Pleasant lighting ranges from 100 lux to approx. 3,000 lux (brightness density of the concerned space is decisive).

At 30 nt pure cone vision is possible, below 3·10^{-3}nt (= 0.01 asb) pure night vision sets in.

Natural radiation sources existing beyond and within the earth's atmosphere are responsible for the actual "seeing" process induced by the human sense of vision. In the former case knowledge of the qualities of extra-terrestrial solar radiation is of primary importance.

Fig.3 compares the spectral distribution of the energy of extra-terrestrial solar radiation and that of a "black radiator" of 6,000°K (photosphere of the sun = 5,800°K). The energy maximum of both "bodies" lies at 0.61 mμ. Furthermore, it may be deducted that in the infrared sector the values of extra-terrestrial solar radiation are very close to those of the "black body's" energy distribution. In the ultraviolet section of the spectrum, however, prominent differences are obvious. This is due to the fact that the sun is not a "black radiator".

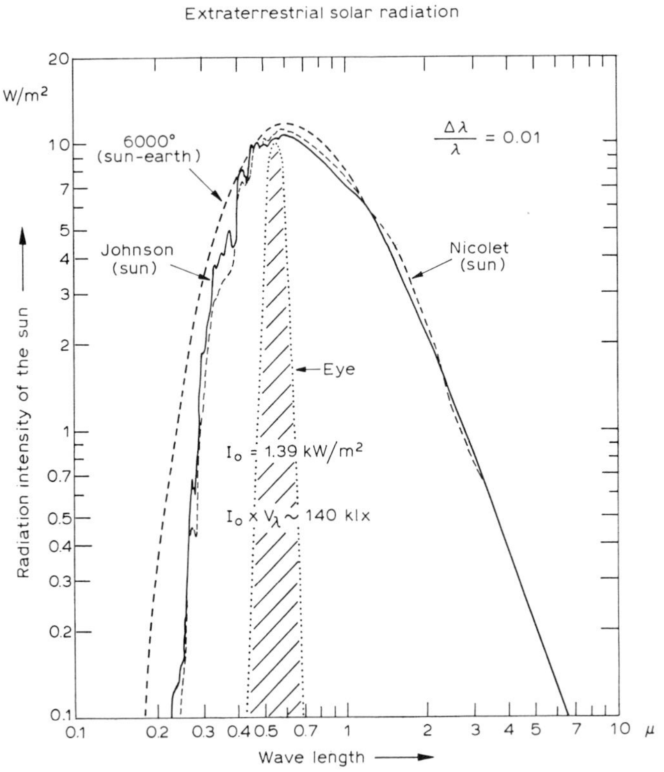

Fig.3. Spectral distribution of the energy of extra-terrestrial solar radiation (radiation intensity in W m^{-2}) and the brightness sensitivity V of the human eye (in relative units) by Johnson and Nicolet (after SCHULZE, 1970).

The energy distribution of extra-terrestrial solar radiation $I_0 = 1.39$ kW m^{-2} (mean value of "solar constant" valid at the moment) in the mentioned three main spectral ranges is of importance to bioclimatology. We have in the:

ultraviolet range ($<$0.4 mμ) approx. 9%
visible range (0.40–0.78 mμ) approx. 48%
infrared range ($>$0.78 mμ) approx. 43%.

The well-known subdivision into the bioclimatically important spectral ranges (cf. also p.10) shows the following percentual values for extra-terrestrial solar energy (SCHULZE, 1970):

UV-C (0.10–0.28 mμ) 0.5%   *UV-B* (0.28–0.315 mμ) 1.5%   *UV-A* (0.315–0.40 mμ) 7.0%

Visible spectral range (0.40–0.78 mμ) 49.8%

IR-A (0.78–1.4 mμ) 29.4%   *IR-B* (1.4–3 mμ) 11.5%   *IR-C* (3–1000 mμ) 2.0%

The main energy discharge of extra-terrestrial solar radiation therefore lies in the spectral range between 0.3 and 3 mμ.

Extra-terrestrial illumination intensity of the sun is of particular interest in the present work and is calculated from the product of radiation intensity I_0 and the values of the effect curve of the human eye (= spectral eye sensitivity V_λ). Thus the value $I_0 = V_\lambda =$ 140 klx $= V_0$ is computed. By means of quotient $V_0 : I_0$ the efficiency value of extra-terrestrial solar radiation for human vision is calculated. Its average for every day lies at 105 lm/W. This value is attained light-technologically for the production of artificial radiators.

Fig.3 gives a further illustration of basic bioclimatological conditions for the explanation of light's existence on the planet Earth in form of a spectral sensitivity curve of the light-adapted eye. The narrow band of light yield at the disposition of the human eye is particularly impressive. Contrary to the spectral distribution of solar energy the eye's maximum brightness sensitivity lies at 555 mμ (details in Table III).

Table II surveys the latitude-dependence of extra-terrestrial solar radiation for the summer and winter semester as well as for the whole year and also supplies the corre-

TABLE II

RADIATION AMOUNTS FOR EXTRA-TERRESTRIAL SOLAR RADIATION S_0 IN 1,000 kWh m^{-2} AND RELATIVE RADIATION T_r (%) IN DEPENDENCE UPON NORTHERN LATITUDE
(After SCHULZE, 1970)

Latitude	Summer semester		Winter semester		Year	
	S_0 extra-terrestrial	T_r relative irradiation (%)	S_0 extra-terrestrial	T_r relative irradiation (%)	S_0 extra-terrestrial	T_r relative irradiation (%)
N Pole	1.55	62	—	—	1.55	62
80°	1.56	63	0.038	50	1.60	63
70°	1.61	66	0.156	60	1.77	66
60°	1.73	70	0.39	64	2.12	69
50°	1.87	73	0.68	68	2.55	72
40°	1.97	76	0.98	70	2.95	74
30°	2.03	77	1.25	73	3.28	76
20°	2.03	78	1.50	75	3.53	77
10°	1.98	78	1.70	77	3.68	78
Equator	1.87	78	1.87	78	3.74	78

sponding indications on "relative global radiation" T_r (in %). The latter is defined by the quotient:

$$\frac{\text{measured global radiation}}{\text{extra-terrestrial radiation}}$$

in consequence: $T_r\,(\%) = T/S_0 \times 100$.

Since S_0 on its way through the atmosphere is attenuated by diffusion on air molecules and aerosols as well as by partial absorption by H_2O, O_3 etc., relative global radiation comprises all influencing dimensions of this attenuation. T_r in Table II refers to cloudless conditions and therefore describes exclusively the effects of specific aerosols on the decimation of extra-terrestrial radiation within the earth's atmosphere. As regards the latitude-dependent behaviour of T_r it appears that in the Equatorial range it is hardly subject to fluctuations at any time of year. Towards the poles the annual amplitude of T_r increases with the change of the solar position. Furthermore, T_r decreases with higher latitudes, especially perceptible in the winter semester.

The sense of light

A few basic facts on the physiology of vision will be useful in the context of natural illumination.

All our sensory organs are the body's "feelers" for its surroundings. They are the only immediate connecting passages between environmental phenomena and the nervous system. Therefore, "seeing" means the reception of information for conscious or unconscious orientation in, and observation of, space. Besides the image, brightness and colour vision of the single eye the sensory coupling of both eyes furthermore enables spacial and stereoscopic vision.

The human eye possesses two separate seeing systems, i.e. one for great light intensities, the *cones*, and one for small light intensities, the *rods*. The cones are therefore used mainly for daylight vision, whereas the rods' function is primarily dusk vision. Both organs are receivers in the human eye's retina. In twilight conditions cones and rods cooperate. Furthermore, the human eye is able to focus certain substances (pigments) at various concentrations, thus having the possibility of distinguishing between great differences in brightness (light and dark adaptation) (Fig. 4).

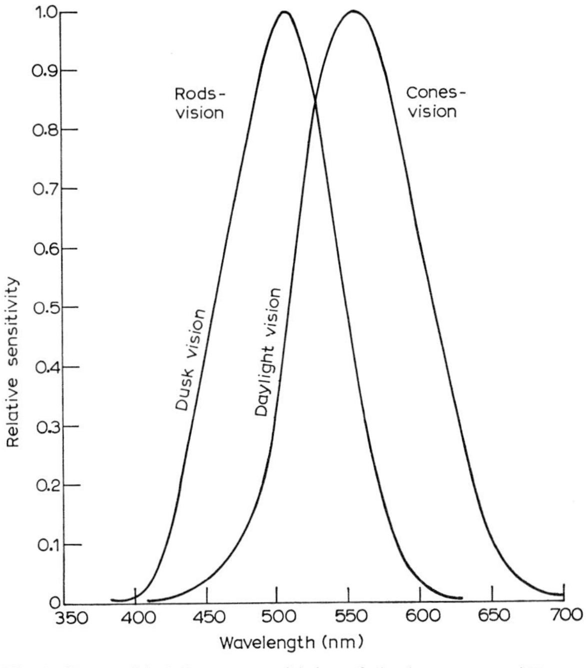

Fig.4. Spectral brightness sensitivity of the human eye (V_λ = curve) (HINZPETER, 1971).

The maximum sensitivity of the light-adapted human eye lies at 555 mμ, while that of the darkness-adapted eye is shifted towards the shorter wavelengths and is found around 507 mμ (cf. Fig.6). Herein the particular feature of the human eye is recognized, i.e., its spectral sensitivity curves lie in the main part of the energy range emitted by the sun. In both cases the curves undergo rapid decline from the maxima to the corresponding minimal values. The minimum darkness adaptation on the ultraviolet side is found at approximately 390 mμ, that of light adaptation on the infrared side around 700 mμ. The exact figures of the human eye's spectral sensitivity (at light-adaptation) are set up according to the CIE dictionary.

TABLE III

SPECTRAL BRIGHTNESS SENSITIVITY OF THE HUMAN EYE
(After CIE-Dictionary)

λ (mμ)	390	408	428	472	510	555	610	652	687	721	753
%	0.01	0.1	1	10	50	100	50	10	1	0.1	0.01

The sense of colour is limited to the cones, and in the close-vision apparatus it is completely absent. Colour sensation is created only by the way in which our eyes function, i.e., by an interaction of light and sense of vision. A normal eye is capable of distinguishing about 160 colour shades, which is of climato-psychological significance. The selection of homogeneous spectral lights leads at certain wavelengths to distinct colour sensation (cf. Table IV).

TABLE IV

PERCEPTION OF COLOUR TONES BY THE HUMAN EYE IN DEPENDENCE UPON THE WAVELENGTH OF HOMOGENEOUS LIGHT
(After TRENDELENBURG, 1961)

Wave length (mμ)	Colour shade
670	red
600	orange
585	yellow
520	green
470	blue
440	indigo
420	violet

From a bioclimatological standpoint it is important to learn some of the basic aspects of sensory-physiology metrics, dealing with the relation between sensation (E) and impulse (R), as well as with the problem of its mathematical presentation. All considerations of these facts lead to the conclusion that not the absolute value of an increasing or decreasing impulse dimension (e.g., $R_0 + 1$, $R_0 + 2$, etc.) but the relative change regarding a given basic value ($R_0 + 1/R_0$ etc.) is responsible for sensory perception. Based on this knowledge Weber and Fechner (see RANKE, 1953) have established the socalled *psycho-physical basic law* in the following general form:

$$E = k \times \ln \frac{R}{R_0} \qquad (k \text{ is a constant})$$

The example of the human eye, however, showed that this law is valid only for medium brightness levels (from $10^{0.5}$ up to about $10^{3.5}$ lx) and otherwise is subject to considerable deviations.

In order to realize improved metrics of sensory physiology STEVENS (1959) based on a comparison of intensities in connection with dynamometric force experiences, was able to discover closer relations between sensation intensity and impulse dimensions. His research results may be mathematically formulated thus:

$$E = k \times \ln \left(\frac{R}{R_0}\right)^n$$

Exponent n is not a constant but depends on the corresponding sensory modality (seeing, hearing, feeling, etc.). By means of Stevens' power function it is possible to create a corresponding mathematical illustration in the double logarithmic coordinate system for all types of sensations (light sense, sense of hearing, pain, etc.).

Fig. 5 shows the appearance of a group of linear curves for various human sensations (pain, pressure, noise, light, etc.). The value of corresponding exponent n indicates the incline of the pertinent modality function. Pain experienced due to electrical shock shows an exponent n value of 3.5, heat sensation 1.6, cold sensation 1.0 and finally

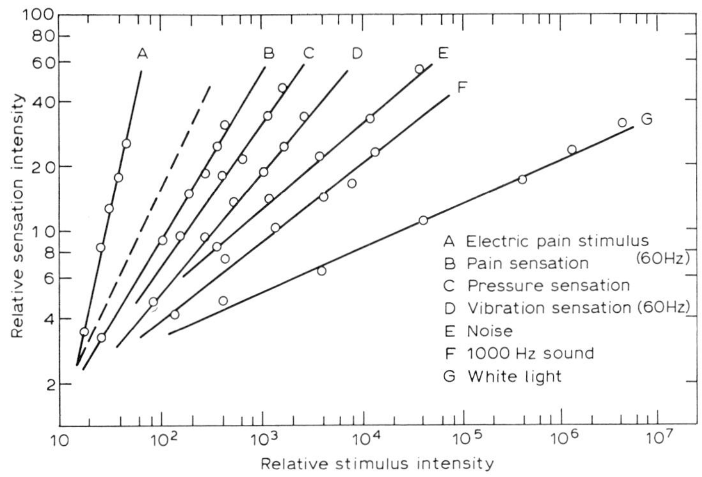

Fig.5. Comparison between an experienced impulse intensity and the corresponding force sensation on a hand dynamometer. The curve was drawn according to average values from a large number of individual measurements. The dotted line indicates the steepness of 1. (STEVENS, 1959.)

brightness of white light only 0.33. These facts indicate that the curves of Stevens' diagram run all the steeper, the smaller the increase in impulse intensity (ΔR) is leading from a sensation just exceeding the stimulus level (ΔE_R) to the maximum possible sensory intensity. Thus for pain we have only few distinguishable impulse levels, whereas on the other hand "light" (i.e., in the modality of the optical sense) reflects in a greatly varied scale of sensations the differences in environment. This means that of our sensory organs the "eye" is the most highly developed one. In Stevens' diagram our sense of hearing is the closest neighbour to our light sense. The dynamometric range is still comparatively large here, which means that it is an important factor in problems of noise protection.

A physiologically significant conclusion may be drawn regarding the sense of light. Considerable power of intervention into vital processes in general is coupled with the fine-graded assimilation of light impulses, such as the rhythmic change of day and night, of light and darkness. Light's effect on the vegetative nervous system (including psychic inductions) as well as numerous hormonal and metabolic processes must also be mentioned. HOLLWICH (1966) pointed out these facts in a series of extensive studies on the connections between eyesight and hormones. The importance for development and functioning of the entire organism by unimpaired light perception through the eye (in humans and animals) can therefore not be underestimated.
Apart from the senses of hearing and seeing, a bioclimatologically noteworthy fact is that, according to Stevens' diagram, in the case of temperature sensations cold impulses ($n = 1.0$), contrary to heat sensations ($n = 1.6$), reveal more distinguishable impulse levels. As to light impulses, however, meteorological and climatological *light changes* acquire, based on Stevens' findings, an increased practical weight. For this reason they deserve additional consideration (cf. p.29).

Climatology of solar and sky radiation, of global illumination intensity, and of circumglobal radiation

Global illumination intensity from a general climatological standpoint and in its relation to global radiation

Continuous measurements and registrations of illumination intensity (global illumination intensity) are carried out at comparatively few points over the earth. This fact has long ago given the impulse to determine the bioclimatologically significant global illumination intensity by means of measurements of global radiation observed in many different places. The mathematical determination of global illumination intensity is facilitated by the fact that measuring units for global radiation and global illumination intensity show practically the same dependence upon the degree of cloudiness. Global illumination intensity is calculated with an accuracy of $\pm 10\%$ by multiplying global radiation with its use effect for human vision (= 105 lm/W) (SCHULZE, 1970). Various investigations have been carried out on the dependence of natural illumination of a horizontal flat receiver surface upon solar elevation, and SIEDENTOPF and REGER (1944) reached the relation given in Table V.
Besides the representation of diurnal variations of global illumination intensities for different latitudes it is particularly interesting from a geomedical point of view to know

TABLE V

REGISTERED ILLUMINATION (klux) OF A HORIZONTAL SURFACE BY SUN AND SKY UNDER CLOUDLESS CONDITIONS
(After SIEDENTOPF and REGER, 1944)

Solar elevation:	5°	10°	15°	20°	25°	30°	40°	50°	60°	70°	80°	90°
Sun	1.1	6.8	15.1	24.6	34.6	44.5	63.2	79.8	93.6	103.5	109.5	111.8
Sky	5.4	7.7	9.4	10.9	12.0	13.2	14.8	16.2	17.3	18.0	18.3	18.5
Global	6.5	14.5	24.5	35.5	46.6	57.7	78.0	96.0	110.5	121.5	127.8	130.3

the daily total illumination intensities for cloudless conditions. They are set up in Table VI for individual latitudes of the Northern and Southern Hemispheres according to calculations by SCHULZE (1970). The conclusion may be drawn therefrom that at winter solstice the greatest global illumination intensities are found, besides at the South Pole (due to the 24-h day), between 20° and 50° southern latitude at about 1000 kluxh, whereas at the same time in the corresponding area of the Northern Hemisphere values of merely 100–700 kluxh are observed. At summer solstice we find the conditions reversed for the Northern and Southern Hemispheres, however, with slightly reduced illumination intensities. In spring and fall the greatest illumination intensities of approximately 850–900 kluxh are measured in the equatorial area when the greatest annual solar elevation is reached.

TABLE VI

DAILY AMOUNTS OF GLOBAL ILLUMINATION INTENSITY (in klx h day^{-1}) UNDER CLOUDLESS CONDITIONS (After SCHULZE, 1970)

	Feb. 4	Mar. 21	May 6	Jun. 22	Aug. 8	Sep. 23	Nov. 8	Dec. 22
N. Pole	—	—	620	950	610	—	—	—
80°N	—	98	600	930	600	97	—	—
70°N	13	230	630	880	630	230	13	—
60°N	104	380	730	900	740	420	103	28
50°N	230	510	820	950	810	500	220	127
40°N	370	630	880	960	870	630	370	250
30°N	510	740	910	960	900	730	510	410
20°N	650	820	910	930	900	810	640	570
10°N	770	870	880	870	870	860	760	690
Equator	860	890	820	770	820	880	850	820
10°S	920	870	740	650	730	860	910	930
20°S	950	820	620	520	620	810	940	1000
30°S	950	740	490	390	490	730	950	1020
40°S	920	630	350	240	350	630	910	1030
50°S	850	510	220	119	210	500	850	1010
60°S	770	370	99	26	98	370	760	960
70°S	660	230	12	—	12	230	660	940
80°S	630	98	—	—	—	96	630	990
S. Pole	650	—	—	—	—	—	640	1010

The information from Table VI has considerable importance with respect to light physiology, particularly for world-wide aviation. Thus a notable change in light intensity occurs for instance on a non-stop flight during the northern winter from northern Europe to corresponding latitudes in the Southern Hemisphere. Global illumination intensity increases from 60°N to 60°S more than 35 times.

A glance at the conditions of global illumination intensity (kluxh) in Alpine areas (eastern Alps), particularly as regards intensity behaviour under cloudless and overcast skies, is instructive (Table VII). This numerical survey makes clear that under both mentioned weather conditions there is always an increase of global illumination at higher altitudes. This fact is primarily related to the increase of global radiation (T in cal. $cm^{-2}d^{-1}$) towards greater heights. Illumination intensities under cloudless skies increase at a relatively small rate compared with that of overcast skies, a fact conditioned by various states of cloud density in lower and higher atmospheric layers. In Alpine regions (Table

TABLE VII

DAILY AMOUNTS OF GLOBAL ILLUMINATION INTENSITY (klx h day^{-1}) IN ALPINE REGIONS (EASTERN ALPS) AND IN THE LOWLANDS, FOR CLOUDLESS AND OVERCAST SKIES
(After DIRMHIRN, 1964)

H (m)	III	VI	IX*	XII
	Cloudless:			
200	538	1056	655	160
1000	585	1110	714	185
2000	609	1150	742	198
3000	619	1171	752	204
	Overcast:			
200	153	275	174	46
1000	195	350	223	59
2000	260	467	295	78
3000	320	576	368	97

VII) the mean global illumination distribution illustrates that under cloudless conditions there is usually an intensity increase of only 5–10% except in midwinter, where an increase of up to 30% may be observed. This latter phenomenon is caused by haze conditions especially in the lowlands, while above 800–1000 m (upper haze level) generally a high light transmittance exists. Under overcast skies the percentage light increase with altitude carries considerably more weight. At elevations of about 3000 m it amounts to more than double the lowlands values. BULLRICH (1948) made very significant observations on this phenomenon. According to his theoretical and experimental investigations it has been proved that diffuse nondirectional light of a spacial expansion is weakened much less by "turbid media" (clouds) than directional light.

Table V already indicated that the illumination of a horizontal surface under cloudless conditions by the sky alone can reach considerable intensities. In order to have some idea of the changes they are subjected to when cloudiness varies in quantity and quality, investigations on the behaviour of diffuse sky radiation (D in cal. cm^{-2}d^{-1}) are pertinent. As early as 1911 DORNO in his *Study on Light and Air in Alpine Regions* dealt extensively also with the influences of cloudiness on the photometric brightness of a town in the Alps (Davos). His results are largely in agreement with those of later investigations of the same nature (THAMS, 1961; BENER, 1963). Fig.6 (THAMS, 1961) relates daily amounts of diffuse sky radiation (D) for the three main types of clouds (C_L, C_M, C_H) and the occurring degrees of overcast. As a consequence, under the sole presence of clouds at a high level (C_H) with a growing overcast, sky radiation (and thus also global illumination intensity) under overcast conditions (8–10/10) increases 3 to 4-fold compared with cloudless conditions. With clouds at a low level (C_L) sky radiation (D) increases rapidly 3 times up to cloudiness of 6/10, and with steadily increasing cloudiness sinks to values equal to those of the lowest cloudiness levels.

For many questions of a practical nature it is helpful to know the influence of weather conditions on the daily and seasonal behaviour of daylight (e.g., switching on and off of street lights, of lighting in factories and offices, etc.). WOERNER (1958) in his contribution to this complex of problems describes the influences of cloudiness levels on the daylight deficit in a city district (Berlin-Rummelsburg) and in non-urban country (Potsdam). These comparisons lead to the conclusion that the length of day (above 7000 lux)

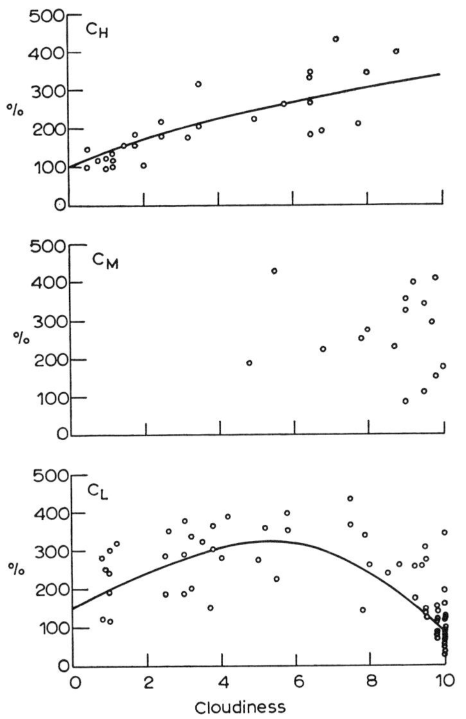

Fig.6. Daily amounts of sky radiation D (cal. cm^{-2} d^{-1}) under different types of clouds and overcast degrees in % of the values under cloudless skies, in Locarno-Monti (Switzerland) (THAMS, 1961).

on an average is shortened during the winter months by clouds, haze, fog or precipitation in the city by 60–70%, in Potsdam by 40–50%. This fact is particularly impressive when considering the percentages of "dark days" (always <7000 lux) in fall and winter:

	X	XI	XII	I	II
Potsdam	1	7	27	22	3%
Berlin	3	26	46	24	4%

Climatology of global radiation T and its components S and D

In addition a few radiation climatological peculiarities shall be pointed out which have practical significance for the characterization of natural illumination intensities. By means of frequency statistical evaluations of global radiation intensities (T, D, etc., in cal. cm^{-2}d^{-1}) together with the current cloud conditions in different climates it is possible to characterize also the often considerable variations in global illumination intensities (B).

In the first place Fig.7 shows the frequency statistical explanation of current daily amounts of global radiation T in intervals of 100 cal. cm^{-2}d^{-1}, whereby characteristic distributions of this element and implicitly the present illumination intensities in differing climates enable interesting insights. In Mawson (Antarctica) as well as in Potsdam (central Europe) the mode of the frequency distribution lies in the two lowest levels of global radiation amounts, whereas the other intensities are represented with comparatively small percentage values. Contrary to the Potsdam conditions one finds in Mawson during the southern summer considerable percentage values of global radiation

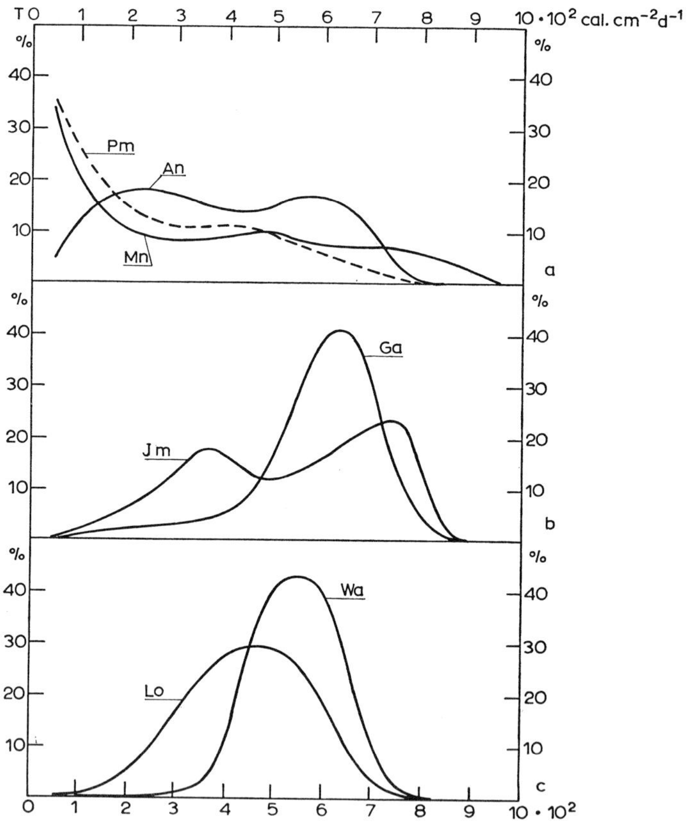

Fig.7. Relative frequency of the appearance of the individual value levels of global radiation (T) in %: (a) Mawson (Mn): φ = 67.6°S, 6 m (1958–1961); Potsdam (Pm): φ = 52.4°N, 81 m (1937–1942); Athens (An): φ = 38.0°N, 107 m (1955–1962). (b) Jerusalem (Jm): φ = 31.8°N, 782 m (1954–1960); Giza (Ga): φ = 30.0°N, 21 m (1957–1959). (c) Wadi Halfa (Wa): φ = 21.8°N, 155 m (1960–1962); Lwiro(Lo) φ = 2.2°S, 1750 m (1957–1959). (FLACH, 1965.)

caused by the "long days" (midnight sun) (cf. p.152). In the eastern Mediterranean area (Athens) the mean intensity intervals show up in a slightly bimodal distribution, whereby the secondary peak represents a characteristic of the Mediterranean radiation climate in summer. Frequency distributions in the middle figure show a noteworthy point in the radiation climates of Asia Minor and North Africa. In Jerusalem again a double peak is observed, but in contrast to Athens this secondary peak is particularly prominent (arid climate). In the extremely arid zone of the eastern Sahara (Giza) only one mode is found with a maximum between 600 and 700 cal. $cm^{-2}d^{-1}$. In the lowest section of the figure two additional frequency distributions are indicated from the arid zone of North Africa (Wadi Halfa) and the African equatorial zone (Lwiro). In Wadi Halfa a true Gaussian distribution of the T values is observed with a maximum between 500 and 600 cal. $cm^{-2}d^{-1}$, which is slightly shifted to the left compared with Giza. The occasional northward movement of the tropical rain belt shows radiation-reducing influences in Wadi Halfa, whereby first of all the S component is attenuated by high and medium high clouds. The maximum in frequency distribution in Africa's tropical zone (Lwiro) is shifted yet further towards the lower levels. Naturally the diffusion range is wider than in arid zones due to frequent high degrees of cloudiness.
Insight into the world-wide spatial distribution of mean annual global radiation amounts (in kcal. $cm^{-2}yr^{-1}$) is given in Fig.8 in the world map of LANDSBERG (1963). Global T

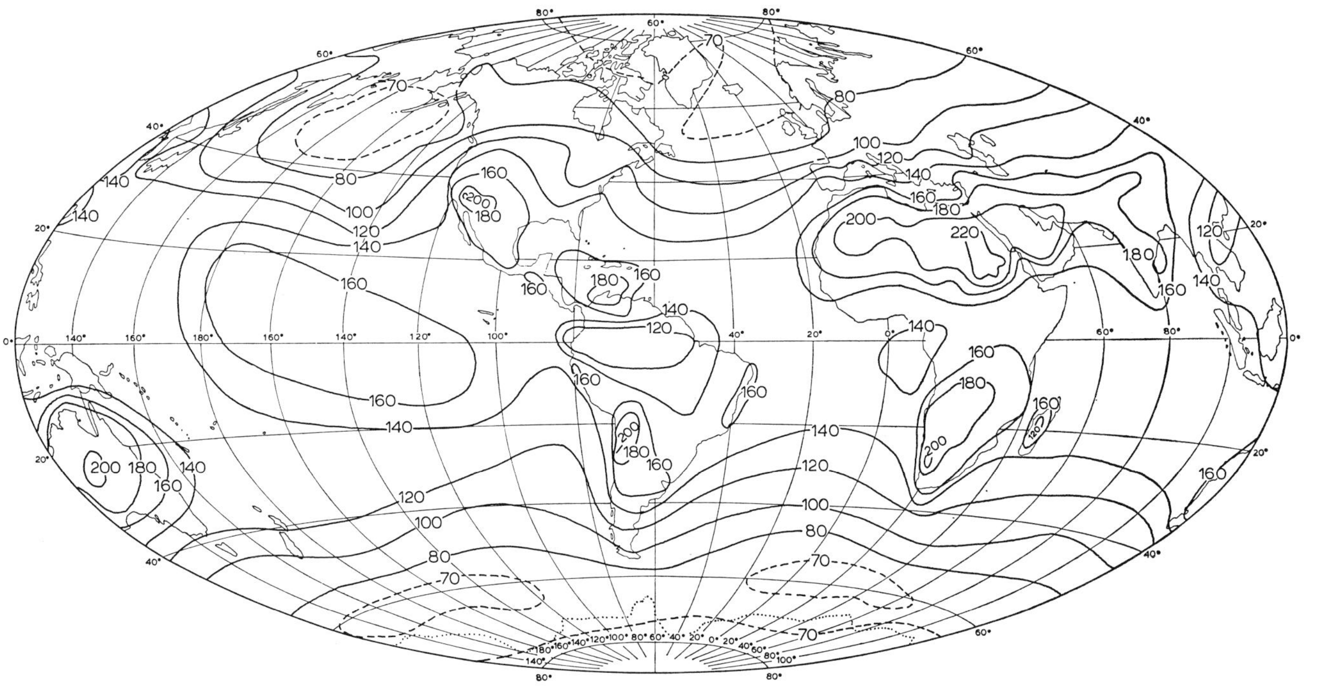

Fig.8. Generalized isolines of global radiation (T), surface, in kcal. cm^{-2} yr^{-1} (LANDSBERG, 1963).

distribution exhibits the decisive dependence of this radiation climatological dimension upon latitude and mean cloudiness conditions. (Details on the difficulties in the charting of T cannot be discussed here.) The map clearly shows the differentiations in the distribution of T given already in the previous figure. Subtropical zones with annual maximum values of 200–220 kcal. cm^{-2} (particularly in the Sahara) as well as the reduced annual T sums in equatorial zones show up conspicuously on the map. In Arctic and Antarctic regions annual global radiation amounts to only about half of that of the Equatorial area.

Temporal and spatial explanations of terrestrial energy fluxes are bioclimatologically important. This is the case not only for questions of the distribution of illumination on earth and of the earth's radiation-induced heating, but also for the evaluation of incoming photochemical energy (ultraviolet radiation) from the sun. Last but not least, they serve geomedical investigations of the most varied kind as well as for efforts made in touristic medicine which are steadily gaining importance.

Here a local climatological outline of lighting conditions at forest edges and in the immediately adjoining stands of trees is in order. These are well-known and not unimportant bioclimatologically. Fig.9 gives an impressive picture of the abrupt changes of horizontal illumination intensity (given here in relative figures) which can occur when passing from open country into wooded areas (with deciduous trees and conifers) under various conditions (with or without foliage, with or without sunshine). Such local climatic conditions are a climate-physiologically "refreshing" experience for a hiker entering a wooded area in summer where a few steps permit him to flee the occasionally excessive heat and light flux of direct solar radiation.

The seasonal structure in the behaviour of global radiation $T (= S + D)$, absorbed by a flat horizontal receiver surface, is dependent upon numerous parameters (latitude,

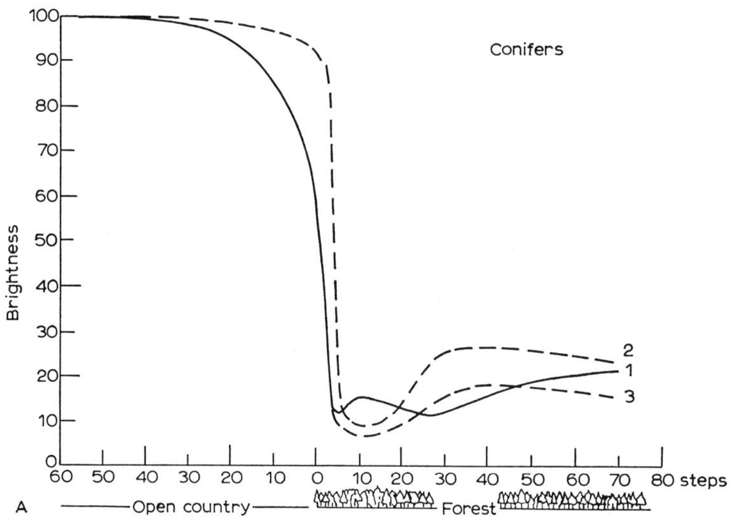

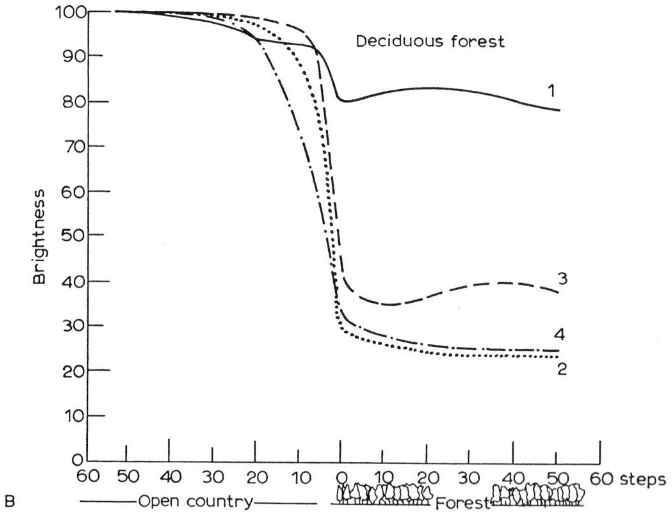

Fig. 9. Illumination intensities at forest edges (brightness measured photometrically on horizontal surface, in %) (LAUSCHER AND SCHWABL, 1934). A. 1 = spring, overcast sky; *2* = spring, full sun, sunny spot; *3* = spring, full sun, shady spot. B. *1* = no foliage, overcast day; *2* = with foliage, no sun; *3* = no foliage, full sun; *4* = foliage, hazy sun.

altitude, air turbidity, and other climatological phenomena such as cloudiness). These statements are sufficiently illustrated in Fig.10, where mean annual variations of T (in cal. $cm^{-2}d^{-1}$) are traced for a subtropical station (Giza, $\varphi = 30.0°N$, $H = 21$ m), a northern European station (Stockholm, $\varphi = 59.3°N$, $H = 44$ m) and for the Alpine valley station Davos ($\varphi = 46.8°N$, $H = 1,592$ m) in the eastern Swiss Alps. The differently shaped annual global radiation variations due to prevalent latitude and climatic influences are evident. Special notice should be taken of the low daily amounts during the Stockholm winter as an immediate consequence of short days and a high degree of cloudiness, as well as of the narrow structure of the annual T variation as a marked contrast to climatological radiation factors of the arid land of the Sahara (Giza). There winter T values correspond with the Stockholm values of spring and autumn, and in Davos the former correspond with the daily averages of the late winter and late fall in

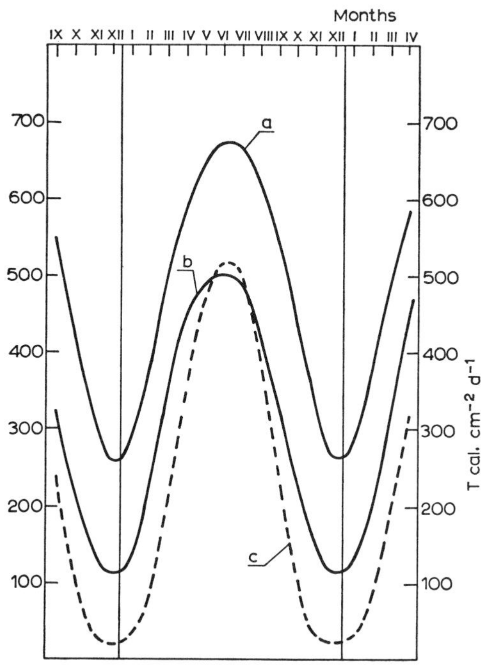

Fig. 10. Mean annual variations of global radiation (T, in cal. $cm^{-2}d^{-1}$) at different latitudes of the Northern Hemisphere: a = Giza, φ = 30.0°N, H = 21 m; b = Davos, φ = 46.8°N, H = 1,592 m; c = Stockholm, φ = 59.9°N, H = 44 m. (FLACH, 1966.)

the Alps. Contrary to the annual T variation of Giza and Stockholm showing on the whole a symmetrical shape, Davos reflects (at 1592 m) an asymmetrical structure of the annual variation of this significant climatological radiation parameter. This illustrates that the late winter and spring values show higher daily amounts of global radiation than the fall months, which are mainly under the influence of clear high pressure areas. This almost exaggerated effect must be attributed to reflected radiation from snow-covered surroundings (slope sites) including also the snowcover-caused back radiation of clouds. This additional "light source" plays a part which should not be underestimated in the natural winter illumination intensities of Alpine valley sites where one frequently finds health resort and winter sports centers.

Bioclimatological significance of the individual components of solar radiation as well as of circumglobal radiation (CGR)

In connection with investigations on global radiation intensities, and indirectly therefrom also global illumination intensity, it has been pointed out several times (first by TOPERCZER, 1931, and more recently by HITZLER and LAUSCHER, 1970) that besides the vertical component of solar radiation calculated from the difference $T - D = S$, also the horizontal component S_h is of distinct bioclimatological importance in particular to man (in standing position). Having measurements of direct sun radiation I at one's disposal (determined with a receiver surface held vertically to the sun) $S = I \sin h$ therefore constitutes the radiation energy incident on the surface of a vertically standing cylinder.

HITZLER and LAUSCHER (1970) made calculations for I, $I \sin h$, and $I \cos h$, for various latitudes and altitudes of countries in the eastern Alps. Fig.11 illustrates the diurnal variation of $I_{\perp} \cos h$ (in mcal. $cm^{-2}min^{-1}$) for various altitudes of the eastern Alps for the months March, June, September and December (only first half of day).

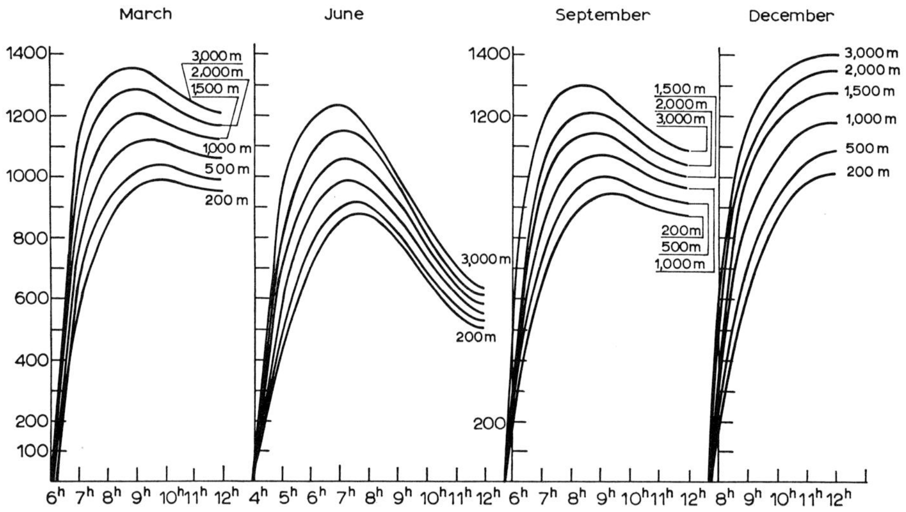

Fig.11. Diurnal variation of the horizontal component of solar radiation (in mcal. $cm^{-2}min^{-1}$) for various altitudes in the eastern Alps (HITZLER and LAUSCHER, 1970).

It may be seen from Fig. 11 that only December shows a simple daily curve for $I_{\perp} \cos h$, whereas this quantity shows up in all other examples as a double curve, specially marked in June. Particularly high values are found in the winter but also during the transitional seasons. This fact is specially noteworthy because in Alpine sites above 1,200–1,500 m in winter and in fall in central Europe the comparative frequency of high pressure weather conditions is expressed in high values of relative sunshine duration.

More than 100 years ago the Italian physicist Bellani developed a principle by which radiation energy coming from natural surroundings was used to change the aggregate state of a filling liquid (alcohol). This type of measurement is being revived and constitutes a comparatively simple means to grasp the bioclimatologically important natural radiation fluxes largely correctly, for instance as they hit a standing person.

The new development of the Bellani apparatus (spherical Bellani pyranometer) by COURVOISIER and WIERZEJEWSKI (1954) at the Davos Observatory enables the determination of short-wave radiation fluxes (300–4000 mμ) of sun (S), sky (D) and of the reflecting ground (R_E) as a complex radiation element of "circumglobal radiation", CGR, by means of a spherical receiver surface (with metal surface). A spherical pyranometer is an integrating instrument and is used exclusively for the measurement of daily CGR amounts (cal. $cm^{-2}d^{-1}$). Several results calculated for the European region with the spherical pyranometer are given in Fig.12.

The top section (a) shows annual CGR variations of towns situated to the north and south of the main Swiss Alpine crest (Basel and Locarno, respectively) as well as for central Alpine valley and crest sites (Davos and Weissfluhjoch above Davos, resp.). In the lowlands the radiation surplus follows essentially the behaviour of sunshine duration, i.e. the amount of weather- and climate-dependent radiation conditions. Therefore, the

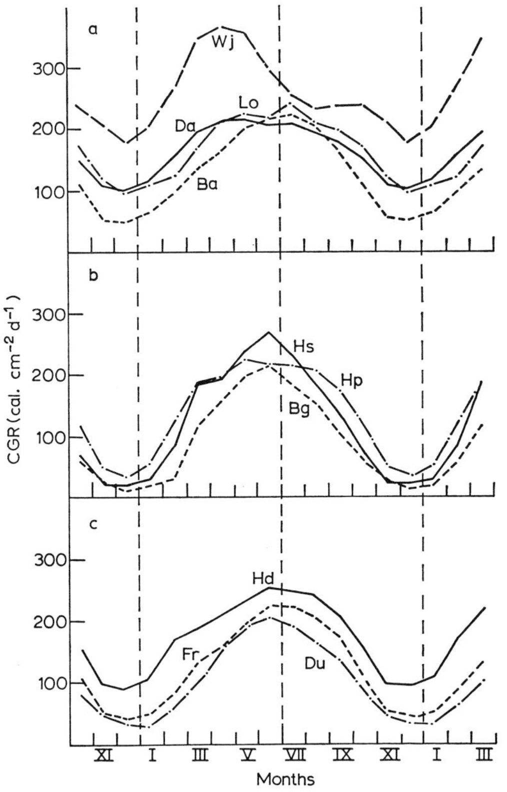

Fig.12. Annual variation of circumglobal radiation (cal. cm^{-2} d^{-1}) according to mean monthly values (for every day)(FLACH, 1968). (a) Swiss towns: $\varphi = 47.5°$, Basel (*Ba*), 318 m, 1954–1966; $\varphi = 46.2°$, Locarno (*Lo*), 379 m, 1953–1959; $\varphi = 46.8°$, Davos (*Da*), 1600 m, 1944–1963; $\varphi = 46.8°$, Weissfluhjoch above Davos (*Wj*), 2667 m, 1955–1962. (b,c) Other European stations: $\varphi = 60.4°$, Bergen (*Bg*) Norway 44 m, 1954–1963; $\varphi = 53.4°$, Dublin (*Du*) Ireland, 14 m, 1955–1966; $\varphi = 60.3°$, Helsinki (*Hs*) Finland, 12 m, 1960–1964; $\varphi = 47.8°$, Hohenpeissenberg (*Hp*) F.R.G. 972 m, 1954–1960; $\varphi = 48.8°$, Freiburg i.Br. (*Fr*) F.R.G., 280 m, 1959–1961; $\varphi = 47.7°$, Höchenschwand (*Hd*), F.R.G. 1,010 m, 1959–1962.

annual maximum is found in midsummer, the annual minimum in early winter. It may be stated, however, that Locarno (Tessin) receiving more sunshine especially in spring and fall is favoured compared to radiation amounts registered in Basel. In fall and winter the north side of the Alps is frequently subjected to fog which reduces considerably the amount of radiation received there. On the other hand, the Alpine valley site of Davos is especially favoured compared with Basel and Locarno in mid and late winter. During this period in high Alpine sites the CGR values rise strikingly compared with those of snow-free areas, due to intensified ground reflex radiation which is more or less guaranteed by a permanently closed snowcover. The crest site of Weissfluhjoch shows the very excess in annual CGR variations. Here the year-round CGR values lie above those of Davos and particularly above those of the lowlands (Basel). The reasons for this are the increase in the intensity of short-wave heat radiation with altitude, but in particular the extremely favourable development of ground reflex radiation R_E by the often renewed snowcover in regions above the timber-line, the greater amount of sunshine in an annual average compared with the lowlands, and finally quite a large number of possible sunshine hours. (As is well-known, the "reflex factor" (albedo) of fresh snow lies between 80 and 90%.) At the measuring point of Weissfluhjoch (2,667 m) the spherical pyranometer used there is placed in such a way that it takes into consideration a state of slope reflection frequently encountered by skiers, a phenomenon particularly intensified during the months February to April due to the quickly rising solar elevations. Later the aging of the snowcover reduces the reflection index which again reduces the entire CGR intensity.

The middle section (b) of Fig. 12 illustrates examples of daily CGR variations of northern European cities (Helsinki and Bergen). Compared with central European stations the high overcast degrees found in the north in winter lead to considerable reductions of CGR radiation, whereas in summer the differences are small. An interesting statement is that the abundant sunshine of early summer in Finland intensifies circumglobal radiation due to the "long days".

Fig.12c gives insight into the CGR radiation conditions of the southern Black Forest (Western Germany) for the examples of Freiburg im Breisgau (Rhine plain) and Höchenschwand in the high Black Forest. This point is characterized by year-round high CGR intensities. The reason is that the dominating site of Höchenschwand is not very confined by the horizon. Snow reflection in mid and late winter and the advantageous sunshine conditions in fall enhance radiation contributions to CGR greatly. In Freiburg i.Br. the fall and winter fog situations and in early summer heavy convective cloudiness on the western slopes of the Black Forest cause a considerable reduction of the CGR element, which improves only in mid and late summer.

THAMS and WIERZEJEWSKI (1963) have dealt with further behaviour of circumglobal radiation CGR in its relations with sunshine duration and global radiation T as well as diffuse CGR. The latter case revealed that diffuse CGR forms a significant part of total CGR, i.e., an annual average of approx. 35%. The percentage is therefore about 3 times higher than the diffuse constituent D of global radiation T.

Many practical bioclimatological questions, particularly those dealing with climatic dosage, aim at the determination of radiation amounts as related to various weather conditions. Very important is the knowledge of daily global radiation amounts to be expected during a year on "clear days". Radiation amounts from so-called "overcast days", however, are bioclimatologically also quite influential. In the absence of solar radiation S the CGR components sky radiation D and ground reflex radiation R_E can occasionally have considerable effects.

In Fig.13 a survey of year-round radiation conditions of the Alpine high valley site of Davos and of the northeastern Swiss lowlands (Basel) are shown. The upper section of the figure represents the even annual variation of CGR daily amounts for Basel and Davos on clear days (based on daily average values collected over years). The increased CGR intensities in the Alpine high valleys from fall through winter into early summer are clearly evident. The greatest radiation condition differences compared with the lowlands (Basel) are the vast contribution of reflected ground radiation caused by complete and often renewed snowcover. After the middle of April the contrasts between "down there" and "up there" diminish noticeably with the disappearance of the snowcover in the Alpine valleys. Subsequently heightened CGR values in the highland are mainly caused by the difference in altitude (1,300 m) between the comparison points and are a consequence of the short-wave heat radiation increasing with altitude. In midsummer and early fall the comparative CGR curves approach each other closely following the high water vapour turbidity of the atmosphere during that period. Only when

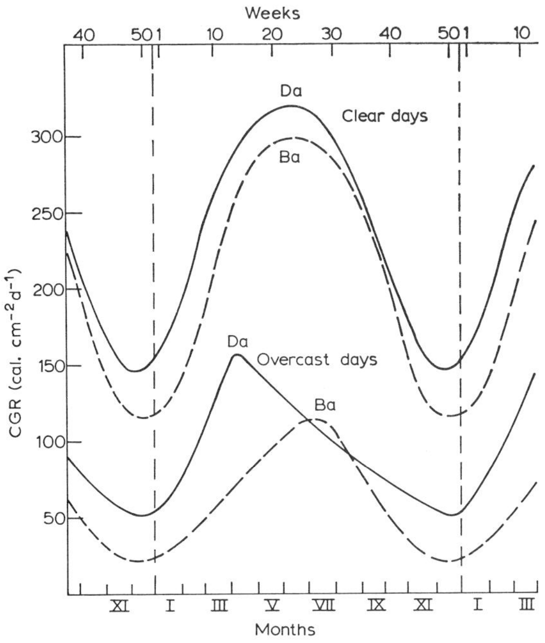

Fig.13. Annual variation of circumglobal radiation (CGR, in cal. cm^{-2}d^{-1}) on clear and on overcast days: Davos (*Da*), 1,600 m, 1944–1963; Basel (*Ba*), 318 m, 1954–1966. Clear days are days with relative sunshine duration $rS \geqq 85\%$. Overcast days are days with relative sunshine duration $rS \leqq 15\%$. (FLACH, 1968).

the autumnal high pressure conditions set in bringing few clouds and clear visibility to the high central Alpine sites as well as with the reappearance of a complete snowcover the Davos CGR values again move farther and farther away from those of the lowland station Basel. At the same time rapidly increasing haze turbidity (a consequence of the formation of low and high fog) in Basel contributes to this phenomenon.

In addition to these mountain climatological considerations on the behaviour of circumglobal radiation CGR it must be pointed out that direct solar radiation (I, in cal. cm^{-2}) as well as natural global illumination intensity (B in klux) in high Alpine sites around 3,000 m H are of the same dimensions as corresponding values for the South Pole (2,800 m H) at a solar elevation of 30° in December (HOINKES, 1961; KUHN, 1973).

The lower part of Fig. 13 illustrates correspondingly the annual CGR variation for "overcast days". A marked maximum of the CGR annual variation in Davos is evident at the beginning of April. The rise from a winter minimum to this peak runs very steeply, whereas the following CGR reduction occurs gradually. This spring maximum is a definite result of reflected ground radiation R_E forming an intensive supplement to diffuse sky radiation (D) and counter-reflection on the underside of clouds. It may be concluded that this reflected radiation constitutes noteworthy portions of the complex element CGR.

Concerning the influence of cloudiness alone BENER (1963) calculated that on overcast days (cloudiness 8–10/10) the mean sky radiation intensities at 50° solar elevation amount according to degree of cloudiness to the 6 to 6.2-fold of the value of a cloudless sky in summer, in winter to the 1.2 to 1.9-fold, however, at lower solar elevations. With regard to the alternating effect between ground reflectivity and sky brightness Moeller pointed out that owing to a complete snowcover under cloudless as well as under overcast skies an intensification of global radiation T and in particular of sky radiation D takes place. In one example for a mean albedo value of 71 % (snowcover) he found an intensification of global radiation T ($= S + D$) by the factor 1.6 (solar elevation $h = 10°$ to 30°) and by 1.46 ($h = 20°$ to 30°) under overcast skies. Under a closed, optically dense cloudcover and, above all, with high snow albedo values the brightening effect can be very extensive. This brightening effect is based not only on the surface albedo and on the current diffusion capacity of the sky but also on the influence of the haze content in the air layer between ground and lower limit of clouds (MOELLER and QUENZEL, 1972). A well-known fact needs to be mentioned at this point. The horizon is brightened up most of all in the presence of a high surface albedo ("ice blink" in polar regions).

In April the vanishing snowcover in Davos (Fig.13) leads to a reduction of reflected radiation and thus to a gradual decrease of CGR in that area. In the lowlands (Basel) CGR reaches the annual maximum on overcast days at the same time as for clear days, i.e., in midsummer. An explanation for such conditions in the lowlands may be the fact that in spring and summer despite overcast skies the cloud structure is loose permitting a certain transparence for light and radiation. In late summer and, above all, in fall this feature disappears steadily, because in that area particularly fog and high fog conditions permit less penetration by light.

The climatology of circumglobal radiation shows the important contributions to knowledge of heat and light radiation made by means of the Bellani spherical pyranometer. The receiver surface of this measuring instrument somewhat simulates human beings, buildings and plants, etc. In this manner the dominating influence of heat and light radiation (besides also as "time indicator") may be determined with great accuracy in its seasonal variations which play a decisive part in climatic therapy (climatic dosage) and in seasonal pathology.

It has been demonstrated that within the range of radiation fluxes of all spectral ranges occurring free in nature the daily and seasonal change of reflected ground radiation and its dependence upon the nature of the ground is of far-reaching significance.

The *reflection index* (also called *albedo*) is the measure describing the relation of radiative fluxes coming from above from sun (S) and sky (D) on the one hand, and on the other radiation R_E reflected by the ground (mostly in percent). Table VIII presents a survey of diffuse albedos for different surfaces for total day radiation (GEIGER, 1961).

TABLE VIII

REFLECTION INDEX (ALBEDO) IN DEPENDENCE UPON SURFACE STRUCTURE (in %)
(After GEIGER, 1961)

Cover of new snow	75–95
Closed cloud cover	60–90
Cover of old snow	40–70
Pure perpetuated snow	50–65
Light dune sand, surf	30–60
Pure glacier ice	20–30
Sandy ground	15–40
Grassland and fields	12–30
Forests	5–20
Water surfaces, sea	3–10

The bottom part of Fig.14 is a graphic illustration of diurnal albedo variations (%) at various altitudes in the Swiss Alps. An interesting comparison between the dimensions indicated for May and November can be made, when on top and on the bottom strongly opposing reflection conditions are encountered. During the summer during complete absence of a snowcover still marked differences are observed in surface albedo characteristics of gravel (valley site) and rocky ground (high site).

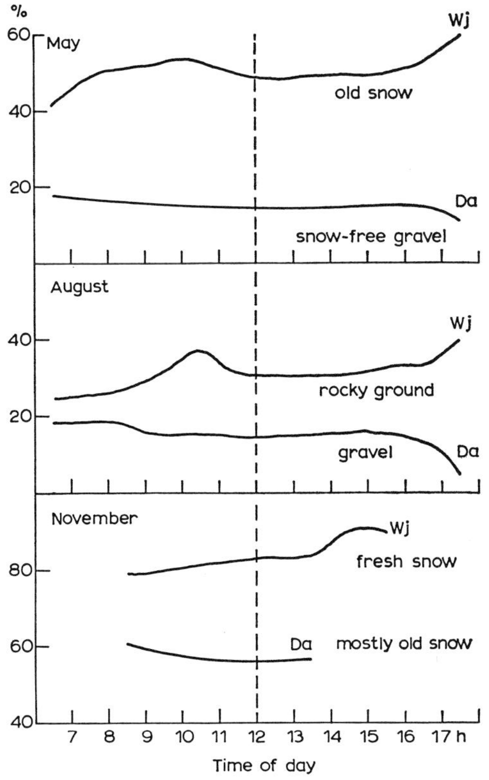

Fig.14. Examples of diurnal variation of reflex number or albedo (%) at various altitudes and under different ground conditions (snowcover, gravel, rock); Davos (*Da*), 1,600 m; Weissfluhjoch above Davos (*Wj*), 2,667 m. (FLACH, 1968.)

Changes of illumination intensity and their relation to reaction times for light impulses

The results of AICHINGER and MUELLER (1953), important for light-physiological questions, are shown in Fig.15. They present the human reaction time in response to light impressions. Based on two years of observations of seven persons, one can see that in all cases, in the annual course, the two elements vary in the same sense. The main maximum occurs from January to May (winter–spring), the main minimum from late fall to early winter. The mean seasonal character of this aspect of the optical sense is shown in the lower part of Fig.16 (small circles). The mean natural illumination conditions have been added (marked by crosses) for a central European lowland station. There the intensities of natural illumination were continuously recorded. Clearly the correlation between human reaction times to light stimuli and the contemporary mean illumination variations

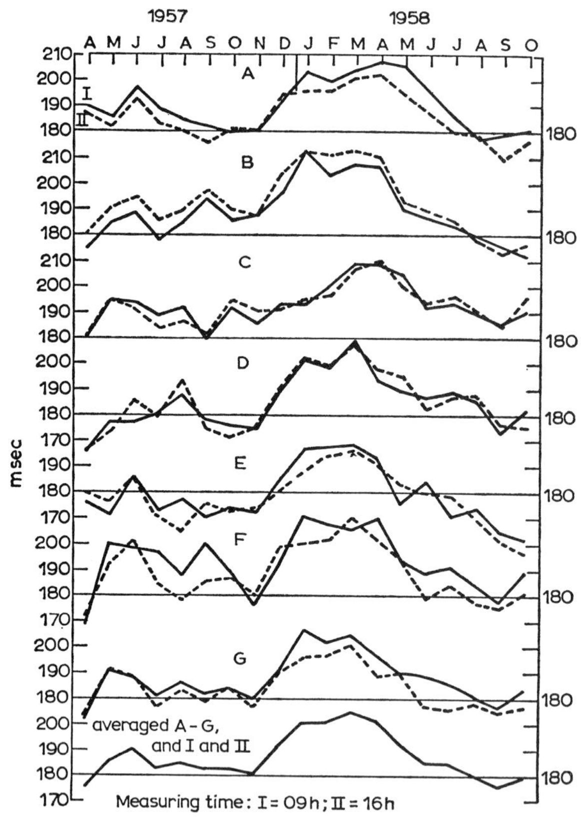

Fig.15. Mean annual variation of reaction time to a light impulse, measured in different test persons, *A–G*. (AICHINGER AND MUELLER, 1953.)

is close. The upper part of Fig.16 indicates for the two variables a function of higher order (FLACH, 1972). It seems justified to assume that with increasing positive illumination changes the reaction times reach a saturation level.

These relations are valid for the climatic zone of the central European lowlands. In the richly illuminated high mountain climate of the Alps in winter different light-physiological reactions can be expected, as studies of radiation climatology lead us to suspect. In the framework of biometeorological evaluation of landscapes this is a systematically important point, especially as it relates to recreation effects.

The close connection between the annual periodical variability of natural illumination intensity and human reaction time to certain light impulses are closely related to the adaptation processes of the retina of the human eye. In the case of brightness adaptation the nervous and photochemical regulations in the retina occur comparatively quickly and the sensitivity of the entire retina is usually reduced abruptly in each case. Thus the adaptation processes in the human eye in a given situation are always subjected to the state of the current basic conditions (cf. STEVENS, 1959, Fig.7). A feature of these complex phenomena is that bright surfaces greatly reduce the retina's sensitivity.

These factual relations are very important for the control of the biological annual rhythm and should not be underestimated. Also in their physiological consequences, as pointed out by HILDEBRANDT (1962), the phases of this rhythm have superimposed additional oscillations which are certainly caused by the seasonal illumination changes.

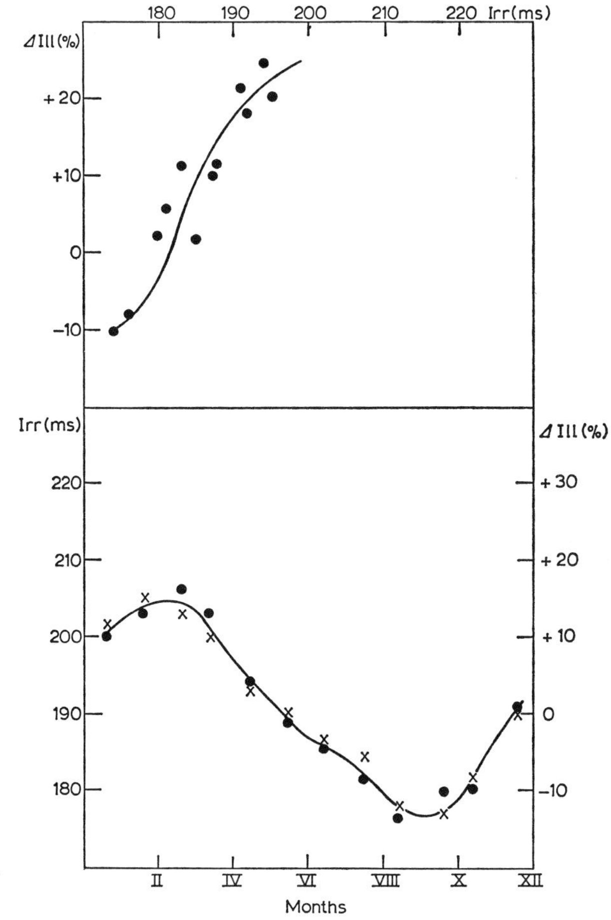

Fig.16. Seasonal response of human reaction time to changes in light stimuli of specific magnitude (in msec), depending on the contemporaneous variations of natural light conditions (%).
Lower section: the seasonal course of reaction times (●) and the natural illumination changes (×).
Upper section: connection between the two variables. (FLACH, 1972.)

Taxonomy of energy fluxes active on the earth's surface and their bioclimatological significance

General taxonomy of radiation fluxes active on earth

A multitude of radiation fluxes is active in the atmosphere, particularly in the habitat of man, animal and plant, with individual energy sources directly or indirectly manifest. A majority of the natural radiations is directed towards the earth's surface, but significant radiation fluxes permanently leave the earth's surface to be lost in the atmosphere or in space. All these radiation fluxes may be distinguished in the following way.

Energy fluxes as part of incoming radiation
Direct solar radiation (*S*)
Diffuse sky radiation (*D*)
Reflected radiation from *S* and *D* (R_E) relating to receiving objects on the earth's surface
Atmospheric radiation (*G*)

Energy fluxes as part of outgoing radiation (directed at atmosphere or into space)
Reflected solar and sky radiation (R_r)
Reflected back radiation of the atmosphere (G_r)
Long-wave radiation of ground (E)

The intensities of the two cited main groups of radiation (each separately) lead as a *difference* of *incoming radiation—outgoing radiation* to the socalled *radiation balance*, a technically measurable parameter which plays an exceptional geophysical and bioclimatological part in the considerations of energy budget in the atmosphere, in the earth and in humans.
The features of biological effects of the radiation fluxes will now be discussed.
In the preceding section we have already discussed, light and brightness acting through the human eye as receiver and transformer. Now the thermic and photochemical effects of natural radiation deserve a closer look, presenting first of all their internationally recognized basic classification in dependence upon their respective wavelengths. Of the direct and indirect radiation fluxes the following ones are of bioclimatological interest:

UV-C	(short-wave ultraviolet)	0.10 –0.28 mμ	*ultraviolet*
UV-B	(medium-wave ultraviolet)	0.28 –0.315 mμ	
UV-A	(long-wave ultraviolet)	0.315–0.40 mμ	
Light and brightness radiation (maximum brightness sensitivity of the human eye)		0.40 –0.78 mμ	
IR-A	(short-wave infrared)	0.78–1.4 mμ	*infrared*
IR-B	(medium-wave infrared)	1.4 –3 mμ	
IR-C	(long-wave infrared	3 –1,000 mμ	temperature radiation of H_2O and CO_2 molecules

These spectral ranges were established mainly according to biological and partially also technical effects and events. The subdivision of ultraviolet radiation was suggested by COBLENTZ and STAIR (1934). The absorption curves of various amino acids from the albumen particle group have their minimal values at approx. 0.315 mμ and are part of the chemical foundation of a living cell. The erythema efficiency curve (by HAUSSER and VAHLE, 1922), the greatest effects of which lie between 0.28 and 0.315 mμ, point strongly to the limits of UV-B. Short-wave ultraviolet ends in the biosphere at 0.29 mμ because stratospheric ozone absorbs all ultraviolet radiation below that wave length. The maximum bactericidal effect of ultraviolet radiation lies in the UV-C range (between 0.25 and 0.27 mμ), i.e. below the end of the surface solar spectrum. Nevertheless, noticeable effects of this type of radiation are yet found in the extreme UV-B (0.28–0.30 mμ) so that the determination of the beginning of the UV-C range is based on biological criteria.

Structure of human skin in relation to influences of natural radiation

Prior to the discussion of the basic effects of natural radiation, the structure of human skin and its bioclimatological absorption capacity for rays of different wave lengths need to be briefly described. First the spectral permeability of individual skin layers must be mentioned.

Fig.17 shows the spectral permeability distribution of the main skin layers. The stratum corneum and stratum germinativum show the maximum permeability in the human eye's greatest brightness sensitivity range, whereas for stratum germinativum and papillary layer ($d = 2.0$ mm) the corresponding maxima are shifted towards the long-wave end of the visible spectrum. In the former case permeability reaches 70–90%, in the latter, however, only 20–40%. At the end of the solar spectrum (290 mμ) as well as beyond short-wave infrared (>1.4 mμ) permeability figures are below 10%. The radiation filtering effect of the stratum corneum (consisting mainly of dead cells) prevents short-wave rays from the UV-C range from reaching the organism.

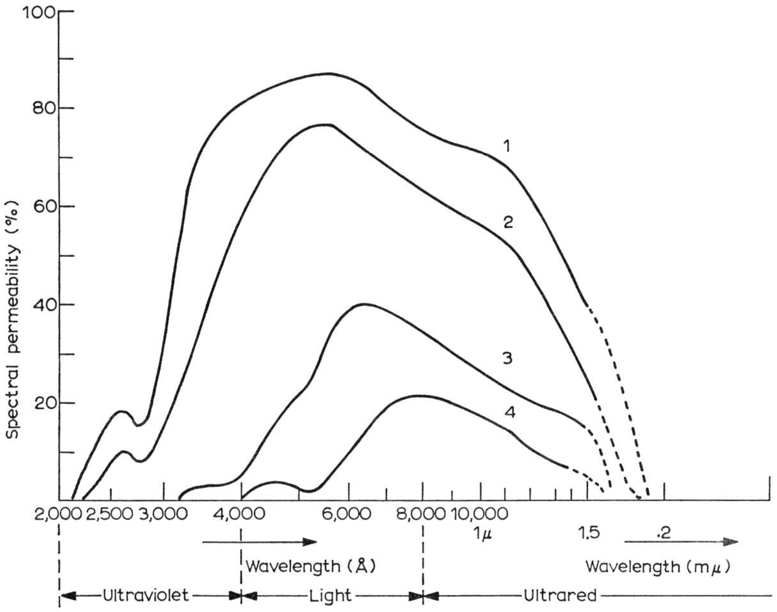

Fig.17. Spectral permeability of different skin layers according to A. Bachem (1931), set up by MEYER and SEITZ, 1949: *1* = for stratum corneum ($d = 0.03$ mm); *2* = for epidermis ($d = 0.05$ mm); *3* = for epidermis and papillary layer ($d = 0.5$ mm); *4* = for entire layer consisting of epidermis and corium ($d = 2.0$ mm).

The main atmospheric filter is the stratospheric ozone which thus protects humans as well as other organisms from the cell-destroying radiation effects of short-wave UV-C (<0.29 mμ). In the infrared radiation above 1.4 mμ the outer skin layers act as filter. This is a consequence of the great amounts of water contained in the tissue, whose permeability limit is precisely 1.4 mμ.

In connection with these special absorption and permeability features of human skin its spectral reflection characteristics deserve the same attention. Fig.18 exhibits the reflection spectrum of human skin (BUETTNER, 1938) under average conditions. It indicates that of the entire terrestrial UV only 1–10% are reflected. Similarly above 2.0 mμ the reflectivity is less than 5%. In the visible range, however, reflection values between 20 and 45% are observed.

These are the values for vertical incidence; at a slant especially on oily or wet skin the

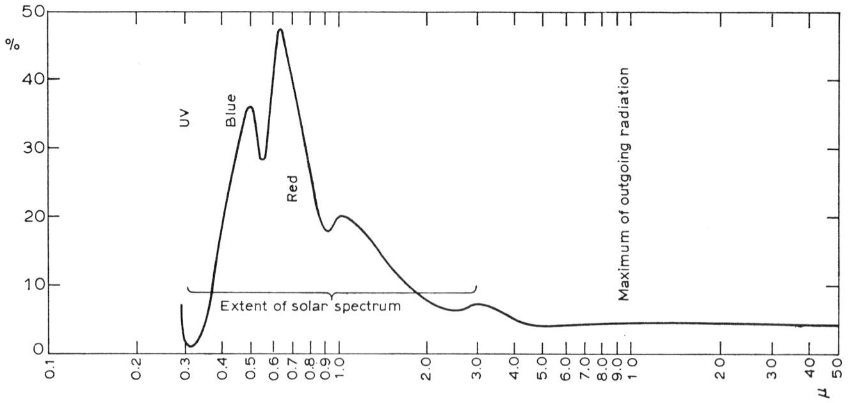

Fig.18. Reflection spectrum of human skin under average conditions for the spectral range 0.3–50 μ (BUETTNER, 1938).

figures lie higher. The skin's spectral reflection is simultaneously the absorption curve for blood pigment, because the reflection originates partly in deeper skin layers (PFLEIDERER, 1938).

It seems obvious that the skin's reflection characteristics depend largely upon the state of its surface. It is therefore understandable that such measurements cannot give a uniform picture of the spectral reflectivity distribution.

Notable differences in skin surface reflection are shown in Fig.19 where besides the so-called normal reflection the curves are shown for reflection with skin erythema and skin pigment. In both cases there is a considerable reduction of reflection by more than 50%, whereby at 55 mμ in the case of the erythema, reflection rises almost immediately to the vicinity of normal values, while for pigmented skin this process is merely doubled (in the erythema case it was a 4-fold increase compared with the minimum of 0.55 mμ).

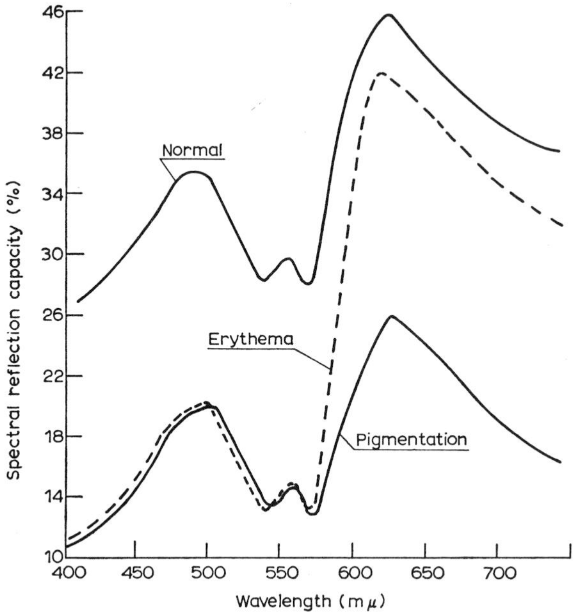

Fig.19. Spectral reflection capacity of human skin in a normal condition, with erythema and with pigmentation (MEYER and SEITZ, 1949).

The normal skin of a central European compared with the skin of black people shows in the visible spectral range a reflection of only 10%, in the IR-A range, however, it is the same for white and black people.

Ultraviolet radiation

Ultraviolet radiation and its biological effects

The notable effects of ultraviolet radiation are most varied in nature. These are caused by the fact that a great many organic compounds (albumen and its component amino acids) have their absorption ranges within the UV. Photochemical reactions take place according to the socalled Grotthus-Draper law that only absorbed radiation can have any effect. Absorption depends on the structure of atoms and their arrangement within the molecules. These only absorb rays having a quantum energy necessary to change an electron to a higher orbit.

One of the best-known effects of natural ultraviolet radiation is the photochemical reaction attributed essentially to UV-B between 0.315 and 0.295 mμ. This reaction takes place in certain substances found in living tissue as an immediate consequence of radiation absorption. Substances thus changed or formed are biologically active. They lead immediately or by diffusion or by transport of body fluids to the physiological effects observed as reactions to UV radiation.

The most familiar photochemical reactions of UV-B is erythema formation (skin reddening) systematically investigated with monochromatic light by HAUSSER and VAHLE (1922) and later by COBLENTZ and STAIR (1934). They elaborated the erythema effectiveness curve given in Fig. 20. This curve has two peaks with a main maximum at 0.299 mμ and a secondary maximum around 0.25 mμ. The main maximum lies within the range of short-wave UV-B occurring in nature, but the other maximum lies beyond the

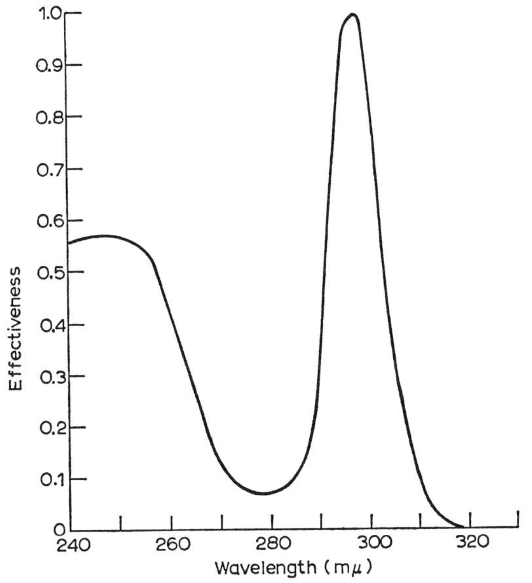

Fig.20. Erythema effectiveness curve proposed by the International Illumination Committee (MEYER and SEITZ, 1949).

terrestrial solar spectrum. Furthermore, the graph shows that above 0.315 mμ (i.e., in the UV-A range) there is no erythema effect.

The term "erythema threshold" is frequently used in connection with applications of UV-B radiation, be they natural or artificial. It describes the weakest but, compared with non-irradiated surroundings, nevertheless clearly defined skin reddening.

Erythema sensitivity varies within wide boundaries. It depends on race, age, sex, season and hair color. Blondes for instance, are 40–170% more light-sensitive than brunettes. Dependence upon season and sex is in connection with hormonal activity of the thyroid, the adrenal cortex and pituitary glands. UV sensitivity occurs in the following decreasing order: chest, stomach, back, face, upper and lower extremities.

Erythema is the most obvious effect of UV-B and has been used for comparison between natural and artificial UV radiators (e.g., medical "sun lamps": mercury vapour lamp or ultravitalux lamps) and as guideline for therapeutical dosage.

Another obvious sign of UV skin irradiation is "pigment formation" (skin tanning). This follows the formation of dark melanine bodies under the influence of radiation at the transition from epidermis to papillary layer. Under the influence of the erythema's inflammation the pigment bodies rise. According to investigations by HENSCHKE and SCHULZE (1939) the pigmentation caused below 0.31 mμ is less a consequence of erythema but a secondary feature of cell decay. This type of skin tanning is therefore called "late pigment". On the other hand, a further pigment is formed between 0.300 and 0.42 mμ in the range of UV-A radiation, which is called "direct pigment". This is an intenser colouring process of the actually present pigment particles. This immediate pigmentation therefore presupposes an earlier "secondary pigmentation" under the influence of radiation with short-wave UV.

Erythema often serves for evaluating UV-B irradiation in various climates and at different times of day and of year. In radiation therapy, particularly in "heliotherapy", the product of radiation intensity and duration of influence play an important part. It is measured in Wsec cm^{-2} and called "irradiation" or "dose". The measure is the incoming energy and not the one absorbed by the irradiated body. In practical cases the erythema threshold dose serves as basis.

A further photochemical effect of UV is the formation of vitamins, mainly the production of vitamin D_3 in human skin. The principal contribution is made by the UV-C range but also by UV-B (up to 0.31 mμ). Vitamin D_3 has special therapeutical effects, such as the cure of rickets and related illnesses, e.g. spasmophilia (cramps in babies and small children), tetany, and osteomalacia (demineralization of bones).

Cell destruction is the basic criterion of the UV-C and UV-B components in therapeutical radiation applications. Epidermis cells of human skin are often involved. The dead cells and the substance issuing from them are resorbed and undergo certain changes. The cell destruction occurring in a UV erythema in the outer skin layers has a different therapeutical effect than the one occurring in an X-ray erythema.

The principal bactericidal effect of UV radiation lies in the UV-C range (maximum between 0.25 and 0.27 mμ); in the outermost UV-B range, however, between 0.28 and 0.30 mμ notable effects exist which may show up under favourable atmospheric conditions (solar radiation with transparent air). Spectral effectiveness curves for individual bacteria types show no particular differences. Essentially the destruction of microorganisms consists in the effect of UV-C radiation on the cytoblast. The effect is statistical

in nature. Hence total elimination of bacteria and viruses is possible only with sufficiently large doses.
UV radiation is detrimental to persons with Basedow's disease (hyperactivity of thyroid gland), for those with nervous over-excitability and for open lung tuberculosis. Otherwise the therapy of bone and joint tuberculosis is the paradigm for the value of a systematic dosis of UV rays from sun and sky when healing sick people (BERNHARDT, 1933; ROLLIER, 1951).
The following effects of UV radiation are observed in a normal healthy person.
(*1*) Influence on metabolism (increase of basic metabolism, increased albumen degradation, etc.).
(*2*) Influence on breathing (deeper breathing leading to a better utilization of oxygen).
(*3*) Influence on blood circulation (lowering of blood pressure).
(*4*) Influence on blood quality (increase of erythrocyte number and of hemoglobin).
(*5*) Influence on glands with internal secretion (e.g., increased activity of thyroid gland).
(*6*) Influence on general state of health (increased performance, reduced susceptibility to colds, influenza, etc.).

Climatology of natural ultraviolet radiation and climatic dosage

Climatology of natural ultraviolet radiation was yet developed just a few decades ago. Historically, at the beginning of this century repeated serious efforts were made to promote this important specialized field of bioclimatological radiation research (C. Dorno, F. W. P. Goetz, K. Buettner, and others). At that time difficulties were encountered in the measurements of UV radiation. These difficulties were successfully overcome only in the years following World War II when a revival of ultraviolet radiation climatology set in (P. Bener, R. Schulze, and others).
The intensity of ultraviolet radiation depends on various parameters whose behaviour is partially even known to the layman. Extra-terrestrial UV radiation is weakened on its way through the atmosphere, like several other types of radiation, by scattering and absorption through aerosol particles (air molecules, haze, other compounds of chiefly foreign nature) and by the ozone absorption. Air molecules participate principally in the diffusion of short-wave radiation but UV is principally absorbed by stratospheric ozone, mostly between 22 and 28 km. Below 300 mμ very little of the incoming radiation penetrates. But the attenuation of short-wave UV radiation increases rapidly towards shorter wavelengths. Depending on the aerosol type the atmosphere's permeability in the UV-C range lies at approximately 70%, and by way of UV-B to the UV-A range it increases by about 10%. Pertinent investigations by R. Schulze and P. Bener confirm that the influence of atmospheric turbidity by haze is relatively small.
Thus the aerosol quality of the lowest atmospheric layers has a powerful influence on the nature of UV radiation. Due to scattering of short-wave radiation on air molecules, sky radiation assumes a considerably greater effect in UV global radiation. Furthermore, the intensity of UV radiation depends greatly on solar elevation and hence in considerable differences in this quantity by time of day and season. A related feature is the dependence upon latitude. Altitude too plays a significant part in therapeutical applications of UV radiation in the mountains.
The bioclimatological importance of UV radiation requires some special climatological

considerations. P. Bener's spectrophotometric investigations furnish valuable clues because they are based upon certain wavelengths. Fig.21 exhibits the diurnal variation of UV global radiation (T_{UV}) and the vertical component of direct solar radiation (S_{UV}) as well as sky radiation (D_{UV}) for wavelength 310 mμ in UV-B (in Davos at 1590 m) on cloudless days in June. Interesting ratios appear between the amount of solar radiation and sky radiation. In the morning and evening hours UV sky radiation outweighs UV solar radiation. Only during the noon hours does solar radiation enjoy predominance. However, UV solar radiation exceeds the corresponding sky radiation only by approximately 12%. These facts are not easily seen from the behaviour of UV-B global radiation, but they are very important to bioclimatology.

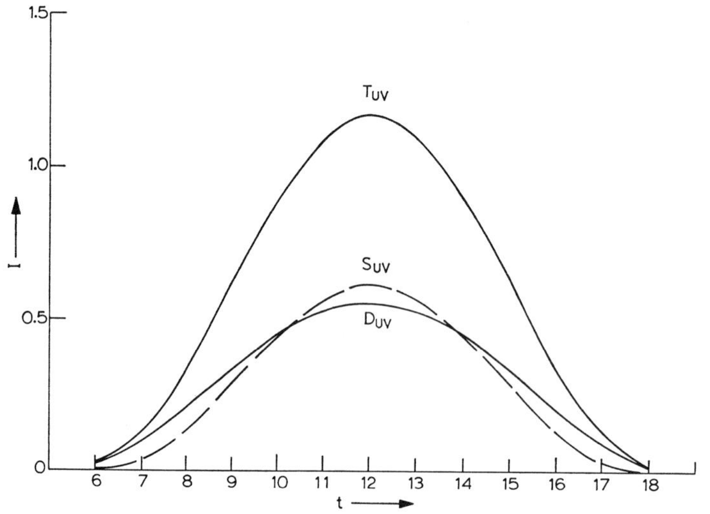

Fig.21. Diurnal variation of UV global radiation T_{UV} (for λ = 310 mμ), of the pertinent UV sky radiation (D_{UV}) and of the vertical component S_{UV} of direct solar radiation in Davos (1,592 m) on cloudless days in June. I = intensity in energetic units (W cm^{-2}); t = true solar time (TST). (BENER, 1964.)

The annual variation of ultraviolet noon intensities of global radiation in Davos for various wavelengths in the UV ranges A and B is instructive. Fig.22 displays the greater annual amplitude of short-wave UV-B (300–315 mμ) compared with those in UV-A (315–380 mμ). This is caused mainly by the strong ozone dependence of UV-B. The annual UV radiation variation is strongly influenced not only by the thickness of the ozone layer and its seasonal change, but also by the seasonal variation of solar elevation and by the changes in albedo (especially in Alpine regions caused by the appearance and disappearance of the snowcover). From the bend in the UV curves during the period April–May the disappearance of the snowcover, its formation at the beginning of November and thus the special effects of ground reflex radiation can be recognized. Maximal UV radiation (with exception of albedo phenomena) is observed not at summer solstice but in July, but the months June through August show on the whole only small differences. At this time of year the influence of the ozone layer's decreasing thickness becomes noticeable.

Table IX gives intensity ratio values (cloudy sky : cloudless sky) to show the influence of sky cover (overcast degree 8–10/10) on UV sky radiation in the presence of different cloud types.

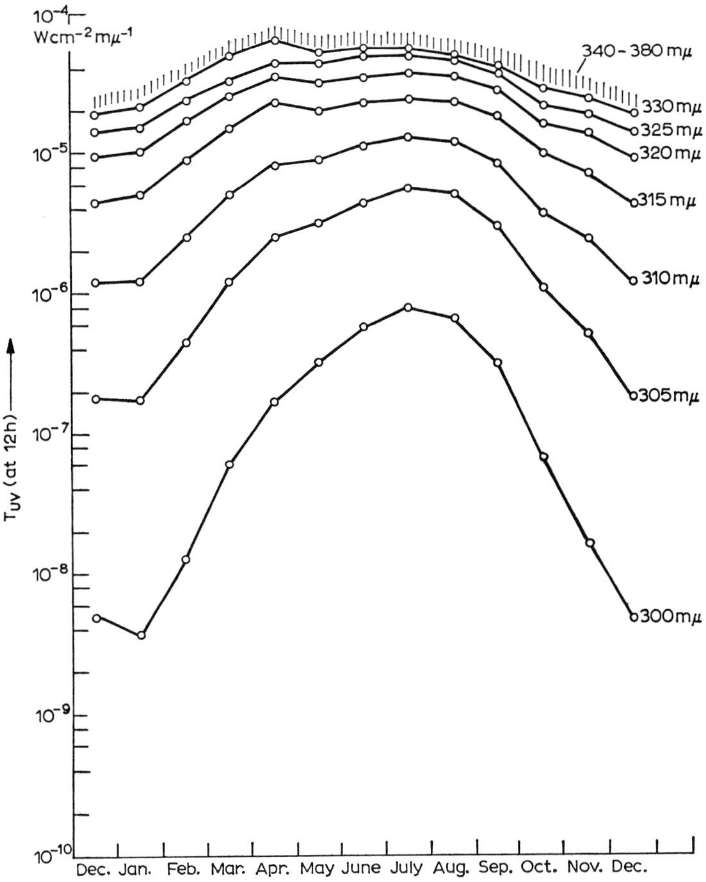

Fig.22. Annual variation of noon intensities of UV global radiation T_{UV} for different wavelengths, in Davos (BENER, 1964).

TABLE IX

MEAN QUOTIENT (Q) OF UV SKY RADIATION CONSIDERED FOR ALL SOLAR ELEVATIONS
(After BENER, 1964)

	λ: 330 mμ Q	370 mμ Q
Lower clouds (incl. Cu and Cb)	0.75	0.86
Medium-high clouds	0.87	0.92
High clouds	1.10	1.23

It can be concluded that high clouds (Ci, Cs) increase sky radiation to an important extent which is also observed partially in the presence of not too dense medium high clouds. On the other hand, very dense and very tall clouds (St, Cb) absorb considerable amounts of energy. This absorption is greater for shorter wavelengths than for longer ones.

For the practical application of UV radiation dosage tables and diagrams have been established (K. Buettner, H. Pfleiderer, R. Schulze, R. Lotmar, and others) and form valuable aids for climatic therapy at sanatoria and hospitals.

The existence of dosage diagrams for the application of UV radiation and also the use of the varied protective means to reduce and divert harmful effects of intensive UV radiation should not mislead from the fact that these deceptive natural forces hide dangers. In the past decades, however, the therapeutic use of UV radiation has lost a lot of its importance except for dermatological cases. Three main reasons for this change in attitude are presented by TRONNIER (1972). For numerous illnesses the newer chemical medicines work with greater certainty and selectivity than the earlier unspecific therapy. But in the last decades there has been a great desire for sunshine, dictated by fashion and cosmetics, often leading to excessive solar irradiation. This may result in: (*a*) a reduction of deficiency symptoms (e.g., rickets); (*b*) an increase of UV-conditioned light damages (e.g., light dermatosis); (*c*) UV carcinogenesis.

The 1st International Conference for the Study of Biological Effects of UV Radiation in Philadelphia (U.S.A.) in 1966 gives reason to take into account a new approach to UV heliotherapy (cf. URBACH, 1969; as well as indications on p.47, etc.).

Light carcinomas are almost exclusively observed on uncovered parts of the body. The genesis of malignancies by "light" lies in the UV-B range below 320 mμ, i.e. in the same range as the erythema effect. In order to produce carcinomas generally intensive irradiation and high radiation doses are required, whereby the thresholds of customary adaptation are surpassed (MIESCHER, 1960). It is interesting to note that the pigmentation of black people protects them against light so that they rarely develop UV-induced carcinomas. Urbach has dealt furthermore with the geomedical aspects of this significant problem complex. Most recently ROBERTSON (1972) furnished research results reached in Australia with a measuring system he constructed for the determination of erythema effective UV radiation*. His observations are based on continuous series of measurements over several years, ranging over an area from 38°S to 53°N.

Certain groups of Australia's inhabitants whose bodies are exposed during long periods of the year to comparatively high UV intensities show frequent skin carcinomas on their heads and upper extremities. Such chronic radiation damages of the skin as irreversible alterations caused by a summation of effective UV-B radiation during an individual's whole life time have been interpreted geophysically by Robertson with regards to the required radiation doses. Therefore, modern medicine warns of overdoses of natural UV-B radiation. This goes also for late winter Alpine climates when the UV snow reflection often considerably increases the corresponding circumglobal radiation. A "suntan" should only serve a true health purpose and not be a social prestige symbol (SCHMIDT-KESSEN, 1965).

All this points up the vital importance of the stratospheric ozone layer. Its diminution by anthropogenic influences could lead to increased skin cancer rates and other detrimental influences in the biosphere. The stratospheric photochemistry is very complex but the introduction of oxides of nitrogen, for example from the exhausts of high flying aircraft and of chlorofluorocarbons by diffusion from the surface furnish chemical species which are apt to destroy stratospheric ozone. Over 60 chemical reactions take place at the level of the ozone layer and the system is not yet fully understood. But as precautionary measure the use of chlorofluorocarbons as propellants in aerosol spray cans is being phased out. A close surveillance of chemical species in the atmosphere and the ozone concentration aloft has been instituted.

* Urbach (Philadelphia) has utilized this UV-B measuring apparatus in a partly further developed form in an extensive network for world-wide information.

Infrared radiation and radiation balance

Infrared radiation and its biological significance

The characteristics of ultraviolet causing effects on humans are absent in the infrared radiation range. IR radiation has no photochemical effect because its quantum value does not suffice for the dissociation or photochemical creation of organic compounds. The main feature of infrared is therefore a more or less unspecific heating effect. It possesses a physiological and therapeutical value which is utilized in artificial IR radiators (heating sun lamps, etc.). IR radiation also can cause an erythema, the socalled "heat erythema". It appears almost immediately together with irradiation and disappears shortly afterwards. A familiar consequence of IR radiation is "hyperaemia", i.e. intensified blood circulation. The natural temperature drop in the skin is reversed and thus stimulates the heat regulating mechanisms.

The following effects are the most important consequences of hyperaemia (especially under artificial IR radiation).

(*1*) Improved nutrition of body organs and thus increased function.

(*2*) Increased resorption together with increased secretion of pathological matter.

(*3*) Bactericidal and inflammation-inhibiting effect.

(*4*) Anodyne effect due to arterial circulation.

Finally, IR radiation covers energy deficits in the organism and therefore has a high biological significance.

Energy deficits in the human organism can occur in different ways, e.g., by environmental temperatures (air or water temperatures) lying far below the mean skin temperature (33°C) or by intensive heat loss of uncovered skin surfaces. The decisive factor is the temperature difference between skin and environment. The extent to which such energy losses can lead is shown in Fig.23. Hence, considerable energy losses are registered for humans already at normal temperature differences.

The mean normal heat loss in humans amounts to approximately 16 kWh per day. By way of nourishment man acquires only 4 kWh per day, so that an impressive energy

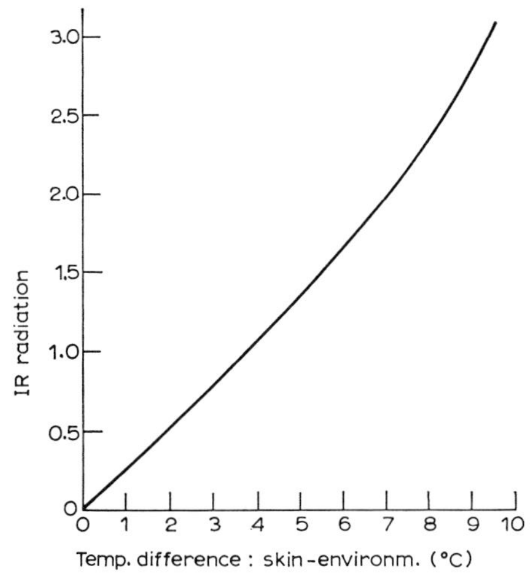

Fig.23. Loss of energy in humans (in kWh d^{-1}) in dependence upon the difference: surface temperature minus radiation temperature of surroundings (SCHULZE, 1970).

deficit becomes obvious and must be covered in some way if human life is to be protected. Energy compensation takes place during the daytime by the heat radiation of sun and sky, and outdoors by the infrared back radiation of the atmosphere (G) effective day and night (indoors it is carried out by radiators, walls, etc.). Furthermore, clothing forms the necessary heat protection (inhibiting loss) which humans need to a greater or lesser extent during the change of climates and seasons. Because the atmospheric back radiation plays a dominant part in covering the natural energy losses its characteristics shall be described here.

Atmospheric back radiation and heat radiation of the ground and of human skin

Atmospheric back radiation constitutes the atmosphere's own radiation. It has its greatest intensities in the IR range. These depend largely on the air temperature, the water vapour content and the degree of cloud cover.

Fig. 24 shows the spectra of the atmosphere's infrared back radiation which plays an extraordinary role also in the earth's heat balance. The top graph shows the minor

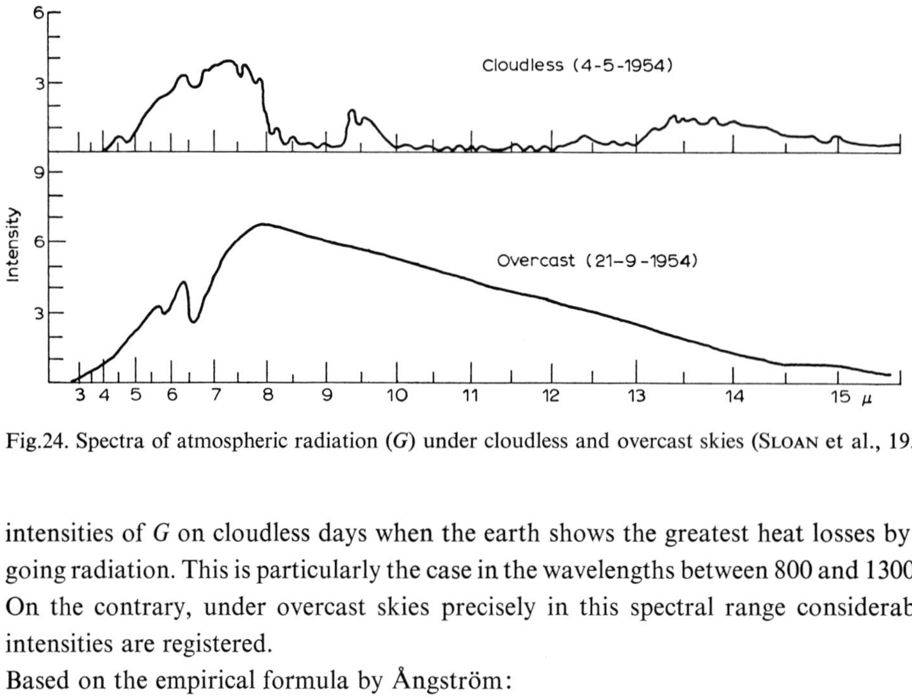

Fig.24. Spectra of atmospheric radiation (G) under cloudless and overcast skies (SLOAN et al., 1955).

intensities of G on cloudless days when the earth shows the greatest heat losses by outgoing radiation. This is particularly the case in the wavelengths between 800 and 1300 mμ. On the contrary, under overcast skies precisely in this spectral range considerable G intensities are registered.

Based on the empirical formula by Ångström:

$$G(w) = \sigma \cdot T^4 \left(u - v \cdot \frac{10-w}{10}\right)$$

it is possible to reach an approximative determination of IR intensities of atmospheric radiation for different air temperatures (T), vapour pressures (e) and degrees of overcast (w). In the following Table X the necessary data are furnished.

The result is that with increasing cloudiness the atmosphere's infrared radiation rises and generally the highest values are reached under a completely overcast sky. The type of clouds as well as the level of their lower limit play a notable part. Atmospheric radia-

TABLE X

CALCULATION OF ATMOSPHERIC RADIATION ($G(w)$) IN DEPENDENCE UPON CLOUDINESS (w) AND VAPOUR PRESSURE (e) ACCORDING TO THE ÅNGSTRÖM FORMULA
(After KNEPPLE, 1956)

e	u	v	e	u	v
0.5	0.946	0.319	8.0	0.960	0.234
1.0	0.947	0.310	9.0	0.962	0.228
1.5	0.949	0.302	10.0	0.963	0.222
2.0	0.950	0.295	11.0	0.964	0.217
2.5	0.951	0.288	12.0	0.964	0.212
3.0	0.952	0.281	14.0	0.965	0.206
3.5	0.953	0.275	16.0	0.967	0.200
4.0	0.954	0.269	18.0	0.967	0.195
4.5	0.955	0.263	20.0	0.968	0.192
5.0	0.956	0.258	25.0	0.969	0.187
6.0	0.958	0.249	30.0	0.969	0.183
7.0	0.959	0.241			

$$G(w) = \sigma T^4 \left(u - v \frac{10 - w}{10}\right)$$

$u = 0.970 - 0.026 \cdot 10^{-0.055\,e}$ (e = vapour pressure, in mm Hg)
$v = 0.180 - 0.148 \cdot 10^{-0.055\,e}$ (w = cloudiness degree, in tenths)

tion with world-wide temperature and humidity conditions is largely dependent upon latitude, as the calculations by BAUR and PHILIPS (1935) demonstrate.

0–10	10–20	20–30	30–40	40–50	50–60	60–90	(°N latitude)
557	542	502	461	410	375	310	(G, cal. $cm^{-2}min^{-1}$)

Furthermore, horizon screenings often have an intensifying effect upon atmospheric radiation conditions.

Due to a decrease of air temperature and vapour pressure with altitude a reduction of atmospheric radiation is found in the mountains. Table XI gives average daily amounts of atmospheric radiation for various altitudes and cloudiness degrees 0 and 10 in the annual variation.

TABLE XI

DAILY AMOUNTS OF ATMOSPHERIC RADIATION G (cal. cm^{-2}) AT EXTREME DEGREES OF CLOUDINESS DEPENDING UPON ALTITUDE (EASTERN ALPS)
(After SAUBERER, 1954)

Altitude (m)	Jan.	Mar.	May	July	Sept.	Nov.
	Cloudiness 0/10:					
200	450	468	627	689	631	511
1000	420	466	571	631	590	473
2000	395	413	501	570	530	430
	Cloudiness 10/10:					
200	600	662	771	830	776	670
1000	577	633	732	792	750	633
2000	550	573	668	738	691	592

These figures show that atmospheric radiation is reduced with increasing altitude, and does so especially under cloudless conditions, whereas under overcast skies this phenomenon is less marked. This example once more underlines the fact of the increased back radiation under a closed cloud cover compared with conditions with few clouds.
Just as the atmosphere emits a long-wave radiation (IR radiation) to the earth and to outer space, the solid and liquid earth surfaces emit radiative energy according to their temperature given by the Stefan-Boltzmann law. In the formula:

$$E\,(= \text{long-wave emission}) = \sigma \cdot \varepsilon \cdot T^4$$

σ signifies the Stefan-Boltzmann constant with a value of $0.826 \cdot 10^{-10}$, ε, the emissivity, a material constant (in the ideal black body = 1.0) and T the absolute temperature in °K of the emitting medium. In nature ε oscillates between 0.90 and 0.99 (in the case of human skin it has a value of 0.54). Therefore, the heat loss of human skin due to emission against a clear sky amounts to the following:

t_L	−30	−20	−10	0	+10	+20	+30 (°C)	
t_H 30°	0.463	.433	.418	.364	.297	.216	.148	(cal. $cm^{-2}min^{-1}$)
t_H 40°	0.548	.521	.494	.447	.384	.303	.234	

These data (Buettner) show that the heat emission of horizontally placed human skin by way of long-wave emission at low air temperatures and mean values of relative humidity (at 50%) causes severe energy losses which at −30°C and +30°C differ by a factor between 2 and 3. As a comparison a few annual variations of daily amounts of long-wave emission of horizontal surfaces are given here.

Radiation balance and its bioclimatological significance

Radiation fluxes occurring in the atmosphere in their changing seasonal behaviour also determine the bioclimatological interplay. Hence a short consideration of the "radiation balance" is necessary. We define:
Short-wave global radiation $T = S + D$.
Reflection of T on earth's surface = R_T.
Long-wave emission of ground $(E) = E + r$ (r signifies the reflected atmospheric radiation the measurement of which is technically related to the determination of E).
Atmospheric radiation = G.
The difference $T - R_T$ is called short-wave radiation balance and the difference $(E + r) - G$ long-wave radiation balance. Total radiation balance Q (called radiation balance for short) is calculated as follows:

$$Q = (T - R_T) - (E - G + r)$$

From the radiation balance equation and the graph in Fig.25 furnishing an annual review of value Q and its individual components for a 10-year period in Hamburg (SCHULZE, 1970) it results that the values $E + r$ and R_T are the loss figures. The ground's

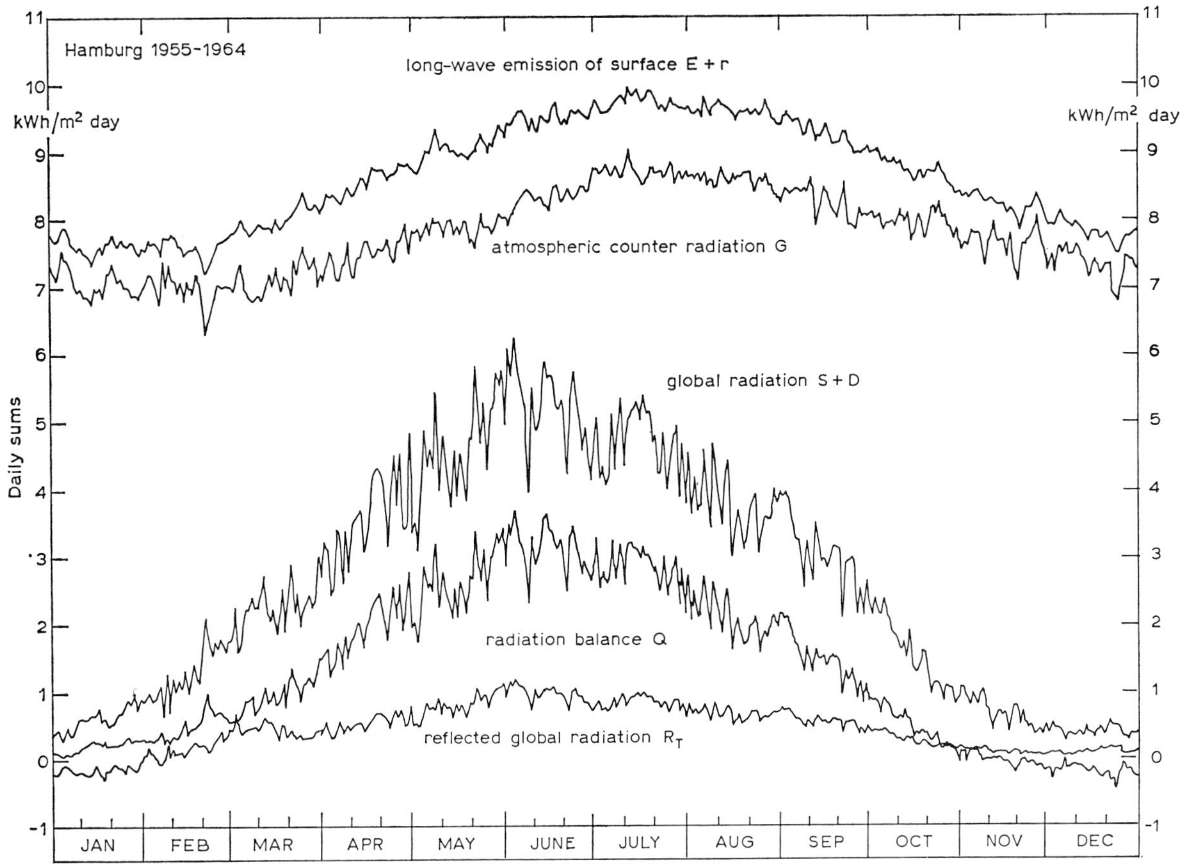

Fig.25. Mean annual variation of daily amounts of the radiation balance and its components, in Hamburg (1955–1964) (SCHULZE, 1970).

long-wave emission carries the main weight. This could not be "neutralized" by day by heat radiation of $S + D = T$ if the highly effective atmospheric radiation did not enter the picture. This circumstance lets the radiation balance Q reach positive values of the greater part of the year. Only in midwinter does Q show negative values when short-wave incoming radiation sinks to very low intensities. Long-wave emission as well as atmospheric radiation have their annual maxima in July, whereas global radiation exhibits its highest figures at the beginning of June. The annual amplitude of the latter is very marked, while the former shows only small annual oscillations due to the almost year-round even course of radiation values. In addition, Fig.25 reveals that the ground's long-wave emission ($E + r$) exceeds global radiation T in all seasons by a considerable measure. The surface loss is four times larger than the global radiation received. But the

TABLE XII

DAILY AMOUNTS OF LONG-WAVE RADIATION $E + r$ AT GIVEN POINTS (in mcal. $cm^{-2}min^{-1}$)
(After DIRMHIRN, 1964)

Month:	I	II	III	IV	V	VI	VII	VIII	IX	X	XI	XII
3,000 m	518	506	525	552	594	622	642	610	620	587	550	525
Tiflis (31°34′N)	611	635	693	740	819	909	939	947	848	765	680	636
Sahara	709	745	843	958	1,028	1,040	1,025	1,016	992	870	765	710

atmosphere's back radiation (G) exceeds global radiation T 3.6-fold. Thus it balances the radiation loss by $E + r$ up to a deficit of 10%.
From a bioclimatological point of view the instructive graph informs us that the atmosphere's IR radiation besides energetic radiation fluxes from sun and sky plays a dominating rôle in heat budget of man and of all organisms, especially since this "dark force" is active at all times and assures the continuation of life as developed phylogenetically on the planet Earth. Atmospheric radiation scatters from day to day based only on weather variations (cloudiness), as is the case with all meteorological elements. This scattering, however, has little weight compared with its average seasonal behaviour.

Further consequences of natural radiation fluxes for practical bioclimatology

Radiation related to clothing

A study on light and life must also include some of bioclimatology's practical side. This includes radiation processes in positions other than horizontal, natural and artificial illumination, and concomitant physiological considerations. Radiation fields in water require special attention.
Clothing is a modification of man's environmental climate. Radiation permeability of the used fabrics and their reflection capacity are of bioclimatological interest. Absorption is to be considered at this point, too. Indications on the effective mechanisms of the most varied kinds of materials show the difference of radiation influence regarding clothing. Table XIII leads to the conclusion that thin materials generally have a high permeability especially in the visible spectral range. Besides fabrics or material, colour and humidity content of clothing play a notable part in the very complex radiation processes. Details cannot be given here and reference is made to the pertinent specialized literature. Work-physiologically for clothing used as "radation protection" it is important that this type of clothing has an efficient protective reflector with an insulating layer between two surfaces (skin and clothing).
Further questions of clothing physiology, of air and perspiration exchange, the assurance of the best cold protection and thus the insulation quality of clothing will be taken up again below.

Radiation on slanting surfaces and on surfaces facing different directions

In geophysics measurements of short-wave heat radiation of sun and sky (T) are based on a horizontal flat receiver surface and several examples have already been given. Regarding human life, however, one must be practical. It is understandable that the knowledge of heat and light radiation incident on surfaces which slant and face many directions is of extreme importance. This goes for bioclimatological problems on the whole and in particular also for the interests of agriculture, construction, road building, as well as ecology, glaciology, etc. A study on human bioclimatology requires a consideration of these problems.
Fig.26 shows pertinent facts from which the conclusion is reached that the surfaces facing south at a 30° and 60° slant receive the relatively highest global radiation values during the entire winter semester. On the other hand, as already mentioned, during the

TABLE XIII

SPECTRAL PERMEABILITY, REFLECTION AND ABSORPTION OF DIFFERENT FABRICS OF VARIOUS COLOURS
(After DIRMHIRN, 1964)

	0.4	0.5	0.6	0.7	0.8	0.9	1.0	1.1	1.2	1.3	1.4	1.5	1.6	1.7	1.8	1.9 μ
(a) Reflection of fabrics in percent																
1*	61	64	65	65	64	64	64	64	63	61	58	53	56	56	55	56
3	22	22	23	23	23	23	23	23	23	23	22	22	23	22	22	22
5	57	58	58	58	59	59	59	60	60	60	58	57	58	58	57	57
6	28	29	29	30	30	30	30	30	30	30	29	29	29	29	29	28
7	45	41	32	35	43	50	55	59	60	59	57	55	54	53	53	52
9	8	9	10	12	21	24	33	38	41	41	40	40	40	40	40	39
13	1	1	2	4	20	31	34	41	44	44	40	38	39	39	38	36
14	9	9	9	9	9	9	9	9	9	9	9	9	9	9	8	8
(b) Permeability of fabrics in percent																
1	25	28	31	31	32	31	31	30	30	30	29	28	29	30	29	27
3	65	70	72	73	73	74	74	75	75	75	75	75	74	74	74	73
5	26	30	34	35	36	36	36	36	36	34	32	30	29	28	27	26
6	54	69	64	67	67	67	67	66	66	65	65	65	65	65	65	65
7	14	15	12	15	21	28	33	34	33	31	28	26	26	27	26	25
9	1	1	2	6	9	10	11	11	10	9	8	7	7	7	6	6
13	0	0	0	1	5	14	21	26	27	27	25	23	25	26	25	23
14	88	90	90	90	90	90	90	90	90	90	90	90	90	90	90	90
(c) Absorption of fabrics in percent																
1	14	8	4	4	4	5	5	6	7	9	13	19	15	14	15	18
3	13	8	5	4	4	3	3	2	2	2	3	3	3	4	4	5
5	17	12	8	7	5	5	5	4	4	6	10	13	13	14	16	17
6	18	11	7	4	3	3	3	3	4	4	6	6	6	6	6	7
7	41	44	56	50	36	22	12	7	7	10	15	19	20	20	21	23
9	91	90	88	82	70	66	56	51	49	50	52	53	53	53	54	55
13	99	99	98	95	75	55	45	33	29	29	35	39	36	35	37	41
14	3	1	1	1	1	1	1	1	1	1	1	1	1	1	2	2

* 1 = artificial silk, white; 3 = cotton etamine; 5 = linen; 6 = nylon; 7 = flannel; 9 = grey sheep's wool; 13 = black wool; 14 = nylon (hose, brown).

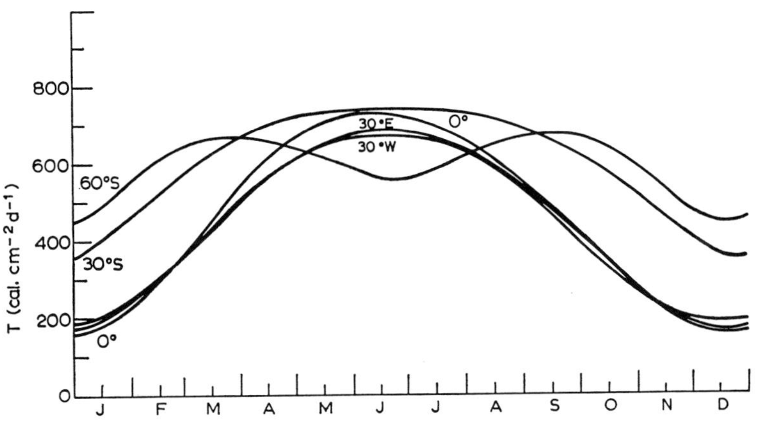

Fig.26. Annual variation of daily amounts of global radiation T (cal. cm^{-2}d^{-1}) on cloudless days on a horizontal surface, on one at a 30 and at a 60° slant facing south, and two surfaces slanting at 30° to east and west, in Locarno (SCHRAM and THAMS, 1967).

summer semester the surface facing south at a 30° slant as well as the ones facing east and west at 30° slants profit most by sky and solar radiation.

Special protective measures against heat and light radiation

Protection against heat and light radiation, such as used for glass housings of measuring or observation points as well as for protective masks and eyeglasses needs mentioning. Frequently radiation reflecting glass is used. In large halls with glass walls a special protective glass should be used in order to assure as great as possible a protection against heat and light radiation especially from the sun. Permeability for various wavelengths of different materials plays a decisive part. Table XIV shows the spectral transmission of common windowpane glass (%). It may be seen that the greatest permeability lies in the visible range and near infrared, as corresponds with general experience.

TABLE XIV

SPECTRAL TRANSMISSION (%) OF COMMON WINDOWPANE GLASS
(After LANDSBERG, 1954)

Wavelength (Å)	3,200	4,000	5,000	6,000	7,000	10,000	20,000	27,000	40,000	45,000
Transmission (%):	0	88	90	90	88	85	85	30	15	0

Transparent glass which must sufficiently absorb heat radiation should have a transmission effective less in the visible range than in the infrared one. Often absorbing glass has a self-heating characteristic which turns it into a heat radiator in its own right. Cooling air flowing past the protective glass can effectively eliminate this phenomenon. On the other hand, reflecting glass heats up much less. For eye protection glasses often nylon is used, whose permeability corresponds with that of windowpane glass in the visible range, but in the infrared range sinks to minor values. Reflecting wire fabrics are effective as protection against heat radiation by reflecting up to 75%.

Protection from ultraviolet radiation (UV-B radiation) is a complex problem. The application of heliotherapeutic cures is a point of decisive bioclimatological weight not only for normal dosage of UV-B radiation but also for the protection against radiation overdoses as well as the abuse of these natural light powers. Based on biological experiments and radiation measurements SCHULZE (1970) determined values (Table XV) of radiation consumption by humans in various spectral ranges, whereby a sunbath was extended until the appearance of the first sun erythema. The results show that under UV radiation at a solar elevation of 50° only half the time is required for a sun erythema than at a solar elevation of about 30°.

Sunburn in its most dangerous forms can be prevented, besides by a sensible dosage, by light-protective means with radiation-absorbing substances, such as lotions and oils, etc. There is, however, a certain adaptation to the influence of ultraviolet radiation by a thickening of the skin's outermost stratum corneum. This leads to the formation of the socalled "light callous" which makes further UV radiation absorption more difficult or completely inhibits it. Hence an adaptation to UV radiation in Alpine regions is not caused by pigment. In the choice of their substance the above-mentioned light-protection means have been tested as to their spectral absorption. In order to have a measure for the efficiency of light-protective substances, adjoing skin surfaces are irradiated with

TABLE XV

INDEX FOR UV RADIATION CONSUMPTION BY HUMAN BEINGS FOR VARIOUS SPECTRAL RANGES UNTIL THE APPEARANCE OF THE FIRST SUNBURN; VALUES FOR 35° SOLAR ELEVATION = 1
(After SCHULZE, 1970)

Solar elevation	(UV-A) :(UV-B)	violet :(UV-B)	blue :(UV-B)	green :(UV-B)	yellow :(UV-B)	red :(UV-B)
	Clear days					
20°	2.4	2.7	2.8	2.8	2.7	3.1
25°	1.6	1.7	1.7	1.8	1.7	1.9
30°	1.3	1.3	1.3	1.4	1.3	1.4
35°	1.0	1.0	1.0	1.0	1.0	1.0
40°	0.9	0.8	0.8	0.8	0.8	0.8
45°	0.8	0.8	0.7	0.7	0.7	0.7
50°	0.7	0.7	0.6	0.6	0.7	0.6
	Half-overcast days					
20°	1.7	1.8	1.8	1.9	1.8	2.0
25°	1.4	1.5	1.5	1.5	1.5	1.6
30°	1.2	1.2	1.2	1.2	1.2	1.3
35°	1.0	1.0	1.0	1.0	1.0	1.0
40°	0.9	0.9	0.9	0.9	0.9	0.9
45°	0.7	0.7	0.7	0.7	0.7	0.7
50°	0.6	0.6	0.5	0.5	0.5	0.5

and without light protection means (Lp) and thus one may determine the duration of irradiation until the first signs of erythema appear (erythema threshold). Light protection factor Q:

$$Q = \frac{\text{duration of irradiation with Lp}}{\text{duration of irradiation without Lp}}$$

indicates by how much time a sunbath must be prolonged when using a particular light-protection means, and what defensive values this substance possesses. Manufacturers of UV light-protective products (for cosmetic application) must take care that they are largely compatible with mucous membranes and skin, non-toxic and preventing allergic effects and that their substance is adapted to the human stratum corneum (adhesive quality despite perspiration and water) (HOPPE, 1973).

Special light effects of a physiological and psychological nature

All complex questions of light protection have yet another aspect in connection with the phenomenon of physiological and nervous fatigue, inasmuch as insufficient lighting can be the cause of industrial accidents and, on the other hand, an increase in illumination intensities may lead to heightened performances. Overtaxation of the visual apparatus may be followed by a tiring of the eyes (burning eyes, reduction of sharp eyesight, headaches, etc.) as well as by nervous fatigue with a reduction of psychic and motor functions. The latter express themselves in lassitude, sleeplessness, etc. Fig.27 shows that work performance rises rapidly with increasingly improved illumination in the lower lux value range, but only slowly in the higher illumination intensity range. This process is connected with a reduction of relative fatigue, however, only up to a certain illumination

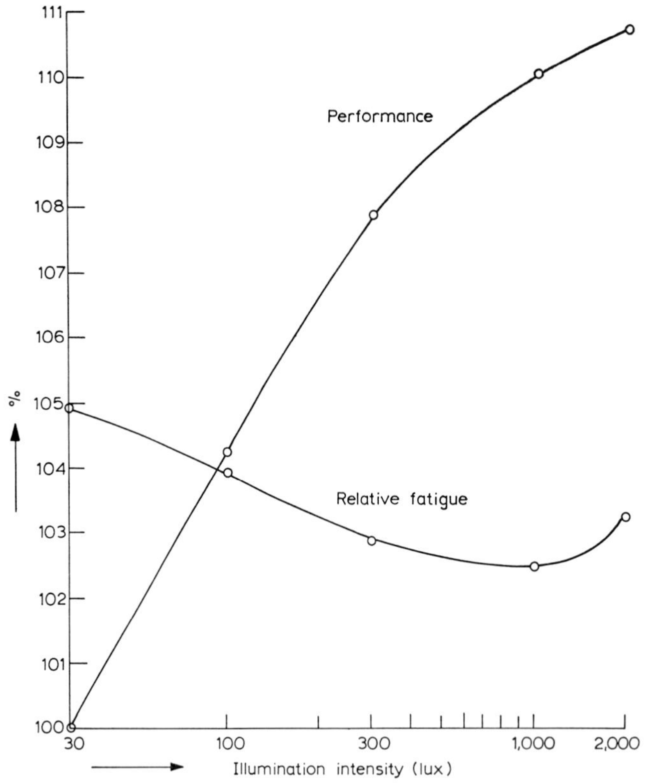

Fig.27. Effects of illumination intensity of work performance and fatigue (GRANDJEAN, 1967).

intensity (about 1,000 lux). At further increased lighting intensity fatigue again increases. Besides the above information on the varied consequences of artificial lighting manipulation for example at working sites, the problem of daylight illumination arises. Neutral daylight in living and working quarters has, besides its function of light supply, a significance in being a connection to the outside world, by allowing a view into the surroundings. For light technology the calculations are based on sky light, i.e., in this case the light intensity of an overcast sky. The daylight quotient is used as a measure for daylight illumination evaluation. It represents the ratio of outdoors illumination intensity to indoors illumination intensity, as follows:

$$TQ = \frac{E_p}{E_a} \times 100\,(\%)$$

where TQ is the daylight quotient, E_p illumination intensity at the measuring point and E_a illumination intensity of a horizontal surface outdoors under an evenly overcast sky. Thus the desired daylight illumination can be expressed by means of the daylight quotient, whereby 5,000 lux are taken as minimal basic values for E_a (cf. Table XVI).
In conclusion a few remarks on the psychological effects of colours are in order (cf. Table XVII).
The most important effects are distance illusions, temperature illusions and influence on the psychological atmosphere. Generally speaking, all dark colours are oppressing and discouraging, while bright colours display a friendly and cheerful aspect. Often a colour

TABLE XVI

ILLUMINATION RELATIONS

TQ at working site	Lux at working site	Illumination requirements
3%	150	low
6%	300	moderate
10%	500	medium
20%	1,000	high

TABLE XVII

PSYCHOLOGICAL COLOUR EFFECTS
(After GRANDJEAN, 1967)

Colour	Distance effect	Temperature effect	Psychic atmosphere
Blue	Distant	Cold	Quieting
Green	Distant	Very cold to neutral	Very quieting
Red	Close	Warm	Very exciting and disquieting
Orange	Very close	Very warm	Stimulating
Yellow	Close	Very warm	Stimulating
Brown	Very close, confining	Neutral	Stimulating
Violet	Very close	Cold	Aggressive, disquieting, discouraging

effect, especially when viewing scenery, depends not only on the absolute colour tone but also on the colour relations and arrangements (HELLPACH, 1950). The same may be said for the contrast in the juxtaposition of light and dark and for the distribution of light and shadow in nature.

Radiation fluxes as aids in the study of atmospheric turbidity

Radiation measurement and air-hygienic orientation

Environmental protection has been promoted at the highest international levels. Hence, it is essential to elucidate such bioclimatological procedures for radiation and light measurements, and monitor the world-wide dimensions of air pollution and promote its control.

Turbidity phenomena at various levels of the atmosphere and total vertical extent have a considerable influence on the reception of natural radiation fluxes on the earth's surface and hence geophysical phenomena are important for the bioclimatology of terrestrial radiation conditions.

Already during the period following the notorious volcanic eruption of Katmai in Alaska (June 1912) American and European scientists noted a considerably reduced radiation transmission in the earth's atmosphere in the years 1912–1914 and partially even later (1916/17). It was DORNO (1912) who, based on his extensive and varied radiation measurement programmes, observed strong decreases in direct solar radiation intensities compared with normal values in Davos, whereas total sky brightness suffered far less noticeable losses because the attenuation of direct sunlight was balanced through an increase of diffuse sky light. At this point a particularly interesting observation by DE QUERVAIN (1912) must be mentioned, who on a passage through central Greenland in the summer of 1912 observed strong colour changes in the scenery. This experience caused fright among the population of Greenland and led him to point out the far-reaching consequences for future comparative actinometry.

The data given in Fig.28 deal with direct air-hygienic measurements (quantitative investigations of the atmospheric nucleus content of air by means of a Scholz nucleus counter) as well as with the consideration of global radiation registrations and of a detailed evaluation of vapour pressure. This example is an introduction to the complex

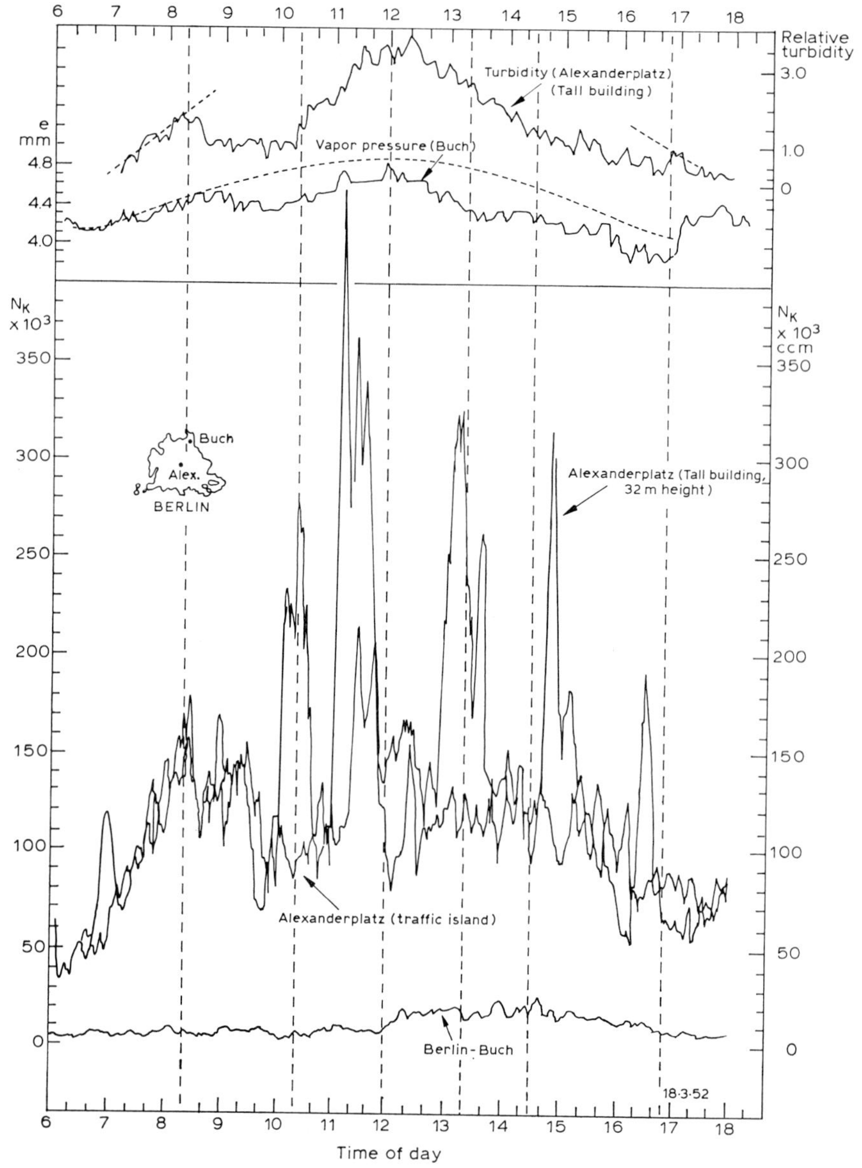

Fig.28. Example of diurnal variation of condensation nuclei content of the air (determined with a Scholz nucleus counter) in the center of Berlin (Alexanderplatz—traffic island; and Alexanderplatz—platform of high-rise building) as well as in the northern metropolitan peripheral zone (section Buch) (FLACH, 1952).

questions of turbidity determinations in the lowest atmospheric layers which emphasize the primary causal relations.

In the upper section of the graph a "relative turbidity state" of the air in the center of Berlin was determined from global radiation registrations. Furthermore, a characterization of diurnal variations atmospheric control factors important in air hygiene, especially of vertical exchange, was achieved by analysis of vapour pressure. The lower section of the graph shows the diurnal variation of the air's nuclei content at a central traffic point in Berlin and at 32 m above it. In comparison the peripheral area of Berlin (Buch to the north) actually shows minimum nuclei counts.

This general survey on a clear March day given clues to the behaviour of metropolitan nuclei aerosol, governed by surface inversions, as convection and vertical air exchange. Apart from "domes" of air pollution caused by sources of exhaust near the upper measuring platform, the behaviour of nuclei counts conforms in every detail with the behaviour of the turbidity (determined from global radiation registrations) and of vapour pressure. This shows clearly that besides the considerable influence by heating, industry and automobile traffic, the atmospheric conditions govern, i.e. formation of nocturnal ground inversion, its dissolution in the forenoon hours and also the variations of vertical air exchange promoted by steeper temperature lapse rates caused by incoming radiation. All these factors are bioclimatologically significant for air hygiene but also for urban construction and for building of health resort and recreational centers. Aerosol research through use of radiation and turbidity measurements give a good insight into general and local variations of air hygienic conditions and readily permit rapid evaluation.

Application of measuring units of atmospheric turbidity and their significance for environmental protection

Since the introduction of the "turbidity factor" (*Tf*) by LINKE (1921) geophysical radiation research has made continuous efforts to improve the procedures of this important sector of atmospheric turbidity research. Historically this was achieved by the introduction of a turbidity coefficient (β) by ÅNGSTRÖM (1929) and later by the turbidity coefficient (B) by SCHÜEPP (1949) which largely enable one to judge air purity by means of direct solar radiation measurements (I). By these means the total atmosphere's turbidity from the ground up to its highest layers may be determined, a subject surveyed by FOITZIK and HINZPETER (1958).

The SCHÜEPP turbidity coefficients α, B and w contain a measure for visible turbidity (haze) B, for the total water vapour content of the atmosphere w and simultaneously for total extinction α (i.e., from α, B and w together). For most investigations however coefficient B (numerically equal with β by Ångström) is sufficient.

In order to follow the behaviour or turbidity coefficient B (by Schüepp) which is of wide interest particular in Alpine regions, Fig.29 displays relative frequency (%) of turbidity coefficient B for points of different altitudes in the Swiss Alpine region (VALKO, 1971). (For details of the calculations the reader is referred to the original literature.) The graph shows that the high altitude stations, Jungfraujoch (3,500 m) and Weissfluhjoch above Davos (2,670 m) have a frequency maximum for very low turbidity coefficients. To a certain extent the valley point Davos (1,590 m) also belongs to this clean class of Alpine sites. Locarno-Monti at 380 m at the south foot of the Swiss Alps in the Tessin

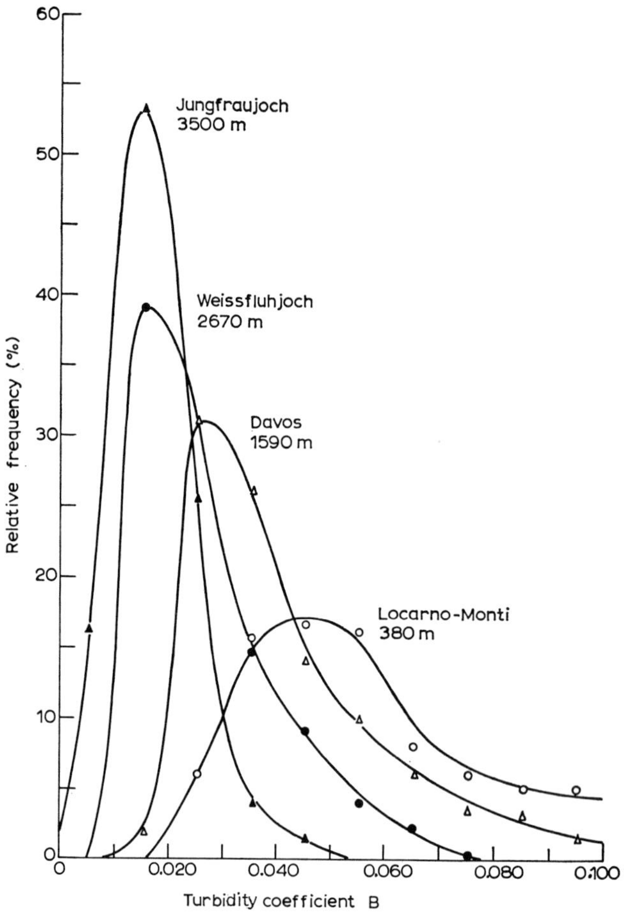

Fig.29. Relative frequency of turbidity coefficient *B* (by W. Schüepp) for four Alpine stations of different altitudes (VALKO, 1971).

shows its turbidity maximum at considerably higher *B* values. At higher *B* ranges strong haze can occur.

In metropolitan areas the turbidity coefficients usually lie above 0.100 and are subject to large temporal and local fluctuations. For this reason, all types of observatories are situated outside of the city haze region.

The mean annual variation of *B* shows that the highest values are registered in midsummer, the lowest in midwinter. One must take into account that the turbidity coefficient is a parameter which indicates exactly the prevailing total turbidity state of the atmospheric layers aloft. This is the case even if turbidity coefficient *B* does not include absorption by water vapour *w*. The relatively high *B* values in summer are explained by the fact that, compared with winter, convection and vertical exchange carry haze and effluents to higher air strata, and the already high water vapour content of the air in summer plays a notable part.

Fig.30 gives a time series of transmission factors worked out by PUESCHEL et al. (1972), based on calculations from measurements of direct solar radiation at the Mauna Loa Observatory (Hawaii). The period 1958–1962 which may be taken as a "normal" period is compared with that of 1963–1971. The latter was under the influence of considerable atmospheric turbidity in the stratospheric range due to the eruption of Mt. Agung (Bali) in March 1963.

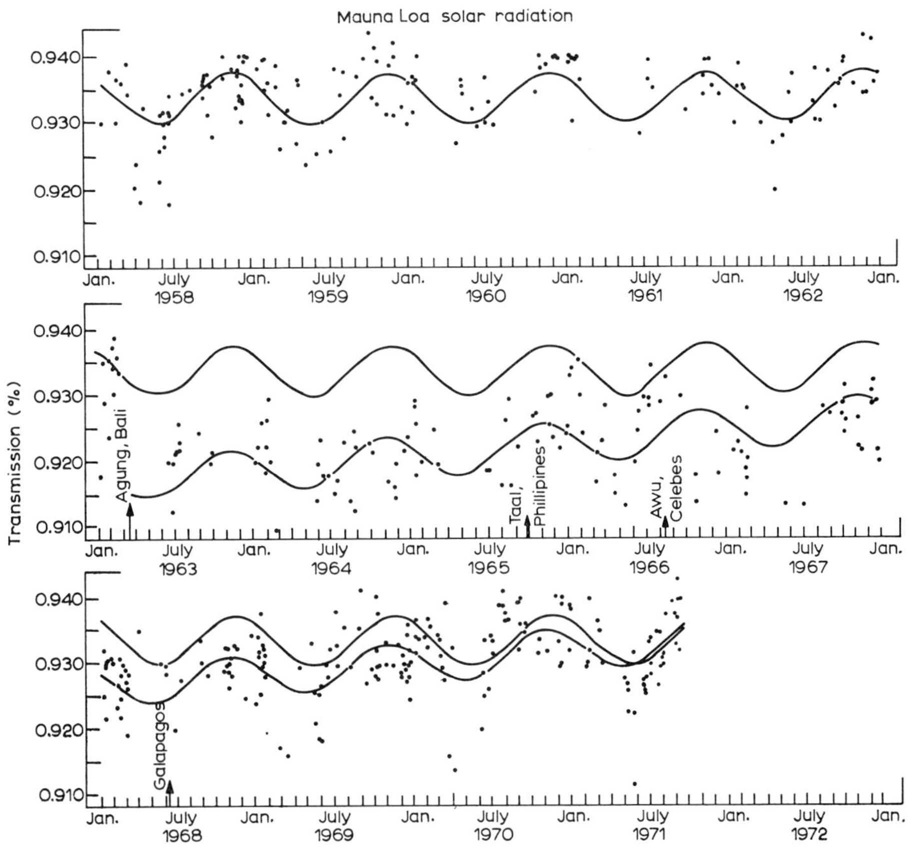

Fig.30. Annual variation of transmission factors on Mauna Loa (Hawaii) for the period January 1958 to March 1963 before eruption of Mt. Agung, Bali, and for the period April 1963 to December 1971 (after volcanic eruption) (PUESCHEL et al., 1972).

After this volcanic event atmospheric transmission was reduced suddenly (from 0.933 to 0.915%) during the interval the volcanic ash reached into the stratosphere, circled the earth several times under the influence of the strong west–east circulation in those strata (cf. similar events of the Krakatoa eruption in 1883 and that of Katmai in Alaska in 1912). Gradually by vertical exchange and diffusion processes there is a gradual reduction of turbidity, so that after 1969 the transmission coefficients largely approached "normal values".

Even at both Antarctic stations, "Roi Baudouin" (coast) and at the South Pole, an influence of the volcanic eruption of Mt. Agung (Indonesia) on turbidity in 1963 was registered (KUHN, 1972). HOINKES (1961) determined at the South Pole a β of 0.015 to 0.020, Viebrock and Flowers found β in 1967 at 0.030 (after KUHN, 1972).

These studies furnish clear proof that radiation measurements, such as determining the direct solar radiation (I) can furnish valuable bioclimatological insight into world-wide atmospheric pollution.

Turbidity determinations have been made in large cities or restricted non-urban areas. With measurements of direct solar radiation MANI and CHAKO (1963) determined total atmospheric turbidity in Poona and Delhi (India), GALINDO and MUHLIA (1970) in Mexico City, KRAMMER (1970) in Basel (Switzerland) and VALKO (1963) in the Tessin (southern Switzerland). Already in 1923 LINKE (1924) carried out measurements of direct solar radiation and air turbidity over the Atlantic Ocean and in Argentina (and also on the Bolivian Plateau) which may be considered as forerunners of determinations carried out today in similar ways. The highest values for turbidity factors were found on the Cap Verde Islands (influence of Sahara dust), the lowest values on the Bolivian Plateau at 3,600 m.

Valuable contributions are made to geophysical turbidity research by determinations of atmospheric transparency over wide areas of the earth made by approximately three dozen astronomic observatories. The investigations are based on the ASTRA project (= Astronomical and Space Techniques for Research on the Atmosphere). Up to present the results (HODGE and LAUTAINEN, 1973) have shown that there is a tendency towards a rising air pollution of the entire atmosphere, using the light transparency of air as a measure for "quality".

The atmospheric aerosol is significant as *absorber*. Aside from H_2O vapour, O_3, O_2 and CO_2, non-selective absorbers for energetic radiation are also active in the atmosphere. The SMIC Report (Study of Man's Impact on Climate, 1971) reports on research to determine the consequences of air pollution by solid and liquid products of anthropogenetic origin for the earth's climates by affecting radiative energy exchanges. Simultaneous measurements of extinction and sky radiation (QUENTZEL, 1967), show that on an average 30% of the solar radiation extinguished in the atmosphere is absorbed by aerosol particles. A change in the aerosol particle concentration, e.g., due to further global air pollution, may lead, at unchanged absorption characteristics of the particles, to an increase of energy reflected into outer space, which would result in a loss of energy on earth and thus cool it (EIDEN and ESCHELBACH, 1973). But the size spectrum and level of the absorbing particles will govern whether there is net cooling or warming.

In this connection satellite meteorology acquires an extraordinaiy significance in the world-wide evaluation of cloudiness, of surface temperature, of total contents of H_2O and O_3, etc.

In relation with atmospheric aerosol as absorber and with the optical scattering behaviour of air it is possible—except for measurements of direct solar radiation—to make certain statements on aerosol near the ground based on measurements of sky radiation (BULLRICH, 1964). Furthermore, UNZ (1969) has shown that by means of measurements of zenith polarization in the diurnal sky (by use of special radiosondes) a basis can be established for the aerosol distribution in the atmosphere. The thus computed turbidity values lead to the deduction of diffusion coefficients and absolute particle concentrations as functions of altitude.

Light scattering and water pollution

Water pollution may also be followed as to its intensity by means of visibility determinations below the surface (e.g., diffusion meter by Petterson). SAUBERER and RUTTNER (1941) presented an impressive example of acute water pollution in an Austrian lake after a storm flood (August 24, 1937), which affected depths down to 20 m. Fig.31 shows that the return to normal turbidity conditions required nearly two weeks until the water layer from 0 to 20 m was cleared by the sedimentation of the suspensions flooded in.

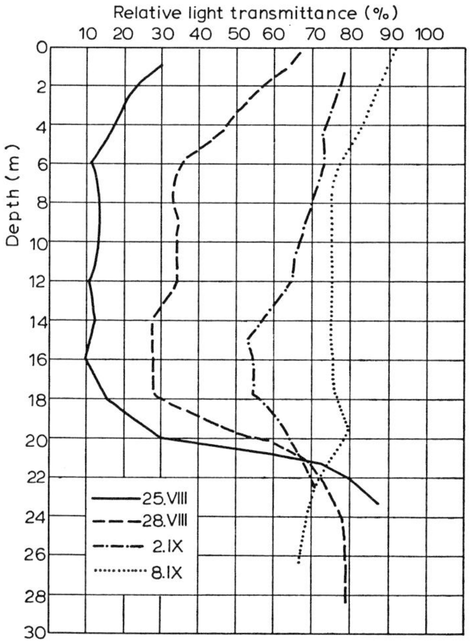

Fig.31. Example of acute water pollution caused by a flood after a heavy thunderstorm rain in the Lunzer Lake (Austria), and its temporal change, determined by means of light transmittance registrations (SAUBERER and RUTTNER, 1941).

Air and life

Basic aspects of atmospheric air

The presence of organic life on earth is predicated on the existence of the atmosphere. Two characteristics are important for bioclimatology.

(*1*) Air as a mixture of so-called "permanent" gases and non-permanent gaseous compounds.

(*2*) Air as a colloidal system (aerosol) comprising gaseous, liquid and solid admixtures, often in tiny quantities, including radioactive isotopes and nuclides.

Changeable weather with variations of pressure, temperature and humidity and also special geophysical influences of electromagnetic radiation processes (light, heat and ultraviolet radiation), constantly transform the atmospheric aerosol in the entire atmosphere and in particulaı in the biosphere. These processes constitute a force which cause characteristic changes in aerosol *sui generis* with usually decisive bioclimatological influence. Furthermore, the constant large and small-scale air movements thoroughly mix atmospheric air with all its constituents. Thus "air" controls many processes and the occasionally diverging impulses develop a continuous interplay of stimuli down into the small-dimensioned molecular range.

A great variety of admixtures (i.e., not conditioned by the basic geophysical nature of the earth's atmosphere) are present especially in the lowest air layers and have their origin from emissions from human settlements, industry, traffic, and from the solid and liquid earth surface. Water vapour (H_2O) is present in greatly variable amounts, especially in its liquid phase, and often governs the colloidal structure of aerosol and the characteristic changes it undergoes by way of water adsorption and absorption.

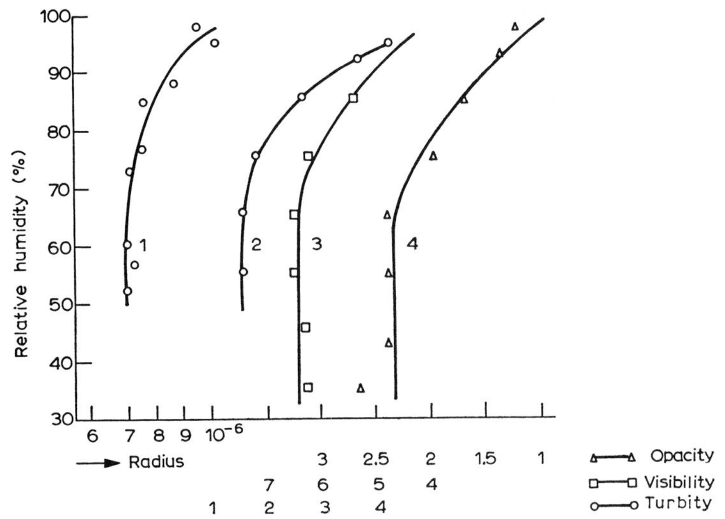

Fig.32. Growth curves for different artificial aerosols in dependence upon relative humitidy RH (%) (JUNGE, 1952). *1* = growth rate of gas flame ions; *2* = turbidity; *3* = visibility; *4* = opacity.

Fig.32 shows the growth of aerosols (often in form of the socalled "condensation nuclei") as function of relative humidity (%). It illustrates that the air's saturation degree influences many characteristics of air (e.g. turbidity, visibility, opacity) and also the conditions of the human respiratory system. Therefore, at relative humidity levels between 60 and 70% the suspensions' growth due to swelling through water absorption and subsequently through water adsorption leads to a rapid increase in droplet radius. Besides such specific influences on the nuclei there are others which are caused by weather changes. In all regions of the earth there is a constant large-scale horizontal exchange of air masses with basically different aerosol characteristics. These show bioclimatologically different effects. Fig.33 shows that fresh polar air masses carry far fewer condensation nuclei than air masses of tropical origin. The greatest quantities of

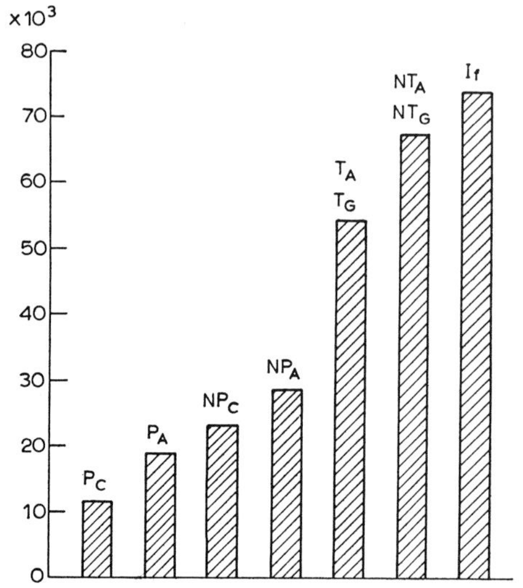

Fig.33. Air's content of condensation nuclei (10^3 cm^{-3}) in dependence upon air mass, in State College, U.S.A. (LANDSBERG, 1938). P_C = Canadian polar air; P_A = Atlantic polar air; T_A = Atlantic tropical air; T_G = tropical air (Gulf of Mexico); N = aged air of tropical origin; I_f = indifferent air locally stagnating.

such aerosol constituents are found in aged or indifferent air which has resided for a longer time over a certain area. (See also the section "Taxonomy of bioclimatological evaluation systems", p.163.)
There are two kinds of influences of the atmospheric aerosol. They illuminate the complexity of processes in the human body, where reactions may be called forth in various organs for generally bioclimatical and particularly air hygienic reasons and probably also meteorotrofic (meteoropathological) ones by way of the respiratory system, by way of the skin surface and the (vegetative) nerve-system (as part of the synaptic relay of information).

Gaseous composition of air near the ground

The *gaseous* composition of atmospheric air is subject only to minor changes in the lowest atmospheric strata (up to the lower stratosphere). Exceptions are carbon dioxide (CO_2) with 0.02–0.04% by vol., ozone (O_3) having a terrestrial atmospheric maximum in an altitude range between 15 and 35 km due to oxygen dissociation by short-wave ultraviolet radiation, and finally water vapour (H_2O) with its pronounced dependence upon temperature. Water vapour may vary between 0 and 4%, which may cause minor changes in the percent-by-volume composition of air. Thus for rising water vapour contents we find for nitrogen:

H_2O (%) 0 0.2 0.9 2.6

minor reductions of:

N_2 (%) 78.08 77.9 77.4 76.05.

Table XVIII is a supplementary survey of the oxygen (O_2) displacement quota in percent of its normal content as related to various values of air temperature and relative humidity.

TABLE XVIII

OXYGEN DEFICIT IN % OF THE NORMAL CONTENT AT DIFFERENT VALUES OF THE AIR'S WATER VAPOUR CONTENT (INDICATED IN VALUES OF TEMPERATURE AND RELATIVE HUMIDITY) UNDER UNCHANGED TOTAL ATMOSPHERIC PRESSURE

Relative humidity (%)	Temperature (°C): 0	10	20	30	40
0	0.00	0.00	0.00	0.00	0.00
20	0.12	0.24	0.46	0.84	1.46
40	0.24	0.47	0.92	1.68	2.91
60	0.36	0.73	1.39	2.51	4.37
80	0.48	0.97	1.85	3.35	5.82
100	0.60	1.21	2.31	4.19	7.28

These fluctuations of oxygen and nitrogen have no biological (physiological or pathological) significance. The variations of the air's vital oxygen content will again be taken up later.

Dependence upon altitude of atmospheric pressure and its climatic effects on man

Reduction of atmospheric pressure with altitude, oxygen partial pressure and its biological consequences

Table XIX shows the gaseous composition of the water-vapour-free atmosphere in man's environment and includes a conversion table for pressure measuring units currently in use.

TABLE XIX

COMPOSITION OF DRY AIR NEAR THE GROUND AND CONVERSION TABLE FOR THE DIFFERENT PRESSURE MEASURES (After BRUENER, 1961)

Gas	Molecular formula and weight		Vol.%
Nitrogen (N)	N_2	28	78.09
Oxygen (O)	O_2	32	20.95
Argon (Ar)	Ar	39.94	0.93
Neon (Ne)	Ne	20.20	0.0018
Helium (He)	He	4.00	0.00053
Krypton (Kr)	Kr	83	0.00011
Hydrogen (H)	H_2	2	0.00005
Xenon (X)	X	130	0.000008
Ozone	O_3	48	0.000002
Carbon dioxide	CO_2	44	0.02–0.04

	bar	atm	at	mm WS	(mm Hg) Torr
1 bar	1	0.987	1.0198	10,198	750.06
atm	1.0133	1	1.0333	10,333	760.00
1 at	0.981	0.968	1	10,000	735.52
1 mm WS	$98.067 \cdot 10^{-6}$	$96.784 \cdot 10^{-6}$	0.0001	1	0.073552
1 Torr (mm Hg)	$1.333 \cdot 10^{-3}$	$1.316 \cdot 10^{-3}$	$1.3596 \cdot 10^{-3}$	13,596	1

There are far-reaching biological consequences for resting as well as for working persons because of oxygen depletion caused by the decreasing partial pressure with height. This is based on Dalton's law, according to which in a gaseous mixture each individual constituent behaves as though it alone were present.

The dependence of the oxygen partial pressure (O_2 pressure) upon elevation (in km) is given in Fig.34, simultaneously for dry outdoors air and for water-vapour-saturated breathing air at 37°C. It results that O_2 pressures are lower for water-vapour-saturated breathing air than for dry outdoors air. Therefore, since breathing air is warmed to 37°C in the respiratory system, a vapour pressure of 47 Torr results for water-vapour-saturated air, which has to be deducted from total atmospheric pressure. Furthermore, dry air has an O_2 partial pressure of 159.6 Torr, and water-vapour-saturated air (at 37°C/RH 100%), i.e. breathing air, a partial pressure of 149.7 Torr at a total atmospheric pressure of 760 Torr (at sea level). Thus at approx. 5,500 m the partial oxygen pressure amounts only to half of that near sea level.

Table XX illustrates the behaviour of total atmospheric pressure in dependence upon elevation. In addition it shows features of other important geophysical parameters (air

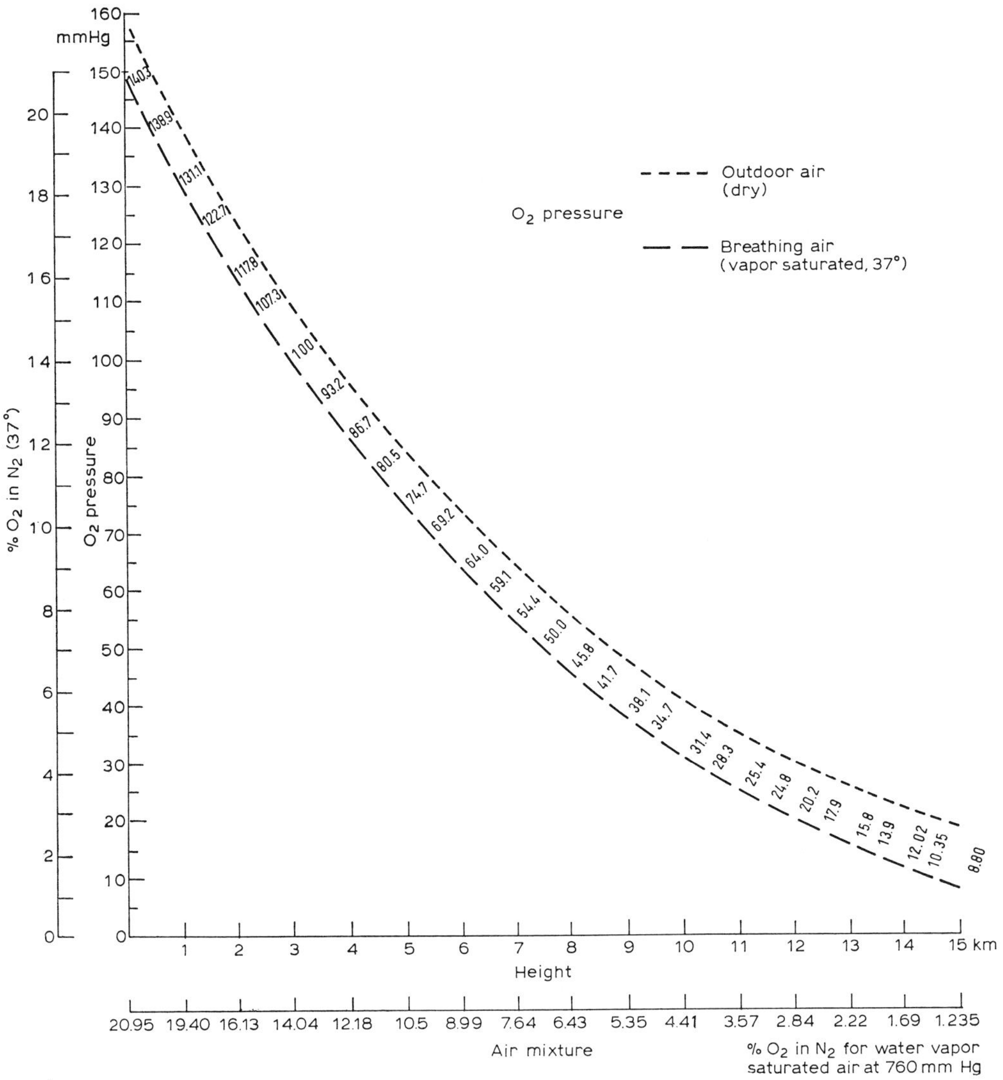

Fig.34. Oxygen partial pressure (O_2 pressure) of dry outdoors air and of respiratory air at 37°C saturated with humidity, in dependence upon altitude. Special abscissa: oxygen content of air mixtures which at inspiration at 760 Torr and water vapour saturation (37°C) possess an oxygen tension resembling the individual altitude levels (OPITZ, 1941.)

temperature, boiling point of water and speed of sound), which will be of interest also in other contexts. Great importance is attributed to the fact that the actual oxygen partial pressure in the lungs' alveoli lies at 15% by volume, while oxygen diffuses into the blood. This corresponds to a local O_2 partial pressure of approx. 107 Torr.

For the human organism an immediate consequence of the reduction of oxygen partial pressure with increasing elevation is a decrease of oxygen saturation of the red blood corpuscles (hemoglobin). Their job is to bind the oxygen inspired into the alveoli and to transport it through the bloodstream into the tissues. When in the breathing air, in the blood or in the tissue a reduction of oxygen partial pressure sets in (due to increasing altitude) the process is called "hypoxia" (a complete lack of oxygen is "anoxia"). In a slightly schematic form BRUENER (1961) distinghuishes the following forms of hypoxia.

(*1*) *Acute hypoxia*, e.g., due to the breakdown of an oxygen apparatus in air with a low oxygen content.

TABLE XX

VERTICAL BEHAVIOUR OF ATMOSPHERIC PRESSURE, AIR TEMPERATURE, BOILING POINT OF WATER AND SPEED OF SOUND*

Height (m′)	t	p mbar	p Torr	p/p_0	Sp	Slg
−1,000	21.5	1139.3	854.5	1.12	103.3	344.2
−500	18.2	1074.8	806.2	1.06	101.7	342.3
0	15.0	1013.2	760.0	1.00	100.0	340.4
500	11.8	954.6	716.0	0.94	98.3	338.5
1,000	8.5	898.7	674.1	0.89	96.7	336.6
1,500	5.2	845.6	634.2	0.83	95.0	334.6
2,000	2.0	795.0	596.2	0.78	93.3	332.6
2,500	−1.2	746.8	560.2	0.74	91.7	330.7
3,000	−4.5	701.1	525.9	0.69	90.0	328.7
3,500	−7.8	657.6	493.3	0.65	88.3	326.7
4,000	−11.0	616.4	462.3	0.61	86.6	324.7
4,500	−14.2	577.3	433.0	0.54	85.0	322.7
5,000	−17.5	540.2	405.2	0.53	83.3	320.6
5,500	−20.8	505.1	378.8	0.49	81.6	318.6
6,000	−24.0	471.8	353.9	0.46	79.9	316.6
7,000	−30.5	410.6	308.0	0.40	76.5	312.4
8,000	−37.0	356.0	267.0	0.35	73.1	308.2
9,000	−43.5	307.4	230.6	0.30	69.7	303.9
10,000	−50.0	264.4	198.3	0.26	66.2	299.6
11,000	−56.5	226.3	169.8	0.22	62.8	295.2
12,000	−56.5	192.3	145.0	0.19	59.4	295.2
13,000	−56.5	165.1	123.8	0.16	56.0	295.2
14,000	−56.5	141.0	105.8	0.14	52.7	295.2

* Height in geopotential meters (m′), in praxi equal to the meter; temperature t in °C; atmospheric pressure p (p_0 at sea level) in mbar and in Torr; boiling point H_2O in °C (Sp); Speed of sound in m/sec (Slg) (normal atmosphere OACI = Organisation de l'aviation civile instrumentale) (From "Scientific Tables", *Documenta Geigy*, 6th ed., 1954).

(*2*) *Rapidly appearing hypoxia*, e.g., when climbing in an airplane under steadily decreasing oxygen content in the inspiratory air.
(*3*) *Chronic hypoxia*, e.g., during a sojourn at elevations above 3,000 m or by inspiration of gaseous mixtures for extended periods with oxygen contents below 14% by vol. Such lack of oxygen leads to the following symptoms (cf. also Fig.35).
Phase I. A sudden reduction of the oxygen pressure to below 100 Torr leads within the shortest time to compensating reactions such as increased frequency and depth of breaths and heightened circulatory activity. Phase I permits practically complete compensation of the effects of oxygen deficiency. Critical judgment and physical energy and intellectual capacity are preserved. Phase I reaches to about 3,000 m.
Phase II. Shows some compensation but reduced performance. Physical strength is steadily reduced with increasing altitude. It starts with the "disturbance threshold" and ends at the so-called "critical threshold" (at 75 Torr oxygen partial pressure corresponding to a height of about 5,000 m).
Phase III. Is identical with the *decompression phase* and essential oxygen metabolism can be held up only for a very short time.
Phase IV. Is the zone of altitude death at approx. 8,500 m.
According to BRUENER (1961) the following point should be made.

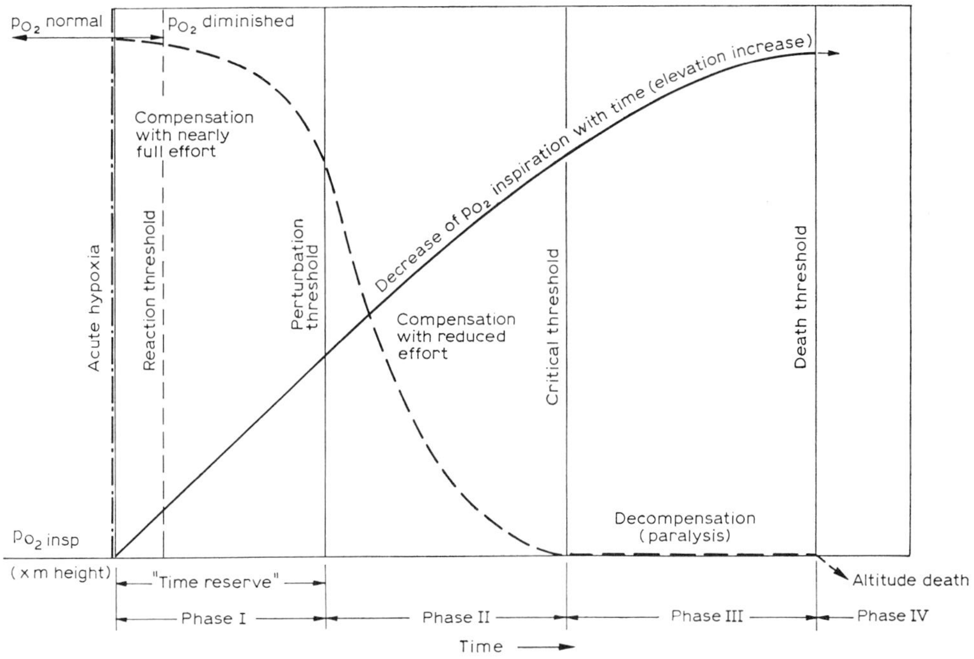

Fig.35. Performance curve (— — —) in connection with the reduction of oxygen partial pressure (p O_2 insp.) when climbing, over period of time (decay of functional performance reserves) (BRUENER, 1961).

Duration of the individual phases depends on the growing oxygen lack. In the so-called creeping oxygen lack the transition from one phase to another (reaction and disturbance thresholds) are often unclear. As long as the hypoxia is not pronounced the transition to higher elevations is expressed in a general increase of vegetative and animal functions (emergency reaction). The senses of feeling and pressure are intensified, smell and taste are sensitized, dark adaptation occurs quicker and furthermore all vegetative functions of sympathetic and parasympathetic innervation are heightened (VON MURALT, 1954). The organisms can, however, reach Phase II subjectively unnoticed, whereby a euphoric condition may detract from an imminent collapse (mountain sickness).

A very slow decrease of partial pressure of oxygen extended over weeks and months leads to altitude adaptation or *acclimatization*. If a change in elevation in the mountains is carried out very quickly (e.g., in a cable car) and if altitudes of over 3,000 m are surpassed a phenomenon may appear which is called "mountain sickness". Its symptoms are: heart palpitations, breathing trouble, headaches, nausea, etc. Sufficient training and carefully chosen climbing speeds can assure the continuation of performance up to relatively great heights (up to 6,000 m). Even at heights between 7,000 and 8,000 m a considerable performance is still possible. Fig.36 gives a striking example of the preservation of muscle strength after adaptation to altitude. The upper limit of any kind of performance lies at approximately 9,000 m (inspiratory oxygen partial pressure 38 Torr). An interesting fact is that the adaptation to mountain climates disappears (after weeks or months) when the "altitude stimulus" (VON MURALT) is absent. Zones and thresholds of altitude adaptation are average values and differ according to age. Individual fluctuations may also be caused by differences in constitution and disposition.

Very impressive alterations are found in the blood in connection with the altitude stimulus, such as the increase of erythrocyte number (erythropoiesis) per volume unit of

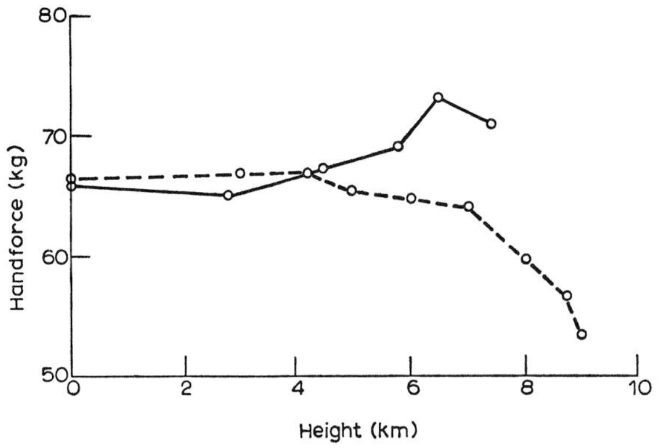

Fig.36. Behaviour of muscular force of a hand flexure before (— — —) and after (———) adaptation to altitude in the Himalayas (HARTMANN and VON MURALT, 1934).

blood and hence increase of hemoglobin and of its oxygen-binding capacity. Therefore, the result is that erythrocyte number and hemoglobin concentration depend upon the supply and absorption of oxygen (influence of hypoxia).

According to VON MURALT (1968) the erythropoiesis sets in above 1,000 m, whereby between 1,000 and 2,000 m greatly varied individual reactions are observed. Only above 2,000 m an erythrocytosis appears in almost all persons. SCHMIDT-KESSEN and PLEHN (1965) observed erythropoiesis stimulations during sojourns in moderately high mountains at altitudes of 700 m. This effect might have been intensified by physical treatment (salt-water bath cure).

In an evaluation of hypoxia influences and acclimatization in Alpine regions one must also always consider the participation of the air's thermal environment. Two examples: in the highlands of central and southern Tibet at various altitudes between 3,000 and 3,500 m, the absolute lowest air temperatures in winter lie between −30° and −40°C (FLOHN, 1970), while the absolute highest temperatures in summer reach values between 25° and 35°C. HOINKES (1961) points to the adaptation difficulties in the Antarctic highland where under an effective oxygen lack (heights between 3,000 and 4,000 m) the air temperatures vary between −50° and −70°C (July) and are often coupled with very strong wind (cooling powers).

ALBRECHT and ALBRECHT (1967) mention that contrary to the simulated climate of vacuum chambers the Alpine climate is influenced by numerous external factors (UV radiation, air movement, extreme temperature and humidity fluctuations, etc.). The same authors indicate the decisive influence of the air's frequently very low water-vapour content predicated on the usually low air temperatures of Alpine climates. Fig.37 gives an idea of the changes of arterial O_2 saturation (%), of oxygen partial pressure (mm Hg) and of vapour pressure (in % of mean ground value) in dependence upon elevation. The often very low air temperatures on the one hand, and the absolute as well as relative humidity on the other, cause considerable water losses at great heights (above 5,000 m), since the respiratory air in the alveoli must be warmed to +37°C and at the same time saturated up to 100% RH.

Concluding the questions of altitude-dependent partial pressure of oxygen, Fig.38 presents an example of the marked changes of blood quality which occurred during a stay on Nanga-Parbat. There is an impressive increase of the erythrocyte number and

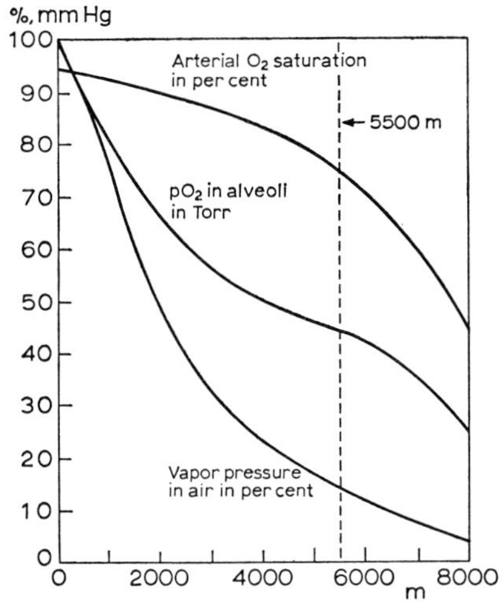

Fig. 37. Changes in arterial oxygen saturation (%), of O_2 partial pressure and of water-vapour pressure (in % of ground values) with increasing altitude (ALBRECHT and ALBRECHT, 1967).

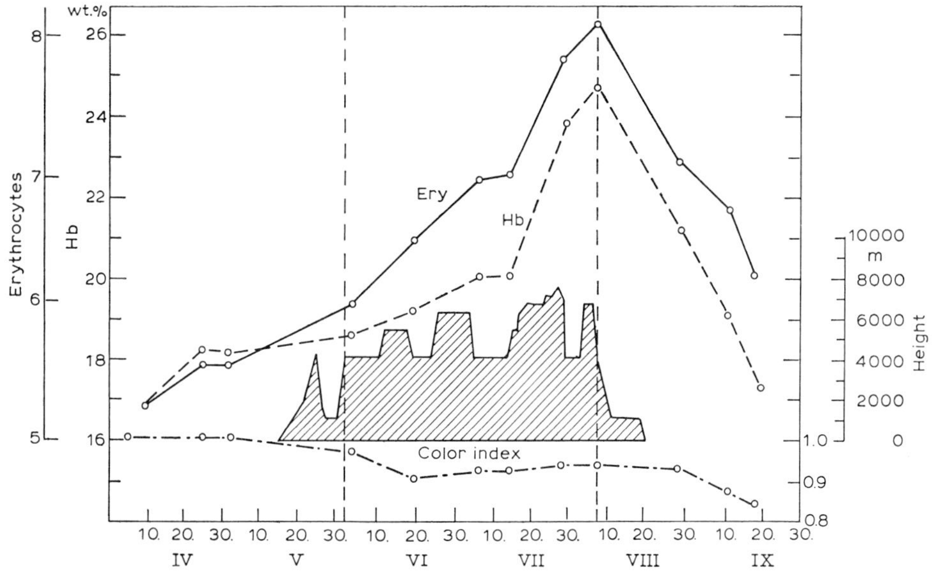

Fig.38. Erythrocyte number, hemoglobin content and colouring index of the blood before, during and after a sojourn of Nanga-Parbat (LUFT, 1941). Mean values from 5 mountain climbers. Hatching: sojourn on the mountain.

of hemoglobin, especially at altitudes between 7,000 and 8,000 m. After the return to the lower mountain areas these blood constituents rapidly diminish.

In aviation the safety of human beings requires for flights at great heights a "pressure suit" or a "pressurized cabin" in order to retain performance. Both types of equipment make use of supplemental oxygen unnecessary as long as no pressure loss occurs. At a sudden drop in the cabin pressure, for example, down to the low outside pressure (especially important for flights above 8,000 m), a pressure drop or an explosive decompression sets in. The resulting pathological symptoms are called "decompression sickness". Its characteristic features are the simultaneous appearance of free nitrogen bubbles in the body and of partially strong pains in the joints.

Influence of altitude on sports

The influence of so-called "moderate altitudes" on the *execution of a sport* in general (mountain climbing, rock climbing, skiing, skating, athletic contests) and on *sports performance* in particular must be mentioned briefly. In the first place one must determine whether at moderate altitudes between 1,500 and 3,000 m lack of oxygen there imposes any risks for athletes (in contests) and for the athletic performance as such. The Olympic Games held in Mexico City (2,300 m) in 1968 led representatives of sports medicine to study this problem more closely. The results showed that with continuous efforts at altitudes of 2,500 m there is no danger of death for the active athlete. In order to develop their full performance capacity, athletes need to spend a week in adjustment (change in altitude) and one in adaptation (accustoming), two weeks in acclimatization and a third or fourth one in the phase of progressive training. In this connection SCHOENHOLZER (1968) stresses that despite certain theoretical possibilities of an emergency reaction, in practice *the performance* in all sports disciplines is endangered by the acute altitude adaptation processes in *unprepared* athletes. Physiological preparations for performance can be carried out in sports centers in the mountains at home. Then, however, care must be taken that between the preparation phase and the actual contest not too much time elapses so that the "return effect" (with signs of fatigue when returning to lower altitudes) can be avoided (CORDES, 1968).

Scientists specializing in sports medicine have reached the following conclusions regarding competitive sports at medium altitudes (REINDELL et al., 1968).

(*1*) Adaptation symptoms (in heart and circulation system, in blood, etc.) do *not* lead to a decisive reduction of physical performance which to a certain degree is dependent upon altitude.

(*2*) An acclimatization period of several weeks is of advantage for avoiding a performance decrease. If possible, an acclimatization phase should last an extended period of time.

(*3*) A reduction of performance of staying power (at 2,300 m) caused by a reduced maximum oxygen absorption, of about 10%, remains also after an extended acclimatization period. Inhabitants native to high altitudes are also subject to these rules. Therefore, no reason appears for an essential change in the measures used up till now for evaluating competitive athletic performance.

Other interesting aspects of sports medicine, such as the influence of time of day and of season, that of thermal and air hygienic conditions, have been presented at an earlier point (FLACH, 1960). Specialized fields of interest such as physical performance capacity, willingness for performance and physical and central fatigue, cannot be discussed here.

Physiology of respiration

Before entering upon air colloidal considerations a few points about the physiology of respiration need to be made.

Respiration is the exchange of the gases oxygen and carbon dioxide between tissue and atmospheric air. The subdivision into approximately 3/4 million alveoli at the end of the finest branches of the trachea, causes the lungs to attain an inner surface of about 50 m^2. Nose, pharyngeal area, trachea and bronchii do not participate in the gas ex-

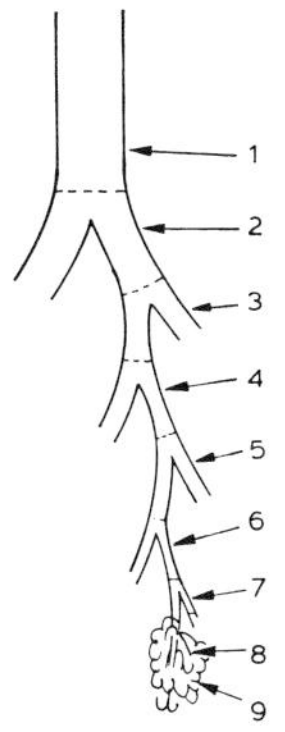

Region	Number	Diameter [cm]	Length [cm]	Surface [cm^{-2}]	Volume [cm^{-3}]	Function type
1. Trachea	1	1.6	12	60	24	Conductive
2. Main bronchii	2	1.0	6	40	10	zone with
3. Lobar bronchii	12	0.4	3	45	5	cilia epithelium
4. Segmental br.	100	0.2	1.5	100	5	
5. Subsegmental br.	800	0.15	0.5	200	10	
6. Terminal br.	$6 \cdot 10^4$	0.06	0.3	3400	50	
7. Respiratory br.	$2 \cdot 10^5$	0.05	0.15	4700	60	Respiratory
8. Alveolar duct	$5 \cdot 10^6$	0.04	0.05	30000	300	zone with
9. Alveolar sacc.	$5 \cdot 10^7$	0.04		250000	2500	alveoli

(Inspiration condition corrected to a lung volume of 3000 cm^3)

Fig.39. Model of lungs, by Landahl, supplemented by Jacobi, during the inspiration phase (Jacobi, 1965).

change. They do, however, have the important job of warming the inspired air to body temperature and to saturate it with water vapour. The surfaces of these spaces are covered with ciliated epithelium, whereby its ciliae move back and forth during the respiratory process. Thanks to this process foreign bodies can be removed from the respiratory system. Reflex actions such as sneezing, coughing, etc. serve the same purpose.

The separation of aerosols in the respiratory system and in the alveolar area takes place essentially by three processes, i.e., diffusion, inertial separation (rebound effect) and sedimentation. Fig.39 conveys an idea of the diameter, surface and volume of the respiratory system's individual regions (Landahl, supplemented by Jacobi, 1966) and is based on the respiratory conditions: minute volume = 15 l min^{-1}, respiratory frequency = 15 min^{-1}.

The separation rate of aerosol particles in the human respiratory system depends initially upon the particle size (in mμ). Fig.40 shows that in the entire respiratory system

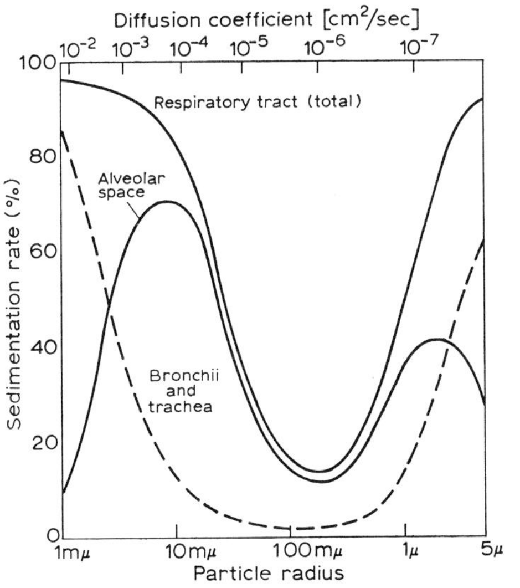

Fig.40. Separation rate of aerosols in the human respiratory system in dependence upon particle size (Jacobi, 1966).

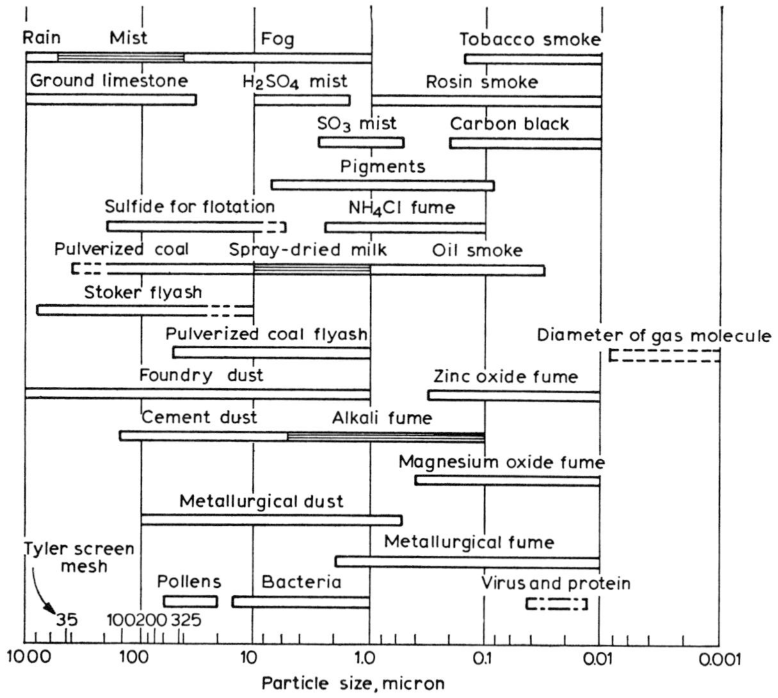

Fig.41. Range of particle size of various materials and organisms (JACOBS, 1960).

the relatively large aerosol particles together with the very small-dimensioned admixtures are generally subjected to accumulation by the above-mentioned processes. Especially in the bronchii and the trachea but also in the alveolar area remarkably high separation rates are found (%). This circumstance deserves particular attention because of atmospheric contamination by radioactive decay products of radon (Rn) and thoron (Tn). Interestingly aerosol particles of a size somewhat largeı than 100 mμ show the lowest separation rate amounting to slightly over 10%. This is essentially due to the strong decrease of diffusion separation in the bronchial tree (wide minimum in the range between 0.01 and 0.2 μ). This minimum is also evident in the alveolar area since a preliminary filtering takes place in the bronchial tree (Jacobi).

A general idea of the dimensions of the anorganic and organic admixtures occurring in polluted atmospheric air is given in Fig.41 (particle size in μ). Excessive movement of all types of air-polluting substances in the respiratory system must go hand in hand with a reduction of the ciliary activity and of ciliary transport.

Atmospheric air as a colloid (aerosol)

Qualities of the atmospheric aerosol

Investigations on the numerous and partially unsolved questions of "environmental protection", especially of air hygiene, require some basic knowledge of the pertinent terms in order to evaluate correctly the often strongly differing characteristics.

The colloidal state of air mixtures of different compositions is characterized by the size distribution, e.g., in liquid substances the drop size, in solid substances the grain size. The degree of dispersiveness is given by the relation $D = O/V$, whereby O means the surface area of the finely distributed substance, and V its volume. When finely dispersed liquid or (and) solid substances are suspended in air or other gaseous mixtures, the resulting system is called "aerocolloid" or "aerosol".

According to whether mainly liquid or solid substances are concerned, one speaks of "fog" or "dust", respectively. The expression "smoke" is used for aerosols which contain chiefly combustion products of a liquid *and* solid nature in form of suspended particles. Dusts carry not only anorganic substances such as stone residues, elements of street pavement (cement, tar, asphalt, etc.) but also organic ones like wood splinters, plant fibres, spores, pollen, etc. Such organic particles are also an essential constituent of house dust. A particularly impressive phenomenon is "smog" (a combination of the terms smoke and fog). This term is used chiefly when the health-damaging effects of smoke and exhaust in the lowest air strata, under the influence of a special atmospheric condition (inversion), are to be emphasized.

The colloidal state of an aerosol and its alterations are determined basically by particle size (cf. above), particle number and shape. Furthermore, their electrical and adsorptive features play a momentous part. In fogs the particles are always spherical, in dusts and smokes, however, they may be of many different shapes. The important point is that the shape of the dispersed particles determines their specific surface and thus also the amount of surface energy. This governs the force interactions between dispersed particles. Thus changes in the shape of suspended substances change a whole series of characteristics of colloidal systems.

The number of particles is controlled not only by effluent processes but also by the characteristics of the suspended particles themselves. Thus smokes and dusts are subjected to *sedimentation*, *aggregation* and *coalescence*. Sedimentation of colloidal substances is a function of size. In a mixture of constituents of different sizes the larger ones fall out first, but the smaller ones stay in suspension. This explains why in stationary air mainly aerosol constituents with smaller radii are found. The fall speed of aero-colloidal constituents can be calculated by the Stokes-Cunningham formula. However, the aerodynamic resistance of gas molecules may not be neglected.

Aggregation (union of solid particles) and coalescence (union of liquid particles) are conditioned by the Brownian molecular movement in gases in which aerosol particles of small and medium sizes participate. This often causes collision of suspended particles and their merging. Thus a continuous reduction of the particle number occurs.

A coalescence of liquid suspended particles, e.g. in fogs, may be conditioned thermodynamically as well as electrically. In the latter case one speaks of *coagulation* when suspended particles lose their repulsion by loss of their electrical charge.

Small aerosol particles can remain suspended for very long periods of time and thus trace substances can often cover large distances before falling out. Well-known examples of this phenomenon are the Sahara dust sedimentations on snow surfaces in the Alps and even in more northerly areas of Europe, and also continental aerosol residues which are observed at sea. Other examples are smoke transport from forest fires and volcanic eruptions.

Many natural and anthropogenic constituents of smoke and dust aerosols are hygroscopic, i.e. they offer a surface for the condensation of water vapour. They are therefore called "condensation nuclei". One measuring technique uses the "nuclei counter" (especially with automatic registration) and is an important aid for determining particle number and range of many effluents. Nuclei sizes range over four powers of ten, starting with a radius of 10^{-7} cm, to sizes between 10^{-4} and 10^{-2} cm. Their substantial and physical-chemical characteristics have been investigated and largely explained especially by JUNGE (1952) and others. A few actual examples of nuclei counts will be referred to below.

Aerosol assessment by nuclei and dust counts in different environments (large cities, medium and high mountains, forests, caves)

The quantitative behaviour of condensation nuclei (also in their characteristic of air hygienic indicators) conveys an idea of the climatic structure of aerosol (cf. Table XXI).

TABLE XXI

SURVEY OF THE FREQUENCY OF CONDENSATION NUCLEI IN DIFFERENT COUNTRYSIDES
(From LANDSBERG, 1938)

Measuring point	Nuclei/cm³: average	average max	average min	absolute max	absolute min
Large city center	147,000	379,000	49,100	4,000,000	3,500
Large city peripheral	34,300	114,000	5,900	400,000	620
Inland area	9,500	66,500	1,050	336,000	180
Coastal area	9,500	33,400	1,560	150,000	0
Mountains 500–1,000 m	6,000	36,000	1,390	155,000	80
Mountains 1,000–2,000 m	2,130	9,830	450	37,000	0
Mountains over 2,000 m	950	5,300	160	27,000	6
Islands	9,200	43,600	460	109,000	80
Oceans	940	4,680	840	39,800	2

An example (Fig.42) shows an air hygienic study of the metropolitan area of Berlin, carried out in a high-pressure weather situation in spring with little wind. On a metropolitan crossing by railway from north to south across the center of the city the quantitative distribution of nuclei and dust aerosol (N_K/ccm and N_S/l) was determined (duration of ride in one direction approximately 2 h). The measurements were carried out continuously with three Scholz nucleus counters and two Zeiss conimeters by several observers. The result reveals a marked "haze hood" over the metropolis, but the two suburban zones have comparatively clean air. The high degrees of contamination in the center of the city and also in an underground tunnel are extremely pronounced. The high N_K and N_S values in the latter section of the traverse are due to the continuous whirling up of dust in the tunnel and to the advection of the nuclei-rich air from the centre of town. The air samples were taken from the engine at the front of the train. The measurements of nuclei and dust conditions were made between 09h and 11h p.m. A traverse accross Berlin (in the opposite direction) on the same line between 11h and 13h revealed many changes. Only in the very centre of the metropolis and, almost exclusively, in the railway tunnels the relatively highest nuclei and dust counts were registered. The improvement in air-hygienic conditions around midnight is caused by the reduction in effluents from industry and traffic, but the metropolitan railway running in the tunnel area leaves its "dust traces" behind.

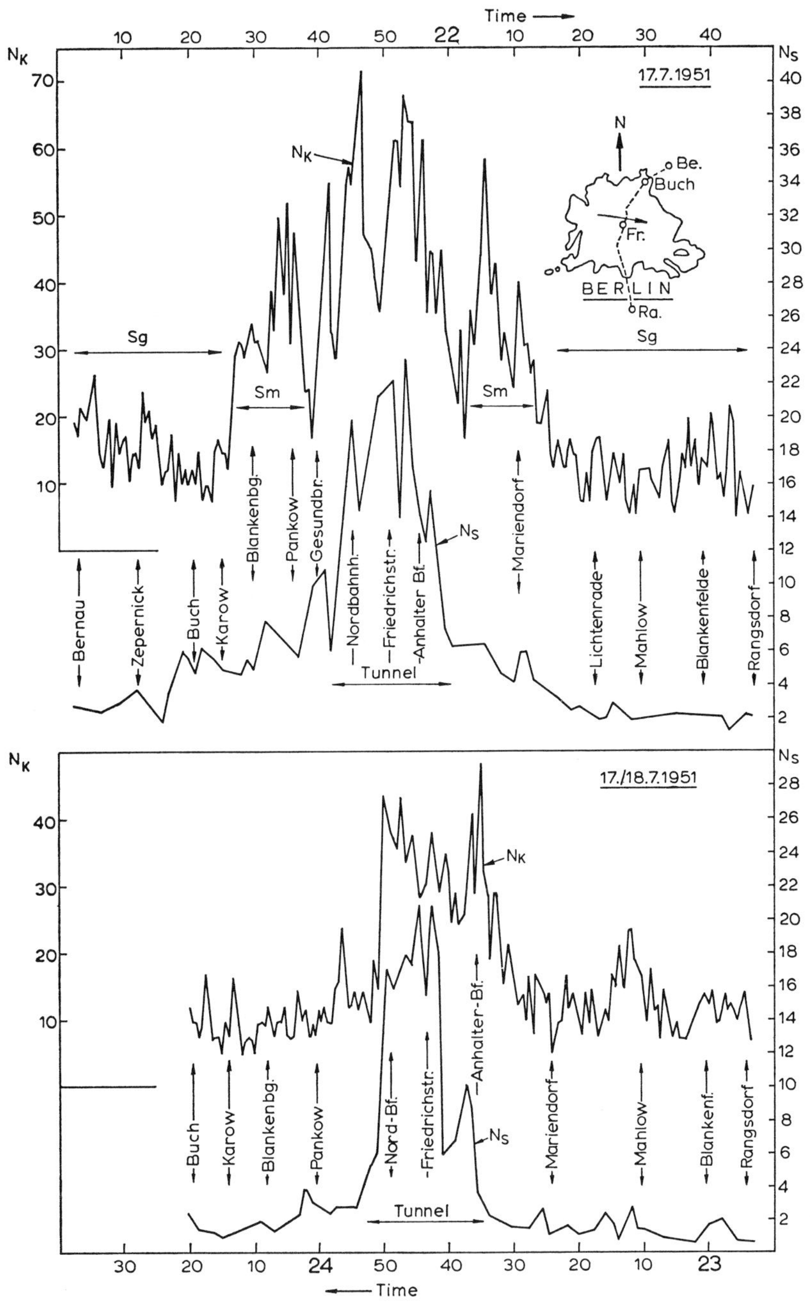

Fig.42. Nucleus and dust particle count profile of a metropolitan railway ride through Berlin and surroundings (FLACH, 1952). N_K = nucleus count cm^{-3} in 10^3; N_S = dust particle count per liter in 10^3. *Sm* = moderate influence of/on municipality; *Sg* = minor influence of/on municipality.

During the two traverses there was no precipitation. Outdoor winds were weak to moderate from the west.

More recently, HOGAN et al. (1973), based on extensive nuclei counts over the Pacific and Atlantic Ocean carried out by various scientists, determined the latitude distribution of nuclei concentrations (cf. Table XXII). In the central areas of these two oceans extremely low nuclei counts prevailed. But they rise quickly in the vicinity of the continents (North America, Europe). Similarity HERPERTZ et al. (1957) as well as BIDER

TABLE XXII

LATITUDINAL VARIATION OF AITKEN NUCLEI CONCENTRATION (cm^{-3}) FOR THE PACIFIC
(After HOGAN et al., 1973)

Latitude	Average concentration	Source and remarks
~90°S*	<200	J. Warburton, DRI single-season data
~70°–90°S*	200–1,000	Readings above 1,000 cm^{-3} may reflect contamination at McMurdo
40°–60°S	<200	I. Allison and U. Radok, University of Melbourne, observations from Nella Dan, single-cruise data
20°–40°S	200–700	Farrell Lines; Thomas Washington, Scripps; C. Owen, ASRC; T. Christian
20°S–30°N	<200–500	C. Owen, ARSC; F. Guenthner; Thomas Washington, Scripps; Moor MacCape, Farrell Lines.
30°–50°N	200–1,000	American Argosy, American Legacy, Moor MacCape, T. Egelston

* Continental site

and VERZÁR (1957) found very low nuclei counts in an Alpine climate (Jungfraujoch, Switzerland at 3,578 m) which differ only little from sea climates and cave climates. Nuclei counts on the Jungfraujoch fluctuated between 50 and 1,000/cm^{-3}.

Persons suffering from diseases of the upper respiratory tract (asthma bronchiale, hay fever, etc.) and skin diseases (neurodermitis) find a certain freedom from relapses in air with little foreign aerosol (sea coast, Alpine and mountain climates) (AMELUNG, 1952). Similar conditions are found for dust aerosol as for nuclei aerosol as shown by NEUWIRTH's (1966) measurings in the Black Forest and in the nearby Rhine plain. These were carried out with an air-sampler. Fig.43 shows the behaviour of the air's dust content in Freiburg i.Br. (Rhine plain under fog) and in the mountain health resort Menzenschwand (1,100 m, sunny, above fog).

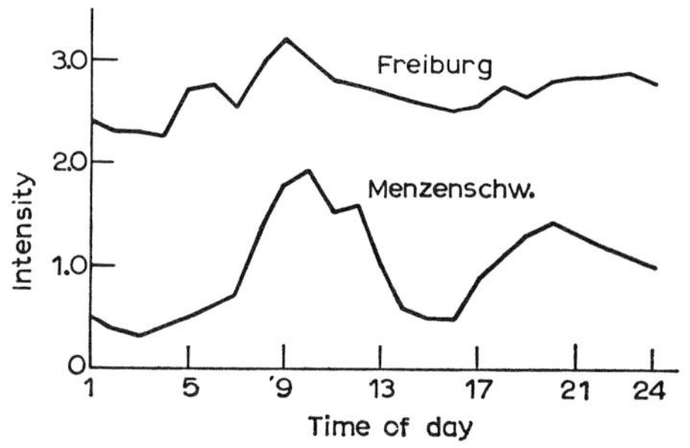

Fig.43. Dust taint during a winter high-pressure weather phase in the high Black Forest (Menzenschwand) and in the Rhine plain below (Freiburg i.Br.) (Jan. 30–Feb. 11, 1959) (NEUWIRTH, 1966).

Because of convectional vertical air exchange, the Black Forest heights with few clouds show definite diurnal variation of the dust content with two maxima (phase of active vertical convection between 12h noon and 17h). But Freiburg i.Br. lying underneath the fog cover does participate in these exchange processes because of the inversion state. In the city the daily periodical course of traffic and effluents from the residential areas' heating is evident, with the comparatively low winds contributing greatly to this phenomenon.

For local climatology and for air-hygienic environmental studies nuclei counts give insight into the great complexity of aerosol behaviour. But for any detailed explanation on chemical air quality and particle concentrations more sophisticated approaches are needed. During the period 1930–1960, when aerosol research acquired an "environmental protection" orientation, nuclei counts were a decisive aid in the establishment of quick geographic air-hygienic evaluations. Such aerosol investigations as were carried out by various sources (cf. LANDSBERG, 1938) threw new light on the depence of natural aerosol upon individual meteorological factors as well as upon complex elements (e.g., air masses). This knowledge is highly valuable for the present and future of aerosol research.

Results of pollen and spore counts

Although the following observations are to a certain extent part of a survey on the occurrence of seasonally conditioned illnesses, a special aspect of aerobiology needs nevertheless discussion here. This concerns the pollen of anemophilous plants, since it constitutes a very important part of bioclimatology and in particular of meteorobiology. Table XXIII shows the blooming times of the main "hay fever" plants which are often the primary cause of allergic sicknesses in spring and summer (moderate northern latitudes). These aliments are the result of an over-sensitivity towards the

TABLE XXIII

BLOOMING TIME OF THE MAIN HAY FEVER PLANTS (CENTRAL EUROPE)
(After DE RUDDER, 1952)

		Feb.	Mar.	Apr.	May	Jun.	Jul.	Aug.	Sep.
1	*Agrostis*/Bent grass					×	×		
2	*Alopecurus pratensis*/Meadow foxtail				×	×			
3	*Anthoxanthum odoratum*/Sweet vernal grass			×	×	×	×		
4	*Arrhenatherum*/Oat grass					×	×		
5	*Bromus*/Brome grass				×	×	×	×	
6	*Chamomilla vulgaris*/Camomile						×	×	×
7	*Convallaria majalis*/Lily of the valley				×				
8	*Corylus avellana*/Hazel	×	×						
9	*Cynosurus cristastus*/Cock's-comb grass					×	×	×	
10	*Dactylis glomerata*/Cock's-foot grass					×	×		
11	*Festuca*/Fescue grass				×	×	×		
12	*Holcus lanatus*/Woolly holcus				×	×			
13	*Hyacinthus*/hyacinth			×	×	×			
14	*Ligustrum vulgaris*/privet					×	×		
15	*Lolium perenne*/rye grass				×	×			
16	*Phalaris arundinacea*/Phalaris grass					×	×		
17	*Philadelphus coronarius*/Mock orange					×	×		
18	*Phleum pratense*/Timothy grass					×	×	×	
19	*Poa*/Meadow grass					×	×		
20	*Populus*/poplar			×	×				
21	*Robinia pseudacacia*/False acacia				×	×			
22	*Salix*/Willow		×	×					
23	*Sambucus nigra*/Elder					×	×		
24	*Secale cereale*/Rye				×	×			
25	*Syringa vulgaris*/Lilac				×				
26	*Tilia*/Lime tree					×	×		
27	*Trisetum flavescens*/Yellow oat-grass					×	×		
28	*Tulipa*/Tulip				×				
29	*Zea mays*/Maize						×		

pollen of grasses, trees and shrubs. Thus always the same persons succumb at the same time and usually with the very same symptoms. Fig.44 (DAVIES, 1969) demonstrates the seasonal differences in maxima of pollen concentrations (grass pollen) measured daily at the same time as they show up in the surroundings of a metropolis (London) and in an Alpine valley (Davos) during the summer.

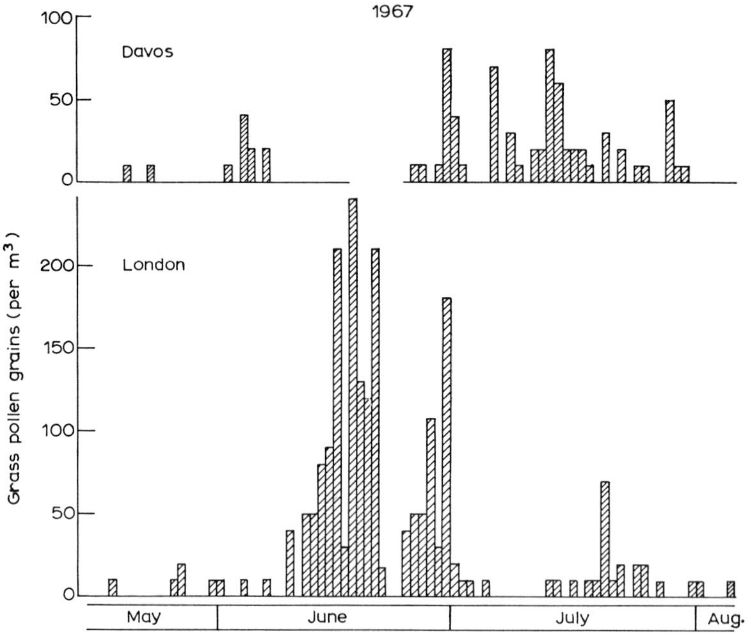

Fig.44. Daily concentrations of grass pollen grains in the air at Davos and London, 1967 (DAVIES, 1969).

A temporal shift in the appearance of the top amounts of grass pollen is very striking. Whereas in London the highest concentrations are observed in June, they appear in smaller quantity in Davos only in July. This is related to the phenology of vegetation and its blooming period which in Alpine regions above 1,500 m sets in a good month later than in central European lowlands. Grass pollen registered in Davos in May and June must be considered as advected and brought into that region from lower ones.

LUEDI and VARESCHI (1935) carried out extensive studies on the distribution, blooming time and pollen sedimentation in the high valley of Davos. Three dozen of the most important grasses were tested. The main emission was found between the end of June and middle of July.

The Deutsche Forschungsgemeinschaft (DFG) has established a measuring network for determinations of pollen and spore distribution. The observations were evaluated by STIX (1971) in various regions of West Germany (cities, open countryside, sea coast, wooded areas) and which he expanded considerably.

Obviously research in this specialized meteorobiological sector—in connection with botany—may greatly expand the knowledge available for the diagnosis and therapy of allergic afflictions of the skin and mucous membranes caused by pollen and fungi. The determination of amount and type of pollen particularly for discovering new pollen (e.g., ragweed pollen) of plants of the genus *Ambrosia* is of high practical significance. For health resort climatology it is very valuable and practical to know the annual periods with the lowest pollen concentrations which vary from place to place.

Special features of atmospheric aerosol

Size spectrum of atmospheric aerosol

As already indicated in Fig.41 (p.66), atmospheric aerosol particle sizes range from molecule complexes (consisting of 10 to 30 molecules) with a radius of $r \sim 10^{-3}\,\mu$ to particles with a radius of $r \sim 10^{+2}\,\mu$, whereby the particle sizes form a more or less continuous spectrum (JUNGE, 1963; cf. Fig.45).
Particles with radii of $r < 0.1\,\mu$ are called "Aitken nuclei", those with a radii interval of $0.1\,\mu \leqq r \leqq 1.0\,\mu$ "large nuclei" and particles with radii $r > 1.0\,\mu$ "giant nuclei".
Based on tropospherical studies chiefly three types of aerosols are distinguished (JUNGE, 1971).

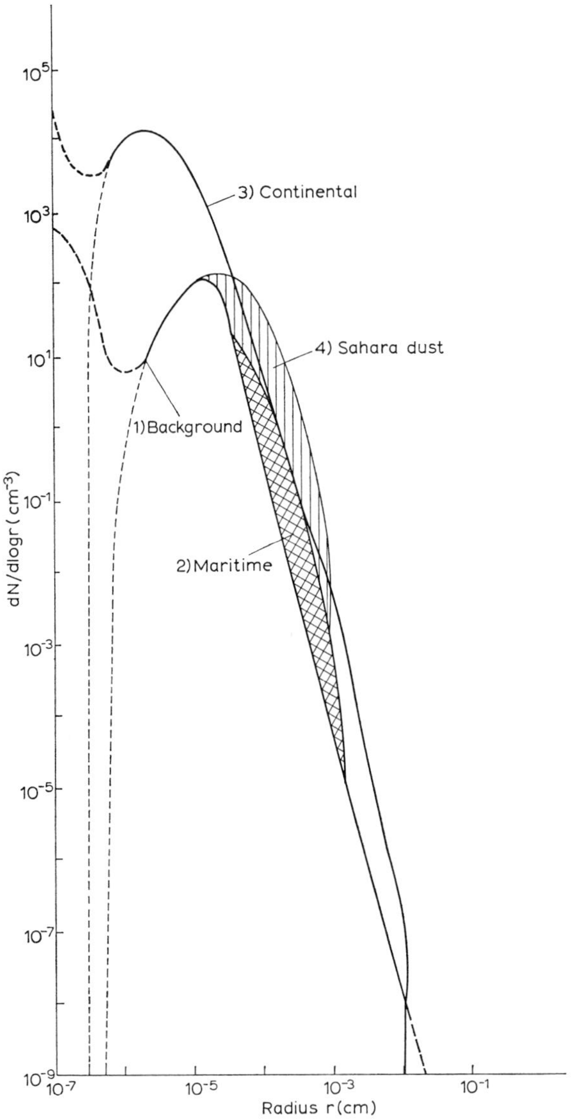

Fig.45. Schematic size distribution of tropospherical aerosol particles with: background distribution; continental distribution; maritime distribution and Sahara dust component (Ch. Junge, from EIDEN and ESCHELBACH, 1973).

Background aerosol. This type fills the middle and upper troposphere and consists of aged, diluted continental aerosol. It fills about 80% of the total troposphere and contains not only continental residue aerosols (mixed with particles of organic and mineral origin) but also sea salt particles. There is continuous production of very small particles of background aerosol (chemical and photochemical production of trace gases). The values of the heavy dotted lines in the left upper section of the figure are yet uncertain.
Maritime aerosol. This type not only contains sea salt particles formed by spray water, breakers, etc. but also admixtures of continental aerosols and very often of desert (Sahara) dust. The maximum of maritime size distribution based on various determinations lies in the Atlantic at $r = 0.2\,\mu$. In maritime aerosols the range above $0.1\,\mu$ to $10\,\mu$ consists largely of sea salt aerosols.
Continental aerosol. On the whole this type consists of mixed nuclei. The maximum aerosol size class lies at approximately $0.05\,\mu$. An Atlantic "Meteor" expedition carried out studies which showed that the size of the Sahara dust component lies between 0.1 and $10\,\mu$.
Thus the aerosol spectrum shows a *size range* in the troposphere of more than five powers of 10. The particle sizes form a more or less continuous spectrum with a radius range from $0.01\,\mu$ to approximately $100\,\mu$. The *particle concentration* in the air layers closest to the earth range from 10^4 to 10^{-7} per cm^3, i.e., about 10 powers of 10. The graph shows in addition that the largest number of aerosol particles which are identical with the so-called "condensation nuclei" has radii below $1\,\mu$. The concentration decrease between 0.1 and $100\,\mu$ is explained as being the result of the statistical distribution of the numerous aerosol sources which contribute to the variety of natural aerosols. The decrease of aerosol particle concentration corresponds with the following power law:
$dN/d \log r = c.r^{-a}$ (JUNGE, 1952)
In the equation c is a constant which depends upon the total mass of the droplet and the vapour pressures over a flat surface of pure water and the surface of the solution droplet. Exponent a fluctuates between 2 and 4, lies most frequently however at 3. Therefore, the above equation is often called the r^{-3} law. N is the number of all particles with a radius smaller than r.
The fact that aerosol particles because of the comparatively long life span of the nuclei in the atmosphere, the coagulation of various particles, the adsorption of gaseous components on the particle surface and condensation and re-evaporation turn into "mixed nuclei", led to the "mixed nucleus theory" (Junge). The dependence lies between the absolutely soluble and the insoluble particles. In aerosol particles an average of 75% water-insoluble substances, 25% water-soluble and about 20% acetone-soluble (organic) components were determined (JUNGE and ABEL, 1971). Continental aerosol contains more water-insoluble particles, oceanic aerosol more water-soluble ones. In general the growth begins with rising relative humidity at about 70%, for mixed nuclei at 60%, and for salt particles (NaCl-nuclei) already between 40 and 45%. The so-called hysteresis phenomenon on aerosol particles can be mentioned here only in passing. It describes their faculty of liberating less water under reducing relative humidity conditions than was absorbed during the reverse process.
For altitude distribution of aerosol valuable information has been collected by direct measurements (from aircraft or balloon) and indirect ones (laser and twilight diffuse light investigations from the ground). EIDEN and ESCHELBACH (1973) have gathered

important facts regarding this subject in a survey which has been referred to several times. Accordingly, the particle concentration falls within the troposphere in an exponential way to a height of about 5 km, and then remains constant up to the tropopause. Above the latter there is again a rise of aerosol concentration in the lower stratosphere, reaching its maximum between 20 and 26 km. Aerosol altitude distribution over the Greenland ice cap showed extremely low particle concentrations immediately above the snow cover. Only above 5–6 km the normal altitude values are attained. Eiden and Eschelbach state that the snow surface is a sink for aerosol particles and must be considered as free of sources.

The fact that the mass of condensation nuclei is small compared with their great number must be considered relative to any possible physiological effects. The importance of NaCl nuclei (giant nuclei) for climatic therapy in a coastal climate or for inhalation cures which are carried out with dispersed sea water (SCHULTZE, 1973), is well-known. There seems hardly a question of specific chemical effects of the natural aerosols, because the amounts of substance in pure air are so small. Sea salt is brought into the air by breaking of waves ("breaker air", Pfleiderer). In the absence of breakers the air is salt-free. In the surf area suspended salt concentrations lie between 0 and 1000 $\mu g\ m^{-3}$. Inland from the shore this finely dispersed salt quickly disappears (after several decameters it is already reduced to half the amount). Suspended salt does not leave the respiratory system again but is absorbed there (JESSEL, 1970).

GEORGII and GRAVENHORST (1972) carried out studies on the constitution of *maritime aerosol* (Atlantic Ocean) during an Atlantic expedition on the research ship *Meteor*. Their results indicated that the sea salt particles no longer deserve the reputation of dominating as condensation nuclei for the formation of maritime cloud elements. Such investigations should be continued regionally; however, the conclusion may be drawn that sulphate particles of which a majority is partly of continental origin and partly due to reactions of gaseous components (from volcanic ash) and which are formed in the atmosphere, exceed the number of sea salt particles also over the open sea. The size range of this aerosol in the lower oceanic air layers shows a maximum at radii between 0.1 and 0.2 μ. The authors have drawn the following conclusions.

(*1*) In the maritime atmosphere the ocean surface is *not*—as previously assumed—the main source of aerosol.

(*2*) Numerically ocean salt particles form only a minor constituent of maritime aerosol (up to altitudes of 1–2 km).

(*3*) Concentration of "background" aerosol in oceanic air layers lies between 300 and 600 particles per cubic centimetre.

(*4*) Sulphate particles and insoluble particles with radii between 0.2 and 1 μ represent the bulk of maritime aerosol.

NYBERG (1972) has furnished additional information on this subject based on the measurements between 1964–1966, in which the mean annual volume distribution of sulphur in the fallout over northwestern Europe was determined. Fig.46 shows clearly that the maximum rates of sulphate (not sulphur dioxide) fallout is found over the industrial centres of Germany, England and southern Scandinavia; furthermore that due to suspensions carried from the neighbouring continents actually a considerable amount of foreign aerosol is encountered over the oceans.

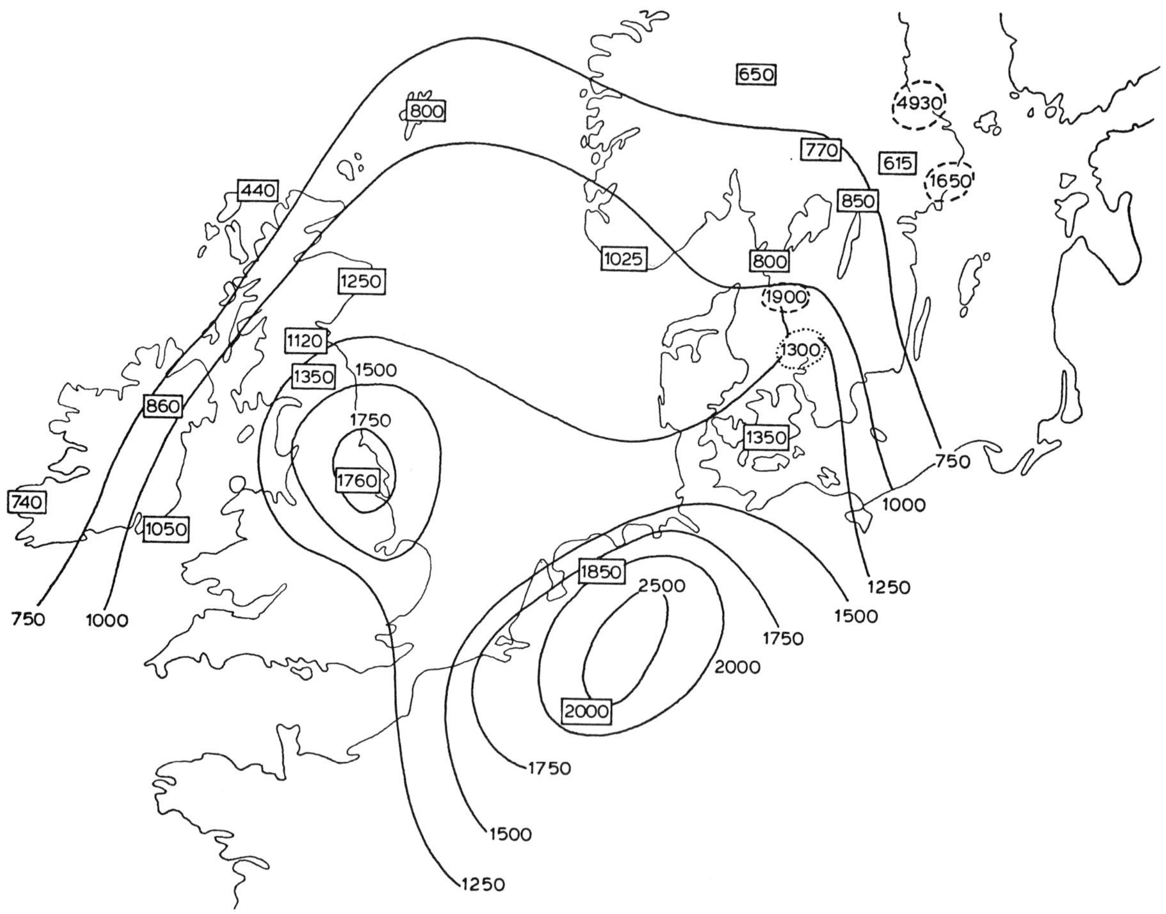

Fig.46. Sketch of sulphur fallout in mg m^{-2} $yr.^{-1}$ in northwestern Europe, 1964–1966 (NYBERG, 1972).

Meteorological influences on the constitution of atmospheric aerosol

At the beginning of the section "Air and life" (p.56) mention has been made of a few factors (relative humidity and the complex element "air mass") which decisively influence the variability of aerosol characteristics. There are two further features which dominate aerosol's composition and distribution. They are the turbulence and exchange processes of atmospheric flow. Eddies thoroughly mix neighbouring air parcels and thus play an important part in the transport of noxious matter from industrial effluents. Equally the extent of washout by atmospheric precipitation and its influence on aerosol particles must be considered.

Very frequently practical aerosol problems confront the theoretician with questions how wind, turbulence and vertical exchange as well as "apparent diffusion" carry and disperse effluents. Meteorological estimates of atomic reactor sites play a critical role because the dispersion of emissions from the nuclear reactors must be determined. A number of authors (O.G. Sutton, M.L. Barad, F. Pasquill, F. Wippermann, W. Klug, R. Trappenberg and M. Diem) set up formulae for the calculation of effluent concentrations by which they evaluated the essential parameters influencing the dispersion. Obviously, air displacements, wind directions and force, degree of turbulence and vertical stability play a very prominent part.

Investigations on the dispersion of noxious substances in the atmosphere by turbulent diffusion (REUTER and CEHAK, 1966) revealed that the concentration field undergoes a very extensive modification due to meteorological circumstances. The dependence upon the existing vertical temperature structure (stable, labile, indifferent) is very marked. Nowadays the lidar technique makes it possible to measure very small concentrations and thus to determine smoke plume profiles. These observations permit better understanding than previously on the local position of a smoke trail and its density. Many investigations revealed that a purely optical impression is often completely insufficient and may lead to an underestimation of effluent dispersal processes.

The mean residence time of aerosol particles in the earth's atmosphere deserve particular attention. FLOHN (1973) found that in the lowest layers of the atmosphere the mean residence time, in general, amounts to only a few days. It must be considered that in these surveys chiefly particles of small dimensions are considered. The medium and larger admixtures settle out quickly. The residence time, however, in the upper troposphere can reach about 1 to 2 months. In the lower stratosphere it jumps to 2–4 months and attains residence times of several years in the middle and upper stratosphere above 25 km. This fact is particularly important for the dispersal of volcanic particulates and aerosol debris from atomic air bursts.

Aerosol dispersal investigations cannot be carried out with the usual wind records. Adequate observations need an aerological network (using sounding balloons or radiosondes) including use of mobile balloon rigs. GEORGII (1956) showed the extent of *pseudo-diffusion* near the ground of aerosol expansion under weather conditions with weak winds. His experiments (cf. Fig.47) show that with total absence of wind or with very small air displacements the relative aerosol concentration undergoes very minor changes (curves *I* and *II*). Under slowly increasing pseudo-diffusion (curves *III* and *IV*)

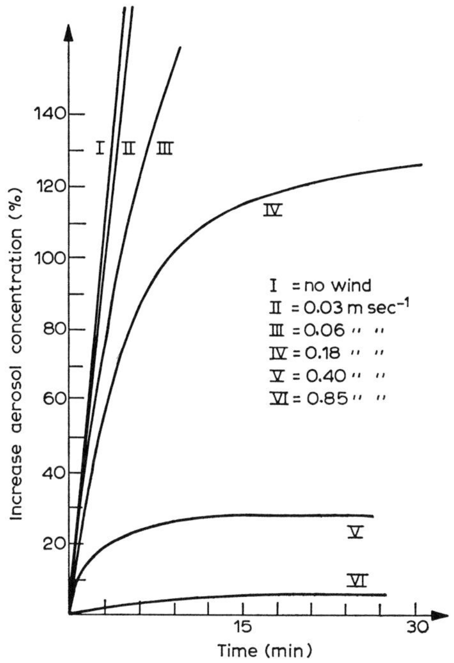

Fig.47. Increase of relative aerosol concentration under constant nucleus production and various intensities of air exchange (GEORGII, 1956).

the aerosol concentration falls rapidly with time (*T*), in order to reach an almost instantaneous reduction of particle density at the onset of convection (*V* and *VI*). Finally a state of equilibrium between nuclei production and aerosol removal is attained.
Besides the removal of aerosol particles by wind, turbulence and exchange, two further phenomena are of special importance for the atmosphere's self-cleansing process, i.e. *washout* and *rainout*. Rainout is the incorporation of trace substances within a cloud, while a washout occurs outside or below the cloud.

Electrical characteristics of natural aerosol

The problems of aerosol's electrical charge are broad and complex. Aggregation and coalescence processes (cf. above) are controlled chiefly by the electrical charge characteristics of an aerosol in a recombination of suspended matter with opposite charges (positive and negative ions). From a practical point the electrical charge offers possibilities for various manipulations of an aerosol, e.g. for its stabilization or quick sedimentation. Based hereon are studies of the physiological effects of atmospheric–electrical fields of suspended substances of unipolar charges on the human organism. However, under the chosen experimental conditions often great uncertainties exist about quantitative electrical charge conditions. The reader is referred to other texts for the theory of electrical charges of an aerosol.
Air electrical climates and their basic behaviour in connection with electrical charge carriers (ions), have as a basic element electrical conductivity which are governed by the content of "small ions", whereby the interaction of ion formation, recombination and accumulation of ions on suspended particles (thus forming medium and large ions) creates a certain equilibrium. Besides the conductivity the vertical electrical field is the second basic characteristic of an atmospheric–electrical climate, and nearest the ground it has the largest value and decreases with elevation above ground. Table XXIV presents the basic values of an atmospheric–electrical climate for the above-mentioned features.

TABLE XXIV

BASIC VALUES OF THE ATMOSPHERIC–ELECTRICAL CLIMATE
(From ISRAEL, 1962)

Measuring point	Field power (Volt m^{-1})	Conductivity (Ohm m^{-1})	Condensation nuclei per m^3
Densely populated areas (large cities, industrial regions)	200–300	$5\text{–}10 \cdot 10^{-15}$	$5\text{–}10 \cdot 10^{10}$
Thinly populated areas (rural areas)	100–200	20–30	$5\text{–}10 \cdot 10^{9}$
Uninhabited areas (oceans, polar regions)	approx. 100	30–40	10^{9} and less

ISRAEL (1961) and REITER (1960) have dealt with the problem of the biological–bioclimatological significance of atmospheric–electrical phenomena. The latter author carried out experiments with individuals who were either weather-sensitive or ill. He checked possible relations to atmospheric–electrical "indicator" elements (longest electromagnetic waves, static field of the atmosphere, ions, aerosols, radioactivity of the air, etc.). As to the significance of atmospheric–electrical elements as possible causal factors Reiter reaches the conclusion that special attention must be paid to the electrical constitution

of aerosol, as expressed by rapid electrical field fluctuation in connection with electromagnetic oscillations or also electrical displacement currents.
Israel also concludes that in this connection only the aerosol state and high-frequency field oscillations can have immediate bioclimatic effects. Regarding the intrusion of atmospheric–electrical phenomena into living quarters, Israel stresses that only the aerosol state in a general sense and in all high-frequency processes can have bioclimatic significance.

Trace gases in the lower atmospheric layers (questions of environmental protection)

Origin of trace gases and their general air-hygienic significance

Besides the aerosol particles present in the atmosphere which either appear as condensation (Aitken) nuclei and the so-called permanent gases, especially the lower atmospheric layers absorb more and more *trace gases* (with mainly very low volume-percentual values). Their presence is linked mainly with emissions from industry, motor vehicle traffic and domestic heating. Hence, they are naturally subjected to great local and temporal fluctuations. Intensified study on the nature and distribution of trace gases was started after 1950 and in the last decade has been accelerated strongly by the demands for intensified environmental protection. Thus the need has also arisen to clarify the trace substance *balance* in the atmosphere (JUNGE, 1971). The biosphere plays an important part in this connection besides purely physical and chemical processes, as is well-known from the behaviour of CO_2.
Trace gases occurring in the lowest atmospheric layers are of largely anthropogenic origin and the immediate sources are domestic heating (oil, coal, wood), motor vehicle traffic and industrial emission. Artificial admixtures to the atmosphere, i.e. not originating from specific atmospheric production, are generally called "air pollution", or "foreign" or "noxious" substances. During the past decades the population increase, technical and economic development and a rising consumption of energy have subjected all organic life (human, animal and plant) to great stress. FLOHN (1973) has pointed out all these problems regarding human activities as climatogenic factors, and he has extensively elucidated all pertinent aspects based on statistical analysis.
These facts gave rise to an intensification of the measures taken for a far-reaching guarantee of "environmental protection" in the specialized field of air hygiene (cf. also Second International Clean Air Congress of the International Union of Air Pollution Prevention Associations, Washington, 1970, and the Environmental Protection Conference, Stockholm, 1972).
The international conference of 1973 at Helsinki supplied further information on the course and technical measurements of foreign admixtures to atmospheric air (Special Environmental Report No. 3, 1974, W.M.O.). At this conference JUNGE (1974) pointed out those categories of atmospheric admixtures (cf. Table XXV) which must be considered as particularly significant and worthy of investigation.
The most important substances from various emissions, as trace substances are: aldehydes (SO_2 and SO_3), nitric oxides (NO_x), hydrocarbons (HC), in particular ammonia (NH_3), nitric acid (HNO_3), sulphuric acid (H_2SO_4), carbon monoxide (CO), carbon dioxide (CO_2), hydrogen sulphides (H_2S), fluorine (F_2), arsenic (As), lead (Pb), etc. They

TABLE XXV

CATEGORIES OF ATMOSPHERIC CONSTITUENTS WHICH ARE OF INTEREST TO MONITORING AT BACKGROUND STATIONS AND TYPICAL EXAMPLES OF SUCH CONSTITUTENTS
(After JUNGE, 1974)

(*1*) Constituents which can influence the climate *directly*:	
CO_2: absorption in the infrared	
Turbidity: scatter and absorption of solar radiation by aerosols	
(*2*) Constituents which may influence the climate *indirectly*:	
Aerosol concentration:	change of albedo and amount of cloud cover due to changes of cloud droplet concentration
SO_2, NH_3, Hydrocarbons: important for production of aerosols from the gas phase	
CO, N_2O, CH_4: possible influence on the stratospheric ozone budget	
(*3*) Deposition rates of various constituents (wet and dry) important for soil and hydrology:	
Acidity, SO_4^-, NH_4^+: soil and water chemistry	
(*4*) Constituents important for the biosphere:	
DDT and other biocides:	transport through the atmosphere
Hg and other heavy metals:	transport through the atmosphere
(*5*) Constituents important for atmospheric tracer studies or as an indicator for pollution:	
O_3, CO, condensation nuclei concentration	

are the chief sources of the very small initial and coagulation nuclei with diameters of 10^{-7} to 10^{-4} cm, as registered in nuclei counters. Precisely for this reason field measurements with nuclei counters have a certain air-hygienic significance. Dispersion nuclei (NaCl) do not enter the pictuie at all as the maritime component of natural aerosols is effective chiefly over the oceans, in coastal areas, and only with very fast air transports (e.g. of maritime polar air to central Europe) affects continental regions. As already pointed out the emissions from settlements, traffic, and industry create the preliminary matrix for the characteristic structures of the "nuclei" as mixtures of hygroscopic, not easily soluble and insoluble substances, and for that reason they deserve the name "mixed nuclei" (Junge). Coagulation and adsorption mechanisms participate in the formation of mixed nuclei. Many dusts also possess "nuclei" characteristics. These generally anorganic substances are joined by "nuclei" from organic ones (for instance house dust), as well as micro-organisms (mites, bacteria, viruses) and vegetable organisms (pollen, spores of higher and lower plants, fungi). The latter may affect humans as "air allergenes" (appearance of hayfever, allergic attacks of asthma bronchiale, or outright diseases such as histoplasmosis).

According to type and frequency, smoke and dust as effluents can provoke unspecific as well as specific diseases. Dust damages to skin, circulation, gastro-intestinal system, but in particular to the lungs, are well-known. Predominant among the latter is the common dust disease (without tissue reaction), inflammatory tissue reactions of the lungs (dust pneumonia) and pneumonoconiosis (lung fibrosis, silicosis, asbestosis). All dusts with free crystalline silicic acid, diatomaceous earth, asbestos, talcum, aluminum and the poisonous substances such as arsenic compounds, cyanogen salts, manganese, morphine, nicotine are considered particularly harmful. Tobacco smoke contains aerosol constituents with diameters of 0.15 to 0.1 μ.

The percentual distribution for the U.S.A. of noxious emission origins is given in Table XXVI. It results that motor vehicles, before heating and industrial sources, constitute the main reasons for air pollution.

Hygienic studies of the outdoor air may not exhaust themselves in the measuring of

TABLE XXVI

PERCENTUAL DISTRIBUTION OF THE ORIGIN OF NOXIOUS SUBSTANCES EMISSIONS IN THE USA (1969)

Sources	CO	HC	NO_x
Motor vehicles	64.7	45.7	36.6
Other means of transportation	9.0	7.2	10.5
Heating, electricity production	1.2	2.4	42.0
Industry	7.9	14.7	0.8
Garbage combustion	5.2	5.3	1.7
Miscellaneous	12.0	24.7	8.4

foreign substances and in the determination of the air-polluting degree, but must also clarify the relations between air pollution and health damages. In principle there are two procedures: toxicological experiments on animals or on human volunteers; as well as the epidemiological course of air-hygienic conditions regarding various population groups. The latter case deals with the study of mortality and morbidity (frequency of diseases). Although the causal relation between air-pollution degree and health damages may not in each case be sufficiently assured, the investigations up to now nevertheless reveal highly suggestive evidence.

Despite smoking's indisputable influence on mortality by lung cancer, statements were made that air pollution due to emissions from industry and motor vehicle traffic form an essential factor for the occurrence of lung cancer. A determination of causal relationships between motor vehicle traffic and mortality through different diseases is difficult. Investigations were also made in connection with air pollution on mortality due to heart and circulatory diseases. However, contradictions arose inasmuch as their fatal consequences also showed a marked meteorobiological dependence upon temperature and humidity conditions (KUHNKE, 1962). In this connection GRANDJEAN (1973) states that also the modern methods of empirical social research may be quantifying and can be guaranteed by appropriate statistical procedures. On the other hand, the circumstance must be considered that the human being possesses sensors for poisonous substances only if they can be seen or smelled. A classic example is carbon monoxide (CO) which is colourless and odourless. An evaluation of noxious effects of air pollution is difficult because of simultaneous presence of several noxious elements in often strongly varying concentrations (cf. also p.86). Experimental investigations on animals and plants no doubt will permit judgements of the dangers of health damages by foreign aerosols. Generalizations are not often possible, either. Epidemiological studies are better suited because the extent of damage is calculated on overall human populations after air-pollution episodes. Such episodes are characterized by: strong increase of pollution; stable inversions under weather conditions with weak winds; rise of morbidity and mortality; irritation of eyes and respiratory system. Relations between asthma frequency and air pollution degree are well-known.

From a meteorological aspect the socalled "inversion states" form a natural favourable phase for accumulation of emissions of all kinds. Fig.48 shows the consequences of a marked *inversion* in London (December 1952) which lasted for over a week. This study has now historical meaning because it was the first time that a large population became aware of the extent and effects of high concentrations of air polluting substances. The excess in mortality was found mainly among the aged and already ill persons. The result,

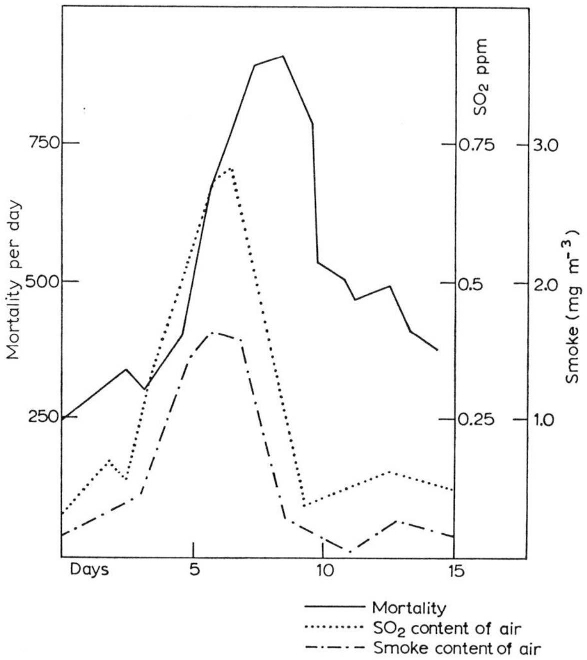

Fig.48. Air pollution and mortality during an inversion phase, London, 1952 (GRANDJEAN, 1973).

however, may be taken as a significant proof of the dangers of local and extensive accumulations of noxious elements. Occurrences of this nature are encountered predominantly during the cold season with its relative high frequency of inversions. Structure and frequency of inversions have been dealt with in numerous specific studies, especially in connection with their characteristics inhibiting dispersal and thus causing rapid deterioration of air purity. Very impressive consequences have been observed in agglomerations of large cities and industrial areas (cf. p.50).

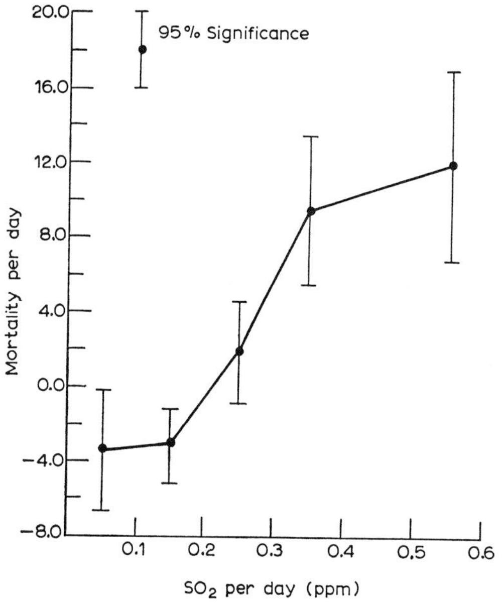

Fig.49. Average number of deaths per day differing from total average and in dependence upon the average SO_2 concentration, New York, 1960–1964 (GLASSER and GREENBERG, 1971).

The result of measurements over several years (1960–1964) in New York must be mentioned, which revealed a significant relation between the number of daily deaths and the air's current SO_2 content (GLASSER and GREENBERG, 1971) (cf. Fig.49). With regards to the far smaller stress on the city air (Vienna) by SO_2 during the *summer semester* STEINHAUSER (1970) stated that during that period the convection phenomena due to incoming radiation (together with turbulence and vertical exchange) are predominant and that thus an intensified vertical replacement of the air layers nearest the ground is taken care of in a natural way (Fig.50).

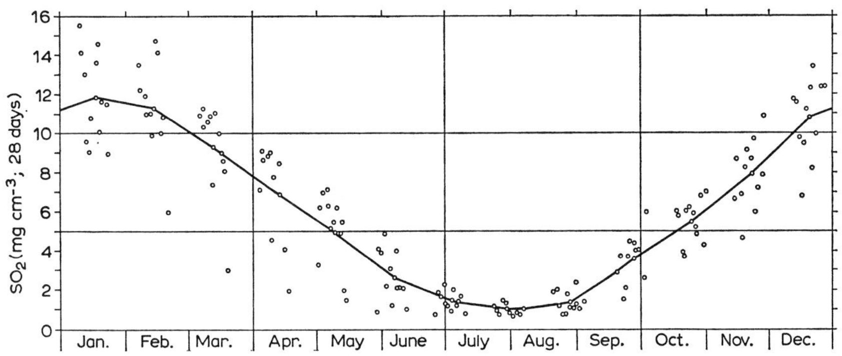

Fig.50. SO_2 course in Vienna Hohe Warte, 1958–1970 (STEINHAUSER, 1970). Points are measured values from the individual exposure periods.

In this context during the midsummer period (June through August) the smallest range of the measured values is observed, i.e. during the season with the highest possible expectancy of an assured reduction of noxious elements in the lower atmospheric layers. It is during this period that they are distributed in a large air space by a far-reaching exchange constellation.

Special meteorological and air-hygienic influences and analyses of aerosol foreign to the atmosphere

The seasonal course of the complex inversions and convections is followed readily by visual range observation. Thereby the main temporal phases can be clearly defined in close connection with processes influencing aerosol (haze or fog and clear visibility).
The lowest atmospheric layers dominate generally by a marked stability, the degree of which may be classified by means of specific stability criteria (TURNER, 1964). On this basis GEORGII and SCHAEFER (1968) determined frequency distributions of SO_2 concentrations (Frankfurt/Main) for the winter phase November–February of the years 1961/62 through 1964/65, expressed in individual stability classes (*7* = extremely stable; cf. Fig. 51).
The increase of SO_2 concentrations with rising stability is easily recognizable.
In order to determine the harmfulness of the air's SO_2 content as function of concentration and duration FORTAK (1971) designed an instructive graph. A particularly interesting feature is that it supplies data not only on the human biosphere but includes the damage limits for the vegetation. Furthermore, the meteorological parameters of stable and unstable air mass structure were duly considered.

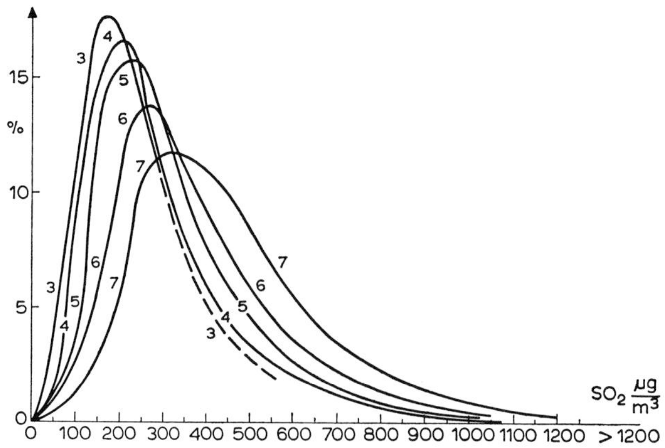

Fig.51. Frequency distribution of SO_2 concentration in dependence upon the atmosphere's degree of stability (7 = extremely stable), Frankfurt, November–February 1961/1962–1964/1965 (GEORGII and SCHAEFER, 1967).

Regarding the stress and stressability of oecosystems (i.e. fields of interaction between living organisms and their environment), especially of the vegetable kingdom, ELLENBERG (1972) emphasizes that for example a deciduous forest in summer is less affected by air pollution than a coniferous forest with evergreen leaves and a comparatively larger surface. In the latter case it must be considered that needles live several years and are thus subjected to higher loads of air pollution also in winter (e.g., by SO_2 emissions) and are thus more endangered.

Among emitting sources motor vehicles occupy the first place. In contrast with heating systems and industrial plants, motor vehicles emit their exhaust directly over the street level and thus contaminate the air layer nearest the ground and especially affect all persons on the street. The latter complain primarily about the evil smell and blackish-brown smoke (e.g., due to inadequate maintenance of diesel motors). All these phenomena justify technical and official measures.

Carbon monoxide (CO) is considered one of the most dangerous exhaust gases as is very well-known from accidents in garages. A concentration of a few parts per thousand (200–500 $cm^3 \, m^{-3}$) may in a short time lead to headaches, fainting and even death. Fig.52 shows the dependence of residence time (h) upon different percent-by-volume values (CO) in a room, together with the relations to individual physiological effects.

CO when inspired is bound in the capillary vessels of the lungs to the blood colouring matter (hemoglobin) and forms carbohemoglobin (COHb). A noteworthy feature is that hemoglobin possesses an affinity 250–300 times larger for CO than for O_2. American scientists have proposed to take a value of 5% COHb as the permissible limit when the endangering substance appears not in form of pure CO but in the mixture of all exhaust gases. This corresponds with an extended CO concentration of 30 $cm^3 \, m^{-3}$. Table XXVII compares CO concentrations in the atmosphere with corresponding maximum COHb values.

Besides carbon monoxide the lead compounds, nitrous gases, SO_2 and soot belong to the noxious substances in motor vehicle exhaust. Lead compounds and sulphur dioxide when originating from motor vehicles appear in concentrations which, at least in general, must not be considered as dangerous to health. Nitrous gases (NO_X) are another matter because they are irritants. Inhalation even of small concentrations (30–40 $cm^3 \, m^{-3}$)

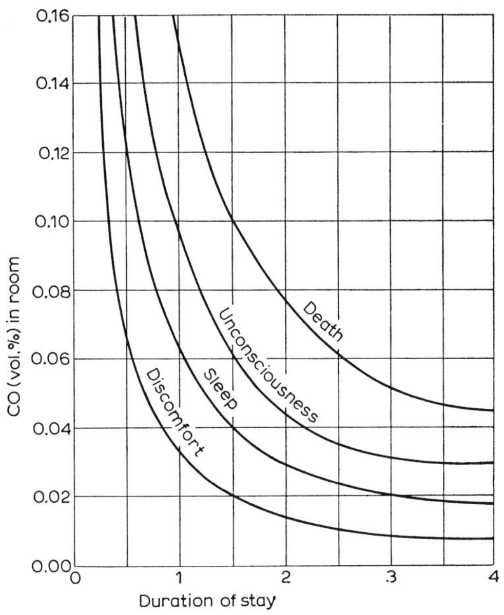

Fig.52. Limits of residence duration (in h) in a room filled with different CO concentrations (DONELLI and KADEN, 1969).

TABLE XXVII

CARBON MONOXIDE CONCENTRATION IN THE ATMOSPHERE AND CORRESPONDING MAXIMUM CARBOHEMOGLOBIN VALUES

CO concentration in air in cm^3/m^3	Corresponding carbohemoglobin values in %
10	1.6
20	3.2
30	4.8
40	6.4
50	8
75	12
100	15
250	36

already leads to irritations of the eyes, the respiratory system and the lungs. Most diesel engine exhaust consists of hydrocarbons. Its health-damaging effect is of considerable significance, because it contains carcinogenic substances and facilitates the progress of irritating substances into the deeper lung sections.

It is of bioclimatological and particularly air-hygienic importance to know the distribution of major emission sources and their respective dangers. The following selection illustrates the contributions of various industries to air pollution.

(1) Anorganic raw materials and heavy chemicals

Chief emissions: (*a*) gases: oxides of nitrogen, oxides of sulphur, ammonia, chlorine, hydrogen fluoride; (*b*) dusts: lime, fluoride, silicon dioxide.

Effects: fine aerosol, large amounts of effluents in low concentrations, heterogenous smells.

(*2*) *Organic substances and finished products*
Chief emissions: (*a*) gases and fumes: chlorine, organic nitrogen and sulphur compounds, carbon disulphide, hydrogen sulphide, formaldehyde, bromine, acetylene, etc.; (*b*) dusts: colouring material dust, detergent dust, etc. Noxious effects: same as under (*1*).
(*3*) *Metallurgical establishments* (foundries and steel works with effluent-producing processes)
Chief emissions: (*a*) gases: CO, SO_2, smokes and acid vapours; (*b*) dusts: sand, silicon dioxide, iron oxides, etc.
Effects: corroding foreign matter, metal oxide aerosols, penetrating smells.
(*4*) *Food and feedstuffs industry* (milling, drying of grass, drying by pulverization, etc.)
Chief emissions: (*a*) gases: hydrogen chloride, nitrose gases; (*b*) dusts: according to manufactured product.
Effects: dusts and smells.
(*5*) *Sewage disposal, garbage combustion* (mud coagulation, combustion)
Chief emissions: dust, smoke, hydrogen chloride and evil smells.

During the past decades the extent of air contamination by effluents and dust has risen greatly so that limits of air pollution concentrations had to be established in order to protect the population. Emission limit values must lie in range which protect man with as large a safety margin as possible from health insults by harmful elements of the most varied kinds. Mainly the following air polluting substances are subject to controls in many countries: particulate pollution (mostly dust); sulphur oxides (SO_X); nitrogen oxides (NO_X); oxidizers (oxidizing smog); hydrocarbons; carbon monoxide (CO)
The limitation of emissions should largely eliminate the noxious effects of air-foreign substances. The present problems are, however, extremely manifold. Air-polluting matter is largely present in concentrations in which harmful effects are often hard to prove. The limiting values lie precisely in those ranges. The objective is to avoid definitely such influences by means of the established air-quality standards. But this is only possible if the required technical remedies are feasible and economically supportable. For many criteria it is often impossible to draw a clear line. It is also not always possible to decide whether in the case of health-damaging effects the air pollution is the sole cause or only an intensifying factor. Indirect effects (on plants, soil, water) must also be considered in the establishment of the limit values.
Many countries have taken steps towards the protection of health. In 1970 Germany's commission for the investigation of health-damaging working materials published a list of the socalled MAK values in the *Deutsche Forschungsgemeinschaft.* The MAK values (maximum working site concentration) is that concentration of a gaseous, vapourous or dust-like working material in the air at the working site which according to present knowledge does in general not damage the health of a person working there for eight hours a day under a weekly schedule of 45 working hours.
Furthermore the VDI (Verein Deutscher Ingenieure) established the socalled MIK values (maximum emission concentrations). MIK values for air polluting substances are defined as the concentrations in the free atmosphere such as according to present knowledge may be considered harmless for human, animal and plant during influencing periods of a given duration and frequency. MIK values are based on the effects and are set up in such a way, if followed, no emission damages are to be expected.

In 1974 the German Federal Republic published a new environmental protection programme in the "Technical Instructions for Keeping the Air Pure". It establishes maximum emission standards for more than 120 gases, dusts and vapours and is compulsory for all plants. The limiting values for SO_2 for instance were greatly reduced, namely from 0.4 to 0.14 mg m^{-3} for long-term effects, and from 0.75 to 0.4 mg m^{-3} for short-term effects. The numerous economic and above all technical measures taken to check damaging emissions from industry, trade, etc, cannot be discussed here.

The necessity of purifying the air in industrial buildings can merely be pointed out here. Installations of such equipment pose thermodynamic, fluid-dynamic and engineering problems. The main job of ventilation installations is the continuous renewal of the air in a room. They include devices to clean or if necessary humidify, dry, heat or cool the air. Environmental problems including air pollution, water and soil contamination, etc. have reached large dimensions. Industrial enterprises have, however, during the last few years made great efforts—partly by law enforcement and partly due to their own initiative—to hold the negative effects on the environment by production as low as possible. Additional important aspects are connected with energy production which, beginning with the primary sources upset the subtle ecological equilibria which nature has built up over millennia. The sum of negative environmental influences is directly proportional to the number of living human beings, and that the main reason for the far-reaching environmental damages is the world's incessantly growing population. Stabilization of population and of industrial production is essential for the future. Attempts are being made by physical and econometric models to establish tolerable limits for survival.

The basic characteristics in the behaviour of the annual periodical variability of coarse and fine aerosols in outdoors air has already been pointed out in Fig.50 (STEINHAUSER, 1970). It became evident that during the Central European winter especially in the large cities and industrial areas the accumulation of effluents was intensified and extended in time due to more heating but also because of the inversion weather conditions frequently present in the lowlands. In summer the usually well developed vertical convection contributes to a periodically strong reduction of foreign aerosol in outdoors air.

The *diurnal variation* of fine aerosol (condensation nuclei) was already illustrated in a striking example in Fig.43 (NEUWIRTH, 1966). The following Fig.53 represents the mean diurnal SO_2 variation. These diurnal variations are caused by interaction of periodical changes in aerosol production and the periodical meteorological influences.

In this example the average diurnal variation shows a maximum between 08h and 12h, and a second smaller maximum between 18h and 22h. The main minimum appears between 03h and 06h, a secondary minimum at 15h. The main maximum is based on the intensified concentration of foreign aerosol and dust with the start of the workday; there is increased production at that time (heating, traffic etc.). Before the vertical convection sets in (on an average between 09h and 11h) these aerosols may permeate the lowest air layers under inversion-like conditions. During the effective convection phase (between 10h and 16h) a reduction of the aerosol concentration at the surface becomes evident. After it ends (between 16h and 17h) a slight rise is observed, which is partially due to a renewed accumulation of foreign particles. Only after the notable reduction of effluent production during the night does the coarse aerosol concentration approach its main daily minimum.

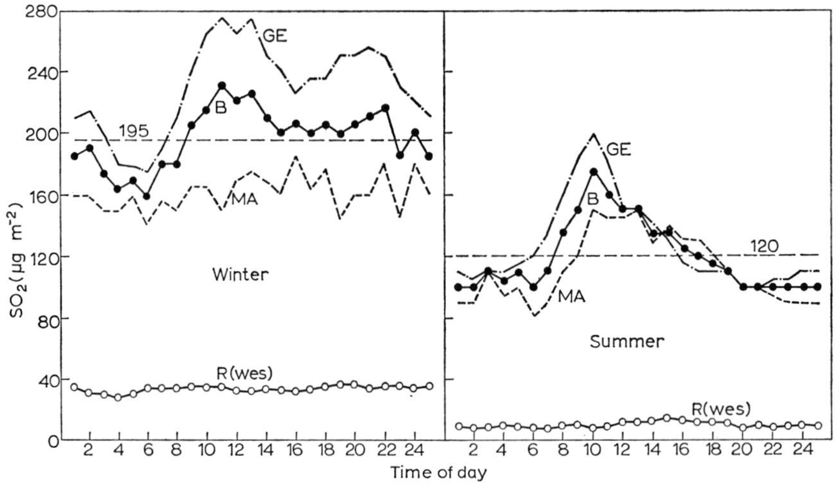

Fig.53. SO_2 diurnal variations at the pure air station Westerland (*R*) and the contaminated stations (*B*) Gelsenkirchen and Mannheim, separately for winter (I/II) and summer (VII/VIII). According to continuous registrations, after Koehler and Fleck (JESSEL, 1970).

STEINHAUSER (1963) obtained the same results by means of albedo measurements of soiled filter papers for determining the diurnal variation of air pollution in Vienna. On days with few clouds during the convective phase between 10h and 17h he observed the air's most intensive self-cleansing process. Maximal air pollution values were found in summer between 05h and 08h and 20h and 23h and in winter between 08h and 11h and 17h and 20h. In all seasons the main maximum is found in the early forenoon.

An extensive specific literature exists regarding the importance of inversions in connection with air pollution and also with the establishment of nuclear reactors, which has been pointed out several times (cf. STEINHAUSER, 1960). REITER (1968) has presented examples of aerosol condition soundings during inversion states in the mountains, carried out from cableway cabins. Convection is hindered primarily by the steepness of the temperature rise in the inversion and not by the actual temperature increase.

These complex problems and the distribution of foreign aerosols in the atmosphere are discussed in Vol.6 of *Advances in Geophysics* (LANDSBERG and VAN MIEGHEM, 1959).

KAYSER (1972) confirmed in a compilation of monthly average values of SO_2 from measuring stations at different sites (cf. Table XXVIII). Compared with a sea climate the very unfavourable conditions of large cities and of industrial centers (Gelsenkirchen in West Germany) are very obvious.

JOST (1973) and BALTRUSCH (1974) carried out in a careful study a three-dimensional analysis of the CO_2 concentration field above the intensely industrialized area of Mann-

TABLE XXVIII

MONTHLY AVERAGE SO_2 VALUES AT DIFFERENT STATIONS (μg m^{-3})
(After KAYSER, 1972)

Measuring point	Oct.	Nov.	Dec.	Jan.	Feb.	Mar.
Vienna Hohe Warte	30	70	60	100	90	60
Westerland	16	28	15	18	17	11
Gelsenkirchen	140	260	270	250	220	250
Mannheim	160	140	200	150	120	180

heim-Ludwigshafen (West Germany). From aircraft they studied the circulation conditions encountered in the upper Rhine valley. The result was that independent of lapse rate class, the CO_2 distributions at different heights over the area were essentially characterized by two groups of stacks with a very strong stack rise (BALTRUSCH, 1974). The geographical climatological considerations of the behaviour of foreign aerosol, show that diurnal and annual variations are similar in the Southern Hemisphere (Pretoria, South Africa) as elsewhere. Fig.54 illustrates these facts in a combined presentation of the diurnal and annual variations (1969) of smoke concentrations (expressed as the "soiling index" and defined as "the darkening potential of the smoke suspended in one cubic meter of atmospheric air, when collected on a Whatman No.42 filter paper having a circular sampling area of 32 mm diameter"). The Pretoria situation shows that the "air-hygienic year" is divided into two periods: during the southern summer (October through March) the lowest concentration values are noted, when the climate is characterized by atmospheric instability and frequent precipitation. During the southern winter (May through August) the highest values of SO_2 pollution are observed. During this period South Africa has a stable weather situation and contrary to summer there is a reduced vertical mixing. The diurnal variation of SO_2 concentrations in the

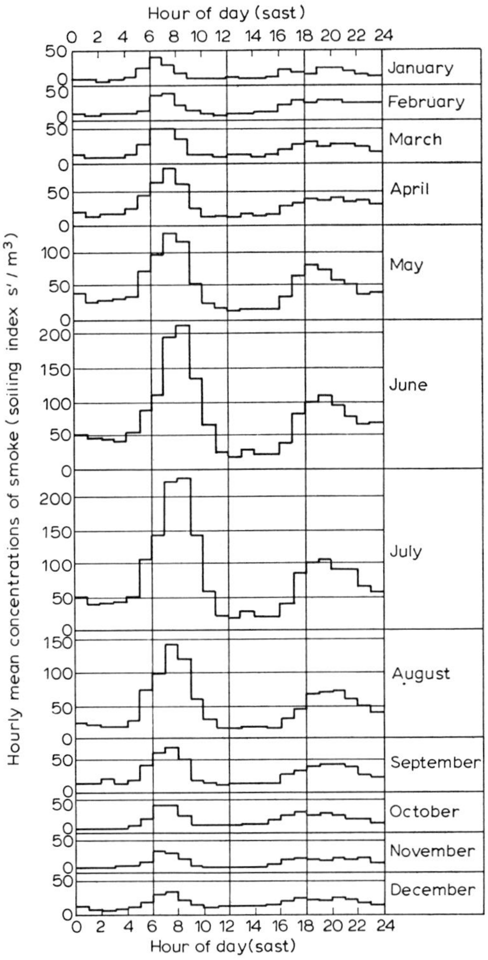

Fig.54. Mean diurnal variation of smoke during the year 1969 at Pretoria (KEMENY and HALLIDAY, 1972).

central area of Pretoria shows their extremes at the same times as found e.g. for Vienna by STEINHAUSER.

GEORGII (1954) made some systematic investigations of air pollution in rooms on higher floors of buildings. He showed that the influence of wind speed is decisive for the air volume exchanged between indoors and outdoors. Air exchange starts only above wind speeds of 6 m sec^{-1}. For cellars pressure effects dominate caused by temperature differences between inside and outside air. An "auto-aeration" takes place through seams and cracks at doors and windows. There is also an "air exchange" through the walls forming an overlapping process. COURVOISIER (1957) attributed an outstanding significance to a "draught" in the indoor–outdoor equalization of aerosol concentrations. LANDSBERG (1964) stressed the particular attention due to air exchange in room-climatological studies, and its importance for rooms with large human groups (school rooms, hospitals, halls, etc.). For the number of Aitken (condensation) nuclei he found the indoor–outdoor relations in non-humidified rooms given in Table XXIX.

TABLE XXIX

INDOOR–OUTDOOR RATIO OF CONDENSATION NUCLEI (base value outdoors 20 to $60 \cdot 10^3$ cm^{-3}) (After LANDSBERG, 1964).

Type of room	Ratio indoor/ outdoor nuclei
Bedroom, unheated, slightly aired	1.3
Bedroom, unheated, not aired	1.5
Bedroom, heated, not aired	2.0
Living room, heated, not aired	2.2
Living room, heated, not aired, people smoking	10.0
Kitchen, heated, not aired, gas stove	20.0

The greatly varied experiences of smell can only be mentioned in passing, because all experiments to create a system of smell assessment have failed (HENSEL, 1966). For the sense of smell (just like for the other senses) a scale of intensity can however be set up. According to STEVENS' (1957) statement (cf. section "Light and life", p.13) the intensity of an olfactory experience (E) may be illustrated by the power function $E \sim k (C-C_0)^n$. C signifies the olfactory substance's concentration, C_0 the threshold concentration, k a constant. Exponent n was determined as 0.5 according to measurements by JONES (1958) and is supposed to be practically independent of the type of the substance concerned.

On the whole the neurophysiology of the sense of smell is yet in its beginnings, and research in this difficult field is for the moment based mainly on animal experiments.

Any influence on smell of air temperature and air humidity is not yet established. It seems, however, that above certain threshold values of "equivalent temperature" (t_e) (cf. also the section "Temperature and life") the smell thresholds sink and thus the smelling intensity increases.

Practical environmental protection will not only require constant checking of air hygienic conditions in large urban areas and industrial regions, but will benefit from areal pollution inventories and maintenance of continuous records, properly published periodically. A good example is given by the Publications de l'Institut Royal Météorologique de Belgique (cf. also the References).

Besides specific local air-hygienic problems the relation of industrial plants to their surroundings must be considered for regional planning. Among the demands for a tolerable environment connected with old and new industrial plants, the following measures are considered important (OLSCHOWY, 1971).
(*1*) Guarantee of the soundness of water, air and soil by proper choice of location, technical installations and biological measures.
(*2*) Planting of a sufficiently wide belt of emission-protecting vegetation.
(*3*) Adaptation of industrial plants to the landscape by the correct choice of the topographic site and of the character of the buildings as well as by sufficient surrounding vegetation.
(*4*) Effective separation of industrial plant and residential, agricultural and recreational sites.
(*5*) Adequate amount of vegetation in and around the works with trees, parks and recreational sites.
Particular importance in these matters is attributed to the statements by A. Baumgartner in which he points out that the meteorological analysis of *vicinity and distance effects of wooded areas* have become a necessity in this age of environmental protection. Under the sole consideration of air-hygienic aspects, wooded areas form extremely important functional constituents with regard to the exchange processes in the atmosphere (cf. below).
At this point further information must be given, also with regards to other effects between forests and surroundings. BAUMGARTNER (1971) carried out a study on "woods as exchange factor"; wherein he states that the intensive division of mass (1 ha spruce woodland has about 2.5×10^9 needles of 5 mg mass each and 1 cm^2 surface) leads to a rapid adaptation to changes in the surrounding climate. This is a consequence of actions of the canopy to intercept radiation of all types, precipitation, aerosols and noise effects. For example, the atmospheric precipitation reaching the forest floor is about 20–40% less than that arriving on the ground of woodless land.
Regarding radiation exchange a forest—in particular a coniferous one—shows the smallest albedo values (approx. 10%), i.e., it has a very high absorption capacity. In this way it influences the light climate of a landscape (cf. also the section "Light and life", p.20). Aircraft measuring has shown that forests act as thermal sinks and thus represent cooling areas in the landscape. Radiation exchange studies lead to the overall result that forests absorb the most energy from radiation budget and thus are—energetically speaking—the most effective elements of the landscape.
Special importance is attributed to the remote effects of forests, based mainly on changes of kinetic energy of atmospheric currents. The horizontal wind component entering a wooded area from the windward side is weakened and to the lee the shelter effect extends up to a distance of approximately 25 times the tree heights. The vertical wind component extends forest air influence up to heights corresponding to half the diameter of the horizontal expanse of a closed wooded area. Fig.55 shows the mechanism of kinetic energy exchange processes. Herein d indicates the height by which a current field is lifted from the ground by a wooded area (= approximately 2/3 of the tree height); D the displacement height; z indicates the aerodynamic roughness parameters connected with the vertical component of the air movement; u is the horizontal component of wind speed. Aerodynamic studies show that z_0 is $3 \cdot 10^4$ times larger over the treetops of a

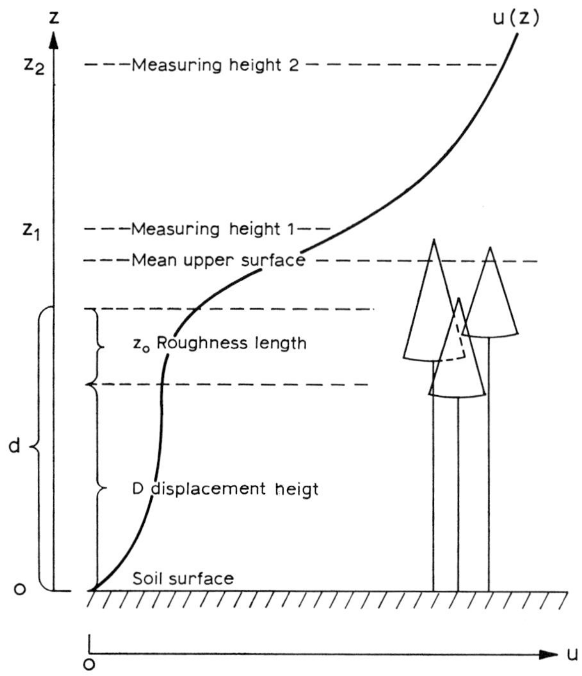

Fig.55. Index of the socalled logarithmic vertical profile of the horizontal component of wind speed in a forest (BAUMGARTNER, 1971).

forest than over a water surface, and 100 times larger than over a meadow. These figures indicate great dispersal capacity of closed wooded areas for atmospheric suspensions (dust, smoke and harmful substance content). BAUMGARTNER (1971) has concluded that, based on these properties, forests are a very active element of the landscape.

The filtering action of forests is very small for foreign *gases* (e.g., SO_2 or CO), as was already shown for the air's content of condensation nuclei. On the whole, however, the social functions of forests and green spaces go beyond pure physical factors.

The problem of water pollution as well as soil and vegetation pollution is not less important, but cannot be discussed here.

Special trace substances in man's environment

Since the beginning of aerocolloid chemistry, ozone (O_3) as allotrope of oxygen with weak binding of an O atom to the O_2 molecule has, among the *trace gases* in the lower atmospheric layers, repeatedly played an important role for suspected bioclimatological influence. Already at the time of its discovery by SCHOENBEIN (1840) ozone was in the center of meteorological and hygienic interests. In recent times CURRY's (1946) meteorological studies again focused attention on the extremely variable trace gas O_3. It participates in the chemistry of the air layers near the ground, and appears often mixed with other oxidizing trace substances in outdoors air.

In the section "Light and life" ozone's vital importance for the world of organisms in the lower stratosphere has already been mentioned. Furthermore it contributes (though in lesser amounts than the gases H_2O and CO_2) to atmospheric radiation (*G*). Whereas carbon dioxide constantly contributes about 1/6 of the intensity of atmospheric radiation the ozone contribution is merely 2% (GEIGER, 1961).

Ozone belongs to those trace substances of the atmosphere which have a strong oxidizing effect. It is the main representative of this group which also includes the oxides of nitrogen (N_2O_3 and NO_2) as well as gaseous chlorine (Cl). Only physical measuring methods (spectral analyses) enable an exact evaluation of ozone alone, whereas many chemical methods determine the totality of oxidizing constituents of atmospheric air, i.e., the total oxidation value. The nitrous gases may under circumstances attain or exceed the ozone values, as demonstrated by EDGAR and PANETH (1941).
Ozone's vertical distribution in the total atmosphere is the result of the photochemical formation and degradation in higher strata (stratosphere), in lower strata (troposphere), of transportation processes (turbulence, convective exchange) as well as of photochemical and catalytic degradation (SCHROEER, 1949).
Chemical and catalytic causes are decisive in the processes reducing ozone near the ground. The reduction value comprises the effects of numerous anorganic substances (sulfate, hydrogen sulfide, carbon monoxide, aldehydes, etc.) as well as of solid and liquid aerosol constituents (Aitken nuclei). Therefore, ozone undergoes a particularly strong reduction in metropolitan and industrial areas, whereas its quantitative presence in outdoors air in thinly populated regions is less subjected to that process. In the past years ozonometry has been greatly promoted and the observations show that primarily weather changes, also in connection with local geographic variations, have great influence on the ozone density in the atmospheric strata nearest the ground.
The lower air layers contain an average of about $2\text{–}4 \cdot 10^{-6}$ percent by volume. Today the ozone contents of the air layers nearest the ground are indicated preferably in $\mu g\ m^{-3}$ (in earlier literature it is expressed in $\gamma = 10^{-6}\ g\ m^{-3}$, γ has been replaced by μg). In general it fluctuates according to measurements by various authors between 0 and 100 $\mu g\ m^{-3}$. The following may be considered as representative.
(*1*) High values of relative humidity ($f \geqq 95\%$), unchanged strong haze or fog as well as continuous precipitation cause low ozone or zero values.
(*2*) Medium and high ozone values are found: (*a*) on clear days with vertical mass exchange; (*b*) during cold front passages; (*c*) in thunderstorm zones; (*d*) in early foehn phases; (*e*) in dispersing haze and fog (dissolution of inversions).
Further observations show that air masses of tropical origin in general have less ozone than polar air masses (especially in summer). This is conditioned by the different stability of these two main air masses of the troposphere, the relatively stable warm mass and the labile cold mass. The former permit as a rule only limited vertical movements, but the latter frequently promote extensive vertical ones.
Measurements of ozone near the ground at the Meteorological Observatory Hohenpeissenberg in rural Upper Bavaria showed that far higher values may exist than have been known so far, and not only as short-lived peak values. Extremely high values of 500–1000 $\mu g\ m^{-3}$ were registered in connection with cold front passages and heavy snow showers, as well as high values of 200–400 $\mu g\ m^{-3}$ on the upper side of inversions and also in the vicinity of Cb clouds (Annual Report of the German Weather Service, 1972) (SUESSENBERGER, 1973).
Fig.56 illustrates details of the annual variation of daily averages of ozone near the ground in Vienna for the year 1958. Except for the relatively high interdiurnal changeability during the summer the sudden decrease of ozone values at the beginning of September is striking. It is caused by accumulation of ozone-reducing anorganic sub-

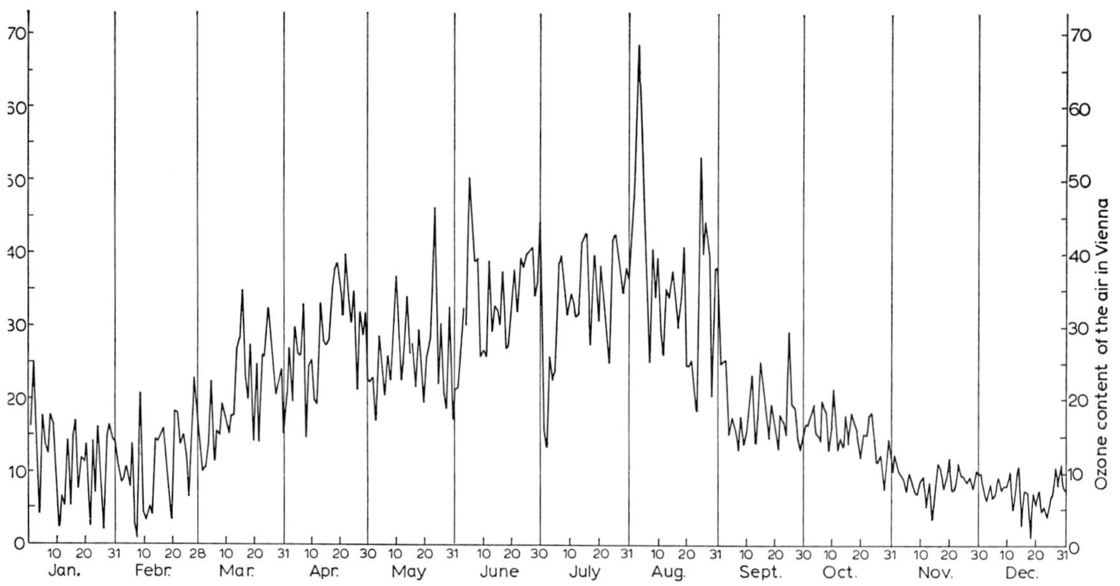

Fig.56. Annual variation of mean daily values of O_3 content of air (O_3/m^3) in Vienna, 1958 (STEINHAUSER, 1960).

stances with increased ground inversion formations. The reduction is even intensified in November and December. The frequently rather high daily ozone averages in summer are closely related to the strongly developed convection and exchange processes at that time of year and possibly to local photochemical formation.

Toxicological studies demonstrate clearly that ozone concentrations of 0.05–0.1 ppm or 0.1–0.2 mg m^{-3} air may cause headaches, nausea, irritation of mucous membranes in nose and throat. The odour threshold for ozone lies at concentrations of 0.01–0.015 ppm. The maximum working site concentration (MAK) was established at 0.05 ppm. Numerous investigations have shown that the concentrations up to 0.05 ppm supportable by humans have no bacteriostatic or bactericidal effect.

The assertion that wooded areas in the medium high mountain resort areas possess "ozone-rich" air is literally "hot air". It has been proved that the air in a forest does not have a significantly higher ozone content and that O_3 is frequently even lower there than in non-wooded areas.

Artificial ozonification of living and working quarters is not recommended since ozone in the concentrations supportable by humans has no hygienic significance whatsoever and in higher concentrations may result in health-damaging effects (WANNER and GILGEN, 1966).

There are some pharmacological findings suggesting that O_3 is capable of building up protective mechanisms in the alveoli against irritating gases (JESSEL, 1970).

The trace element iodine (I_2) is of considerable bioclimatological importance. Like ozone it is an extremely variable element and appears on an average of only $3.5 \cdot 10^{-9}$ vol. % in fresh air. It originates for the greatest part in iodine-accumulating types of sea-weed, the burning of which frees I_2, and in far smaller amounts from the sea itself. In Europe mainly the coastal areas of northwestern France, Ireland and Scandinavia act as iodine sources in maritime air mass transports to the continent. Coastal values during the scorch of sea-weed amount up to 6,000 $\mu g\ m^{-3}$, otherwise only to 0.6–1.0 $\mu g\ m^{-3}$.

As a rule in central Europe the average values fluctuate around 0.5 μg m^{-3}. In regions with minimum iodine concentrations the formation of goitres is promoted and constitutes a marked bioclimatic phenomenon (influences of metabolism and hormones in man). Since in coastal climates the air's iodine content lies 20–30% higher than in the interior, goitres are more rarely observed there.

Studies by TEUB (1970) demonstrate that in air-iodine investigations at the German North Sea coast (Westerland on Sylt) the sea water forms an important iodine reservoir as a consequence of iodine liberation from sea-weed and algae. An evaluation of aerosol-iodine in dependence upon wind direction showed a significant increase of iodine concentrations with maritime air currents compared with the content in air masses of continental origin. Atmospheric iodine in its natural genesis and in its natural effective form seems to be non-toxic (JESSEL, 1970).

Fresh air's content of carbon dioxide (CO_2) outside of industrial areas and concentrated motor vehicle traffic reaches an average of 0.02–0.04 vol.%. According to KREUTZ (1941) the daily fluctuations amount to 0.05% (between 0.025 and 0.075%). The CO_2 *diurnal variation* shows—as illustrated in Fig.57—a definite noctural maximum and a main minimum by day (between 10h and 18h) (HUBER, 1952). The CO_2 diurnal variation runs largely inverse to that of ozone near the ground. Furthermore, during the night within the ground inversion the air enriches itself with CO_2 (much of it from humus and plant decay). During the day, however, CO_2 decreases rapidly when the convective exchange sets in and active CO_2 assimilation by plants starts.

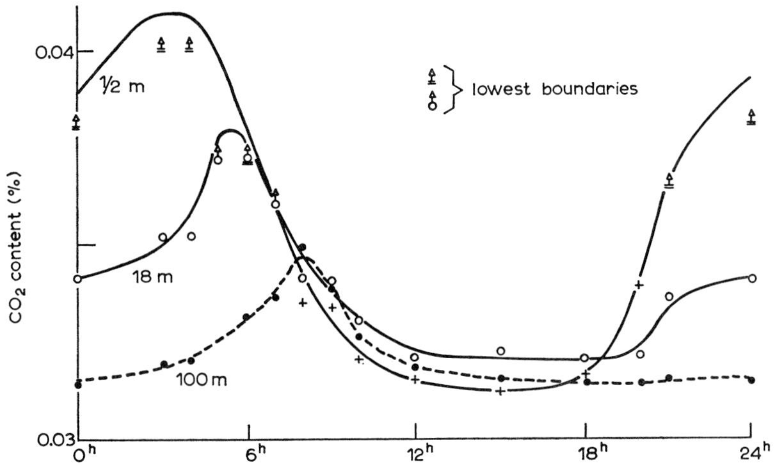

Fig.57. Diurnal variation of CO_2 content in the lowest air layers (summer observations) (HUBER, 1952).

The annual variation at least in the Northern Hemisphere reveals the greatest CO_2 content of the air during the cold season (main maximum between March and May), the phase of the lowest CO_2 contents in midsummer (July–September). The summer minimum in the annual CO_2 variation is obviously a result of the photosynthetic use by plants in summer.

During the last decades the world use of fossil fuels and the anthropogenetic destruction of vegetation has led to steady increases of CO_2 in the air, illustrated in Fig.58. This graph is based on CO_2 observations at the Mauna Loa Observatory (Hawaii), which clearly shows the annual variations from 1959 to 1971 (mean CO_2 increase between

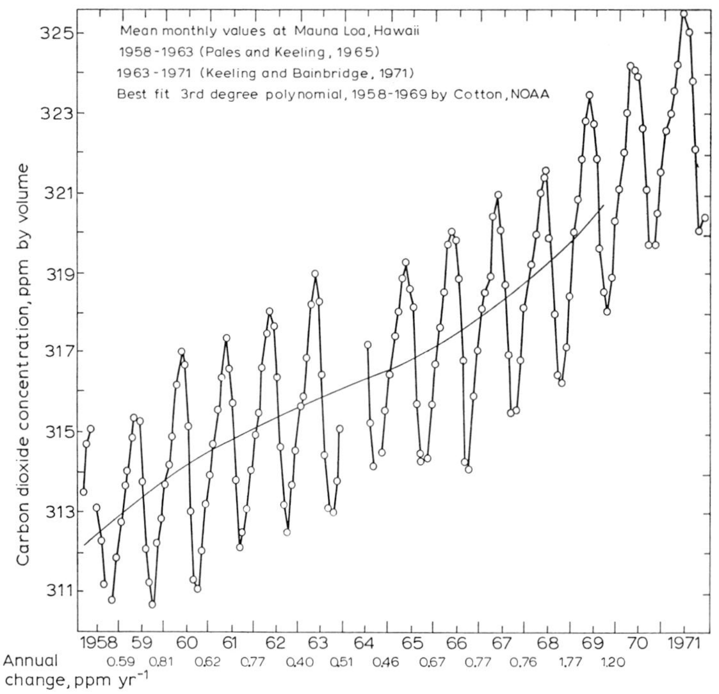

Fig.58. Mean monthly atmospheric carbon dioxide concentrations at Mauna Loa (MACHTA, 1972).

approximately 312 and 323 ppm). A world-wide increase of CO_2 concentrations is important for the atmosphere's energy balance, as pointed out by FLOHN (1973) in his studies on large-scale climatic fluctuations and influences. The CO_2 increase has presently only minor human-bioclimatological significance but its potential future effect on climate needs watching.

Radioactivity of air and its significance for the biosphere

Although research on the atmosphere's radioactivity has been going on since the beginning of the century, the advent of artificial fission has accelerated it enormously.
The origin of natural radioactivity with respect to the human biosphere is based on two sources: the first is found in the earth's outer crust which furnishes in gaseous form (emanations) decay products of the three radioactive elements from uranium^{238}, uranium^{235} and thorium^{232}. The second source is found in the lower stratosphere (cf. Fig.59). This schematic illustration (KREBS, 1968) indicates the atmospheric circulation of the emanations. They emerge from the earth's outer crust through cracks and capillaries and enter the atmosphere, or are brought to the surface dissolved in spring water. Then they are carried to the higher atmospheric layers by convection, exchange and diffusion. In this way they decay into daughter products and return to earth with aerosol particles and precipitation. In the case of precipitation-free return, the main means of emanation transport are sedimentation and downward convection. Table XXX gives a general survey of natural atmospheric radioactivity.
Interesting observations of a meteorological and bioclimatological nature regarding the mean behaviour of radioactive emanations (including the conditions in the lower air layers between 0 and 16 m above ground) have been furnished by PEARSON and JONES

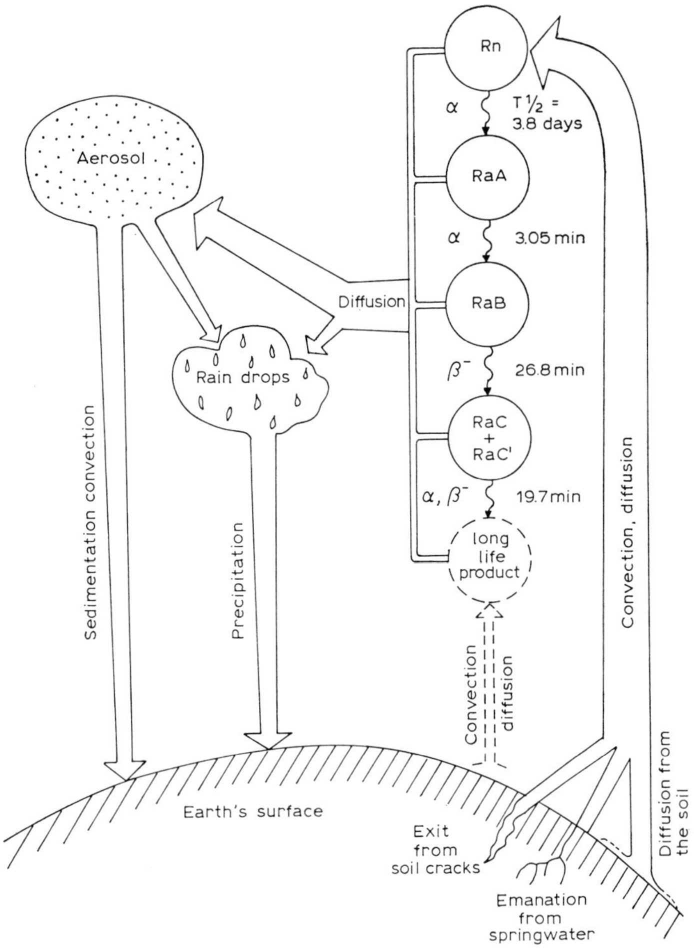

Fig.59. Circulation of radium emanation (radon) and its consequent products in the atmosphere: *Rn* = radon; *RaA* = polonium^{218}; *RaB* = lead^{214}; *RaC* = bismuth^{214} (KREBS, 1968).

TABLE XXX

GENERAL SURVEY OF NATURAL RADIOACTIVITY
(After ISRAEL, 1962)

Standard substances		Emanations	
Ground		Radioactive period depth	Deficit (curie)
	Uranium^{238}: $3\cdot10^{-6}$ g g^{-1}	Radon: 105.5 cm	$3.1\cdot10^{-11}$
	Thorium^{232}: $1\cdot10^{-5}$ g g^{-1}	Thoron: 1.35 cm	$0.034\cdot10^{-11}$
Sources	Radium^{226}: 10^{-14} – 10^{-13} C cm^{-3}	Radon: 10^{-14} – 10^{-12} C cm^{-3} (maximum values up to 10^{-9} C cm^{-3})	
Atmosphere		Condentration near ground	Radioactive period height
		Radon: $1.58\cdot10^{-18}$ C cm^{-3}	1,350 m
		Thoron: $1.74\cdot10^{-18}$ C cm^{-3}	17.5 m
		ThB:	457 m

(1965). Fig.60 shows in an isopleth illustration the radon^{222} concentration based on eight 24-h mean values of time-of-day fluctuations between 0 and 16 m above ground. It also shows the high concentration values during the evening and night hours in the lower and medium ranges of these air layers. During the day the atmospheric convection and

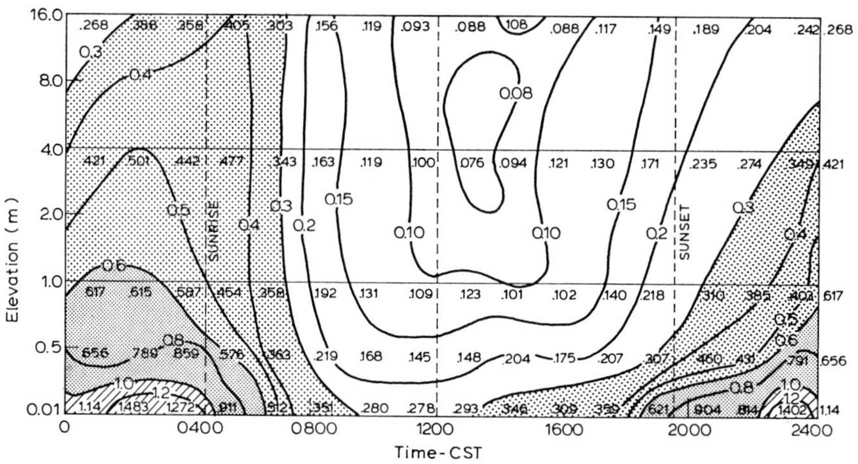

Fig.60. Radon^{222} concentration (in 10^{-12} C l^{-1}) variation with elevation and time at Argonne National Laboratory. Average values for eight 24-h periods (PEARSON and JONES, 1965).

exchange effects on the vertical aerosol structure of air show a decrease of the Ra concentration by a factor of 4 to 6, so that the lowest values are found between 3 and 9 m height. This may be explained by the fact that even under an intensive exchange during the day the exhalation from the soil continues and therefore its minimum may not be expected to be found directly at the ground (0–1 m).

It is of interest that the Rn contents in coastal areas is affected by wind direction. During on-shore winds only very small Rn concentrations are observed. This is caused by the lacking or very small exhalations over water. An off-shore wind, however, greatly raises the content of Rn (ISRAEL, 1962).*

Biological interest in natural radioactivity is based mainly on questions of radiation contamination and of transport, distribution and incorporation processes. Radiation contamination of animal and human organisms comes from: (*a*) radiation of ground and surroundings; (*b*) α, β and γ radiation of incorporated radioactive substances. Incorporation takes place by respiration and food consumption.

Radiation from external sources amounts to approximately 50–150 mrad year^{-1}. However, besides places with low radiation intensities there are areas in which environmental radiation may be very high, such as the Kerala region in India (2,800 mrem/year) and the Monazite areas of Brazil with up to 12,000 mrem per annum (KREBS, 1968).

In outer space the doses to which astronauts are exposed vary greatly and especially in connection with changing solar activities. Radiation doses coming from "inside" depend on the amount of radioactive elements which an individual carries inside since birth and that which is absorbed during life through food, water and air. The amounts vary with the radioactive state of a person's domicile and the corresponding conditions of

* Before discussing the biological effects of radioactive radiation, the currently used units need mentioning. *Radioactive decay.* The international unit of radioactive decay is the curie (C). 1 curie is that amount of a radioactive nuclide which shows an activity of $3{,}700 \cdot 10^{10}$ disintegrations per second. The curie may be subdivided decimally: 1 curie = 10^3 millicurie (mC) = 10^6 microcurie (μC) = 10^9 nanocurie (nC).
Radiological dose units. Irradiation dosis is the dosis emitted and absorbed by air. Unit: röntgen (r). The *absorbed dosis* of any kind of ionizing radiation is the energy which is transmitted to matter (air, tissue, etc.) by ionizing particles per mass unit of the irradiated material at a given point. Unit: rad (rad) = radiation absorbed dose. The equivalent of relative biological effects of different ionizing types of radiation is the REB (relative biological effectiveness). The product REB $\times$ rad is the rem (rad equivalent man).

the atmosphere, the natural "fallout" from aerosol and precipitation. In areas where fish and sea food are the main nutrition sources considerable amounts of radioactive substances may be absorbed. No living creature can escape natural environmental radiation because there is no spot on earth completely free from radiation. An evaluation of radium poisoning cases showed that 0.1 μg (= 10^{-7} g) Ra^{226} may be taken as the tolerance limit. In connection with Fig.40 (p.65) which represents separation rates (%) for aerosol particles in the human respiratory system, the special studies by JACOBI (1966) revealed that the greatest radiation influence can be expected in the bronchial area. This result corresponds with the experience that most lung carcinomas of uranium mine workers start in the bronchial tree area.

According to recommendations by the International Committee on Radiation Protection (ICRP) the following doses are the maximum permissible ones.

(*a*) For persons professionally in contact with radiation: after the 18th year the biological equivalent dosis of 5 rem per annum shall not be exceeded; from the 30th year on the dosis is set at 60 rem.

(*b*) For persons considered not professionally endangered by radiation the dosis is set in general at 1/10th the dosis for radiation workers (0.5 rem/year);

(*c*) For the entire population a genetic radiation dosis of 3 rem in 30 years by natural environmental radiation plus a dosis of 5 rem in 30 years from other radiation sources (excluding medical treatments) are tolerated.

Nuclear arms tests, mainly carried out from 1950 to 1958, have brought artificial radioactivity into the air. Those explosions caused artificial radioactive particles in form of fission products to reach great altitudes. Air currents distributed the material to all regions of the earth. The larger particles fall out quickly in the vicinity of the explosion site, the smaller ones accumulate on unspecific aerosol particles and often reach the ground in a comparatively short time due to precipitation in rain-out and washout processes. Fig.61 shows the course of long-lived artificially radioactive fission products in the atmosphere from 1953–1961 as determined at Heidelberg. It is very evident that the increasing number of atomic bomb explosions heightened the radioactive contamination of the air, but after the nuclear air burst ban (1958) it reduced rapidly. A second set of tests in 1961 led again to increases.

There were many studies on radioactive fallout. STEINHAUSER's (1974) work on strontium-90 deposits from precipitation in the area between North Africa and the Arctic (cf. Fig.62) are very instructive. The figure exhibits the cumulative curves of Sr-90 deposits from precipitation between 1954 and 1970. They reflect the effects of the first

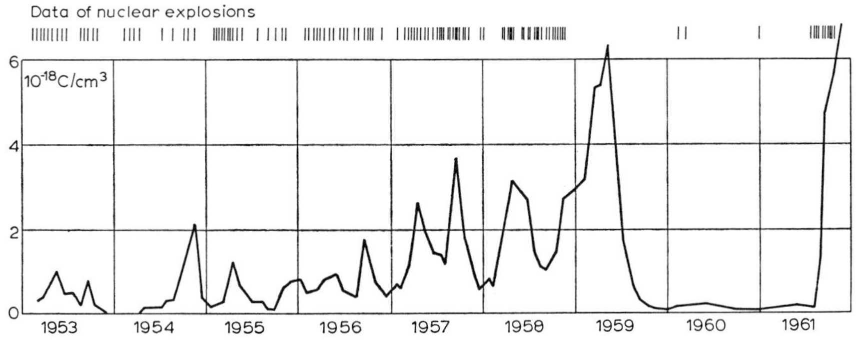

Fig.61. Concentration of long-lived artificially radioactive decay products in the atmosphere (1953–1961) at Heidelberg (ISRAEL, 1962).

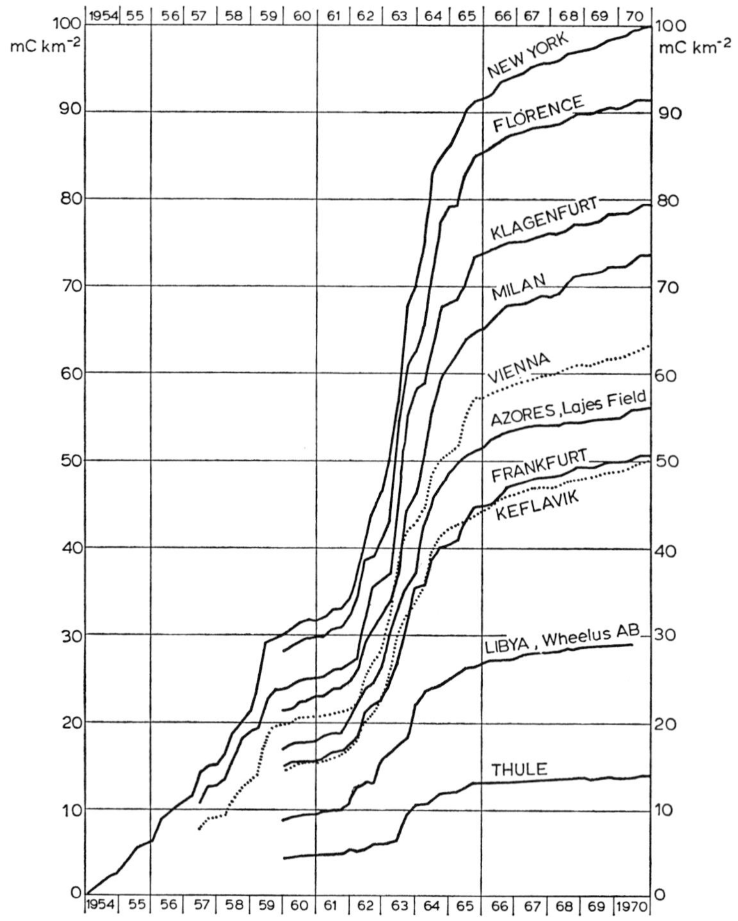

Fig.62. Sum curves of strontium-90 deposit by precipitation water from January 1954 till December 1970 in the area between North Africa and the Arctic (STEINHAUSER, 1974).

atomic air bursts series till 1958 and of the second one till 1962. The observations were published in the *Quarterly Summary Report of the US Atomic Energy Commission, New York* (Health and Safety Laboratory, Fallout Program). Considering the decay time (Sr-90's radioactive half-life is 28.1 years) one sees that approximately 73% of this fallout has not yet decayed. The greatest difference in the Sr-90 activity in precipitation appears between the second and fourth quarter. This activity must therefore be fed from a Sr-90 reservoir in the stratosphere. Thus a seasonal variation of the transport processes from the stratosphere to the troposphere are an important part of the process. Radioactive nuclides can also get into the environment from nuclear power plants and possibly from other potential applications of nuclear explosions.

It needs further emphasis that the mean radon content of European continental air moves between 130 and $1{,}500 \cdot 10^{-12}\,C\,m^{-3}$. In areas with specially high radon contents in the ground air such as in the vicinity of Bad Gastein (Austria) $1{,}300–3{,}000 \cdot 10^{-12}\,C\,m^{-3}$ were observed. This is not only gaseous radon but also aerosol particles with adsorbed emanations. These enter the human body through respiration.

Of the biological effects of radioactive radiation influences it is known that the range of γ-rays is 17,000 times larger than that of α-particles and 350 times larger than that of β-particles. Ionization of the body tissues is carried out to 90% by α-rays. Furthermore, the effect of α-rays is concentrated—due to their short range—in a small area near the

radiation source. In contrast, the other 10% (β- and γ-rays) distribute their ionizing effect more deeply into tissues.
When inhaling radium emanation one must assume, based on diffusion laws, that it immediately comes into contact with all tissues through the circulation system. Animal experiments showed that 90 min after the incorporation of radon-containing air there is exact proportionality between the radon content of the blood and that of the respiratory air. On an average the blood's radon content amounted to approximately 29% of that of the respiratory air. The effects of radioactive substrates on organs and organic systems (metabolism, circulation, nerve-system, etc.) cannot be elucidated here.
Yet it should be pointed out that in balneology and balneotherapy by means of mineral water cures radioactive substances are put to use and seem to manifest themselves in various and significant "cure effects".

Temperature and life

Basic aspects of a synthesis of the complex terms temperature and life

At the beginning of this review, in the introduction to bioclimatological aspects, it was pointed out that for heat physiological reactions in all different kinds of ambient climates, terms from the general walks of life will be used as expressive descriptions for bioclimatological environmental systems. This does not mean, however, that in the consideration of thermal environmental factors attention will be focused on individual elements. Rather in the heat physiology sector only the knowledge of all the complex causal relations can accurately reflect the working system.
It is, therefore, understandable that repeated theoretical and experimental attempts have been made to interpret and measure the integrative effect of climatic environmental factors which influence thermal sensations of all possible shadings. However, the complex comprehension and application of physiologically sensible and appropriate relations is beset by a number of difficulties, although climatology has the possibility of comprehending quite accurately the main features of thermal effective complexes. Considerable differentiation in the definitions of climate-physiological phenomena must also be reckoned with. WEZLER (1950) distinguishes the following: form and type of a stimulus; intensity of the stimuli; duration of the stimuli's influence; temporal variability of the stimuli; point of the body where the stimulus is effective in connection with the sensor density; individual differences in stimulus sensitivity and types of reactions.
On purely physiological grounds difficulties are encountered in the interpretation of climate-conditioned influences on the heat and energy balance of the human body and on the loss of energy (heat emission or heat loss). Nevertheless, a large variety of experiments has been carried out in order to find the most objective basic features and to provide theory and practice with extensive application possibilities. All these problems will be discussed in detail at a later point. For practical reasons a few basic facts on the circumstances vital to the human body shall be presented here.
In Fig.63 HARDY (1972) presents a survey in accordance with today's state of knowledge on the human thermoregulatory phenomena which depend upon terrestrial and extraterrestrial thermal conditions. It is of particular interest to bioclimatology that the vital

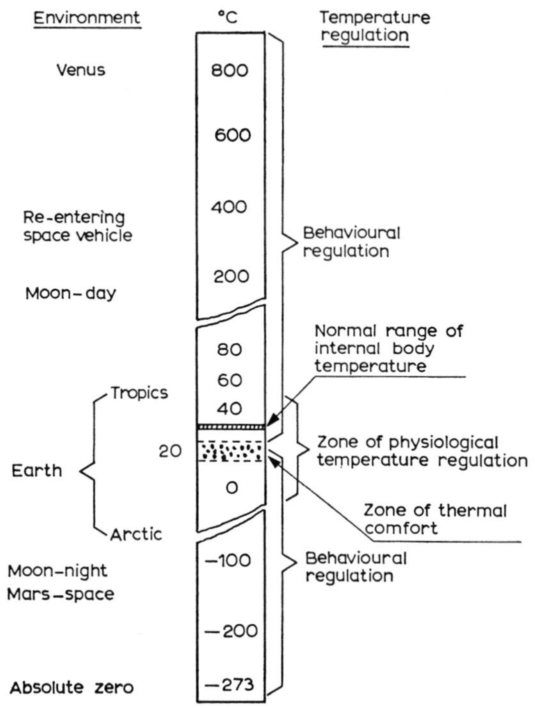

Fig.63. Thermal environment and human thermoregulation (HARDY, 1972).

conditions for our life and existence provided by the earth's atmosphere constitute only a relatively small thermal range which guarantees physiological temperature regulation. In extra-terrestrial regions "non-physiological regulation" is necessary in order to form a foundation for life and work.

General thermo-regulatory phenomena in man with regard to the terrestrial and thermal conditions

One of human life's special characteristics is that the organism has its own specific temperature within a fairly narrow range. For warm-blooded creatures, the so-called "homeotherms" (man, mammals, birds) this temperature band comprises only a few degrees centigrade. Fig.64 shows that within a normal range between 36° and 38°C, temperature regulation is still effective. Beyond those levels considerable damages occur in temperature regulation. Above 43°–44°C body temperature reaches its upper lethal limit,below 25°–24°C its lower lethal limit. Without going into details on the effects of extreme body (rectal) temperatures it must merely be mentioned that already around 40°C there is a definite tendency towards collapse (HENSEL, 1955), and at 41°C as a rule the blood circulation reaches the limits of its capacity. However, this goes only for exogenous *hyperthermy* (high air temperatures in connection with radiating heat) and not for a fever condition where there is the possibility of a change in the regulating centers.
When strongly cooled down, an unclothed person soon reaches the limits of regulatory capacity. The mean tolerance time for a voluntary exposure to cold (+1°C and slight wind as presupposition) which is limited by discomfort, cold pains, fatigue from shivering, etc. amounts to approximately 1 h. The limits of man's body temperatures in *water*

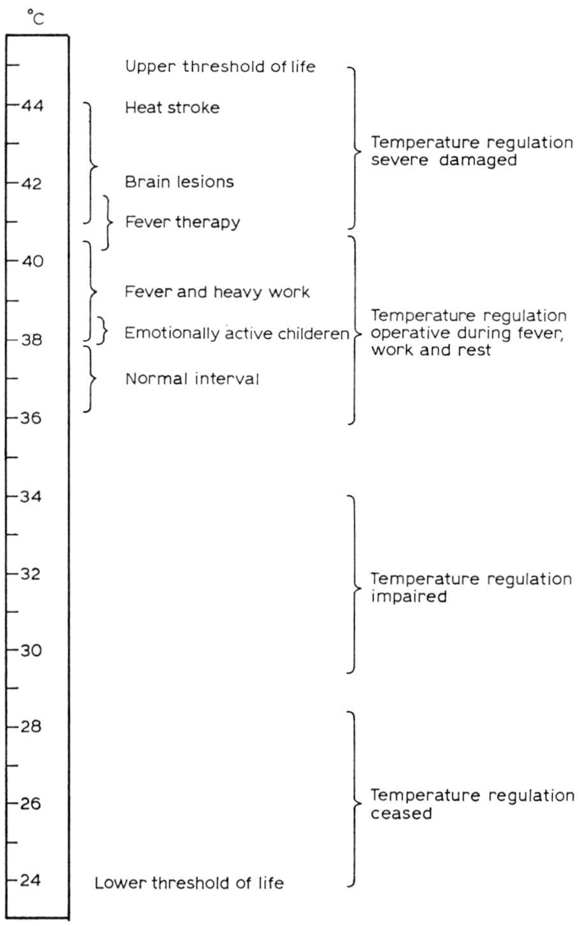

Fig.64. Total range of rectal temperature and of temperature regulation in man (DU BOIS, 1951).

are of great importance. For instance, at water temperatures of about −1°C survival is possible up to 1 h (ADOLPH and MOLNAR, 1946) at body temperatures between +15° and 21°C mean survival times of approximately 10 h were observed. Medium degree swimming performances often signify a considerable increase of the tolerance times. Fig.65 gives the main stages of human supercooling according to GROSSE-BROCKHOFF (1969). It distinguished three phases, i.e., the first so-called stimulation increase (approximately to 34°C body temperature), the second of stimulation decrease to 24°–26°C, followed by the paralysis phase which is very similar to apparent death. During these three *hypothermy* phases the thermal regulation capacity is maintained to a certain degree; this process may be observed during the warming up of a super-cooled organism in hot surroundings.

A further aspect of bioclimatological consequences for the evaluation of influencing intensities of different ambient climates is the physiological phenomenon of a constant heat current from the interior of an organism towards the outside. The reason for this is that the *body shell* together with the skin surface always indicates a lower temperature than the *body core*. These circumstances presuppose a homeotherm body core and a body shell subjected to more or less extensive temperature fluctuations. Fig.66 shows the core of the human body consisting of torso and head as organs of high consumption. The shell is formed chiefly by the extremities. The blood is warmed in the core area,

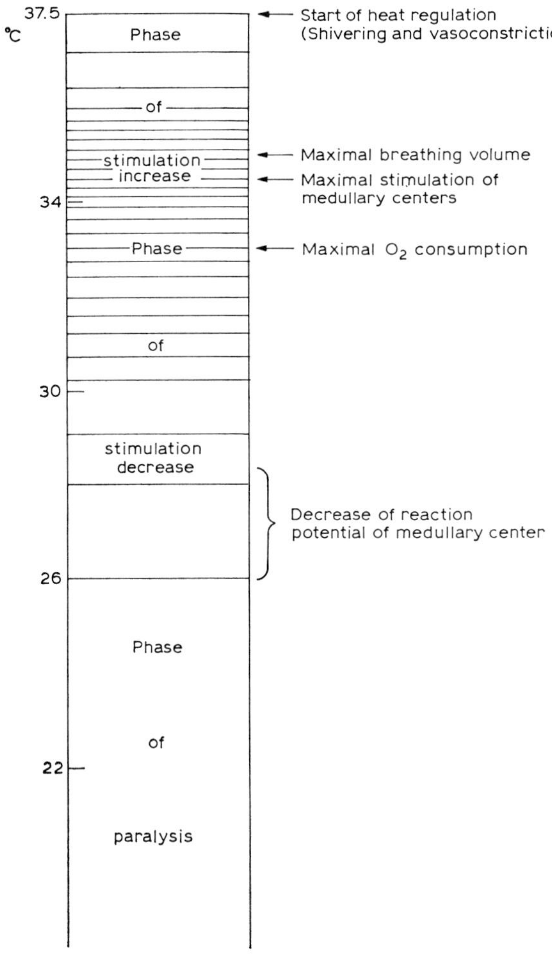

Fig.65. Thermal environment and heat tolerance adapted from FOLK (1966) (HENSEL et al., 1973).

while in the shell it is cooled. Temperature differences between core and shell can be very great, although sufficiently high ambient temperatures reduce them considerably. Thus high external temperatures cause certain body areas to become part of the core which at lower temperatures are part of the shell. Strictly speaking one can therefore neither distinguish *the* body temperature nor a *mean* body temperature. In practice, as a rule rectal temperature is taken as "body temperature".

General aspects of heat production and mean tolerance limits

Bioclimatology frequently bases its measurements of thermal complex units on teh term "body temperature" or "skin surface temperature". Often this inadequate distinction has led to controversies regarding the determination of cooling processes and their physiological applicability. It is therefore useful to assess heat production and body temperatures of homeotherms at differing air temperatures.

Animal experiments have clearly shown that heat production in dependence upon external temperature is subject to considerable variations; this fact is also of considerable importance for man's bioclimatically controlled behaviour. Fig.67 gives a schematic survey of these relations. Two aspects predominate: in homeotherms it is mainly the

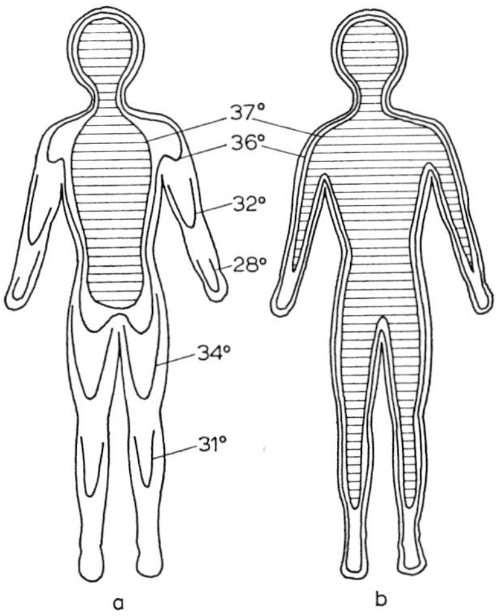

Fig.66. Definitions of body core and body shell: a. low outdoors temperature; b. high outdoors temperature (ASCHOFF and WEVER, 1958).

external temperature's *indirect effects* on heat production which cause a contrary course ($t_2 - t_3$). They must be distinguished from the *direct effects* which appear within extreme temperature ranges when the interior body temperature can hardly be regulated any longer. At a certain external temperature (t_3) which also varies with the acclimatization degree, heat production reaches its lowest point. In a smaller or larger external temperature range a thermally neutral zone ($t_3 - t_4$) appears. In the *hyperthermal* phase in connection with increasing core temperature the heat production in man increases 13% per degree temperature increase. Below the range of the so-called neutral zone the heat

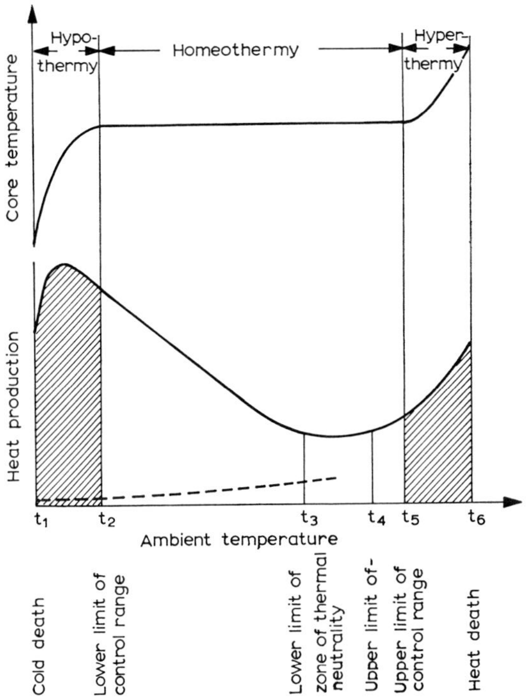

Fig.67. General course of heat formation and of body temperature of homeotherms at different external temperatures. Dashed line: heat production of poikilotherms. (HENSEL et al., 1973.)

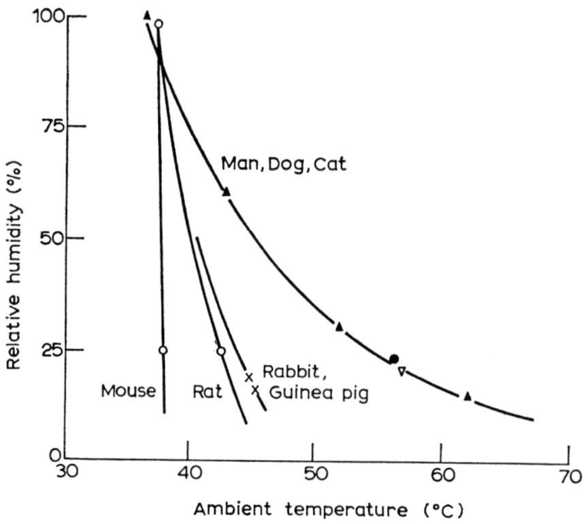

Fig.68. Mean tolerance limits of some mammals during a 3-h stay under different air temperatures and relative humidity conditions (ADOLPH, 1947).

production grows in an almost linear fashion with sinking temperatures. Below the temperature value t_2 in the *hypothermal* range after a short further increase of heat production it falls off quickly due to exhaustion of the reserves and paralysis of the regulating centers. At the same time the core temperature drops rapidly. The indication t_1 means cold death, t_6 heat death.

The preceding has dealt with the temperature limits of life; now the limits of heat regulation shall be pointed out, because it is also a problem of considerable physiological interest for industrial medicine. Fig.68 conveys an idea of the extent of the mean tolerance limits under exposures to different air temperatures and relative humidities. One can conclude that man in his tolerance limits is decidedly dependent upon the temperature-humidity environment. The highest known temperatures on earth, measured in the shade, of about 55°C require minimal humidities of below 25% in order to avoid the immediate danger of hyperthermy (understood under elimination of incoming heat radiation from sun and sky). Important pertinent factors such as performance capacity of sweating and evaporation as well as skin temperature and skin humidity will be discussed later.

Processes of heat production and heat extraction in man

General dependence and processes of heat production and heat emission

Heat production by homeotherms, in particular by man, depends upon a large number of factors which are also significant for the core temperature regulation. Roles of various importance are played by size of body, food consumption, age, sex, activity, function of glands with internal secretion, acclimatization, and above all the varying environmental temperatures and the continuously effective elements of heat radiation (cf. Light and life, p.7), of air humidity, of air displacements, etc.

Essentially there are three factors controlling heat and energy production in the human body.

(*a*) Consumption of food and its combustion by means of inspired oxygen.

(*b*) Physical activity (work, sport, etc.).

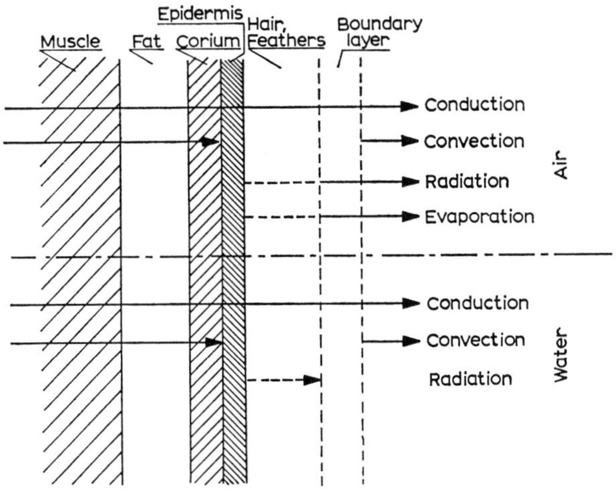

Fig.69. Heat transport between body and environment (HENSEL et al., 1973).

(*c*) Incoming heat radiation from natural and artificial radiators (sun, sky, room walls, heaters, etc.).

They form the assets in the heat balance which are necessary for the maintenance of a constant core temperature. The losses are those of removal of sensible heat by conduction and convection, and by radiation governed by the temperature of the body and of the surroundings as well as by evaporation. All of them are closely connected with weather and climate fluctuations. Dominating roles in these intricate influences are played by the characteristics of the human skin surface with its often greatly varying behaviour, and by the lungs and respiratory system. Comprehension of the changeable circumstances and significant bioclimatological situations in the weather courses and climate structures are facilitated by a review of heat physiological facts. These deal with transport methods and the heat fluxes from body core towards body shell and beyond to the surroundings.

According to estimates by HENSEL et al. (1973) in a human body at rest the greatest heat production (approximately 50%) is concentrated in the abdominal organs (particularly in the liver); under medium work loads the production is naturally taken over by the muscles (75%). The sketch in Fig.69 shows the processes of heat transport from the interior of the body to the skin surface. The body's heat emission and evaporation through the skin surface and the respiratory system including the lungs, besides conduction and convection are the main mechanisms for heat exchange with the environment.

Components of the inner and outer heat currents

Three factors are important in the heat transport from the interior to below the skin surface and above all within the skin system itself, namely: W_B = *blood stream* (progress of arterial blood to below the outer skin); W_L = *heat conduction stream* conditioned by the drop in temperature from the inside towards the outside; W_A = *heat exchange* (disorganized movement of liquid constituents of the skin). Thus the sum $W_B + W_L + W_A$ constitutes the *internal heat stream* which plays a decisive part in the principal behaviour of physical heat regulation. Total heat emission Q_B, i.e. the *external heat stream*, corresponds to the sum of the three factors:

$$Q_B = Q_{KL} + Q_{SH} + Q_{VH}$$

Q_{KL} is heat exchange by conduction and convection (KL); Q_{SH} is heat exchange by long-wave emission (S); Q_{VH} is heat exchange by evaporation (V) by way of skin surface (H).

The following relations are valid for these individual factors:

(*1*) $Q_{KL} = \alpha(t_H - t_L) \cdot F_{KL}$

(*2*) $Q_{SH} = \sigma \cdot \varepsilon(T_H^4 - T_U^4) \cdot F_{SH}$

(*3*) $Q_{VH} = \beta(e_H - e_L) \cdot F_{VH}$

(*1*) Q_{KL}. Here t_H and t_L signify the temperatures of skin surface and air, respectively; F_{KL} is the size of the surface participating in the heat exchange; α is the coefficient of heat transfer (cal. cm^{-2} min^{-1} $degree^{-1}$).

According to posture F_{KL} varies in size. The heat loss is the larger, the more the surface is bent. The surface of open skin spaces amounts to approximately 50% in a reclining person.

Thermal conduction coefficient or coefficient of heat transfer α (in cal. cm^{-2} min^{-1} $degree^{-1}$) is a complicated function of conductivity, air density and wind speed (v), of the viscosity of air and finally of body length. For a forced stream Buettner placed it at $\alpha = 0.021\sqrt{v}$. The strong dependence upon wind of the coefficient of heat transfer (under conduction and convection) in the case of an adult human being is emphasized by the following values: $\alpha = 0.006$ ($v = 0$ m/sec), $\alpha = 0.020$ ($v = 1.0$ m/sec) and 0.065 ($v = 10.0$ m/sec), with 760 Torr atmospheric pressure and an temperature of 30°C and $(t_H - t_L) = 10$°C postulated. For a nude person reclining on his back, α amounts to 0.006 at a temperature difference of 10°C (skin to air), or generally, for calm conditions:

$$\alpha = 0.0034 \cdot \sqrt[4]{(t_H - t_L)}$$

Among the components of the human heat balance none is more dependent upon weather than the conduction and convection flux. It has only been possible to measure α and to obtain useful information on this factor after the introduction of the laws on the boundary layer with the help of the Nusselt similarity theory. Fig.70 supplies

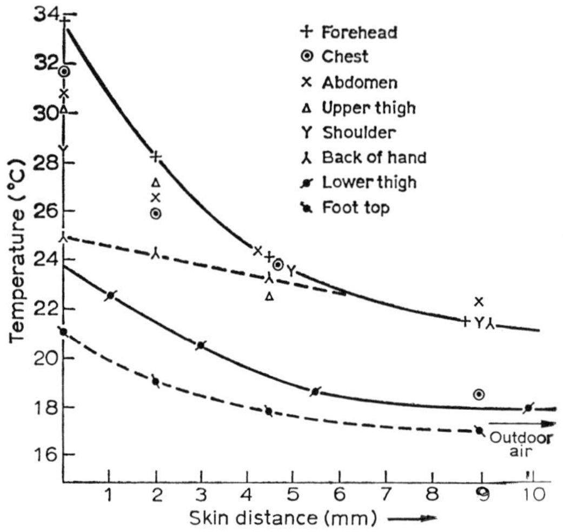

Fig.70. Measurements of the temperature limit layer over the uncovered skin of a reclining person (BUETTNER, 1938.

information on the behaviour of the temperature boundary layer over the bare skin of a reclining person. The greatest heat loss by conduction and convection occurs on the forehead, chest and stomach.

(2) Q_{SH}. In this case F_{SH} is the active body surface. T_H is the absolute temperature (°K) of the skin and T_U of the surrounding surfaces (e.g., walls). σ is the Stefan-Boltzmann constant. Unit ε_H is called radiation coefficient of the skin. It is the relation of the emission of a given surface to that of a black body at the same temperature. Reliable measurements by Buettner (1938) set it for human skin at 0.954, which is a deviation from the black body radiation of only about 5%. This means that within the energy range spectrum of human skin (5–50 μ) only $1 - \varepsilon_H$ parts are reflected from vertically incoming radiation, and the greatest percentage (approx. 95%) is absorbed (cf. Table XXXI).

TABLE XXXI

RADIATION FIGURES FOR TEMPERATURE RADIATION
(After BUETTNER, 1938)

Black body	1,000
Snow	>0.995
Frost	0.985
Water	0.965
Human skin (ε_H)	0.954
Bronze colour	0.50
Fur	0.99
Wool	0.96
Pine needles	0.96
Wood	0.89
Clod with grass	0.98

ε_H varies for dry and perspired skin.

The radiation factor for skin is very close to that for water and other substances of importance in the skin structure. Obliquely incoming radiation shows an ε_H of 0.893 (BUETTNER, 1938).

GAERTNER and GOEPFERT (1964) again investigated the radiation characteristics of live human skin. For the back they found ε_H at 0.976, for the forearm at 0.960 and the sole of the foot at 0.941. Incidentally the average lies very close to the value determined by Buettner: $\varepsilon_H = 0.954$. Based on the studies by the above-mentioned authors it must again be stated that live human skin is not a "black radiator".

The heat stream of infrared temperature radiation Q_{SH} starting from the skin surface impinges on surrounding bodies of the temperature t_U, outdoors on the air's water vapour or on clouds (fog). The skin's radiation loss may also be defined as follows:

$$Q_{SH} = \alpha_{SH}(t_H - t_U)$$

whereby α_{SH} signifies the coefficient of heat transfer due to radiation. It is of the same dimension as α_{KL} (in conduction and convection), where:

$$\alpha_{SH} = 4\sigma \cdot T_H{}^3 \cdot \varepsilon_H \quad \text{(Buettner)}$$

In connection with bioclimatological evaluations of outdoor air conditions one frequently encounters questions on the extent of skin radiation in relation to different weather and climatic conditions. The actual heat loss from the skin through emission compared with ambient air amounts to:

$$Q_{SH} = \varepsilon_H(\sigma T_H{}^4 - G)$$

σ signifies the Stefan-Boltzmann constant, ε_H the skin's radiation factor and G the atmospheric back radiation which is strongly dependent upon t_L, e_L and the amount of clouds (cf. p.40 and Fig.24).

(3) Q_{VH}. Evaporation or water vapour transfer is a process of diffusion–convection. The coefficient of heat transfer α is replaced by the evaporation coefficient β. It can be calculated from α because both figures deal with the same boundary layers and the same flow conditions. Hence $\beta = 2.18$.

Furthermore:

$$\beta = 0.046.\ \sqrt{v}\ \text{cal. cm}^{-2}\ \text{min}^{-1}\ \text{(mm Hg) with wind}$$
$$\beta = 0.013.\ \sqrt{v}\ \text{cal. cm}^{-2}\ \text{min}^{-1}\ \text{(mm Hg) without wind}$$

Finally, e_H and e_L signify the vapour pressures over the skin surface and the ambient air, respectively.

Fig.71 expresses the boundary layer behaviour due to vapour pressure over human skin (e_H) and related to different parts of the body. Here, too, it is clear that the greatest evaporation amounts are encountered on the forehead, chest and stomach. Incidentally, the quantity Q_{VH} is the heat loss by evaporation (cold due to evaporation) and not the water loss. This is proportional to $\beta(e_H - e_L)/580$ (g cm^{-2} min^{-1}). It must also be pointed out that the heat transfer coefficient (under conduction and convection) grows with $\sqrt{p}$ (p = atmospheric pressure in Torr), i.e., in the lowest air layers it is the biggest. On the other hand, the evaporation factor β behaves inversely, it changes with $1/\sqrt{p}$. This fact is of importance in Alpine bioclimatology.

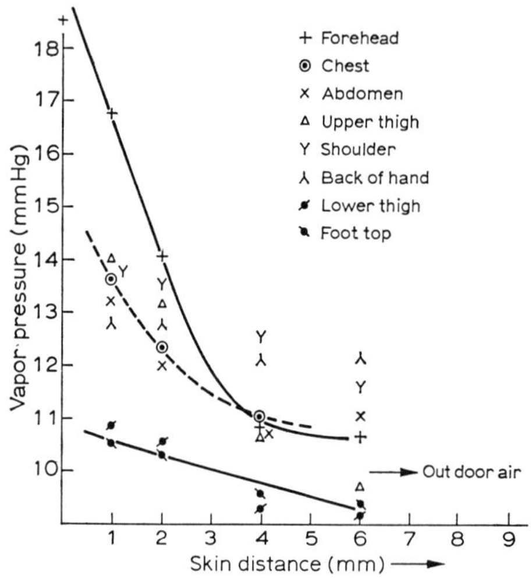

Fig.71. Boundary layer of humidity (vapour pressure) over human skin according to point registrations of air temperature and relative humidity (point hygrometer) (BUETTNER, 1938).

Evaporation does not take place only through the outer body surface (*H*) but in part also through the lungs and respiratory system (*A*). Radiation and wind have no immediate influence on this type of heat loss. Heat loss through respiration is composed of: (*a*) *actual respiration* (A), whereby the difference ($t_A - t_L$) plays a decisive part (t_A = temperature of expiratory air);
(*b*) *heat loss by additional expiration of water vapour*, VL; here the difference $e_A - e_L$ is concerned, where e_A is the vapour pressure of the expiratory air.
Heat loss due to additional expiration of water vapour, i.e., quantity Q_{VL}, becomes the natural additional value to the heat loss by evaporation through the skin surface Q_{VH}. Thus we have a total heat loss of $Q_{VH} + Q_{VL}$. Both terms of this sum are always positive, while in warm climates Q_{SH} and Q_{KL} and occasionally also A can be negative. Contrary to the sum $Q_{VH} + Q_{VL}$ the sum $Q_{KL} + Q_{SH}$ is called the *dry heat loss*, according to Buettner. In moderate climates it constitutes the main part of the heat balance. In the case of conduction–convection (KL) air temperature is the decisive factor, in the case of the skin's heat emission (SH) it is the ambient temperature. These facts are of notable physiological significance, as it makes a great difference if a cooling process is caused by air movements or primarily through infrared emission.

Evaluation of losses in the heat budget, and physical and chemical heat regulation

An evaluation of all losses in the heat budget led to the following relations (BUETTNER, 1938; BRADTKE and LIESE, 1952).
Q_{SH}: cooling through heat emission by the human body due to lower ambient temperatures, approx. 43%.
Q_{KL}: cooling through conduction and convection in the border area between skin surface and air, approx. 27%.
Q_{VH}: cooling through evaporation by way of skin surface, approx. 18%.
Q_{VL}: cooling through evaporation by way of lungs and respiratory system, approx. 12%.
Thus skin and respiratory system are the areas of contact for atmospheric ambient influences, in particular by thermal factors. They regulate the physical heat regulation and cause the described cooling processes.
Physical heat regulation is subjected to the following control factors (THAUER, 1939).
(*1*) *Body posture* and its changes. In order to determine the extent of cooling processes the concerned body's surface size is of fundamental importance. The size of the geometrical surface of a human being (*F*) can be calculated with the empirical formula (Dubois):

$$F = g^{0.425} \cdot l^{0.725} \cdot 71.84 \text{ cm}^2$$

wherein *g* is the weight and *l* the length of the body. In a standing nude human about 80% participates in radiative emission, also in losses through conduction and convection.
(*2*) *Blood circulation in the skin* and its changes. Under the influence of cold the blood vessels in the skin contract, whereas under the influence of heat the opposite normally takes place. One must distinguish between local reactions (e.g., in the extremities) and distant reactions. Connected with the latter are for instance cold reflexes in body parts which are not near the spot of the cold effect.

(*3*) *Water secretion* through the skin.

(*4*) *Blood circulation in the mucous membranes* of the respiratory system.

Among the gains in the human heat budget the oxidation processes (intake of food and its combustion as well as physical activity in work and exercise) form the foundation for *chemical heat regulation.* Its extent depends upon the value of metabolism (under basal conditions), upon the amount of muscle work and on the combustion of food. The external influences on the chemical heat regulation and thus the changes in the internal heat production are closely linked with the cooling and heating of the body shell.

In the organism certain mechanisms participate in the effects of these two basic forms of heat regulation, namely the nervous and hormonal controls which together with the hypothalamus, regulate temperature in homeotherms. Heat-physiological details cannot be given here, but a few remarks shall emphasize the importance of thermoreceivers. The sensory organs for temperature perception in warm-blooded creatures have a double significance for temperature regulation, namely by an immediate release of reflex processes and by temperature sensations. We must assume a reflex control when it appears at a distance from the stimulus and the central temperature does not change. In temperature control we find beside this nervous control also hormonal processes. They are, however, slow and of significance in connection with seasonal fluctuations, acclimatization and hibernation of animals.

Responses of skin and tissue temperatures and of skin humidity

In all the preceding considerations skin temperature t_H plays an essential role. The predominant part of the heat produced in the body core passes through the body shell in order to reach the environment through the skin. Hence there is a temperature drop in the skin tissue. Fig.72 shows that this varies regionally and depends upon clothing, ambient temperature, etc.

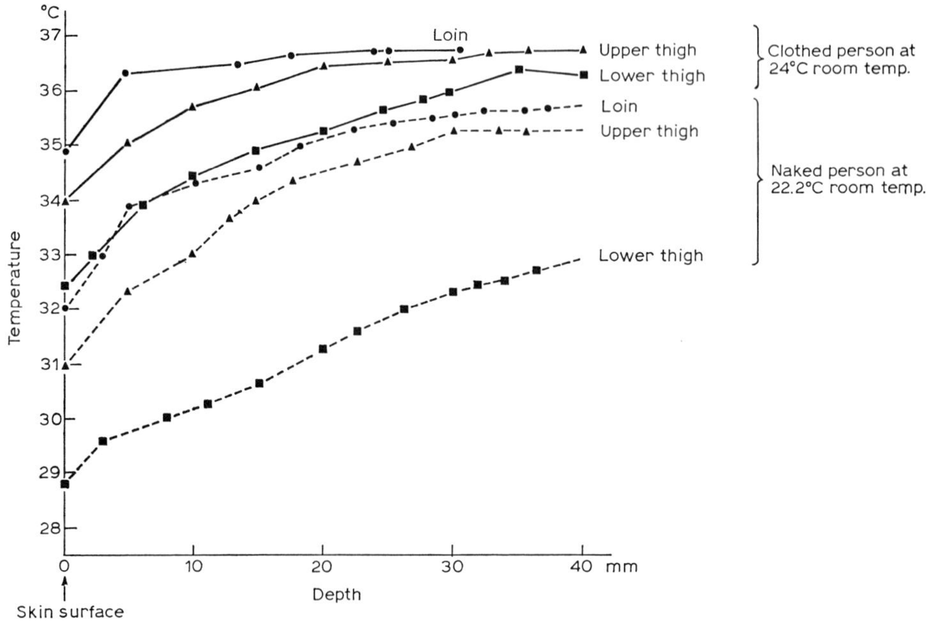

Fig.72. Tissue temperature at different depths below the skin surface of a resting person (READER and WHYTE, 1951).

Fig.73 shows the dependence of rectal temperature and skin temperature upon ambient temperature t_L. Skin temperature t_H exhibits the greatest exposure differences. WEZLER and NEUROTH (1949) found local differences in skin temperature up to 15°C, whereby the ends of the extremities revealed the lowest temperatures. Torso and forehead on the other hand are the warmest.

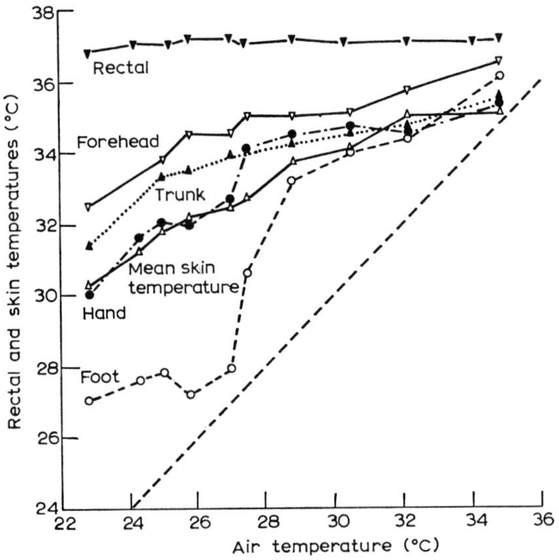

Fig.73. Skin temperatures and rectal temperature in man at different room temperatures (HARDY and DU BOIS, 1938).

Mean or integral skin temperature t_{Hm} fluctuates (according to HENSEL, 1955) between 22° and 37°C, the temperature drop between core and skin surface lies on an average between 15° and 1°C. With normal clothing and room temperatures between 15° and 21°C the mean skin temperatures vary between 32° and 35°C.
The condition of skin humidity e_H is influenced by the fact that with increasing skin temperature not only the heat stream by conduction, convection and emission from the skin surface grows, but that simultaneously evaporation rises due to the increase of e_H. e_H can be measured indirectly by way of the skin temperature. By means of the vapour pressure differences (hyperhumidity) $e_H - e_L$ and the emitted heat Q_{VH}, e_H can be determined. One can also measure skin temperature t_H and the relative humidity of the skin surface f_H and thus determine e_H. According to measurements by PFLEIDERER and BUETTNER (1940) e_H fluctuates as a rule between 20 and 60%, while f_H is always smaller than f_L. When the skin sweats f_H approaches 100%.

The process of perspiration in man under various climatic conditions, including its dependence upon work

Even at the smallest deviation, in form of overheating, from the thermal conditions considered "comfortable" by man, the skin's sweat glands begin to secrete "sweat". This type of water secretion is called sensible perspiration in contrast to that by means of "diffusion" (previously called insensible perspiration). "Diffusion" on the whole is considered dependent upon skin temperature t_H and the ambient humidity e_L. It is not, however, linear in all circumstances with hyperhumidity ($e_H - e_L$). Diffusion, which takes place without the participation of sweat glands, represents a very constant figure (approx. 25% of total heat emission), as long as the external climatic conditions remain within a normal range. The extent of this insensible perspiration varies according to external conditions between 20 and 30 g/h.

Normal wet sweating (i.e., active secretion process) is controlled in such a way that according to the adjustment of the heat balance smaller or larger skin surfaces are completely wet while the others remain in the normal diffusion state. When 70% of the skin surface are kept wet to retain the heat balance, BUETTNER (1938) declares that the state of sultriness perception has been reached, and when the water cover approaches 100% one speaks of "unbearable" conditions. The so-called steam sweating sets in under dry-hot desert winds with temperatures of 40°C and more, as well as under intensive incoming heat radiation. In that process the skin dries and leaves a salt crust because the evaporation is greater than the water supply (BUETTNER, 1938).

The thermoregulatory adaptations of the human body under uncomfortable climatic conditions are of great bioclimatological significance. A clarification of the relations is demanded not only by bioclimatology but even more so by industrial medicine. In the present context it is not, however, possible to discuss details of the pertinent heat physiological processes and phenomena.

The results of heat-physiological studies, illustrate limiting temperatures under which naked test persons did not produce sweat for different levels of heavy work (Fig.74). The linear dependence appearing between gross energy transformation and room temperature t_L separates the areas with and without sweat production. The temperature values from the indicated curve are to be taken as the "comfort values" for the corresponding degrees of work loads. In the combinations lying to the right above the line, where the body produces sweat, there was a warmth sensation, in the range of the curve itself the thermal sensations were neutral.

For a normally clothed person doing light housework the hatched zone in Fig.74 indicates the comfortable range. One may describe that ambient temperature as "comfortable" in which there is neither an increase in metabolism as sign of chemical heat regulation, nor sweat production as sign of physical heat regulation against overheating of the body.

Heat and cold acclimatization

Man's heat acclimatization is quite remarkable. It plays a very important part in seasonal climate variations and for life in the tropics or in desert climates, and particularly for the adaptation when working under great heat stress. Fig.75 shows the time course of man's acclimatization in a desert climate. An interesting fact is that on the first day the rectal temperature is quite high and during the immediately following phase is reduced gradually. After approx. 10 days close-to-normal conditions are attained (ADOLPH, 1947). Regarding the time course of heat acclimatization, as measured by sweat production, man's water consumption does not begin to cover the water loss during the first days of the heat stress. As a consequence the body becomes dehydrated. During the course of acclimatization man becomes always more thirsty while his water secretion remains constant. His water loss is generally connected with a feeling of thirst (cf. Fig. 76).

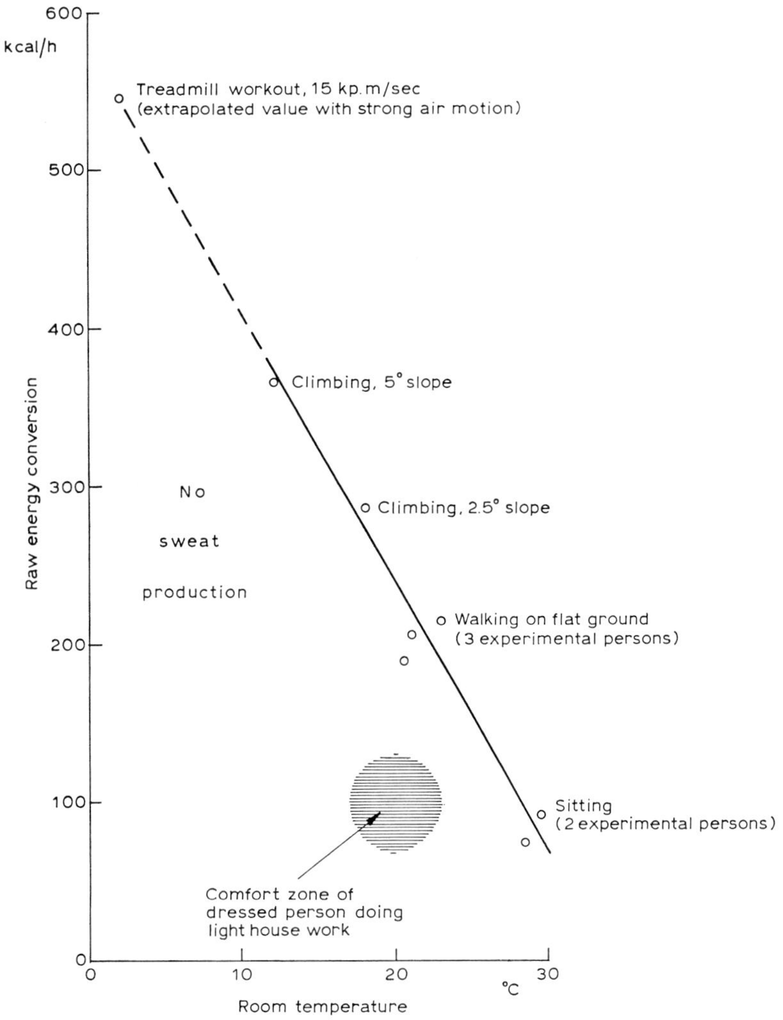

Fig.74. Maximum possible energy output of naked person without producing perspiration, at different room temperatures and air displacements of 0.1 to 0.5 m sec^{-1} (WENZEL, 1961).

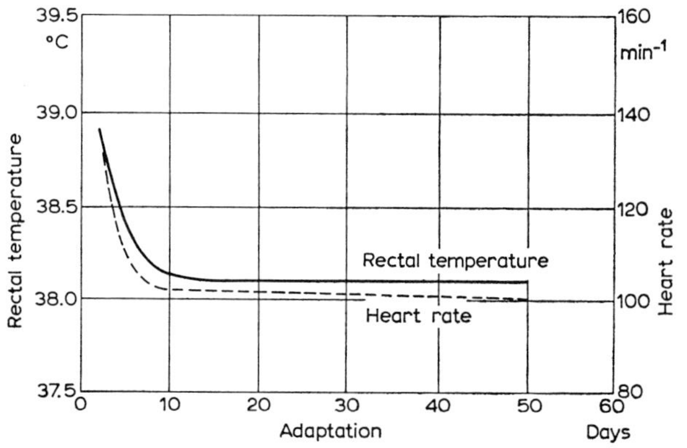

Fig.75. Acclimatization course for man in a desert climate. Rectal temperature and pulse frequency after a standardized work performance (ADOLPH, 1947).

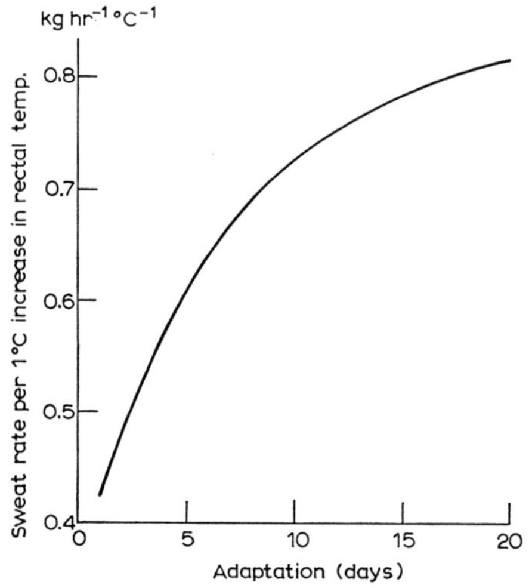

Fig.76. Temporal course of heat acclimatization in man measured from the perspiration production per degree of rectal temperature increase. Mean values from several persons, work 87 kcal. m^{-2} h^{-1}, room temperature 37.8°C, relative humidity 78% .(LADELL, 1951, cit. HENSEL, 1955.)

While in the *tropics* the main danger is inadequate water evaporation, in the desert it consists in the body's loss of water which reduces the resistance against heat. Harmful dehydration can only be avoided by drinking more than one's thirst indicates, contrary to a previous frequently encountered opinion.

Heat-physiological studies state the following circumstances as responsible for a breakdown of the water and heat balance in man.

(*1*) Breakdown due to hot-humid sultriness. Vapour pressure e_L and air temperature t_L are so high that despite extensive wetting of the skin not enough sweat can evaporate. Lack of air ventilation worsens these conditions.

(*2*) Breakdown due to dry hot overheating, whereby dry hot air, wind and heat radiation exceed the water supply to the skin.

Pathological manifestations will result in such cases in form of heat cramps, heat exhaustion, heat stroke and sun stroke.

The dominating circumstance for *heat cramps* is the NaCl loss. A person is particularly subjected to heat cramps when exposed to radiative heat while simultaneously carrying out heavy physical work (blast furnace stokers, ship stokers, mine workers, firemen, etc.).

Heat exhaustion (heat collapse) is that state of heat stress on the organism in which the regulation of circulation breaks down.

A *heat stroke* is the most dangerous form of heat damage. Prerequisite for it is the inhibition of heat emission and an abnormally large heat supply from outside. This combination causes an increase in body temperature. Beside high ambient temperature, a high humidity content of the air is important, only light air movements and strong atmospheric back radiation. The decisive criterion is a surge in body temperature to 42°–43°C (final condition: cardiac insufficiency).

A final condition is *sun stroke*. It occupies a special place because intensive incoming heat radiation (S,D) on the head is the important factor. The characteristic feature of a sun stroke is its sudden appearance together with a marked increase in body temperature (like in a heat stroke).

Regarding *cold acclimatization* HENSEL (1955) states that neither during the course of the seasonal changes nor in the inhabitants of polar zones noteworthy physiological cold acclimatization signs could be found. Acute supercooling occurs usually during stays in water of 10°–15°C (ship or airplane accidents). It is also encountered during blizzards if a person is not sufficiently protected by weather proof clothing. The "cold damages" met with in such cases are to be clearly distinguished from the so-called "cold symptoms". They include chill blains, frost bite, and catarrhs of the upper respiratory system (inappropriate reaction of the organism towards cooling) in connection with an infection caused by the common cold virus groups.

Temperature sensations

General aspects of climatic measures

Human temperature sensations are centered around two entirely different quantities, warm and cold, within each of which there are intensity gradations. In a rough classification the terms: indifferent—cool—cold—freezing, on the one hand, and: indifferent—moderately warm—warm—hot, on the other, are well-known. On both sides the intensity scale ends in pain (heat pain and cold pain). In the human being the temperature sensors are discontinuous due to their distribution over the skin in form of "sensory points". The distribution of "cold-sensitive points" is considerably denser than that of "warm-sensitive points". Furthermore, the distribution of the sensory points varies as to the body parts. The greatest density of cold- and warm-sensitive points are found in the face. In all temperature sensations the size of the stimulated surface and thus the number of activated thermal receivers plays a decisive part. The larger the surface, the narrower is the indifferent range of a temperature sensation. For temperature influences affecting the whole body the thermal indifference range lies only between 32° and 35°C skin surface temperature.

Beside sensations of a general type and in extreme temperature ranges with reactions of cold and heat pain (which may appear below 17°C and above 45°–50°C) there are sensations which are difficult to localize and are partly of a judgmental nature. They concern in one realm the "freezing" sensation, in an other "sultriness". The middle range where these sensations are absent is the thermal "comfort zone". At a normal temperature distribution on the surface it is found at 32°–34°C mean skin temperature (THAUER, 1939). In this connection Hensel stresses that comfort and sultriness are integrative quantities. They reflect totally the general thermoregulatory state of the organism in a given environment.

Major interest centers on numerous practical questions (in work, sports, therapy, recreation, etc.) which require heat physiological evaluation of weather and climatic factors which govern the thermal ambient climate. In this connection the integration of physically measured climate factors in form of climate attributes (complex quantities, comfort indexes, etc.) becomes important. It is not possible, however, to list all experiments made for supplying useful definitions and observing methods of integral effects of thermal environmental factors. Great difficulties have been encountered, for example, when trying to develop a test body whose heat balance could be taken as a congruent model of the human body's heat balance. BRUENER (1959) in a survey pointed out 18 different measures for total climate quantities and mentioned 27 test bodies whose size varied between a few centimeters and normal body length.

Measurements or physico-mathematical constructs of thermal environmental phenomena of the terrestrial atmosphere must start from the fact that parameters which also fulfil the requirements of physiology to some extent, can only be achieved by way of climatic integrating parameters. Thus Bruener stresses that such an integrating parameter is best suited for the evaluation of heat-physiological reactions if it corresponds with the current stress by different climates on the human organism at rest and at work. Up to the present there are only very few test body methods and calculations which permit adequate heat physiological evaluations. Comprehensive method is hardly amenable to all situations in life, because it makes a great difference if a measuring method is adapted to outdoor or indoor conditions (in buildings, at working sites, etc.).

Meteorological influences on skin temperature

The difficulties to find a true, heat-physiological acceptable solution in this important sector of bioclimatology are great. Fig.77 shows the air temperature and wind influence on forehead temperature. It may be concluded that at an air flow above 0.3 m/sec the air temperature influence on forehead temperature is the greater, the lower the current

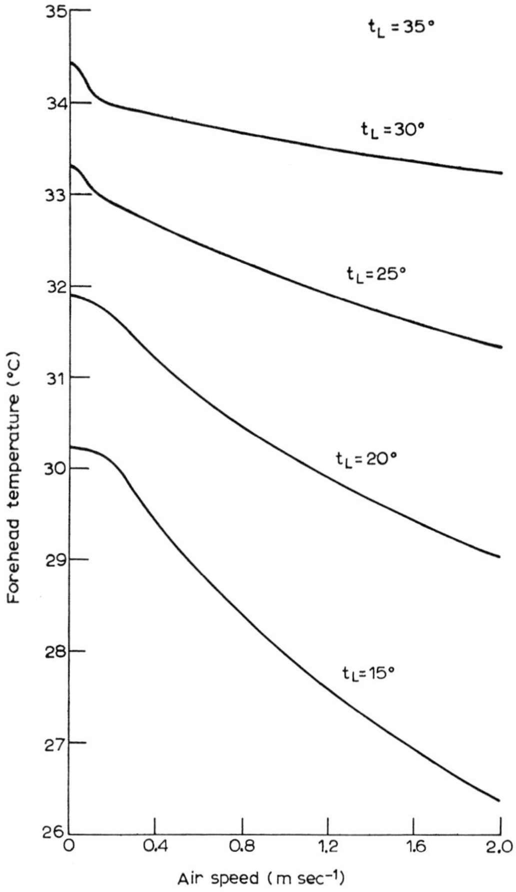

Fig.77. Dependence of forehead temperature upon air speed at different air temperatures (BRADTKE and LIESE, 1952).

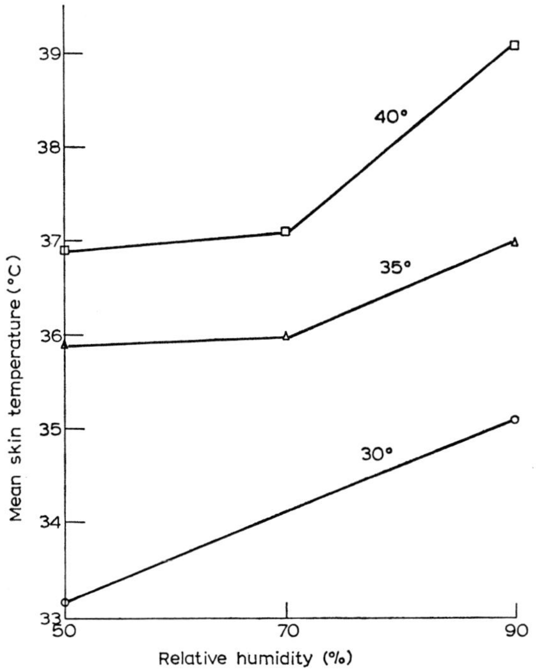

Fig.78. Mean skin temperature as function of relative humidity. After 2.5h subjection at 30°, 35° and 40°C (NEUROTH, 1948).

air temperature and the larger the wind speed is. Fig.78 indicates that with growing relative air humidity (%) the mean skin temperature rises the quicker, the higher the air temperature lies. Both graphs show for the case of tropical climate conditions that wind influence on forehead temperature vanishes above an air temperature of 34°C. However, high air humidities (above 70%) at tropical air temperatures lead to a rise of the mean skin temperatures. These processes throw light on the influences of *individual* climatic factors with regard to the different levels of heat emission (conduction, convection, evaporation, etc.). In this connection some experiments by BRADTKE and LIESE (1952) interesting. They studied the influence of heat radiation on skin temperature (cf. Fig.79). Under an intensive incoming heat radiation of 1.5 cal. cm^{-2} min^{-1} in a short time considerable skin temperature rises of 6°–7°C can be observed, that may cause very unpleasant heat sensations. On the contrary, radiant energy flux of 0.5 cal. cm^{-2} min^{-1} produces mild and pleasant sensory reactions and involve only minor skin temperature rises even for extended duration.

Fig.80 illustrates how temperature influences alone can affect processes in the interior of the body.

Above the comfort range of air temperature (18°–22°C) the blood circulation in the skin rises rapidly. Circulation values at approximately 40°C air temperature are 8–10 times higher than at 20°C.

Characterization of thermal environmental influences as well as special measuring techniques for determining "cooling intensity", or "cooling power" and "cooling temperature"

Complex thermal environmental quantities enjoy a certain preferred position because their effects on man are most obvious. They are the total effects of short-wave heat

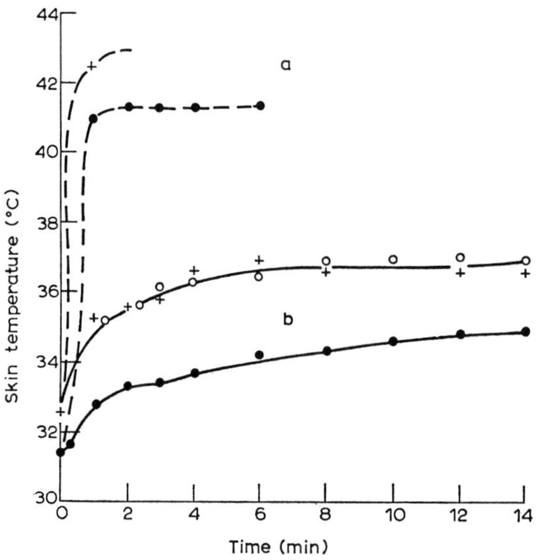

Fig.79. Skin temperature and incoming heat radiation: a = incoming heat radiation 1.5 cal. cm^{-2}min^{-1}; b = incoming heat radiation 0.5 cal. cm^{-2} min^{-1}, at the same ambient temperature of 19°C (BRADTKE and LIESE, 1952).

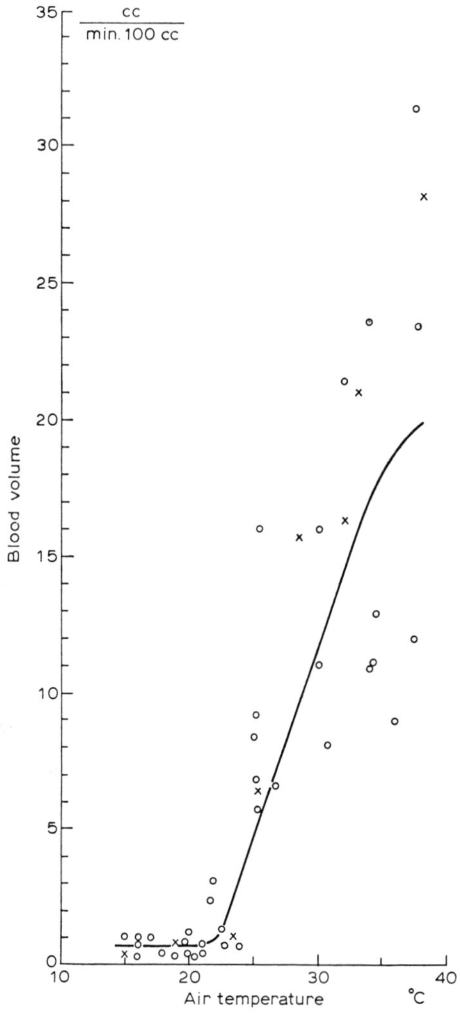

Fig.80. Blood circulation in a hand at different air temperatures (3 clothed test persons)(FORSTER et al.,1946).

radiation from sun, sky and ground reflection ($S + D + R_E$) as radiation complex, as they are comprised in the form of circumglobal radiation (CGR) (cf. section Light and life, p.22); furthermore of dry-bulb temperature of air (t_L) and air displacements (wind speed v). Here air humidity, e.g., in form of vapour pressure (e in mm), relative humidity (f in %) or wet-bulb temperature (t_f) are not initially incorporated into the thermal environmental dimensions. Later it will be shown that such a procedure—exclusion of the very important heat physiological and meteorobiological phenomenon of air humidity—cannot represent the true bioclimatic circumstances.

As long as 80 years ago VINCENT (1890) described an empirical relation for the common reaction of air temperature, wind and radiation of the sun on human skin. As physiological point of reference he took the outer surface of the thumb ball (t_{HD}). Radiation was determined with a black spherical thermometer, whereby the radiation intensity was taken as "excess temperature" (t_{Ue}) compared with air temperature. (This procedure is, however, extremely inaccurate; it harbours many sources for errors.) v signifies wind speed (m/sec). The empirical Vincent formula follows:

$$t_{HD} = 26.5 + 0.3\,t_L - 1.2\,v + 0.2\,t_{Ue}$$

Later Vincent established a simplified formula (without the radiation term):

$$t_{HD} = 30.1 + 0.2\,t_L - v\,(4.12 - 0.13\,t_L)$$

Both formulae are valid only for air temperatures above 17°C. This may be considered as the beginning of research on thermal effectiveness complexes for acquisition of heat physiological knowledge. Experiments on test bodies of heat loss yielded other empirical relations, during the past 50 years. Such procedures will be described later on p.122. In connection with the 80-year-old experiments by Vincent more recent investigations by HOESCHELE (1970) need mentioning. He used skin surface temperature t_H as a measure for the thermal state and found the optimal range between 32° and 34°C t_H. This corresponds with results by other observers. Hoeschele's studies are remarkable because his results are based on a model calculation which includes air temperature, air humidity (!), heat radiation, wind, heat production of the body, and clothing.

Already 60 years before Vincent's experiments instruments were used in order to comprehend "cooling effects" due to thermal environmental factors (e.g., HEBERDEN, 1826). During the last 5 decades of this century well-aimed experiments were carried out. Their structure, physically and also physiologically better founded and orientated, permitted acquisition of further knowledge in the field of temperature sensations.

Thus the *katathermometer* by HILL (1923), THILENIUS and DORNO's (1925) *frigorimeter*, and PFLEIDERER and BUETTNER's (1935) *frigorigraph* have found a wide-spread application on a regional as well as on local and small-scale climatological level. Usable observation series have been set up also with other measuring devices for integral climatic parameters.

BRUENER (1959) made an informative but incomplete survey of the efforts made during nearly two centuries for the establishment of bioclimatological devices for such parameters with information on the construction and functional principles and the date of their introduction (cf. Table XXXII). The measuring principles given therein are based mainly on Newton's cooling law which in its simplest form can be stated as follows:

$$\frac{\mathrm{d}W}{\mathrm{d}t} = \alpha\,(T_{\mathrm{K}} - T_{\mathrm{U}})$$

TABLE XXXII

SURVEY OF SUM VALUE APPARATUSES
(After BRUENER, 1959)

Group	Description	Proposed by	Year	t_L	f	v	S	Definition and principle
Unheated apparatuses	Wet-bulb thermometer	Cullon	1777	×	×	(×)	(×)	Temperature of a thermometer with humid sphere
	Ball thermometer	Vernon	1923	×		(×)	×	Temperature in the interior of a blackened copper hollow sphere of 20 cm ∅
	Black sphere actinograph	Linke-Poschmann	1932	(×)		(×)	×	Surface temperature of a blackened copper hollow sphere (measured thermo-electrically)
	Resulting thermometer	Missenard	1935	×	×	×	×	Interior temperature of a blackened copper hollow sphere the surface of which is covered to 36% by wet felt strips
	Rubber globe thermometer	Brunclaus	1941	×		×	×	Temperature in the interior of a blackened inflated rubber balloon
	Ice calorimetric measuring method	Siple and Passel	1945	×		×		
	Wet ball WWSt	Linsel	1951	×	×	×	×	Surface temperature of a hollow copper sphere covered with wet black material
Apparatuses heated once	Heated thermometer	Heberden	1826	×		×	(×)	Measurement of cooling time of a heated thermometer
	Heated thermometer	Osborne	1835	×		×	(×)	Time required by a heated spirit thermometer to cool from 32° to 27°C
	–	Krieger	1876	×		×	×	Cooling time of a water-filled copper cylinder
	Homeotherm	Frankenhäuser	1911	×	(×)	×	×	Water-filled copper cylinder as above. May be supplied with wet cover
	Heated thermometer	Grosse	1914	×		×	×	Warming of a common thermometer by 10°C in t_L, measurement of time for cooling
	Dry kata-thermometer	Hill	1913	×		×	(×)	Time required by a preheated alcohol thermometer with large vessel (at 5°C) to cool from 38° to 35°C; converted by means of a factor into mcal. $cm^{-2}\,s^{-1}$
	Wet kata-thermometer	Hill	1928	×	×	×	×	As above but with wet cover

TABLE XXXII *(continued)*

Continuously heated apparatuses	Reichenbach apparatus	Reichenbach	1922	×		×	×	Heating current consumption of a heated copper cylinder filled with oil
	Calometer	Hill	1923	×		×	×	Heating current consumption of a nickel spiral heated to 37°C
	–	Phelps	1923	×	×	×	×	Heating current consumption of an apparatus covered with wet sheepskin
	Davos frigorimeter	Dorno and Thilenius	1925	×		×	×	Heating current consumption of a solid blackened copper ball of 36.5°C surface temperature
	Coolometer	Weeks	1931	×		×	(×)	Heating current consumption of a copper cylinder heated to 37°C
	Cooling apparatus	Pfleiderer	1931	×		×	×	Surface temperature of an evenly heated copper hollow sphere
	Eupatheoscope	Dufton	1932	×		×	×	Heating current consumption of a black copper cylinder heated to a constant temperature
	Frigorigraph	Pfleiderer and Buettner	1933	×		×	×	Surface temperature of a double-walled hollow copper sphere, evenly heated
	Thermointegrator	Winslow and Greenburg	1935	×	×	×	×	Surface temperature of an evenly heated copper cylinder
	Trifrigorigraph	Pfleiderer	1943	×		×	×	Surface temperatures of three spheres differing in absorption and heating
	Copperman	Hall	1946	×		×	×	Metal man whose individual body parts may be heated separately
	Metal man	Pedersen	1948	×		×	×	Metal man heated evenly with 73 kcal. h^{-1}, measurement of surface temperature
	Climate sound	Riedel	1958	×	×	×	×	Wet sphere (test body = skin when at work) interior temperature (preheated water) measured in comparison with external temperature

Here dW/dt means the heat emitted per unit time, T_K the temperature of the test body, T_U the ambient temperature, α a factor (heat transfer coefficient) which depends on the shape of the test body, its position in a room and the total state of the ambient air (e.g., air calm oı in motion, with laminar or turbulent current).

As results from Bruener's survey, most of the indicated measuring methods for the comprehension of integral climatic parameters deal with "dry" atmospheric ambient factors excluding the contribution by air humidity. If this factor is taken into account, it is determined in a psychrometric way. Merely in two cases (Phelps, Riedel) did the experiments tend towards more immediate acquisition of physiological knowledge. Measuring devices for the determination of "katavalues", "cooling power", and "cooling temperature" are still in world-wide use. Hence, their technical and empirical formulations must be briefly discussed.

In 1923 HILL introduced the "katathermometer" which is a climate integrating device. Its actual function is merely to evaluate the influences of air temperature and air movements (wind). This measuring device (alcohol- or mercury-in-glass thermometer with large liquid bowls) does not take into consideration the influence of incoming short-wave heat radiation ($S + D + R_E$). There is, however, a heat loss through conduction and convection as well as by radiative emission which, under low ambient temperatures, can be quite large. In order to have a katathermometer with less reaction, silver-plated katathermometers have been introduced. For a normal katathermometer the relation originally given by HILL is:

$$A_{HILL} = \alpha_K\,(36.5 - t_L) \text{ (in mgcal. cm}^{-2}\text{ sec}^{-1})$$

whereby A_{HILL} is called "cooling power of air" or "katavalue". (The definition "cooling power" should be reserved for measuring devices which also take into account the component of incoming heat radiation.) BRADTKE and LIESE (1952) shaped Hill's basic formula in the following way in which the values for the heat transfer coefficient were determined by including wind influence:

$$A_{HILL} = (0.105 + 0.485\sqrt{v})\,(36.5 - t_L) \text{ for } v \geqslant 1.0 \text{ m/sec}$$
$$A_{HILL} = (0.205 + 0.385\sqrt{v})\,(36.5 - t_L) \text{ for } v \leqslant 1.0 \text{ m/sec}$$

In these equations in the first bracket the first term of the sum indicates the influence of free convection, and the second one that of air flow. Considering heat emission processes on the katathermometer, LEHMANN (1936) found a relation which later was simplified by Buettner:

$$A_{HILL} = (0.26 + 0.34\, v^{0.622}))\,(36.5 - t_L)$$

As only temperature and wind influences appear in these formulae, corrections must be made for incoming heat radiation from the sun (S) and for atmospheric pressure (p). Both of these (incoming heat radiation and increasing altitude) cause a reduction of the katavalues. When including incoming heat radiation from the sun (without contribution by D and R_E) the product goes $S \cdot \varepsilon_K \cdot q_K$, whereby S signifies the intensity of solar radiation (mgcal. cm^{-2} sec^{-1}), ε_K the absorption factor 0.54 and $q_K = 0.287$. Atmospheric pressure is taken into account by the factor $\sqrt{p/p_0}$.

Later THILENIUS and DORNO (1925) developed the *Davos frigorimeter*. This is a climate-integrating apparatus which includes besides temperature and wind influences also the radiation intensities of sun (S), sky (D) and ground-reflected radiation (R_E) in the determinations of thermal complex effects on a test body (blackened massive copper sphere with a mean surface temperature of +36.5°C). This instrument is a measuring apparatus (with and without registration devices) for the determination of "frigorimeter values" (contrary to the "katavalues" determined with the katathermometer) or of "cooling power". It is defined as that heat amount in millicalories which 1 cm² of the test body (+36.5°C) emits per unit time (sec) due to the cooling effect of atmospheric influences. The frigorimeter's measuring principle is a registration of the cooling effect by way of an electric heating device. Therefore the cooling power determined with the frigorimeter is equal to the heating energy required for the compensation of the heat loss, divided by the necessary exposure time. Under subtropical and tropical conditions, in connection with little wind, warming effects may appear in the frigorimeter. Attempts have been made therefore to include also the "warming effects" with the frigorimeter (THAMS, 1962). AMBROSETTI (1965) in a climatological study has dealt extensively with the warming factor on the southern side of the Alps.

An example of the bioclimatological value of "cooling power" is given in Fig.81 by comparing the annual variation of air temperature with the cooling factor for Basel (318 m) and Davos (1,592 m). In the upper section of Fig.81 the corresponding annual variations of air temperature show that there is a marked parallelism in the elemental behaviour at both stations. The mean difference (due to the difference in altitude) amounts to 7°C. If one would attempt a bioclimatological evaluation of the northern Swiss lowlands and the central Alpine region based on these temperature comparisons alone, one would be led far astray. The lower section of the figure shows the simultaneous annual variations of cooling power (*CP*).

The comparison shows that during the winter in the lowlands the frigorimeter values lie considerably above those of the Alpine valley sites for quite extended periods. In summer the situation is reversed. In winter the behaviour of *CP* is predicated—in Alpine climates (Davos at 1,592 m)—on the frequently large amounts of radiation and relatively calm

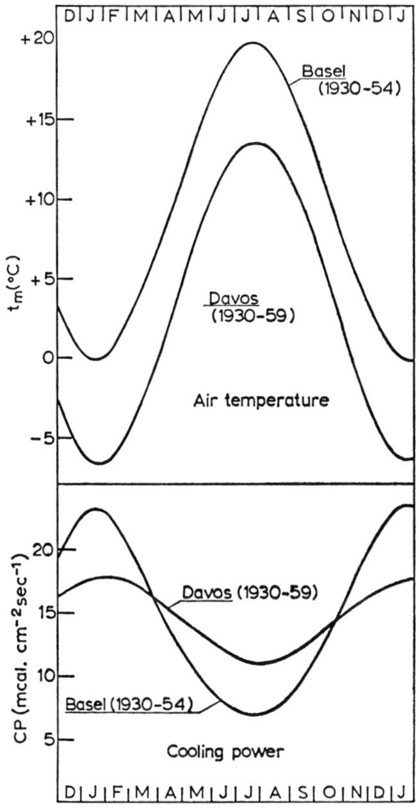

Fig.81. Mean annual variation of air temperature (t_m, in °C) and of cooling power (*CP*) measured with the Davos frigorimeter in Basel (318 m) and in Davos (1,592 m), period 1930–1959 (FLACH, 1964).

conditions. On the other hand, during the same time of year the lowlands (Basel at 318 m) show the greatest wind speeds of the year and much reduced sunshine duration (very frequent fog and high fog conditions). During the summer semester these climatic conditions are, in comparison with winter, largely reversed, mainly with regard to air movements.

The obvious usefulness of the term "cooling power" in evaluations of bioclimatic or regionally conditioned phenomena led to the introduction of this complex factor into practical applications of bioclimatology, such as climate therapy and research on health resort climate.

For Pfleiderer and Buettner's frigorigraph a test body in the shape of a hollow copper sphere of 15 cm diameter is used. In adaptation to the absorption quality of human skin it is painted reddish-yellow and is warmed continuously with a heat stream of 0.16 cal min (about basic turnover of the human body under comfortable conditions). The heat loss caused by climatic factors cools the frigorigraph sphere whose surface temperature is defined as the "cooling temperature". Incidentally this is not identical with skin temperature.

For the "frigorigraph temperature" or "cooling temperature" (t_{Ph}) BUETTNER (1938) determined the following relation:

$$t_{Ph} = t_L + \frac{0.160 + 0.60\left(\frac{S}{4} + \frac{D}{2}\right)}{0.008 + 0.020\sqrt{v}}$$

(Under calm conditions the denominator in the second term has the value 0.014.)

In order to survey the performance level of the two best-known types of instruments (frigorimeter and frigorigraph), WIERZEJEWSKI (1963) experimentally determined their differences. In both types of equipment (beside the evaluation of the humidity influence) the regulable resistance between body core and body shell for adaptation and reaction to environmental stress is missing. Wierzejewski established the following relations of frigorimeter and frigorigraph functions:

$$CP = k \cdot (t_{Fr} - t_L) - (S + D + R_E) \quad \text{(frigorimeter)}$$
$$t_{Ph} = t_L + (Hz + (S + D + R_E)) \cdot k \quad \text{(frigorigraph)}$$

Besides the already known quantities in the equations, Hz signifies the heating current, held constant in correspondence with the cooling temperature, and k a heat transfer coefficient characterizing wind influence and heat emission. From both equations one may deduct a relation between cooling power CP (t_{Fr}) and cooling temperature t_{Ph}:

$$CP - Hz = k \cdot (t_{Fr} - t_{Ph})$$

The conclusion is that the same temperatures have different cooling power according to wind force. Only under conditions with weak winds does one find a close relation between the two measured quantities.

Sensation scales and difficulties in their interpretation

These considerations (cf. also p.112ff.) demonstrate that by means of a physical test body exclusively determining heat extraction one can hardly obtain true physiological information on the presence or absence of an equilibrium between heat production and heat emission. Thus all these devices for determining the cooling process of metallic surfaces are at best able to provide some measure for the demands on the heat regulatory capacity of an organism.

Many difficulties are encountered when trying to characterize the climate and its physiological impact by means of katavalues, cooling power and cooling temperatures. Problems in interpretation arise when trying to classify differently gaged sensation levels which are to be considered as reliable. The reasons are as follows.

(*a*) Heat extraction values (measured or computed) relate to differently obtained integral effects of the pertinent factors.

(*b*) Most pertinent investigations completely exclude the humidity influence in the atmospheric environment. This circumstance can hardly enable an objective evaluation of a given heat-physiological condition. This can easily be seen from the fact that rising air temperature increases the influence of air humidity on the human heat sensation, whereas decreasing air humidity lets the "cold" compensability rise if the wind influence is of a minor extent (e.g., in winter in an Alpine climate of the moderate latitudes).

(*c*) The influence of clothing cannot be defined in every case.

(*d*) In comparisons one often uses averages of heat loss values determined at different times of day.

(*e*) In a sensation classification the individual psychosomatic and especially the thermoregulatory behaviour and the current state of acclimatization is not taken into account.

These circumstances are presented in Fig.82 in a selection of "sensation scales" currently in use and based on the best-known heat loss regulations.

Mathematical determination of cooling effects

BRUENER (1959) has provided a historical survey (Table XXXIII) of heat loss values determined experimentally with physical test bodies and by calculation, such as have been presented in the preceding section. Regarding the purely meteorological development it requires a supplement: Assmann's internationally known aspiration thermometer for the determination of dry-bulb and wet-bulb air temperature was introduced in 1887; equivalent temperature as a measure for the air's heat content (cf. below) was established in 1894 by VON BETZOLD.

Complex explanation methods for the determination of the sensation climate of extreme climatic zones

The above table (BRUENER, 1959) can make no claim to completeness since in the following two decades (1950–1970) further and new efforts were realized which correspond with the demands for specific climate evaluations. A very useful compilation by LANDSBERG (1972) illustrates these partially new methods of research.

The frequently extreme climatic conditions to which visitors to the Arctic or Antarctic

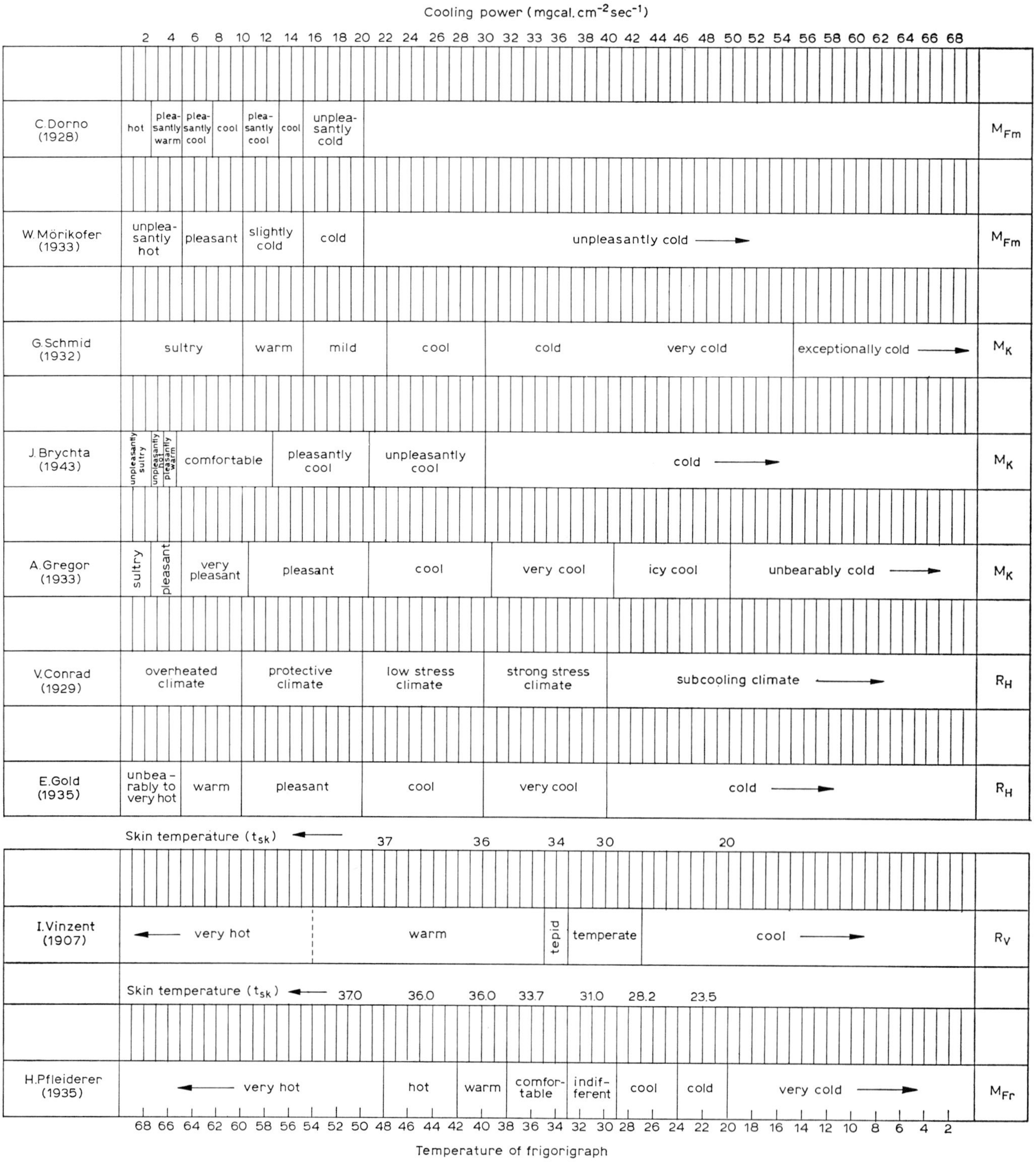

Fig.82. Survey of "sensation scales" in use based on determinations of cooling power and cooling temperatures: M_{FM} = frigorimeter measurement; M_K = katathermometer measurement; R_H = calculations based on the Hill formula; R_V = calculations based on the Vincent formula; M_{FR} = frigorigraph measurement. (FLACH, 1957.)

TABLE XXXIII

SURVEY OF CLIMATE SUM VALUES TO BE DETERMINED MATHEMATICALLY AND GRAPHICALLY
(After BRUENER, 1959)

Description	Proposed by	Year	t_{tr}	RF	w	Str.	Definition
Wet temperature	Haldane	1905	×	×			Temperature of a wetted thermometer at w 2 m s^{-1}
Absolute humidity	—	—	(×)	×			Water vapour content of air if g kg^{-1} or g m^{-3} for dry air
Saturation deficit	Flügge	1905	×	×			Difference between maximum humidity and absolute humidity at a given temperature
Thaw point	Bruce	1916	×	×			Temperature at which air of a given humidity shows RF = 100%
Physical saturation deficit	Dorno	1925	×	×			Difference of the air's vapour pressure and vapour pressure of expiratory air
Equivalent temperature	Von Betzold	1894	×	×			Addition of condensation heat of air humidity to t_{tr}
Prött temperature	Prött	1919	×	×			Equivalent temperature calculated by approximative formula
Total temperature	Linke	1922	×	×			Equivalent temperature calculated as $t_{tr} + 2\,e$
Perceived temperature	Linke	1926	×		××	×	$= t_{tr} - 4\,w + 12\,J$ (J = incoming radiation, in cal. cm^{-2} min^{-1})
Comfort lines	Frank	1941	×	×	×		Table with desired combination of climatic values
Sultriness indices	Ruge	1932	×	×			Measure for an excess of the sultriness curve of 14.08 g H_2O m^{-3}
Sultriness values	Scharlau	1943	×	×			Excess of sultriness limit of 14.08 Torr vapour pressure in °C
Dry comfort index	Bradtke	1951	×	×	×		t_{tr} divided by dry katavalue
Wet comfort index	Liese	1935	×	×	×		T_f divided by wet katavalue
t_{effA} (A = American)	McConnell, Houghton, Yaglou and Miller	1924	×	×	×		Climatic conditions causing the same sensations have the same t_{eff}. At RF = 100% and $w = 0$, $t_{tr} = t_{effA}$
Operative temperature	Winslow, Gagge and Harrington	1937	×		×	×	Temperature that a room should have for the same physiological effectiveness as with wall temperature = t_{tr} and w = 1 m s^{-1}
Corrected t_{effA}	Bedford	1946	×	×	×	×	t_{effA} which instead of t_{tr} contains the bulb temperature according to Vernon
t_{effB} (B = Belgian)	Bidlot and Ledent	1947	×	×			$t_{effB} = 0.9\,t_f + 0.1\,t_{tr}$

regions are exposed require special bioclimatological methods for the determination of the often enormous cold stress. To this end SIPLE and PASSEL (1945) deteımined by means of a plastic cylinder filled with water the amount of hourly ice formation and designated the heat extraction calculated therefrom as "wind chill" (kcal. m^{-2} h^{-1}). This is a special kind of "cooling power" which can be of high utility in the characterization of the human heat stress by cold climates. These authors deducted the following relation from their observations:

$$H = (\sqrt{100v} + 10.45 - v)\ (33 - t)$$

COURT (1948) later presented an improved version:

$$H = (10.9\ \sqrt{v} + 9.0 - v)\ (33 - t)$$

WILSON (1967) then used the wind chill index to investigate the point at which exposed skin areas begin to freeze under natural conditions (Antarctic). He found close relations to this quantity (according to the Siple-Passel formula) whereby the formation of chilblains was observed under different temperature and wind combinations (Fig.83).

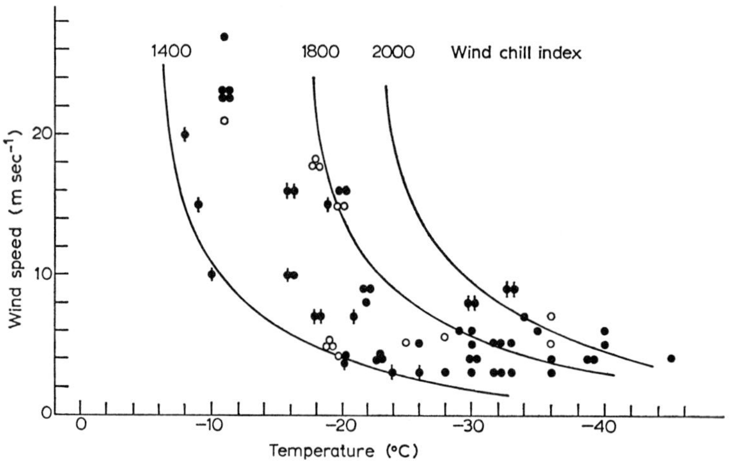

Fig.83. Recorded cases of frostbite in relation to temperature and wind speed. Open circles mark cases of frostbite that have occurred during shifting meteorological conditions (the mean of the extremes in temperature and wind is plotted). Dots with a vertical bar indicate cases of frostbite observed during sledging (wind speed estimated). Black dots are frostbites with established data for temperature and wind speed. The curved parameter lines correspond to a wind chill index of 1,400, 1,800 and 2,000, respectively. (WILSON, 1967.)

Table XLII (p.152) informs on the temperature and wind conditions encountered at the South Pole (p.152) which is of interest in connection with the complex element of "wind chill". Fig.83 conveys an idea of the level of the wind chill index under various temperature and wind conditions. The right side shows the sensation scale corresponding to the sensory impression under strong and strongest cold stress.

For the bioclimatological characterization of the earth's tropical and subtropical zones special "discomfort indexes" were introduced, e.g. by THOM (1957), WEBB (1960), TENNENBAUM et al. (1961) and by KAWAMURA (1965). Tennenbaum set up a simple relation:

$$DI = \frac{t + t'}{2}$$

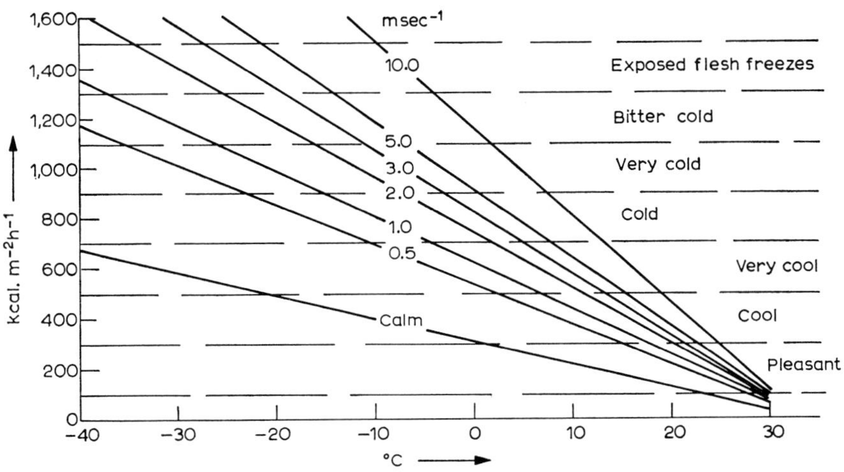

Fig.84. Wind chill index in kcal. m^{-2} h^{-1} as related to temperature (°C) and windspeed (m sec^{-1}) (LANDSBERG, 1969).

Based on this relation JAUREGUI and SATO (1967) carried out a bioclimatological landscape evaluation of Mexico.

Special bioclimatological interest focuses on the atmospheric environmental conditions of the *arid zones*, for physiological as well as psychological aspects. A research survey on these problems published by UNESCO in 1963 (*Environmental Physiology and Psychology in Arid Conditions*) presents the varied opinions on this subject by specialists (GLASER, 1963; LEE, 1963; MACFARLANE, 1963; FUHRMANN, 1963; SARGENT II,1963; and others). Similar objectives are found in *Climatologie* (UNESCO, 1958) and *Bioclimatic Maps of the Mediterranean Zone* (UNESCO-FAO, 1963), of which the latter deals mainly with ecological problems.

LEE (1963) compiled the characteristics of dry-warm climates and their impact on the human heat balance (cf. Table XXXIV). TERJUNG (1967) employed some climatophysiological indexes for the bioclimatological characterization of tropical and sub-

TABLE XXXIV

IMPORTANT CHARACTERISTICS OF HOT DRY ENVIRONMENTS
(After LEE, 1963)

Characteristic of environment	Magnitude	Effect on human heat balance
Direct solar radiation	High, with little natural shade	Marked addition to heat gains
Solar radiation from clouds, etc.	Moderate	Some addition to heat gains
Solar radiation reflected from ground, etc.	High to moderate	Marked to moderate addition to heat gains
Thermal radiation exchanged with ground, etc.	Moderate towards body	Some addition to heat gains
Thermal radiation loss to sky	Moderate	Moderate addition to heat losses
Air temperature	Often above skin temperature	May represent moderate addition to heat gains
Air vapour	Usually low	Important channel for heat loss essential to restore balance
Air movement	Variable, often high	Promotes heat loss when vapour pressure high but gain when temperature very high

tropical conditions. For the example for Africa (by using observations from 800 stations) he presents informative distributions of typical characteristics of its greatly differing atmospheric ambient conditions. Figs.85 and 86 illustrate a selection of bioclimatological sketches valid for the warmest time of year (July).

They deal with effective temperatures and air temperature under calm conditions during the night. This quantity can be specially recommended for characterizing the nocturnal

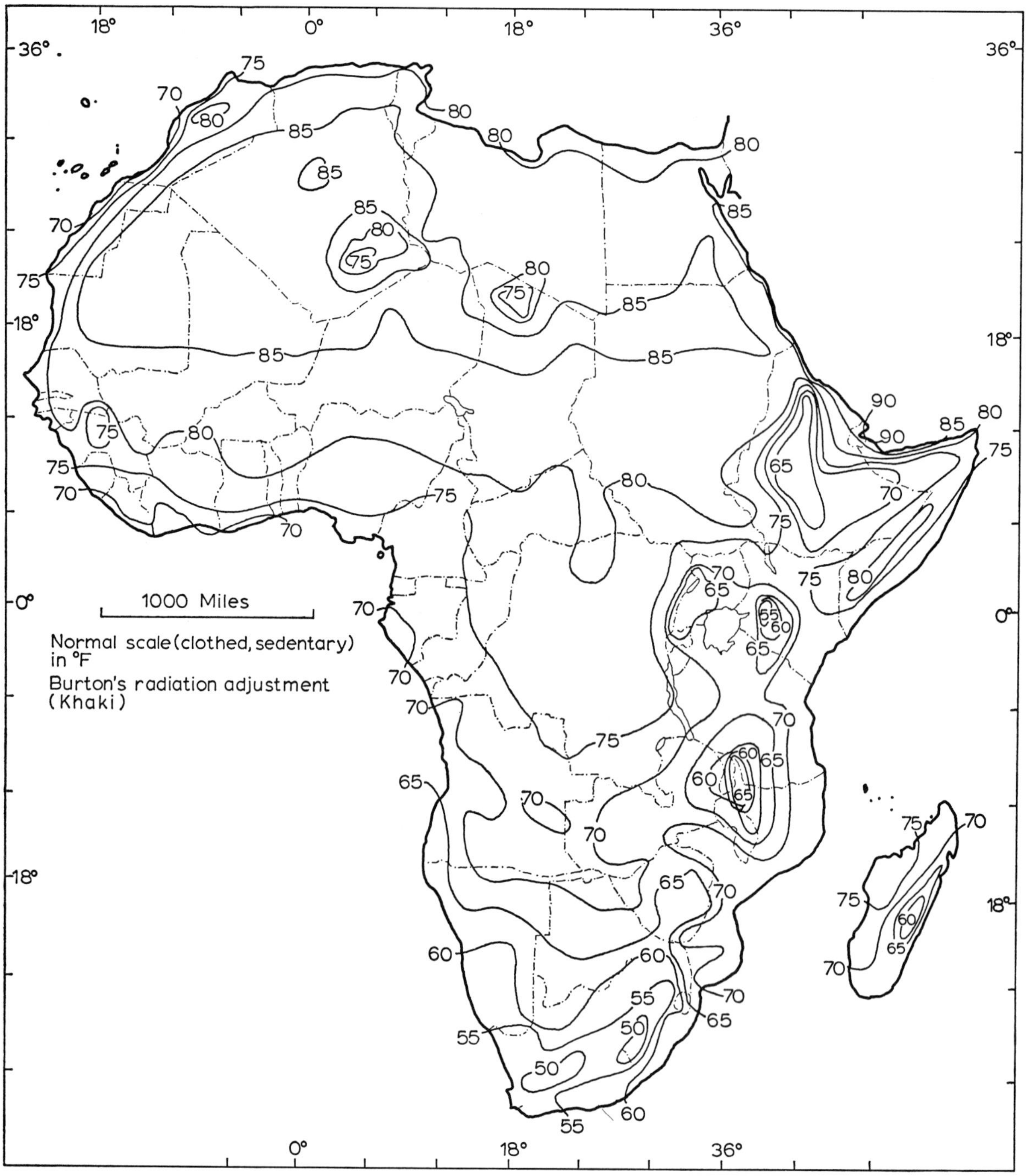

Fig.85. Effective temperatures (*ET*) of °F in Africa; July, mean maximum (TERJUNG, 1967).

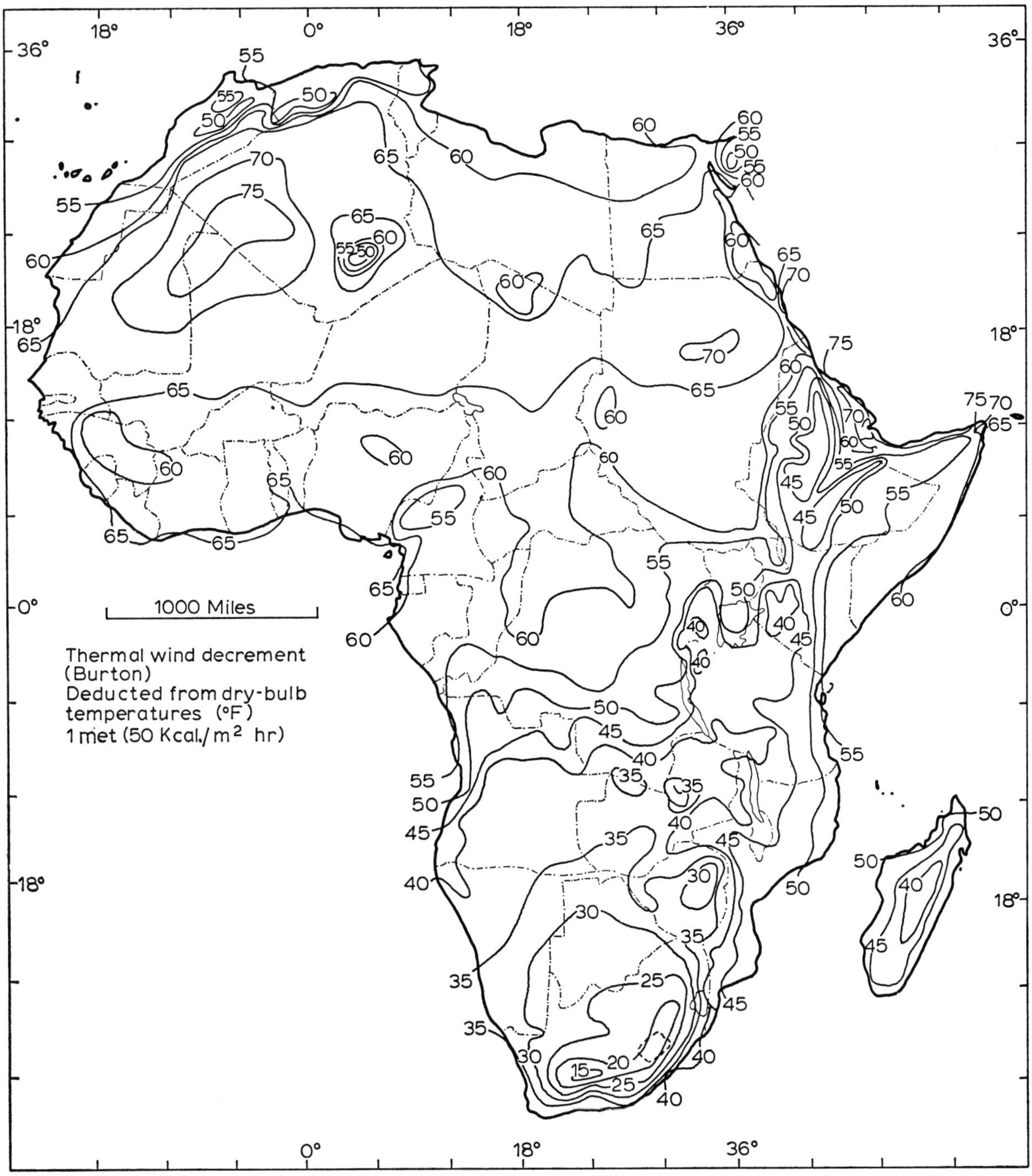

Fig.86. Still-air temperature (SAT) deducted from dry-bulb temperatures (°F) in Africa; July, nighttime. (TERJUNG, 1967.)

climate conditions in the tropics and subtropics with their profound unfavourable influences on sleep.

It must be noted, however, that regions classified in terms of "tropical stress" actually have far less oppressing conditions than the dry-hot arid zones or the transition zones between dry-hot and humid-hot areas. The result is that from a heat physiological standpoint the North African summer is by far the most stressful season. The localities with the greatest stress are in the area along the southern Red Sea and in parts of southern Somalia. In comparison the mildest zones are in the eastern highlands of South Africa and the eastern Atlantic coastal areas which are influenced by the cold ocean currents (upwelling zones).

For heat physiological and aerosol–climatological (meteoro-biological) assessment it is of interest to obtain the seasonal fluctuation of the humidity conditions (vapour pressure and relative humidity) in the Sahara and its northern and southern boundary zones. DUBIEF (1971) has presented an instructive section between 6° and 37°N at a longitude of 2°E.

Fig.87 shows distribution values of water vapour content in winter (January) not only in the Sahara itself but also in its immediate boundary zones, which closely approach those of Alpine climates even of winter. This range of comparatively low water vapour content below 8 mbar reaches at this time of year from 17° to 33°N at a given longitude of 2°E. In the zone between 10° and 6°N as well as along the Algerian coastline vapour pressure quickly rises to very high amounts and thus characterizes highly tropical climatic conditions. In the summer (August) this southern boundary zone (Sudan)

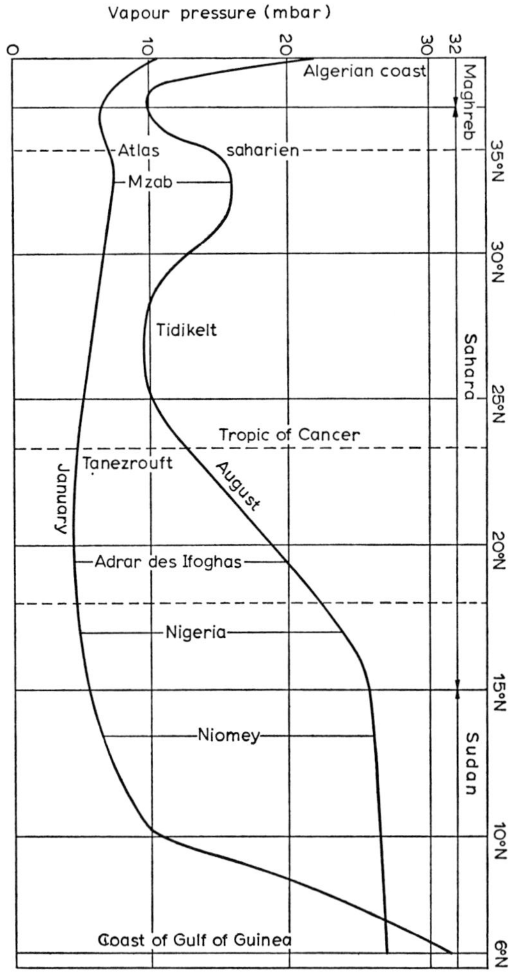

Fig.87. Annual fluctuation of vapour pressure (e in mbar) in North Africa between 6° and 37°N (2°E) (DUBIEF, 1971).

extends to approximately 18°N because of the northward shift in the innertropical convergence zone. Between 25° and 35°N the vapour pressure behaviour reveals the influence of the northern part of the subtropical zone, when the sky aspect at this time of year causes a strong vertical exchange and therefore a reduction of vapour pressure.
Corresponding behaviour is shown in summer (August) by relative humidity, while in winter the Sahara lies for the greatest part in humidity ranges below 35%, often even below 25% (Fig.88).

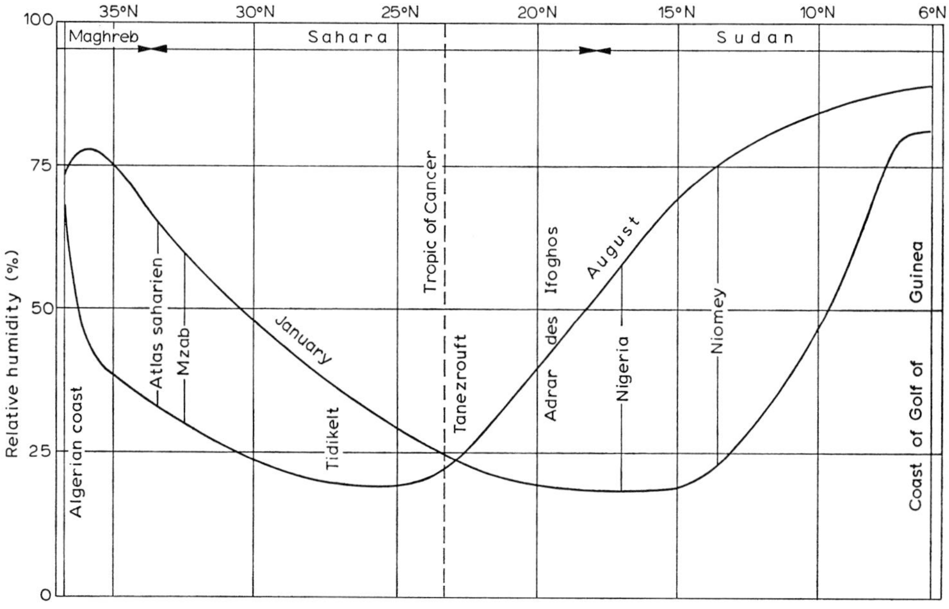

Fig.88. Annual fluctuation of relative humidity (*RH* in %) in North Africa between 6° and 37°N (2°E) (DUBIEF, 1971).

The July illustration is based on the effective temperature (*ET*) according to the Missenard formula:

$$ET = t - 0.4\,(t - 10)\cdot(1 - f/100)$$

where t = air temperature in °C; f = relative humidity in %.
Here areas under heat stress with *ET* 20°C appear between 40°N and 20°S. They more or less correspond with the areas with average July vapour pressure values (map according to LANDSBERG, 1964) and the mean July values of relative humidity (map according to SZAVA-KOVATS, 1938). The isolines of $ET > 20$°C comprise the areas with discomfort stress, while the ones with $ET < 15$°C encircle the cool and cold ones.

"Sultriness" and its common meteorological interpretation

Figs.74–80 have dealt with the problem of heat physiology in man with its significance for climate-conditioned levels of work performance, for phases of sweat production, "spectral distribution" of climate sensations, etc. They have thus shown that these perspectives have important consequences for applied medical climatology, but in particular for the different branches of industrial medicine. If medical climatology is above all interested in climatic therapy (climate dosing) and of bioclimatological evaluations of localities, then industrial medicine's main objective is to determine the limits of human performance under more or less normal conditions and especially under extreme bioclimatic conditions in outdoors and indoors circumstances (e.g., in factories, mines, etc.). Admittedly the latter points are far more important for practical medicine, as in the former case local empirical experiences are doubtlessly able to furnish useful information on the absolute sensation ranges.

For assessment of outdoor air conditions, bioclimatology must first of all determine the large-scale and regional differences in order to describe the integral heat physiognomy of a locality, if possible by incorporating the significant values of air humidity. In indoor working climates, however, one should find the bioclimatic and heat physiological values for evaluating performance and limits of endurance of a working person and the extent of the climatic effects.

The term "sultriness" commonly defines an atmospheric condition incorporating the simultaneous effects of certain temperature and humidity levels in the ambient air. In numerous investigations and experiments for climate therapy and indoor hygiene the behaviour of certain value pairs t_{tr} and RH (%) have been singled out. In a graphic survey BUETTNER (1938) demonstrates (cf. Fig.89) that especially in the temperature range between 25° and 30°C (t_{tr}) with humidity levels of approximately 50–70% (RH) the evaluations of sultriness conditions coincide very well. Below and above these ranges one finds definite deviations from the temperature and relative humidity value groups which cause sultriness sensations (outdoors). Far better agreement is encountered in comparisons based on the air's equivalent temperatures ($t_e = t_{tr} + 2e$) (e = vapour pressure in Torr). By using this equivalent temperature (t_e in °C) as a measure for the air's heat content LINKE (1938) and independently also von Dalmady established the upper limit for the comfort zone at t_e = 56°C, while ROBITZSCH (1931) defines the sultriness range in the temperature values between 37° and 70°C (t_e).

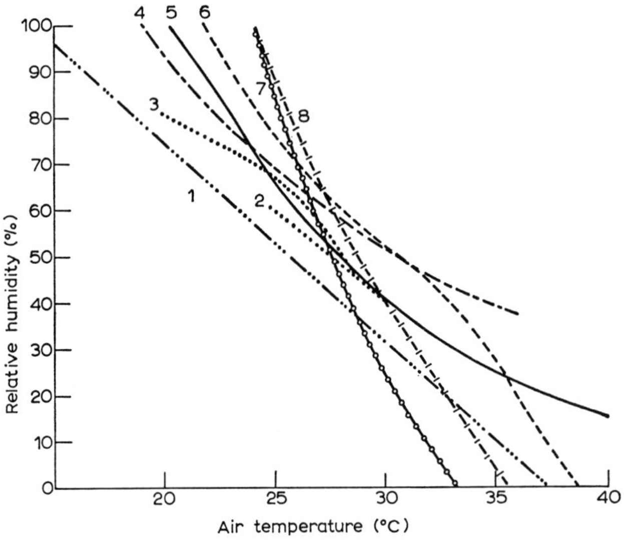

Fig.89. Site of sultriness limit according to various authors: *1* = Proett-temperature = 37.5°C, "comfortable"; *2* = Rubner; *3* = Lange, Lancaster; *4* = Fleischer, thaw point 19°C; *5* = Linke-von Dalmady, equivalent temperature 56°; *6* = Tyler, limit of comfort zone; *7* = effective temperature of 24°C; *8* = effective temperature of 24°C (BUETTNER, 1938).

Later SCHARLAU (1943) tried to give a climate physiological explanation to the work results from investigations by various authors (RUBNER, 1896; LANCASTER, 1898; LINKE, 1922; CASTENS, 1925; RUGE, 1932; etc.). As basis he took the sultriness limit known as the "Lancaster-Castens curve" (in the temperature–humidity diagram) which was checked by Ruge, whereby all air states above the curve are defined as causing sultriness. If one figures out the pertinent vapour pressures from the corresponding value pairs of temperature and relative humidity of the Lancaster-Castens curve, one finds a surprising constancy of the water vapour contents which on an average amount to an e value of 14.08 Torr. (The premature conclusion has been drawn herefrom that a vapour pressure of approximately 14 mm has a true physiological significance.) In a diagram of the so-called "hygrotherms" SCHARLAU (1943) therefore introduced the vapour pressure of 14.08 Torr as a straight line (parallel to the abscissa), above it the "sultriness range" is to be found, and below the "comfort range". Such a representation of sultriness and comfort limits is a gross simplification of the actual outdoors conditions and their relation to sultriness, because physiological reactions appear almost exclusively not in linear but in logarithmic dependence.

Difficulties of this natuıe in dealing with problems of "sultriness" have been pointed out by SCHARLAU himself (1952). Several other authors have also pointed out the concession to be expected in a regional and world-wide climatological treatment of this ambient climate sensation which becomes often extremely unpleasant for people of all ages. In a geographical survey regarding local and tempoıal distributions of sultriness TROLL (1969) has made the same interpretation. As already mentioned one will be obliged to search for approximate solutions which meet the practical demands on bioclimatology.

Climatological aspects of equivalent temperature in relation to "sultriness" and "comfort"

Historically VON BETZOLD (1894) is to be thanked for the term "equivalent temperature" (t_e) which permits close description of the sensation value of "sultriness". Later KNOCHE (1905) urged by von Betzold, tried to give mathematical proof of the climatological significance of this complex quantity. In 1919 PROETT, a sanitary engineer, dealt with the assessment of the air's total heat content and created the "Proett meter". This enables one to obtain this important bioclimatological quantity without the use of psychrometric tables. LINKE (1922) stressed the significance of the "Proett theorem" for hygiene, because "nowadays no science exists without an ulterior motive". Somewhat later LINKE (1938) again pointed out the advantages of equivalent temperature for bioclimatological questions. ROBITZSCH (1951) then gave a new impulse to the use of t_e, and according to him two definition equations are distinguished:

$$t_e = t + \frac{1510}{p} \cdot E_w \qquad \text{(for water)}$$

$$t_e = t + \frac{1730}{p} \cdot E_e \qquad \text{(for ice)}$$

In these equations t signifies the temperature of the dry-bulb thermometer, p atmospheric pressure at the observation point and E_w and E_e the saturation vapour pressures for water and ice, respectively, with regards to the measured air temperature. For an atmospheric pressure of 755 Torr the factor in the second term becomes 2, a fact which simplifies the calculation of equivalent temperature at sea level if the air temperature is above 0°C (in the case of sub-zero temperatures the factor is 2.3). In the following the necessary conversion factors are given for:

$H =$	100	400	800	1,150	1,450	1,800	2,100	2,450	2,750	3,000 m	
$2\frac{755}{p_m}$	2.0	2.1	2.2	2.3	2.4	2.5	2.6	2.7	2.8	2.9	
$H =$	0	350	650	1,000	1,300	1,600	1,850	2,100	2,400	2,650	2,900 m
$2\frac{755}{p_m}$	2.3	2.4	2.5	2.6	2.7	2.8	2.9	3.0	3.1	3.2	3.3

the most important altitudes (H) related to the mean atmospheric pressures (p_m), appearing in the equations:

$$t_e = t + 2\frac{755}{p_m} \cdot E_w \qquad \text{(for water)}$$

$$t_e = t + 2.3\frac{755}{p_m} \cdot E_e \qquad \text{(for ice)}$$

From theoretical considerations based on the psychrometer formula, the following relation is also valid:

$$t_e = t' + 2\frac{755}{p_m} \cdot E'_w$$

This is valid as long as the ventilated wet-bulb thermometer is not coated with ice. Thus equivalent temperature depends solely on the indications (t') of the wet-bulb thermometer. Based on these relations Proett as well as Robitzsch introduced "equivalent thermometers", which unfortunately are no longer available commercially.

The above-mentioned definition equations for equivalent temperature indicate that with this quantity the total heat content of the air can be determined through sensible heat with addition of latent heat (as found from the current water vapour content).

A closer look at the nature of equivalent temperature is desirable, because it regulates the manifestation of sultriness. Furthermore this thermal complex element—as will be shown at a later point—has yet a greater meaning than "only" that of its nature as climato-physiological index.

Table XXXV shows the internal structure of equivalent temperature (t_e) in its relation to the constituent elements of vapour pressure (latent heat) and of air temperature (sensible heat). Not only the vapour pressure groups suspected of sultriness ($e \geqslant 14.1$ mm, after Scharlau) were used as starting point, but also lower e values were related to equivalent temperature. Nature makes no jumps. Hence equivalent temperatures below 55°C contain considerable contributions from e values between 12 and 14 Torr. Thus it is justified to take into account also the t_e groups 50°–54°, 45°–49°, 40°–44° as regards their bioclimatic influence.

Air temperatures and vapour pressures represented by the t_e group of 35°–39°C correspond, in the majority of cases, for central European acclimatization conditions, to the "comfort" sensation ($t_e = 37.5$°C is according to Proett the temperature of well-being).

TABLE XXXV

FREQUENCY DISTRIBUTION OF VALUE LEVELS OF AIR TEMPERATURE (t in °C) AND VAPOUR PRESSURE (e in Torr) IN DEPENDENCE UPON EQUIVALENT TEMPERATURE (t_e in °C), ALWAYS AT 14H00, BASEL (318 m), PERIOD 1930–1959, IN PERCENTAGES OF THE CASES APPEARING IN THE INDIVIDUAL t_e GROUPS
(After FLACH, 1972)

t_e (°C):	35–39	40–44	45–49	50–54	≧55
t (°C):	(%)	(%)	(%)	(%)	(%)
10–12	—	—	—	—	—
12–14	1.6	—	—	—	—
14–16	17.4	2.3	—	—	—
16–18	24.2	12.1	1.2	—	—
18–20	27.1	22.0	7.8	0.7	—
20–22	20.2	31.0	20.1	4.5	0.4
22–24	7.7	21.3	32.6	15.9	2.3
24–26	1.8	9.1	24.3	34.3	10.5
26–28	—	2.1	9.9	26.4	27.9
28–30	—	0.1	2.9	10.9	27.7
30–32	—	—	1.2	5.2	18.2
32–34	—	—	—	1.5	8.8
34–36	—	—	—	0.6	4.0
36–38	—	—	—	—	—
e (Torr):					
4–6	0.5	—	—	—	—
6–8	15.3	2.2	0.2	—	—
8–10	50.8	32.2	7.2	2.5	—
10–12	33.1	51.4	46.5	14.3	3.3
12–14	0.3	14.2	40.2	56.8	23.2
14–16	—	—	5.9	25.5	50.8
16–18	—	—	—	0.9	18.8
18–20	—	—	—	—	3.7
20–22	—	—	—	—	0.2

The two following t_e groups (40°–44° and 45°–49°C) include obvious transition sensations. They range from comfort if air temperature is the dominant factor to a slight sultriness as the vapour pressure contribution increases. The t_e group 50°–54°C with regard to air temperature level must already be considered as the initial phase of the true sultriness sensation. With 25% of the vapour pressure values between 14 and 16 Torr, t_e values fall into the 50°–54°C group. Equivalent temperatures ⩾ 55°C finally give rise to strong and very strong sultriness, with about 74% of the vapour pressure values between 14 and 20 Torr. This result coincides to a certain extent with previous statements by Linke and von Dalmady (sultriness above a t_e value of 56°C).
Table XXXV gives a simple survey of equivalent temperature at midday with respect to its two components. It is indeed an influence factor, noteworthy for global, regional, and local climatology.
In the bioclimatological analysis of atmospheric sultriness the components of heat radiation and particularly of long-wave atmospheric back radiation (G_A) as contributing factors should be considered. Table XXXVI deals with these questions. It compares the behaviour of the *degree of overcast* (in tenths of the sky cover) during midday period and also the daily *sunshine duration* (in hours), with the chosen intervals of equivalent temperature. The following remarks apply.

TABLE XXXVI

FREQUENCY DISTRIBUTION OF VALUE LEVELS OF OVERCAST DEGREE (tenths) AT 14H00 AND OF THE DAILY AMOUNT OF SUNSHINE DURATION (h) IN MIDSUMMER (JUNE THROUGH AUGUST) IN DEPENDENCE UPON EQUIVALENT TEMPERATURE (t_e in °C), BASEL (318 m), PERIOD 1930–1939, IN PERCENTAGES OF THE CASES APPEARING IN THE INDIVIDUAL t_e GROUPS
(After FLACH, 1972)

Cloudiness:			Sunshine duration:			t_e group
0–5	6–7	8–10	⩽5	6–8	⩾8	
27	9	64	65	14	21	35–39
36	9	55	51	13	36	40–44
42	10	48	40	16	44	45–49
60	7	33	25	17	58	50–54
67	11	22	8	20	72	≧55

(*a*) The t_e group 35°–39°C has the relatively highest percentage of large cloud cover and therefore short sunshine duration (64 and 65%). Correspondingly, the highest percentages for small cloud cover and high sunshine duration amounts are 27 and 21%.

(*b*) In the t_e group ⩾ 55°C one notes a marked reversal of the conditions described under (*a*). 72% of daily sunshine duration are 8 h and more, and 67% of cloud covers are below 5 tenths (the latter fact signifies—for Basel—approximately 90% relative sunshine duration). ZENKER (1971) confirmed the higher frequency with few clouds on sultry days by observations made in moderately high mountain areas.

(*c*) Medium cloud cover (6/10–7/10) and corresponding sunshine duration occur exclusively as low percentages in the individual t_e groups.

The statements (*a*) to (*c*) indicate that in the case of sultriness conditions (t_e ⩾ 55°C) the *short-wave incoming heat radiation* from *S* and *D* participates decisively in approximately 70% of the cases as an "accompanying element". Therefore, long-wave atmospheric back radiation which does not vary much in intensity, does not exhibit much importance, not even during the daylight hours (cf. sultriness studies by KNEPPLE, 1948, and KING, 1955). One can deduce this from the fact that on days with true sultriness (t_e ⩾ 55°C) the sky cover is in two-thirds of all cases under 5/10 and thus atmospheric back radiation G_A contributes only 5% or less, for all cloud types. Similar conditions hold for values of 6/10 and 7/10, where they furnish caloric supplements of 5–7% (G_A). In the latter case one must probably include a considerable increase of short-wave *sky radiation* (*D*), especially in the presence of high clouds (Ci, Cs). At the summer solar elevations in central Europe, according to calculation by BENER (1963), the short-wave heat radiation at 9/10 Ci and Cs cloudiness represents 2 to 3 times the value of cloudless condition values, and corresponds to approximately 0.3 cal. cm^{-2} min^{-1}. Such cases are not rare prior to the end of a sultriness situation (pre-thunderstorm situation).

For emphasis of the above statements the components decisive foı radiation balance are presented for a clear day (Hamburg), in Fig.90. The graph shows that during the greatest part of the daylight hours global radiation $T(S + H)$ by far exceeds atmospheric back radiation G_A (called *A* in Fig.90). Between 10h00 and 14h00 it reaches more than the double amount.

Furthermore the facts given under (*b*) state that precisely the long-lasting incoming radiation during the day simultaneously causes above-normal air temperature values and comparatively high vapour pressures (at least during the morning and evening

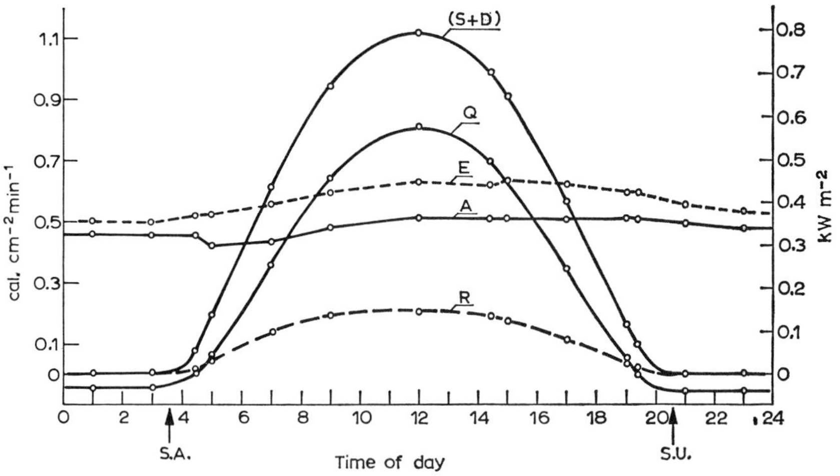

Fig.90. Behaviour of radiation balance Q and its components on a clear summer day (Hamburg); Q = radiation balance; E = long-wave ground radiation; A = atmospheric back radiation; R = reflected global radiation; $S + D$ = global radiation. (FLEISCHER and GRÄFE, 1954.)

maximum). An advective contribution to the appearance of a sultriness sensation within a current of subtropical air masses (meridional air circulation) at t_e values under 60°C is negligible. HERRMANN (1959) also points out the limited effect of such advection for outdoors sultriness. On the other hand, DAMMANN (1964) would attribute sultriness situations almost exclusively to the advection of tropical air.

The same author (Dammann) cites a unique case of sultriness which is restricted to desert areas and their immediate vicinity. During dust storms in the Sahara the desert air is enriched with very fine dust particles; this leads to an intensification of mainly long-wave but partly short-wave diffuse atmospheric radiation. Thus sultriness sensations can appear although at high air temperatures the relative humidities are extremely low. In this context a common observation on mid-summer evenings in large cities is an intensive heat emission of long-wave radiation from *stone masses* (buildings, asphalt roads, etc.). Together with the evening vapour pressure maximum the sultriness sensation is thus often considerably heightened (cf. quantity E in Fig.90).

It results from the preceding Table XXXVI (comparison of t_e with cloudiness and sunshine duration) that from the t_e group 35°–39°C to that of $\geqslant$ 55°C the phases with few clouds and much sunshine increase in number, while the opposite ones decrease. The daily phases with much sunshine increase very rapidly above $t_e \geqslant$ 50°C, and thus also the contribution of short-wave incoming heat radiation (mainly of S), in connection with rising values of air temperature and vapour pressure, i.e., with an increase of sultriness frequency.

In order to check on further climato-physiological consequences the *cooling power* (*CP* in mcal. cm^{-2} sec^{-1}) for the *daylight hours* was measured in Basel with the Davos frigorimeter. Table XXXVII relates the *CP* level of 1 to 4 mcal. cm^{-2} to the given t_e groups. The table (CP/t_e) provides an interesting view of the midsummer behaviour of *CP* and its percentual appearance within the individual t_e groups. In the case of marked sultriness ($t_e \geqslant$ 55°C) the lowest *CP* values (1–4 mcal.) are by far most frequent. But in Basel one must expect a certain number of daily phases with high heat parameters (THAMS, 1962). AMBROSETTI (1965) stated for Locarno that the hours with hot conditions are 80% fully sunny.

TABLE XXXVII

FREQUENCY DISTRIBUTION OF VALUE LEVELS OF THE COOLING POWER (*CP* in mcal. cm^{-2} sec^{-1}) MEASURED WITH THE DAVOS FRIGORIMETER DURING THE DAYLIGHT HOURS OF THE MONTHS JUNE, JULY AND AUGUST IN DEPENDENCE UPON EQUIVALENT TEMPERATURE (t_e in °C) IN BASEL (318 m), PERIOD 1930–1939, IN PERCENTAGES OF THE CASES APPEARING IN THE INDIVIDUAL t_e GROUPS
(After FLACH, 1972)

1–4	5–8	9–12	13–16	17–20 *CP*		t_e group (°C)
16	50	24	8	2	June	
18	23	33	17	9	July	35–39
17	29	33	12	9	Aug.	
47	36	16	1	—	June	
38	30	25	7	—	July	40–44
37	40	15	5	3	Aug.	
68	25	7	—	—	June	
51	31	15	3	—	July	45–49
43	38	14	5	—	Aug.	
69	26	3	2	—	June	
80	19	1	—	—	July	50–54
70	29	6	—	—	Aug.	
89	11	—	—	—	June	
71	29	—	—	—	July	≧55
92	8	—	—	—	Aug.	

The summer percentage of the *CP* level (1–4 mcal.) in the t_e group 35°–39°C already amounts to 17%, in the t_e group $\geqslant$ 55°C it climbs to 84% (as an average for the three summer months). This signifies a five-fold growth. The *CP* levels (5–8) and (9–12) at the same time decrease rapidly in their relative frequency by 25–30% (cf. Fig.91).
Thus the thesis that growing sultriness, i.e., rising t_e values, is accompanied by linear reduction of the cooling power, is incorrect.
In addition to this type of climatic data evaluation for sultriness studies, corresponding analyses of equivalent temperature with sunshine duration, cloudiness and cooling power were also carried out for the central European stations of Karlsruhe and Dresden. They confirm the statements for Basel, Locarno, and Davos, considering the special climatic features of these locations.
Table XXXVII has already made it quite clear that under sultriness conditions as indicated by the determining element "equivalent temperature", cooling powei as such has only a limited capability of mitigating the heat-physiological stress. Effects of wind speed to lower the sultriness sensation (especially with high air temperatures) is small when considered as a purely scalar dimension. If one includes the vectorial components of air movement, additional aspects result.
Each high-resolution record of air temperature, relative humidity and wind (direction and force) shows that particularly on days with few clouds (also in connection with sultriness) with turbulent convection and vertical exchange pulsing rises and falls of temperature, vapour pressure and relative humidity occur.
Fig.92 documents the fact that temperature and humidity fluctuations in the boundary layer, i.e., in the human life's sphere, can be of quite considerable magnitude under

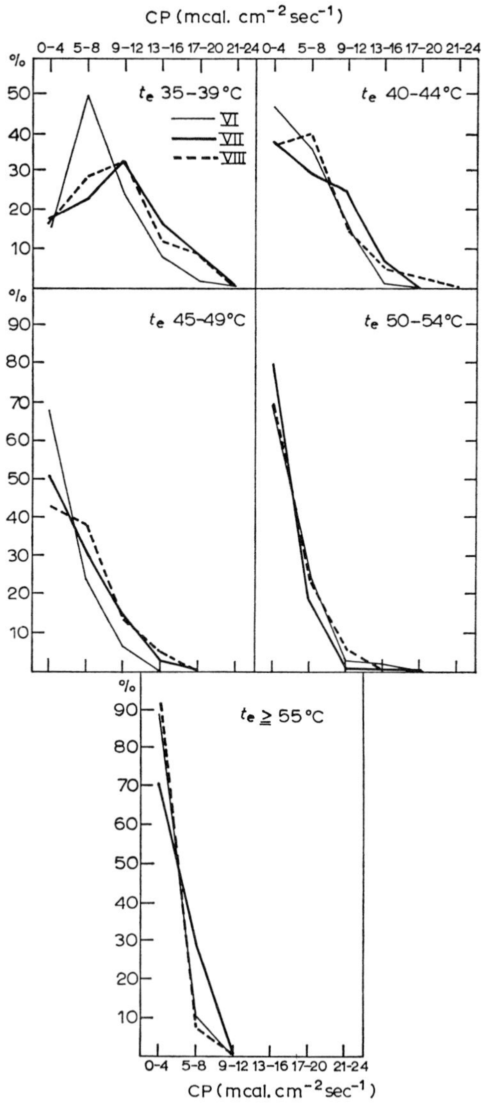

Fig.91. Frequency distribution (%) of cooling power (*CP*) measured with the Davos frigorimeter in dependence upon individual t_e groups during the summer months June through August. Basel (318 m), according to average values collected over 15 years. (FLACH, 1972.)

strong incoming radiation and intensive ground-heating. The figure illustrates the reduction of vapour pressure which is strongly felt during the day as a consequence of the vertical exchange taking place at that time. The process is all the more evident, the greater the distance is from the ground (at only 5 cm height it is not very pronounced because of constant supply of water vapour from the soil). Corresponding processes are found in relative humidity. This daily periodicity and its peculiarities play a noteworthy role for climato-physiological aspects of sultriness. This type of specific dynamic processes in the atmosphere cause an attenuation and mitigation of the sultriness sensations.

Basic meteorological facts for a sultriness analysis have been established for Basel; in order to acquire further information, Table XXXVIII presents the percentage frequency of relative humidity (f in %) as a function of daily sunshine duration and related to the various t_e groups. The latter have been arranged in two main ranges: between 35° and

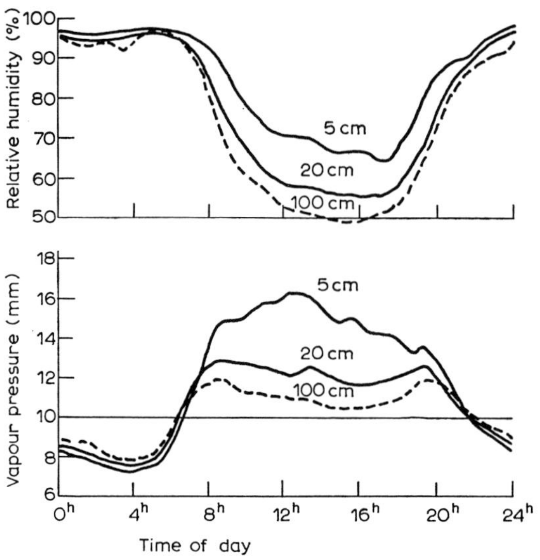

Fig.92. Diurnal variation of air humidity (vapour pressure and relative humidity) in the lowest air layer (1 m) (M. Franssila, from GEIGER, 1961).

49°C, with none or little sultriness, and in t_e values $\geqslant$ 50°C. The table shows foı the days not subjected to sultriness that the f values show their highest frequency during clear episodes in lower ranges than for the case of sultry days. This statement as such is not new, but a close study of the frequency values of f at t_e values of the $t_e \geqslant 50$ group reveals that frequently rather dry air masses take part in sultriness situations. This means that convection and vertical exchange provide heat-physiological relief also on the so-called sultry days, which are quite numerous.

A simpl ecalculation based on the relation $Q_{VH} = \beta(e_H - e_L)580$ g cm^{-2} min^{-1} demonstrates that the simultaneous vertical aiı exchange always leads to a certain rise in evaporation of water from the human skin. This is moreover coupled with an increase of air movements, even if only small.

This kind of air exchange in unison with pulsing reductions of vapour pressure (and thus also with relative humidity) and an increase of wind speed is an exclusive feature of the outdoors climate. This fact prohibits the indiscriminate transfer of heat-physiological experimental results from an artificial climate to outdoor conditions, because in a closed room the above-mentioned phenomena are not present.

In addition, the higher f values in sultriness situations, i.e. under heightened t_e values, are also accompanied by moderate to strong haze. Synoptically seen, this is frequently observed in the lability of pre-thunderstorm situations even if there is no longer a high-reaching convection as in anticyclonic weather phases. It is therefore unnecessary to consider the intensity of haze for the evaluation of sultriness intensity (King).

In true sultriness conditions ($t_e \geqslant$ 55°C) there is a primary participation of significant *short-wave* heat radiation intensities of S and D. This fact weakens the hypothesis (Knepple, King) of a decisive participation of long-wave atmospheric back radiation (G_A) (with approximately one third of the total impact) in the formation and promotion of sultriness sensations.

To date all sultriness hypotheses have excluded the heat effects of short-wave heat radiation (particularly by S) in its *daily periodical* behaviour as well as in its *cloudiness-*

TABLE XXXVIII

FREQUENCY DISTRIBUTION OF VALUE LEVELS OF RELATIVE HUMIDITY (f in %) AT 14H00 IN DEPENDENCE UPON THE DAILY TOTAL SUNSHINE DURATION (h) AND OF EQUIVALENT TEMPERATURE (t_e in°C) IN SUMMER (MAY–AUGUST) IN BASEL (318 m), PERIOD 1930–1939, IN PERCENTAGES OF THE CASES APPEARING DURING A SUNSHINE DURATION INTERVAL

(After FLACH, 1972)

	Daily amounts of sunshine duration:							
f(%)	0–2	2–4	4–6	6–8	8–10	10–12	12–14	>14
t_e values (35–39) + (40–44) + (45–49):								
96–100	—	—	—	—	—	—	—	—
91–95	15	1	—	—	—	—	—	—
86–90	19	4	—	—	—	—	—	—
81–85	9	5	—	—	—	—	—	—
76–80	16	6	4	—	—	—	—	—
71–75	13	11	1	5	—	—	—	—
66–70	17	16	10	10	2	—	—	—
61–65	7	26	21	8	2	3	3	4
56–60	3	19	30	22	4	10	5	4
51–55	1	10	17	30	13	22	13	4
46–50	—	2	10	15	26	30	22	31
41–45	—	—	3	7	38	23	22	35
36–40	—	—	2	3	9	10	16	13
31–35	—	—	2	—	6	2	10	9
26–30	—	—	—	—	—	—	9	—
21–25	—	—	—	—	—	—	—	—
t_e values (50–54) + (⩾55):								
96–100	—	—	—	—	—	—	—	—
91–95	4	—	—	—	—	—	—	—
86–90	17	—	—	—	—	—	—	—
81–85	12	4	3	—	—	—	—	—
76–80	4	3	9	—	—	—	—	—
71–75	—	11	9	—	—	—	—	—
66–70	17	25	17	8	6	2	—	—
61–65	17	29	14	18	16	7	4	—
56–60	21	18	9	18	18	25	2	14
51–55	8	10	23	35	22	25	18	21
46–50	—	—	8	16	20	17	31	14
41–45	—	—	8	3	13	14	22	21
36–40	—	—	—	2	5	7	10	14
31–35	—	—	—	—	—	3	10	14
26–30	—	—	—	—	—	—	3	—
21–25	—	—	—	—	—	—	—	—

conditioned intermittent "*pulsations*" and ensuing *consecutive psychic effects.* As already pointed out, in the same way the pulsatory fluctuations of the air's equivalent temperature (with components t and e) as brief alternating mitigating and intensifying conditions for sultriness have not been considered.

The efforts by LEISTNER (1964) and ROBITZSCH (1931) for the solution of the sultriness problem provide a physiological aspect to the purely meteorological determinations in this field. This is done by the introduction of an equivalent temperature for skin ($t_{e(H)}$). This parameter may be approximatively inferred by electrical resistance thermometers and special hair hygrometers. Linke denied the exact determinability of this dimension and stressed that it would be more sensible to speak of an equivalent temperature of the air layer near the skin. On that basis Leistner set up a physioclimagram for the *windy North Sea coastal climate* including equivalent temperature of the ambient air and wind speed for his determination of climatic sensation steps from "cold" to "comfortable" to "sultry".

Equivalent temperature as a regular climato-physiological unit for the study of sultriness distributions (central Europe)

The atmospheric causes of sultriness are of remarkable bioclimatological significance. But the multiple causes combining to specific values limit the use as a *true aid to medical practice* (touristic medicine, industrial medicine). For this purpose one should supply only the simplest arrangements of pertinent climatological data in order to be truly useful. One can also only recommend against combinations of differently founded *complex elements* with heat-physiological objectives into one single parameter because different weights attached to double individual elements, in such combinations (e.g. air temperature) may lead to erroneous interpretations.

The preceding again points out the necessity of finding a clear solution by means of the basic thermal elements t and e now for various problems including sultriness. With the help of the complex element "equivalent temperature" (t_e) this seems feasible for several reasons.

(*1*) t_e is a physically founded quantity defining the air's heat content. Its two main components (t and e) are easily comprehensible. Its constituents are continuously measured at all climatic stations over the world, usually including atmospheric pressure. In contrast, the climatological element t' (indication of wet-bulb thermometer) used frequently in other thermal complex factors, that is linked with the equivalent temperature by the relation $t' = t_e + 2\,E'$, and that thus can also be used for a definition of the air's total heat content, t_e has the advantage of incorporating the significant and primary components t and e. These two factors, besides radiation component, control the bioclimatic aspects of "temperature and life".

(*2*) The t_e grouping used in the preceding paragraphs for the characterization of the warm range (comfortable—warm—hot—sultry) due to its inclusion of t, e and of short-wave heat radiation (S, D, R_E) (the latter by way of indications of sunshine duration or also of the midday cloudiness degree) is sufficiently justified for physiological consideration.

The inclusion of absolute or relative sunshine duration (*AS* in hours and *RS* in percentages) and thus of a characterization of incoming radiation conditions, has the advantage of significant discrimination (cf. Table XXXIX).

(*a*) In moderate zones there is the frequent sensation "comfortable" by marked involvement of radiation parameter $T\,(S + D)$ or CGR $(S + D + R_E)$; e.g., in the presence of polar maritime air masses.

In sea-coast climates (moderate zones) as well, and particularly in Alpine climates during all seasons the inclusion of relative sunshine duration with equivalent temperature t_e is appropriate for the characterization of such comfort conditions. The groupings of t and e are not alone responsible for these conditions (e.g., t_e 35°–39°C, 30°–34°C, etc.), because vapour pressure values below 8 Torr do not contribute greatly to the comfort sensation. The limiting value of *RS* (= 75%), on the other hand, includes clear days as

TABLE XXXIX

CHARACTERIZATIONS OF SENSATIONS IN THE COMFORT AND SULTRINESS RANGES (MODERATE LATITUDES) BY MEANS OF EQUIVALENT TEMPERATURE LEVELS AT MIDDAY (t_e °C) AND OF RELATIVE DURATION OF SUNSHINE (*RS*) FOR THE EXAMPLE OF THE BASEL SUMMER CLIMATE*
(After FLACH, 1972)

t_e group	t, e and *RS* parts (frequency percentage)		Characterization of sensation	
35–39°C	18 t 24	50%	*Cf* ($t + T$)	Comfortable
	8 e 12	80%	*Cf* (t_e)	
	RS 75%	30%		
40–44°C	18 t 24	80%	*Cf* (t_e)	
	8 e 12	80%	*Cf* ($t_e + T$)	Comfortable–warm
	RS 75%	30%		
45–49°C	18 t 24	65%		
	24 t 28	30%	*Cf* ($t_e + T$)	
	8 e 12	50%		Warm–hot
	12 e 14	40%	*Dcf* ($t_e + T$)	
	RS 75%	40%		
50–54°C	18 t 24	20%		
	24 t 28	65%	*Dcf* ($t_e + T$)	
	8 e 12	10%		Hot–sultry
	12 e 14	20%	*Ss* ($t_e + T$)	
	e 14	70%		
	RS 75%	75%		
⩾55°C	18 t 24	0%		
	24 t 28	40%		
	t 29	60%		
	8 e 12	0%	*Ss* ($t_e + T$)	Sultry–warm
	12 e 14	20%		
	e 14	80%		
	RS 75%	75%		

* t = air temperature in °C; e = vapour pressure in Torr; *Cf* = comfort; *Dcf* = discomfort; *Ss* = sultriness; T = global radiation.

well as partly clear days, a fact which is generally valid for the moderate latitudes (FLACH, 1963). Comfort sensations can be felt even at t_e values below 35°C with sufficient incoming radiation.

(*b*) When including the daily sunshine duration with equivalent temperature (midday value) one takes into account the short-wave light and heat radiation fluxes contributing most persistently to the sultriness effect with all its physiological and psychological consequences. The radiation parameters, i.e., duration of sunshine or midday cloudiness implicitly indicate convection and vertical exchange and thus the pulsations of t and e and also of t_e during the daylight hours.

(*c*) Using relative sunshine duration and midday cloudiness it is possible to gage the diurnal variation of sultriness sensations. Often long sunshine duration on sultry days should not be neglected including the effects of reflection radiation from clouds and ground. Together with the global radiation they are of far greater importance than atmospheric back radiation alone which is a rather uniform factor.

(*3*) It is a secondary requirement to include *air displacements* as a scalar quantity or as cooling power in all comfort and sultriness studies. Both characteristics (wind speed v and cooling power *CP*—the latter as complex element) show extremely large local fluctuations mostly from micro- or mesometeorological circulations. However, these parameters are not decisive for the thermal sensation range of sultriness.
In the windy coastal climates (e.g., European coasts of the Atlantic and North Sea) regular high wind speeds prevail (above 4–6 m/sec). There wind breaks, forests, beach baskets etc. help to reduce ventilation to comfort ranges described by equivalent temperatures. Yet in summer at t_e values between 35° and 45°C frequently sea and sun baths are taken in the boundary layer near the ground, where there are usually low wind speeds. With the exceptional higher equivalent temperatures ($>$ 45°C) in these regional climates the relatively high wind speeds are often a very welcome bioclimatic feature.

Countryside-climatological behaviour of equivalent temperature and cooling power

In order to convey an idea of the seasonal sultriness states in the tropics (example of San Salvador at 13°43′N and 89°12′W at 698 m) a comparison is made between the average amounts of sultry hours (based on the sultriness limit value with e = 14.08 mm—SCHARLAU, 1943) at San Salvador (Central America) and Potsdam (central Europe), calculated by DIETERICHS (1958) (cf. Table XL).

TABLE XL

COMPARISON BETWEEN AVERAGE AMOUNTS OF SULTRY HOURS AT POTSDAM AND SAN SALVADOR

	I	II	III	IV	V	VI
Potsdam (1924–1938)	—	—	—	—	5.3	23.4
San Salvador (1952–1956)	103.6	175.6	308.8	528.2	662.4	707.6

	VII	VIII	IX	X	XI	XII	Year
Potsdam (1924–1938)	62.8	44.9	9.6	0.1	—	—	146.1
San Salvador (1952–1956)	716.2	726.0	713.8	613.2	388.8	214.8	5858.0

The greatest danger in long-distance travel is a climatic lag. Although the traveller enjoys a fully air-conditioned jet aircraft, he may well be "exhausted" upon arrival in the tropics because of the stress of an instantaneous change into a humid-hot environment. Similar conditions are encountered there in air-conditioned hotel rooms which create a stress by continuous changes with outdoor conditions. Besides the influences of temperature and humidity the often high intensities of heat and ultraviolet radiation create additional heavy loads. All these circumstances (especially in the tropics) make prophylactic and therapeutical care through touristic medicine (MOHRING, 1971) in this age of mass tourism very important.
But even without such "jumps" to distant climate areas, regional differences in one single

region often give rise to impressive climato-physiological contrasts. In Fig.93 LEISTNER (1964), based on values determined by HERRMANN (1959), set up a sketch illustrating the regional sultriness grades of the German North Sea coast and the adjoining northern German inland areas to the south.

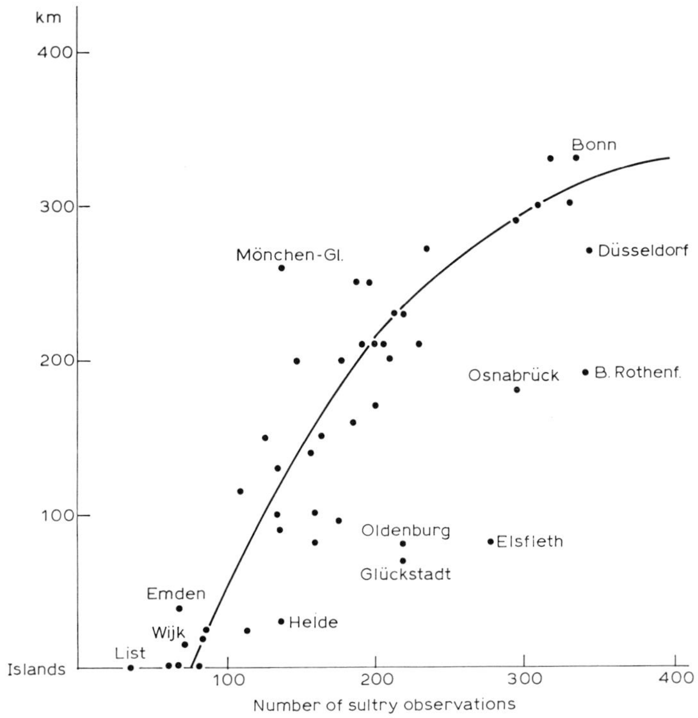

Fig.93. Increase in sultriness in northern Germany from the North Sea coast towards the interior (for altitudes below 100 m) (H. Herrmann, according to LEISTNER, 1964).

The maritime climate (Atlantic and North Sea) extends its influence approximately 150 km inland; then it falls off rapidly. In Leistner's sultriness definition wind force has considerable weight; hence Fig.93 not only expresses the increase in sultriness towards the interior but also the higher sultriness frequency. The latter indicates that there is no definite advantage in over-evaluating air displacements in sultriness studies. Each locality has its own peculiar wind structure (some of which depends on the varied placement of wind measuring devices).

Purely from a bioclimatological point—direct illustrations of the sultriness distributions within a region or over the whole earth deserve no doubt our primary interest. Yet the causal relations for such maps as they relate to heat physiology should not be neglected. General climatology offers a variety of world maps of interest such as the distribution of air temperature (t), vapour pressure (e), sunshine duration (Sn), etc.

Fig.94 represents the water vapour partial pressure distribution (e in mbar) as determined by LANDSBERG (1964) and is also valid for July. It can be seen that the continents of the Northern Hemisphere in this case show comparatively high vapour pressure values. In contrast the vapour pressures below 10 mbar are prominent in South Africa and Australia where despite the low latitude they cover wide areas. In this context, the analyses of the details of environmental radiation undertaken by VERNEKAR (1975) are

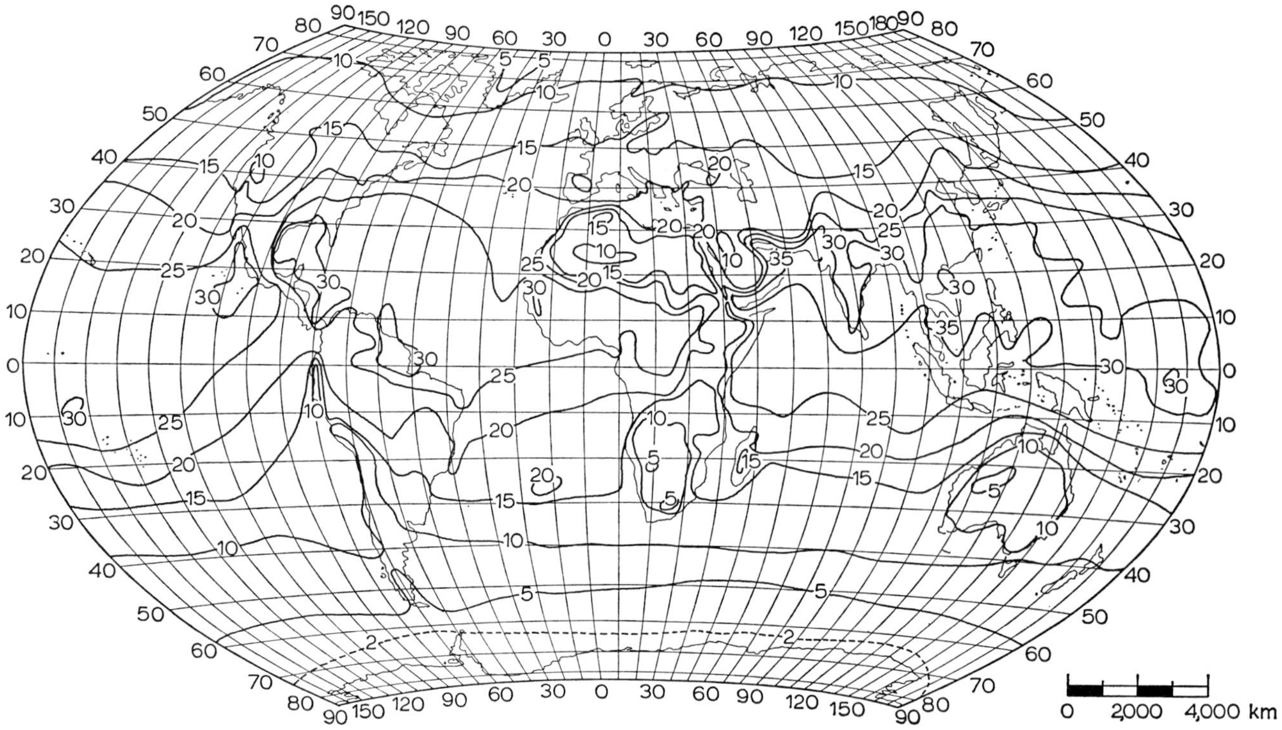

Fig.94. Mean distribution of water vapour pressure (mbar) in July on earth (LANDSBERG, 1964).

of some bioclimatic significance. They encompass the radiation conditions on earth for average cloudiness, the mean surface albedo, etc.

Particularly vast expanses with *e* values $>$ 30 mbar are observed in Arabia, around the Indian subcontinent, in Southeast Asia and in the central Pacific. The areas above approximately 20 mbar correspond with the distribution of permanent and periodical sultriness on earth. The absolutely lowest vapour pressures below 2 mbar are, of course found at that time of year in the Antarctic. The earth's arid zones reveal the lowest monthly average values of vapour pressure (between 5 and 10 mbar) despite the often high air temperatures. As a rule vapour pressure shows a continuous increase from the polar areas towards the equatorial regions in strong dependence upon the atmosphere's absorption capacity for water vapour dictated by air temperature.

During the past few decades cooling power registrations have been carried out in various zones of the earth by means of the Davos frigorimeter. They are of special bioclimatological interest because the contributions of short-wave heat radiation of S, D and R_E have been included in the system. Diverse climates can serve as examples for a bioclimatological assessment of cooling power measured with the frigorimeter. The frigorimeter stations extend from northern Greenland to the middle latitudes of North America (FLACH and MOERIKOFER, 1967) (cf. Table XLI).

Fig.95 presents examples of the *mean diurnal variation of* CP_{Fri} for the main seasons. Interesting comparisons and impressions of the basic bioclimatic structure of areas of northern Greenland, of the maritime parts of Scandinavia, of Alpine sites and climatically moderate zones of North America can be obtained from the figure. The relatively "mildest" conditions are encountered in the Alpine sites (Davos) and in the North American lowland station Belmar with its, at times, subtropical character. In contrast,

TABLE XLI

LIST OF FRIGORIMETER STATIONS USED*

Country	Station	φ (N)	λ	A	H	E
U.S.A.	Belmar N.J.	40°11′	74°04′W	5	2.5	meadow
N. Greenland	Tuto-West	76°28′	68°40′W	305	?	above ground
N. Greenland	Camp Century	77°10′	61°08′W	1,920	?	above snow
Norway	Bergen	60°23′	05°21′E	40	20	tower
Switzerland	Davos	46°48′	09°49′E	1,592	8	flat roof

* Abbreviations: φ = geographical latitude; λ = geographical longitude; A = altitude of the station in m a.s.l.; H = height of the frigorimeter sphere above ground; E = place of exposure

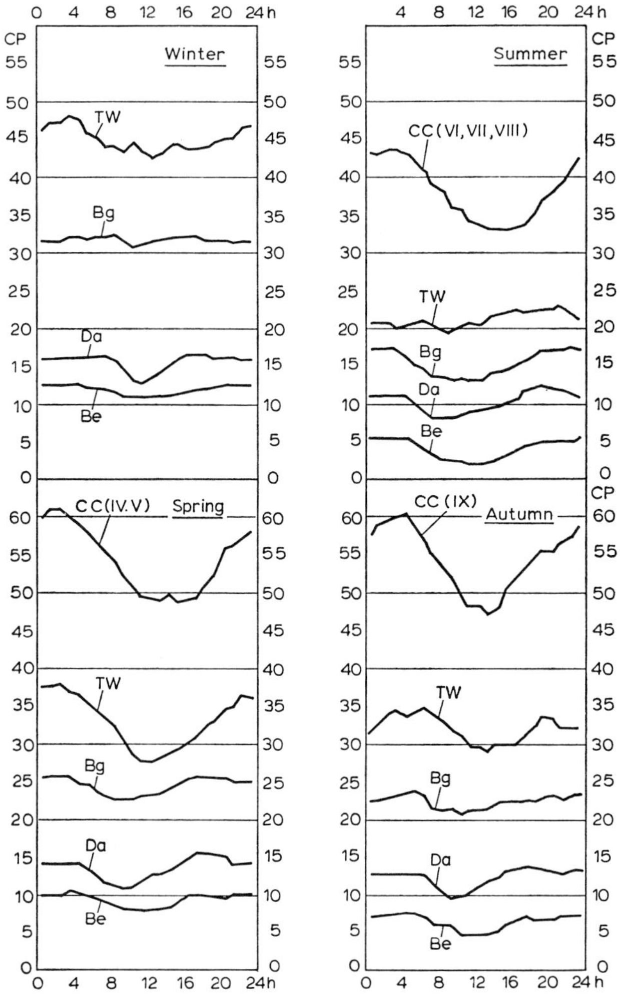

Fig.95. Diurnal variation of cooling power (CP in mcal. cm^{-2} sec^{-1}) at Tuto-West and Camp Century (Greenland), Bergen (Norway), Davos (Switzerland) and Belmar, N.J. (U.S.A.) for the different seasons. Tuto-West (*TW*), 305 m, 1961; Camp Century (*CC*), 1,920 m IV–IX, 1962; Bergen (*Bg*), 40 m, 1954–1960; Davos (*Da*), 1,592 m, 1941–1945; Belmar (*Be*), 5 m, 1957–1960. (Flach and Moerikofer, 1976.)

there are conditions of extreme intensity in northern Greenland (Tuto-West and particularly Camp Century) which is undisturbed in all seasons. In the summer there is a notable reduction of *CP* in Camp Century. During that time the influence of the continuous incoming radiation may be felt. In Tuto-West, in the vicinity of the coast, the

TABLE XLII

MONTHLY AVERAGES OF THE MAXIMUM AND MINIMUM AIR TEMPERATURES (°C) AS WELL AS MEAN MONTHLY WIND SPEEDS (m/sec) AT THE SOUTH POLE (US), $\varphi = 90°00'S$, $H = 2{,}800$ m

	I	II	III	IV	V	VI
t_{max} (1957) °C	−21.1	−27.8	−37.2	−32.0	−34.5	−41.3
t_{min} (1957) °C	−34.3	−56.2	−63.6	−67.3	−73.6	−71.7
Wind speed (m s^{-1}) (1957)	5.0	5.0	5.8	7.7	7.7	8.6

	VII	VIII	IX	X	XI	XII
t_{max} (1957)	−40.4	−43.0	−49.8	−43.0	−19.0	−18.9
t_{min} (1957)	−72.7	−73.2	−74.5	−65.6	−48.1	−29.7
Wind speed (m s^{-1}) (1957)	7.9	8.1	6.5	7.1	4.7	3.5

TABLE XLIII

DAILY TOTAL RADIATION $T (= S + D)$, SOUTH POLE (in cal. cm^{-2}d^{-1}; 1958)
(After DALRYMPLE, 1966)

1958	Jan.	Feb.	Mar.	Sep.	Oct.	Nov.	Dec.
1	925	446	249	—	59	526	940
2	896	701	294	—	71	426	950
3	886	485	276	—	129	456	912
4	M	M	247	—	171	454	920
5	875	578	226	—	89	599	938
6	910	522	252	—	167	639	952
7	956	594	130	—	226	M*	954
8	938	632	147	—	234	642	969
9	874	588	126	—	209	698	982
10	685	564	135	—	133	668	984
11	810	605	87	—	M	683	993
12	662	597	92	—	M	725	963
13	766	579	109	—	222	M	895
14	686	576	M	—	267	735	857
15	645	559	M	—	321	757	755
16	625	536	80	—	353	773	832
17	782	273	66	—	245	780	737
18	797	267	41	—	333	664	801
19	808	301	30	—	384	633	914
20	857	283	23	—	357	M	951
21	860	401	17	6	270	M	980
22	853	314	15	5	349	646	882
23	752	311	13	8	375	764	M
24	692	355	12	12	372	768	M
25	683	225	7	14	326	792	919
26	587	226	4	20	288	866	910
27	752	227	2	35	364	828	919
28	650	240	—	46	320	819	954
29	596	—	—	54	337	870	968
30	729	—	—	59	375	908	936
31	575	—	—	—	527	—	937

* M = missing data.

incoming radiation is hindered by cloudiness and fog. At the same time the *CP* values at Belmar vary only between 0 and 5 mcal. cm^{-2} sec^{-1}.
The comparatively low *CP* values in winter in Belmar and Davos are noteworthy. In the latter case (cf. also Fig.81) weather conditions prevail with little wind and often much sunshine in the high Alps augmented by a considerable amount of ground-reflected radiation (snow). This demonstrates the mild bioclimatic winter conditions compared with the Scandinavian station Bergen (with mild sea air but strong winds and abundant cloudiness).
For the knowledge of the structure of polar climatic zones, the following example for the annual variation of the components contained in cooling power CP_{Fri} (temperature, wind force and global radiation) is presented, and expressively describes the fundamentals of the bioclimate of the South Pole. In Table XLII the monthly averages of temperature maxima and minima as well as of wind force (1957) are presented; furthermore in Table XLIII the daily amounts of global radiation $T(S + D)$ for the year 1958 according to DALRYMPLE (1966). This reflects the extreme bioclimatic conditions (caused also by the seasonal contrast of the incoming radiation conditions) very well. Thus during the southern summer the very high daily amounts of global radiation are predominant, when the continuous solar radiation has values which approach those for the arid zones. In this connection one must consider that the albedo of the South Pole's snow- and ice-covered ground in midsummer (1957 and 1958) was found at 88%. This reflection value would reduce the cooling power there, had it been recorded by the Davos frigorimeter, and have considerable weight. But the low air temperatures and the comparatively high mean wind speed considerably increase the cooling power. The albedo value indicated by Dalrymple largely coincides with the values found by HOINKES (1971) at the South Pole (mean values on overcast days 87.9%, on clear days 89.0%). The absolute range was between 93.4 and 84.3%.

Other methods for characterizing thermal sensations, in particular in connection with physiological measures for physical performance and indoor climatological observations

Already earlier the interactions of "sultriness and work" were pointed out. The determination of sultriness in its effects upon work performance, and related to air conditioning at working sites, is of particular interest in industrial medicine and hygiene. For instance, not only the t_e or *CP* conditions play a part, but also the adaptation to a given climate (acclimatization), the change to a new type of climate (adjustment), and work training. It is mainly the latter which determines the establishment of endurable limiting values. BRUENER (1959) made a survey of integral climate values which could be determined by calculations or nomographs. Besides the thermal evaluation factors such as those discussed in the preceding part, e.g. in form of "cooling power" and "equivalent temperature", also other complex elements must be considered in the evaluation of a thermal ambient climate. This was often done for evaluating working conditions under severe climatic conditions (e.g., in mining, steel industry, plantation work in the tropics, etc.). In this connection Bruener points out that, for example, at a dry working site normal work can be carried out at temperatures far above 28°C (previous upper limit in German mines around 1900) without reaching or even exceeding the limit of physiological possibilities. The situation changes when the ambient humidity at a working site reaches values which cause a sensation of intense sultriness. Fig.74 already has shown the gross energy turnover with regard to climatic situations (t, f and e) and with different work loads.

Even before PROETT's (1919) ideas HALDANE (1905) has pointed out that with the dry-bulb thermometer alone (i.e., by indication of dry tempeıature) one cannot obtain a measure for the suitability of a climate in relation to physical work loads. He advocated that readings of the wet-bulb thermometer (which as is known are proportional to equivalent temperature t_e) would be more suitable for this purpose. More than 20 years later YAGLOU (1927) suggested that for a more objective evaluation of a climatic situation dry-bulb temperature, wet-bulb temperature and wind speed should be used.

In a series of mass experiments YAGLOU and MILLER (1924) established which air temperature and air humidity (determined by means of dry-bulb temperature t and wet-bulb temperature t') is felt as equally comfortable as a state with water-vapour saturated air ($t_L = t'_L$). A certain thermal sensation in the comfortable range, for example, was attributed to an effective temperature $t_{effA} = 20°C$, if the majority of test persons agreed. This corresponds to a humidity-saturated air of 20°C. Fig.96 gives graphically the results of these American investigations, from which details may be read. It is a network of lines which contains the governing elements for characterizing a climatic *indoor situation* with lines of the same t_{effA}. The "comfort zone" resulting from the tests is shown as a closely hatched area between 17° and 21°C t_{effA} and relative humidities between 40 and 70% (later they were reduced to the range between 30 and 60%). For persons normally clothed and only doing light work the following sets of values of t_L and f (%) are valid for calm air at $t_{effA} = 19°C$:

$t_L = 20.3°C$ $f = 70\%$
$t_L = 21.2°C$ $f = 50\%$
$t_L = 22.3°C$ $f = 30\%$

In Fig.96B the behaviour of human rectal temperature (°C) in dependence upon effective temperature (t_{eff}) is referred to. The result is that at approximately 31°C t_{eff} body temperature begins to rise. At 35°C effective temperature the upper tolerance limit is reached (cf. descriptions by Bruener on p.158). Later YAGLOU (1927) included air displacements (wind speed) in these considerations and created nomograms as shown in Fig.97. They are suitable for indoor and outdoor air if the radiation influence can be neglected.

In order to take into account, within an indoor climate, the effective radiating heat, mainly of infrared origin, the so-called "globe thermometer" was used. VERNON (1930) introduced this apparatus (black dull copper sphere with Hg thermometer) which, by determining the difference between the indications of this instrument and that of a common Hg thermometer, establishes an "excess radiation temperature". (A disadvantage of this instrument is the great inertia with which it adjusts.) If in Yaglou's diagrams one sets instead of the dry-bulb temperature the so-called "globe temperature" (*GT*) one obtains the "corrected effective temperature" (VERNON and WARNER, 1932).

Besides this term Yaglou also introduced the "wet-bulb globe temperature" (*WBGT*), a complex parameter expressed by the equation:

$$WBGT = 0.7\,t'_L + 0.1\,t_L + 0.2\,GT\ (°C)$$

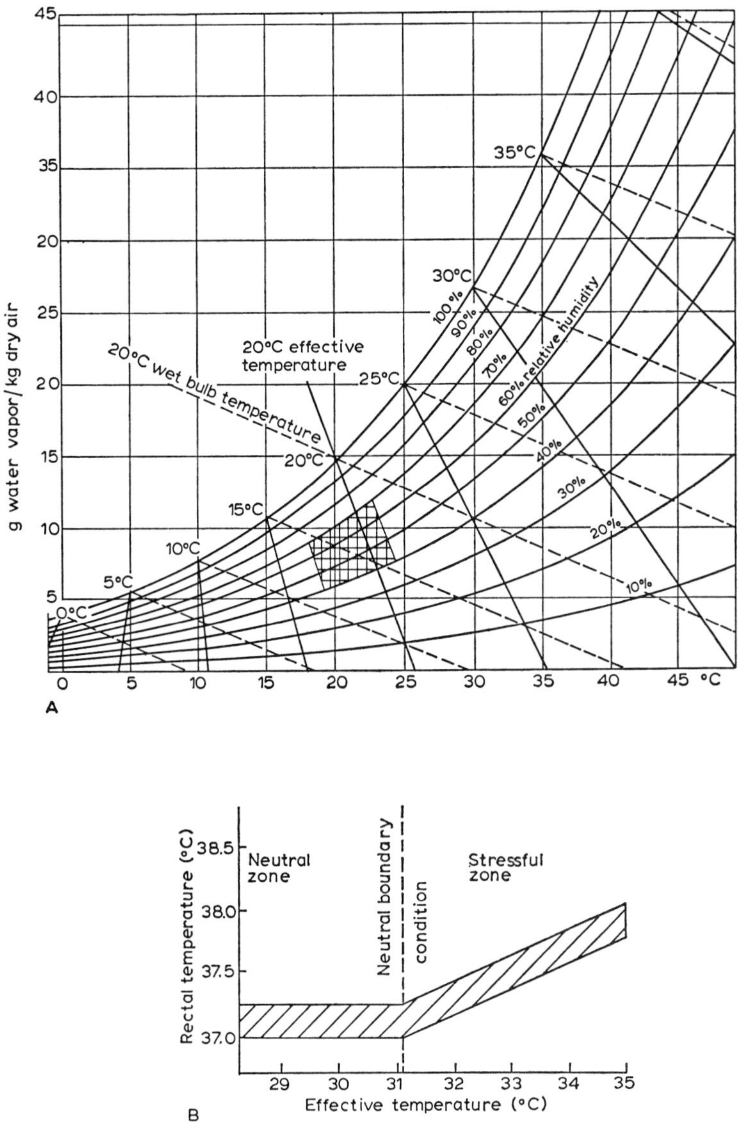

Fig.96. Psychrometric diagram (mixing ratio vs. dry-bulb temperature). Also indicated are relative humidity (curved lines), wet-bulb temperature (gently sloping lines), effective temperature (steeply sloping lines). Stippled area indicates an accepted area of indoor comfort for a majority of healthy persons in appropriate clothing (HOUGHTEN and YAGLOU, 1924). Relation of rectal temperature to effective temperature, in two subjects (after MACPHERSON, 1962) (from LANDSBERG, 1972).

BIDLOT and LEDENT (1947) defined the so-called "Belgian effective temperature" (t_{effB}) which is in use in Belgian mines. It is calculated according to the formula:

$$t_{\text{effB}} = 0.9\, t'_{\text{L}} + 0.1\, t_{\text{L}}\ (^{\circ}\text{C})$$

The definition, however, neglects variable wind speed and instead assumes a constant wind speed of 1 to 1.5 m/sec.

Much medical effort has been devoted to the elucidation of physiological reactions to a given climate. These have led to noteworthy successes and practical usefulness. CARLSON and BUETTNER (1957) have made a graphic compilation of the physiological requirements

as they appear through heat stress in connection with the main meteorological factors (temperature, humidity, air movement, heat radiation, etc.).
Chiefly pulse frequency, skin and rectal temperature, sweat production, etc. have been used as physiological measures for the performance capacity and performance limits in a defined climate. The increase that these factors were subjected to led to establishment of comparable climate combinations. Certain increases in the physiological parameters were taken as limits for the acceptability of a climatic situation or for its changes.
Physiological indications have to be taken into account for the problem of "acclimatization", i.e., man's adaptation to a climatically unfavourable working site, just as they are also considered for the determination of physiological measures, e.g., for underground work possibilities in mines under aggravated climatic conditions.
The following climatic ranges can indicate suitability of a climate by use of the American

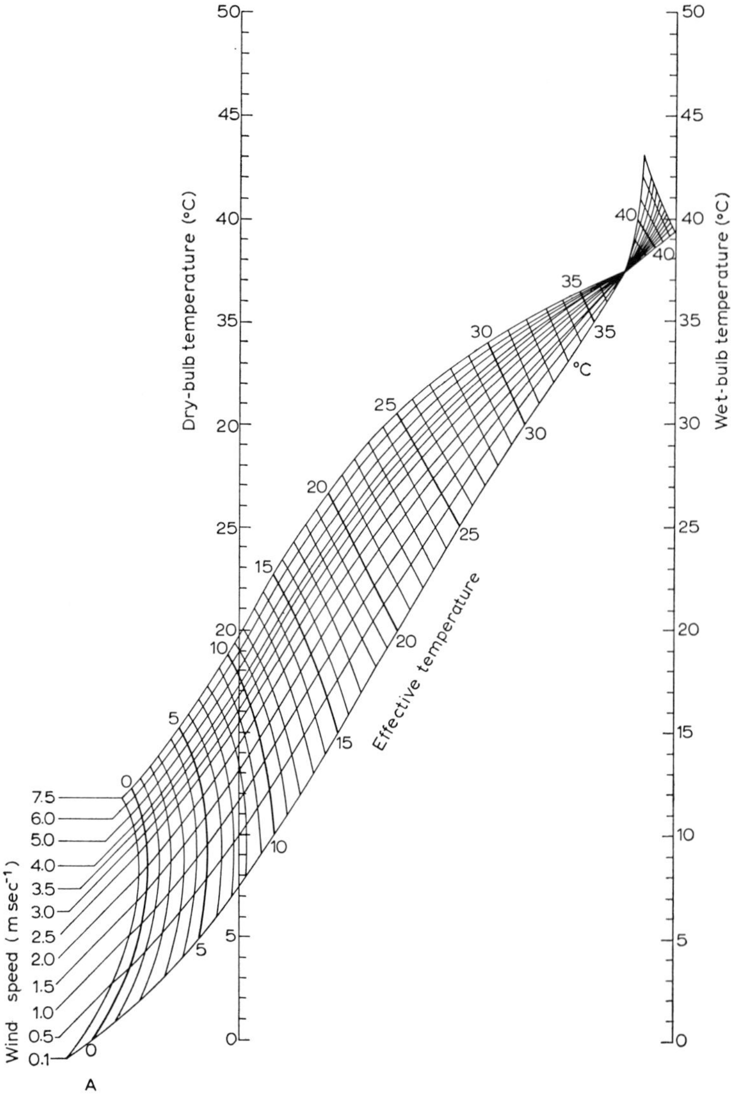

Fig.97A. Effective temperature for a clothed person in dependence upon dry-bulb temperature, wet-bulb temperature and wind speed. B. As for A, for an unclothed person. (YAGLOU, 1927.)

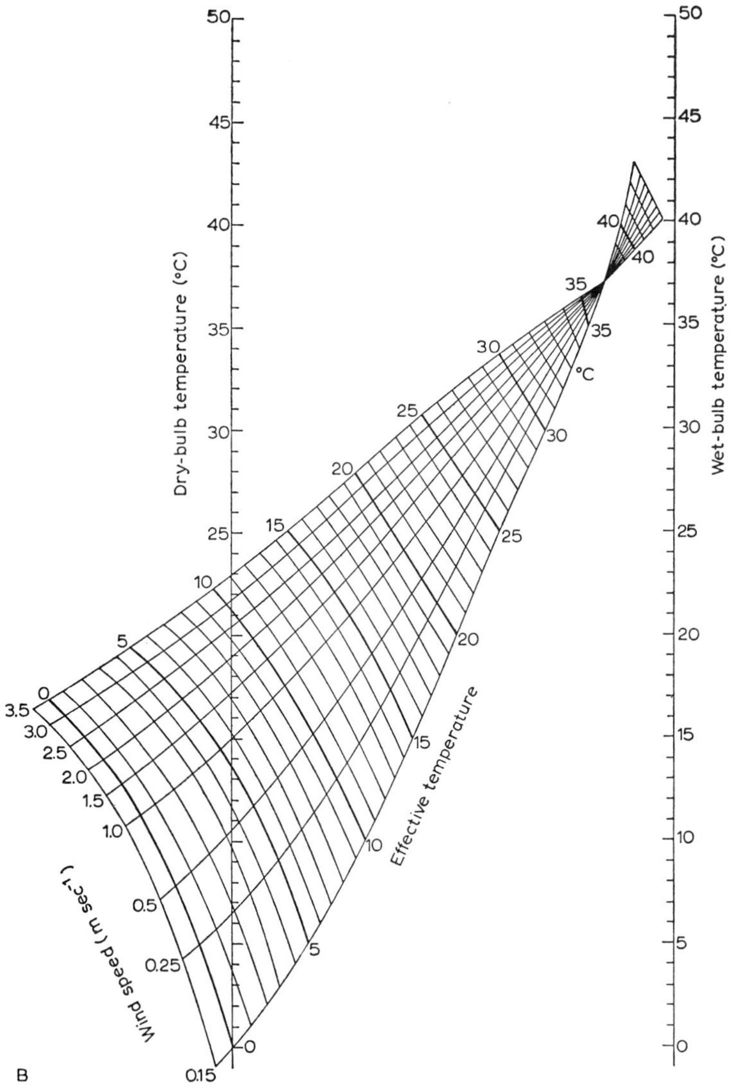

Fig.97B, legend see Fig.97A.

effective temperature (t_{effA}) (cf. Fig.96), according to investigations by BRUENER (1959) and others which correspond with results by other authors:

(*1*) Up to 28°C range of full working time and full performance.

(*2*) 28°–32°C range of 6-h working time with restricted performance.

(*3*) 32°–35°C range of short-period working time with greatly restricted performance.

(*4*) Above 35°C only exceptional work under close technical and medical control.

Fig.98 presents Bruener's version of work performance and performance capacity in relation to effective temperature (t_{eff}) and consideration of the acclimatization process. It can be seen that before acclimatization is completed already at t_{eff} values of 20°–22°C there is a notable decrease in performance which rises yet further above 24°C. When acclimatization is completed, the hatched curve section shows that only above 28°C is there a decrease in performance which rises still further above 32°C. In the range 34°–35°C work performance approaches the values "zero" (cf. Fig.96).

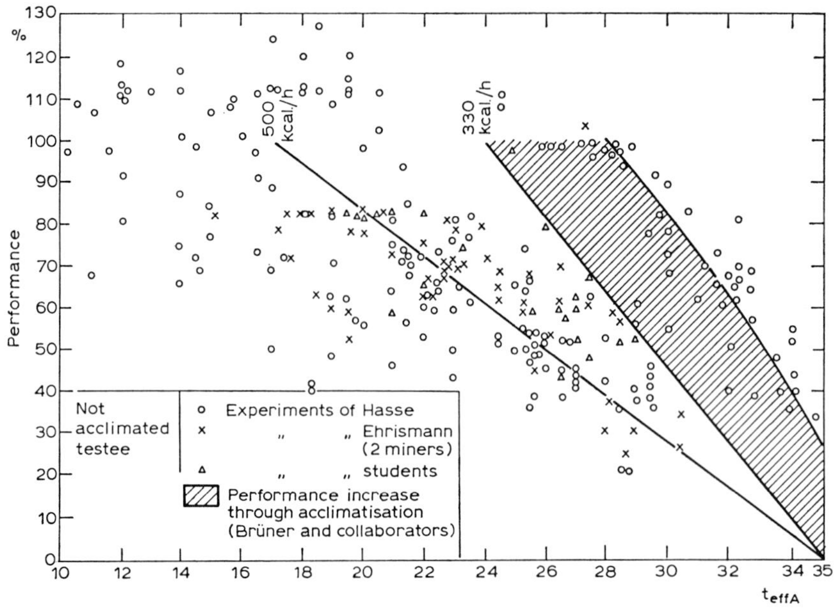

Fig.98. Performance capacity in dependence upon the effective temperature with consideration of acclimatization (BRUENER, 1959).

In Fig.99 the curves—only employing the individual elements t and $f(RH)$—convey an idea of the behaviour of the threshold ranges which are important for working persons regarding the effects of certain climatic indoor conditions (WEZLER and THAUER, 1948). American effective temperature (t_{effA}) is still in wide use. Most work physiologists con-

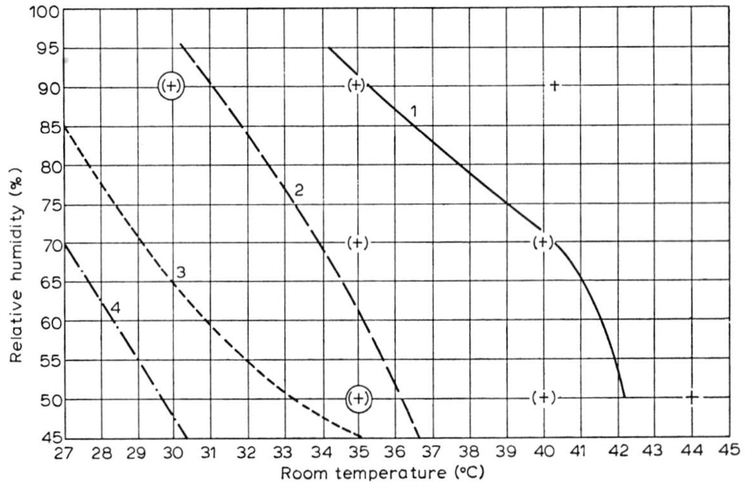

Fig.99. Supportability limits in dependence upon room temperature and relative humidity (WETZLER and THAUER, 1948).
Curve 1 = critical threshold, above which body temperature rises quickly. Organic stress and disturbed functions lead after a certain time to the imminent danger of collapse. Equivalent temperatures t_e belonging to curve *1* are in the range between 115–125°C.
Curve 2 = disturbance threshold, body temperature also rises noticeably, but later there is a new equilibrium with constant body temperature.
Curve 3 = above this curve approximately a 40% reduction of physical performance is observed.
Curve 4 = reaction threshold in which there is probably a certain incrimination of physiological functions and of full physical performance.

sider it the integral climatic parameter which is the simplest as correlate to physiological stress. (An Assmann aspiration psychrometer, a normal anemometer and possibly a precise wind meter (katathermometer) are sufficient.)
Within the framework of "Temperature and life" some remarks on the bioclimatic conditions in normal living and working quarters are in order, because man spends a great deal of time in "closed surroundings".
In this connection McConnell and Spiegelmann (1940) made a summer inquiry based on effective temperature with the help of 745 normally clothed test persons. Their results are given in Fig.100. It shows that there is no optimum indoors climate covering everybody. There will always be someone who in a generally "optimal climate" will be "too cool" or "too warm". Thus the creation of thermally comfortable indoor conditions for an assembly of people will always be a climate-technical and a psychological problem (Wenzel).

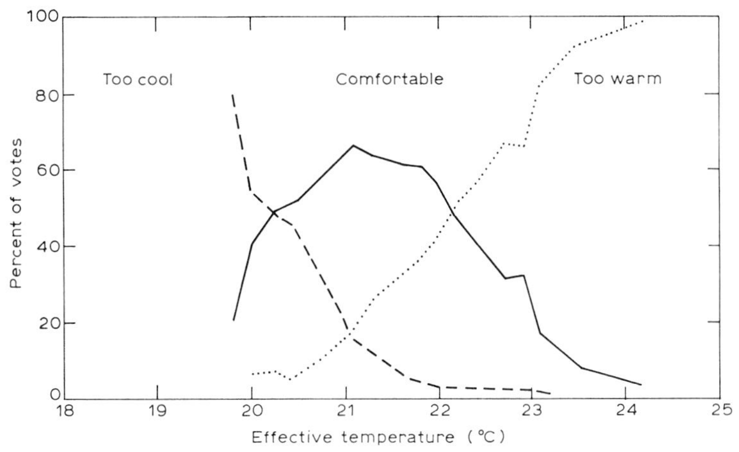

Fig.100. Sensitive evaluation of different climatic conditions by 745 normally clothed persons during office work, in dependence upon effective temperature (°C) (McConnell and Spiegelmann, 1940).

During the past years interest in radiative heating, especially from ceilings, has risen. Results have shown that the radiation of a 40°C warm ceiling at air temperatures of 10°–11°C is found pleasantly warm, at over 18°C air temperature, however, as increasingly "unpleasantly oppressing". Radiative temperature according to investigations by various authors should as a rule not exceed 37°C.
Beside the thermal indoor conditions the air exchange (between living or working quarters and outdoors) with normal closing surfaces (doors and windows) play a great hygienic role.
Fig.101 (Flach, 1954) provides further information on this specialized theme. In an experiment temperature and ventilation conditions were measured near an open and a closed double window (the weak external wind blew approximately parallel to the window front which led to suction effects). The upper section of Fig.101 gives the diurnal variation of air temperatures at 180 cm (head height of a standing person), at 120 cm (head height of a seated person) and at 60 cm (knee height of a seated person) above the floor as it appeared at 50 cm distance from a temporarily opened and closed double window. With open windows there are considerable temperature contrasts between head

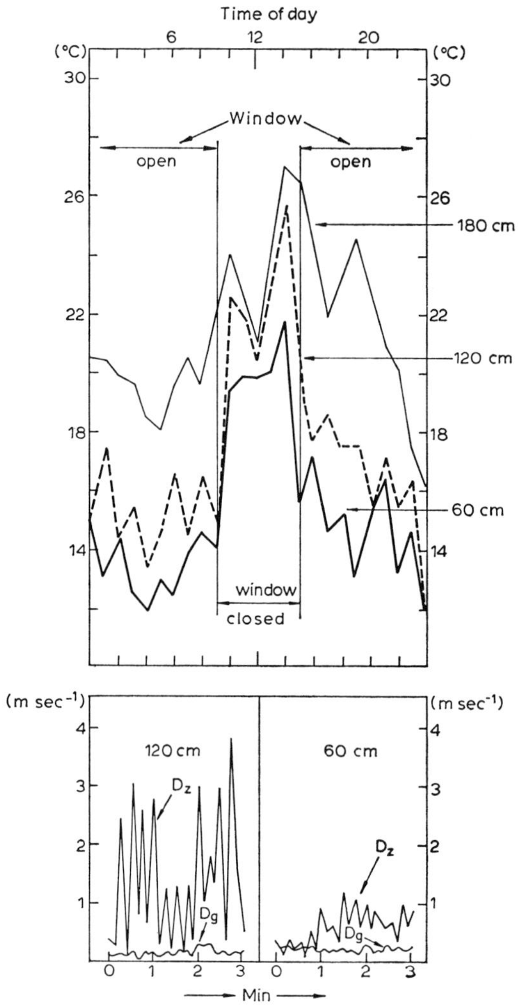

Fig.101. Air temperature and ventilation dimensions in an office (resistance electrical measurements) at different levels above the floor with opened and closed windows. For details cf. text. D_z = draught; D_g = passage. (FLACH, 1954.)

and knee height of a standing person. Thus at external temperatures from 8° to 11°C in an immediate effect of the external air the vertical temperature differences lie between 6° and 15°C. With closed windows the temperature differences are reduced to only a few degrees and thus underline the physiological monotony of climatic conditions indoors.

The lower section of Fig.101 shows that in a "draught" (D_z) there is a very intensive air exchange (determined by a sensitive hot wire anemometer with ventilation levels up to 4 m/sec. In the case of a "passage" (D_g) with closed windows the ventilation figures reach only a few centimetres per second.

Protection against unfavourable climatic effects especially in purely thermal ranges

Considerations of human behaviour in different temperature climates require some attention to the complex questions of protection against unfavourable climatic effects. The following survey shows the type of protective measures against stressful climatic effects (at work) in heat and cold, and reach a high degree of productivity in the work to be carried out.

The "natural behaviour" indicated in Table XLIV points to such measures carried out without any technical aid, purely by instinct and experience. The section "conscious adaptation of work technique to climate" deals mainly with "acclimatization" processes.

TABLE XLIV

THE PROBLEM OF PROTECTION AGAINST STRESSFUL CLIMATIC EFFECTS AT WORK*
(After MUELLER, 1961)

Type of protection	For heat	Evaluation	For cold	Evaluation
Natural behaviour	Use of natural protection (shade, cool in forests, caves, water)		Use of natural protection (sun, earth-warmed caves, windbreakers)	
	Reduced food consumption during greatest heat of day		Increased food consumption	
	Cooling food and drinks		Warming food and drinks	
	Economic work motions		Extensive work motions	
Conscious adaptation of work technique to climate	Acclimatization	+	Acclimatization	+
	Reduction of work load	+	Increase of work load	+
	Intervals to cool down	+	Intervals to warm up	+
	Reduction of working time	+	Reduction of working time	+
	Correct intake of salt and liquids	+	—	
Technical influences of working climate (macro- and micro-climate)	Cooling of air	++	Warming of air	++
	Air movements	+	Stopping of draughts	+
	Reduction of radiation temperature (screening)	++	Increase of radiation temperature	++
	Drying of air	+	Humidification of air	+
	Lighter clothing	+	Warmer clothing	++
	Taking off clothing	+	Heated clothing	++
	Special clothing (aired, reflecting)	+		

* For details see text.

The body reacts to the same climatic stimulus with a more intense counter-regulation after acclimatization. Technical measures with an immediate influence on climate are to be considered as specially effective. Alteration possibilities regarding external climatic phenomena are manifold (e.g., radiation and wind protection, specialized clothing, etc.). Therefore clothing, as is known by experience, as a personal protection against adverse environments plays a dominant role for "artificial air conditioning". (In MUELLER'S (1961) table the signs for evaluations of protection (+ or ++) mean that they are effective either in a limited or pronounced way.)

In the section "Light and life" it was mentioned that there are special modifications of the ambient climate of man with respect to the radiative permeability of clothing fabrics and reflection characteristics of the used materials. Each piece of clothing which prevents a direct irradiation of the skin has protective qualities.

Protection against cold by means of clothing is not predicated on the heat conductivity or the thickness of the fabric, but depends on the amount of air retained in the material. This factor is decisive.

The "insulation" of a certain piece of clothing is expressed in units of "clo": 1 clo is that amount of insulation which conveys a feeling of comfort in a room at $t = 22°C$ with air movements not more than 3 m/sec and relative humidity not over 50%. A man's ordinary business suit corresponds with the insulation of approximately 1 clo (Mueller). The best possible insulation is, according to the same author, in practical use 1.6 clo per cm thickness, but the theoretical ideal case is 1.85 clo. Incidentally the clo is an empirical unit which is valid only for clothing under quite specific conditions. The insulation of the same layer in flat condition is far larger than when curved.

Immobile "dead air" in clothing is decisive for a good heat insulation. Hence outdoor wind reduces the insulating quality. Table XLV demonstrates the reduction of heat insulation (in clo units) under increasing wind forces (m/min).

TABLE XLV

HEAT INSULATION IN DEPENDENCE UPON WIND FORCE
(After MUELLER, 1961)

Wind force ($m\ min^{-1}$)	Heat insulation (clo units)
6	1.00
11	0.80
15	0.70
23	0.60
37	0.50
64	0.40
130	0.30
320	0.20
610	0.15
1,370	0.10

At a wind force of approximately 10 m/sec only 1/7 of normal insulation is effective. In this connection the air permeability (in $m^3\ m^{-2}\ min^{-1}$), depending upon the material used in a piece of clothing, is of particular interest. Table XLVI lists the restricted air permeability of uniforms or special wind-proof fabrics. Normal uniforms with an insulating value of 3.5 clo (at 1 m/sec wind force) retain only 2.0 clo in the case of wind forces of 11 m/sec. Wet clothing, e.g. after avalanche accidents, retain very little heat insulating capacity. This is due to the fact that the so-called "dead air" is replaced by liquid water. Pelts of polar animals have the highest insulating values with 6–8 clo.

Dominating importance is attributed to the total heat insulation of clothing which is used at different external temperatures dependent on the degree of activity. Interest in clothing for the polar zones is steadily growing (cf. Fig.102).

Activities are indicated in MET values, i.e. in multiples of the basic metabolism of 50 kcal. $h^{-1}\ m^{-2}$ of a person sitting quietly. The MET values include the total metabolism (rest and work metabolism). The clo values in Fig.102 are valid for insulation by air and clothing. The amount of clothing required is calculated by subtracting the air insulation, as expressed in Table XLV, dependent on wind speed.

TABLE XLVI

AIR PERMEABILITY OF VARIOUS CLOTHING MATERIALS
(After MUELLER, 1961)

Material	Air permeability*
Nude (wind speed causing a pressure of 12.6 mm WS)	850
Mosquito net	400
Loosely woven shirt	120
Normal shirt	average 28 (8–70)
Tropical material	13–18
Uniform twill	4
Poplin	2
Windproof special fabric	0.3

* In m^3 m^{-2} at 12.6 mm WS pressure difference.

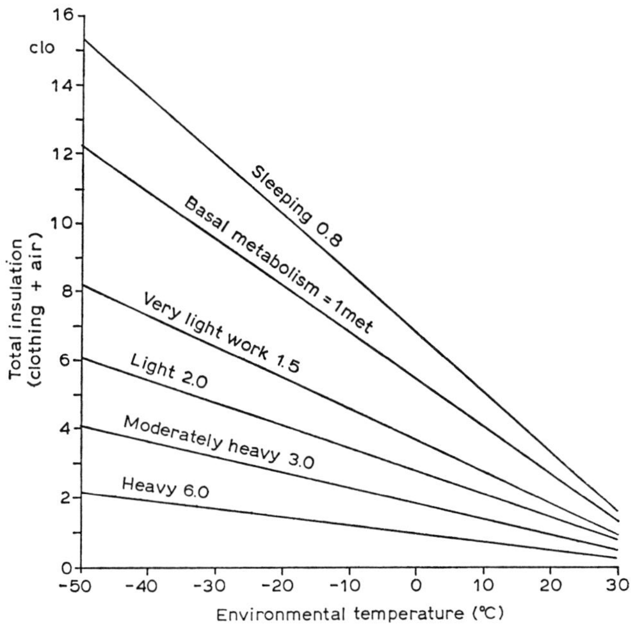

Fig.102. Necessary cold insulation in clothing (BURTON and EDHOLM, 1955).

Bioclimatological evaluation systems

Basic aspects of spatial evaluation and its scientific implication

A study on human bioclimatology such as the present one can only partially elucidate and classify the varied problems and tasks dealing with bioclimatological evaluations of specific sites. The abundance of results collected on this subject would make any such undertaking difficult. Several biological and physical disciplines contribute to this subject. It is as yet premature to develop guidelines for practical and objective use. Some aspects of bioclimatological local site evaluations have been dealt with in the main preceding sections.

Yet the entire problem of such site evaluations with all their specific aspects is of international importance. It should therefore be included in a survey of world climatology. Like any problem regarding systems relations, the search for values, e.g. in a recreational area, must be based on the actual atmospheric conditions. Environmental protection actions have shown clearly how difficult it is to obtain a reliable overview of present circumstances. With environmental protection as an important target of site evaluations, a brief outline will be given on the procedures.

Such a survey includes assessment of impact on environment by: (*a*) human living habits (refuse in cities and villages, garbage, detergent residues, etc.); (*b*) industry (chemical, physical, thermal); (*c*) agriculture (fertilizers, pesticides, etc.); (*d*) traffic (private automobile traffic, truck traffic, railways, aviation, etc.); (*e*) energy producers (power works, district heating plants, etc.).

Even this short list shows need for cooperation between biological, specially medical, physical, chemical, climatological, legal and other specialized disciplines in order to assess the present state of a region and participate in operational research.

Important background information has been collected by "International Biological Programme" (IBP) which emphasized the use and development of natural food sources and man's adaptation to natural and artificial habitats.

UNESCO's environmental protection research programme "Man and the Biosphere" (MAB) aims to project the effects of today's actions on the world of tomorrow. One purpose is improved husbanding of the biosphere's natural resources. Climatology, in particular bioclimatology, has important contributions to make to MAB's four main subjects, namely of natural environment, industrialized and urbanized regions, damages to environment and of the biosphere's pollution.

Urban planning and the establishment of recreational areas presuppose bioclimatological site evaluations.

Recreational site suitability depends upon the given natural and culture-geographical factors. Often the terms "recreation" and "tourism" are inaccurately used as synonyms. Change of locality is a dominant topic in recreation but it is a large and problematic sector (KLOEPPER, 1972).

Health resort climatology and its problems

The terms and institutional programmes, particularly of MAB, concern a special field of bioclimatology which comprises "health resort climatology" and specifically the problems of "health resort climate classification". This has, for almost five decades, been a lively topic in central Europe, especially in the Alpine countries (Switzerland, Austria, Germany).

In a critical summary (Problems of health resort climate research as a basis for climate therapy) KNOCH (1962) points out the problems which have to be dealt with not only by climatology but to an even greater extent by medicine. For many years numerous pertinent questions have been studied and have doubtlessly contributed notably to classifications (Amelung, Jungmann, Fleisch and Von Muralt, Hoegl, Moerikofer and others). But there is still much territory to be covered before a practical solution of these intricate problems has been found.

Noteworthy suggestions have been made by German, Austrian and Swiss specialists for health resort climate classifications (survey by SCHNELLE, 1961) in which mainly the comprehensive term "climatic health resort" (Flach, 1954) was used. Frequently static aspects were employed as basis for attempts to elucidate local geographic and average local climatic factors according to their endowment which "promote health" and "promote recreation" in the general sense of "stimulation" and "relaxation" as well as "stress". The *dynamic* structures of the "weather variability" encountered chiefly in the moderate zones were not always properly taken into account. Any recreation, balneological or climatic cure is subject to the greatly varied nuances of weather which can either inhibit or promote the cure. Therefore "fair weather" does not contain exclusively positive weather sensations, just as "bad weather" does not necessarily remain a purely dreary memory. In a climatic cure just as in any therapeutical programme the personal individual drive towards recovery is just as important as the instructions by the physician.

Often a lack of mutual consensus may be observed which leads to negative consequences instead of a beneficial application of the programme aspired to in practical health resort climatology.

In addition there are omissions in climatological health resort analyses regarding the weather changeableness and the climate therapeutical classification of the *seasonal* structure of regional and local climates. It is essential to define bioclimatologically the seasons and transition. SCHMIDT-KESSEN (1965/1973) emphasized that a local climate requires evaluation of each element apt to affect man.

The *Medical Problems of Vacations* (WACHSMUTH, 1973) has been a subject for considerable discussion. The report on the first ADAC Doctors Congress (1973) presents not only the problems of fatigue and recreation as physiological and psychological fields of interest but also locally conditioned influences in the use of climate and the subject of vacation illnesses and health risks.

For the climatological aspects DAMMANN (1968) and BECKER (1972) have carried out experiments to help medical interests and endeavours to evaluate climate and climatic classifications. These authors limited their studies to recreational areas in Western Germany. In his cartographical survey Dammann presents geographically and climatologically useful distinctions which he describes as "physiological". The idea is less the use of true physiological relations between locations, climate and man in a purely medical sense but rather the desired aids for climatic therapy as useful additions to geographical maps of spas, therapeutical and recreational sites.

In the map of the bioclimatic zones of Western Germany BECKER (1972) takes yet another step forwards and presents a differentiated explanation of the levels of bioclimatic "stress", "relaxation", and "stimulus". This climatic map is based on these three terms which are in use since several decades in climatic therapy. Becker employs the prevailing conditions of incoming radiation, cooling power and in particular of the daily fluctuations in the temperature–humidity environment, sultriness, etc. This bioclimate map places the main emphasis on the summer values of the thermally effective complex. The transitional seasons were not considered, although the stressful phases of winter in the lowlands (frequent inversion phases with fog and haze near the ground) have been partially included in the evaluation.

There is a need for bioclimatological illustration not only of a comparatively favourable season (summer) but also of the numerous change and transition phases within the seasonal structure of a climate foı use in the medical practice of climatic therapy. It is, for example, wrong if Alpine regions are classified generally as areas of "strong stimuli". During the frequent high pressure weather situations in winter only a mild stimulus may be present in valley sites as well as in the ski run areas if one excludes an *overexposure* to ultraviolet radiation. The levels of cooling power lie far below the values recorded in the often fog-covered lowlands (with a lack of incoming heat radiation and much stronger air turbulence in the annual variation).

The autumnal Alpine climate is noted for its abundance in sunshine at altitudes above 1,200 m, and for clear visibility and usually unhindered incoming solar radiation with little wind. In a bioclimatological classification these circumstances constitute a striking example of the heat-physiological condition of "relaxation". As an annual phase it occupies an essential place in the course of the Alpine seasons, not only from a thermal standpoint. In a meteoro-biological sense the autumnal lowland weather conditions with much haze and fog constitute—in comparison with Alpine conditions—sources for various disturbances (catarrhs, bronchitis, etc.).

With regard to the numerous endeavours for classifications one must agree with PFLEIDERER's (1961) statement that it has not yet been possible to develop a complete picture from the mosaic of individual climatological elements which would permit valid conclusions on the responses of the human organism.

Thus the bioclimatological site evaluation, much needed for pıactical purposes, requires still further research.

Methods of research in the thermo-hygric environment for bioclimatological requirements

Seasonal differences in thermo-hygric behaviour

In the absence of bioclimatological guidelines in the desired fashion only a few partial aspects can be briefly elucidated.

The basic behaviour of the thermo-hygric ambient climate can be discussed in terms of mean annual variations of a few complex bioclimatological elements but also some individual meteorological elements. They are founded on simple symmetry observations and their objective is to consider the genetics of the elements t, e and in particular of f as it affects aerosol quality and its changeability.

The opposing balance of the annual course of meteorological elements at summer solstice underlines the actual separation of two halves of the year with season-pathological significance. At the beginning of the second half of the year (end of June), first gradually and then with increasing effect the ergotropic branch of the vegetative influence of the atmosphere sets in and reaches its (central European) climax in August/September, and then decreases.

In contrast, the annual trophotropic phase appears chiefly in the first half of the year (from February on) and reaches its seasonal end in the second half of June.

Fig.103 represents the seasonal course of two well-known and previously discussed complex elements, namely of equivalent temperature t_e as a measure for the air's heat content, and of cooling power *CP* measured with the Davos frigorimeter as a measure for heat loss (for the example of a representative central European lowland station). Annual curves reflected at summer solstice reveal that the second half of the calendar

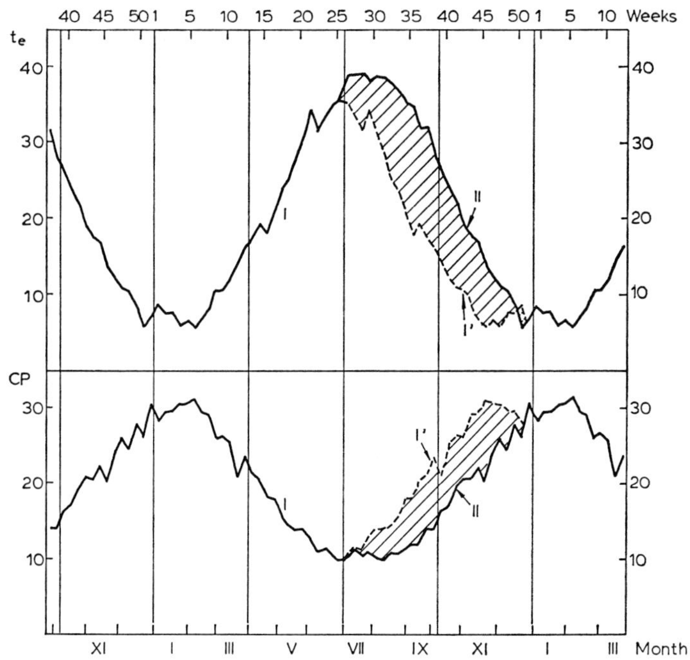

Fig.103. Mean annual course of equivalent temperature t_e (°C) and of cooling power *CP* (mcal. cm^{-2}sec^{-1}) in a representative example of a central European lowland station. *I* and *II* = first and second semi-annual course; *I'* = reflected elemental course of the first half of the year. Mean weekly values from 15-year period.

year—considered climato-physiologically—is in many ways the thermally favoured phase. In the case of equivalent temperature the phenomenon of a comparatively heightened heat content of the air during the second half of the year is determined by the participation of the basic elements air temperature and water vapour content. The behaviour of cooling power which compared with the first half of the year shows more favourable climate-physiological conditions in the second half, is determined by the increased air temperatures and the partially improved conditions for incoming radiation and also by the relatively low wind speeds in midsummer and fall.

Fig.104 deals separately with the individual elements *e* and *f* in the same way. Here too it may be recognized that in the second half of the calendar year the water vapour supply (*e*) and the saturation degree of the air with water vapour (*f*) as formation forces of natural aerosol show values several times higher than in the first half. The illustration of the differences reveals in the case of *e* a dominant climax during the period from August till September with a continuous change process. But the difference amounts of relative humidity exhibit periodically increasing or decreasing courses. They are divided into four periodically appearing excess phases, whereby the periods with relatively low difference amounts correspond with the average appearance of well-known central European anticyclonic singularities. On the other hand, the periods with the largest positive *f* deviations correspond with the stages of increased precipitation probability. These illustrations of the seasonal behaviour of *e* and *f* are a first indication of the seasonally bound influences of the annual weather course on atmospherically controlled aerosols and their changeability. As shown at an other point, these phenomena in the humidity behaviour of air have, mainly during the second half of the year, far-reaching meteorobiological significance for the chronopathology of infectious diseases.

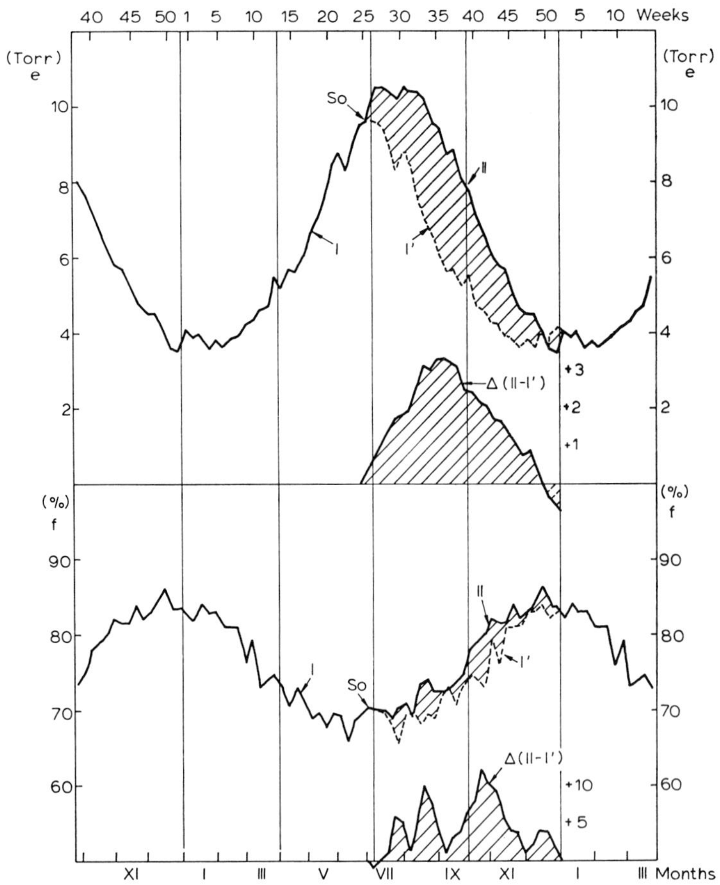

Fig.104. Mean annual course of water vapour content *e* (Torr) and of relative humidity *f* (%) as factors influencing the quality of atmospheric aerosol, for the example of a central European lowland station. *I* and *II* = first and second semi-annual course; *I'* = reflected elemental course during the first half of the year (*II* − *I'*) = difference between courses of *II* and *I'*. Mean weekly values from 25-year period.

Air mass climatology and its relevance to formation of water vapour aerosol

It is of special aerocolloidal interest to analyze the behaviour of the changing parameters Δe and Δf in connection with weather phenomena, in particular as related to air masses (according to Linke-Dinies) of different origins and qualities.

Fig.105 shows first the mean annual variation (1901–1940) of the relative frequency in the appearance of the main types of air masses and air mass groups present in central Europe (*E*). It shows the regional differences in the incoming flow of specifically differing air masses. The following points are of interest.

(*a*) Sharp decrease of polar maritime (*PM*) and polar (*P*) air masses at the transition from eastern (*e*) and northern (*n*) regions to western (*w*) and southwestern areas of central Europe.

(*b*) While western and northern regions of central Europe are supplied with approximately equal amounts of continental (*C*) air masses, the eastern *E* (*e*) region is definitely favoured in this respect.

(*c*) When comparing region *w* with *e* there is a striking reduction of maritime (*M*) and tropical maritime (*TM*) air masses. Already in northern central Europe there is a notable reduction in the frequency of the mentioned air mass groups.

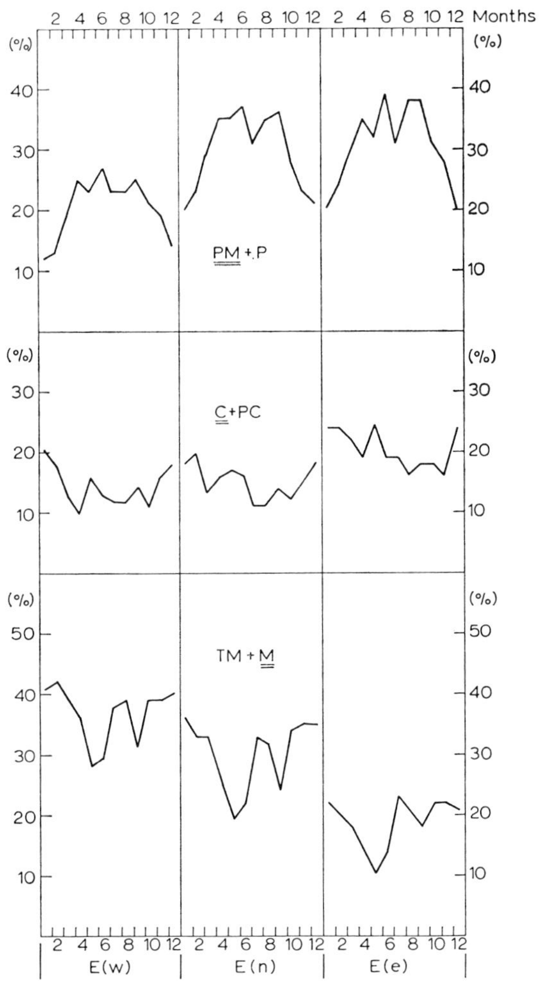

Fig.105. Relative frequency (%) of the main air masses and air mass groups (*PM* + *P*), (*TM* + *M*), (*C* + *CP*) in Central Europe (groupings of monthly mean values according to seasonal course, period 1901–1940). *P* = polar air (origin in Arctic areas); *PM* = maritime polar air (origin in northern Atlantic); *M* = maritime air (origin in Atlantic between approx. 45° and 55°N); *TM* = maritime tropical air (origin in Atlantic between 30° and 40°N); *C* = continental air (origin in Eurasian continent between 48° and 58°N); *PC* = inland polar air (origin in inner Eurasia and northern Fennoscandia). (Underlined air mass has natural climatological predominance). *E* (*w*) = western central Europe; *E* (*n*) = northern central Europe; *E* (*e*) = eastern central Europe.

The seasonal and regional distinction illustrated by these frequency statistics of air masses of different origin and of different basic aerosol qualities furnish *one* of the necessary diagnostic tools of a European aerosol climatology.

In addition to the preceding, Fig.106 provides a more detailed survey of the seasonal behaviour of the most important air mass groups and is also based on a 40-year observation period with statistics from Potsdam (northern Germany). In the air mass combinations in the group in question the underlined air mass is climatologically predominant.

The seasonal courses show marked annual variations in their mean behaviour. The winter extremes in their position near solstice exhibit a more definite character than their summer counterparts. The latter are temporally and numerically rather distorted caused

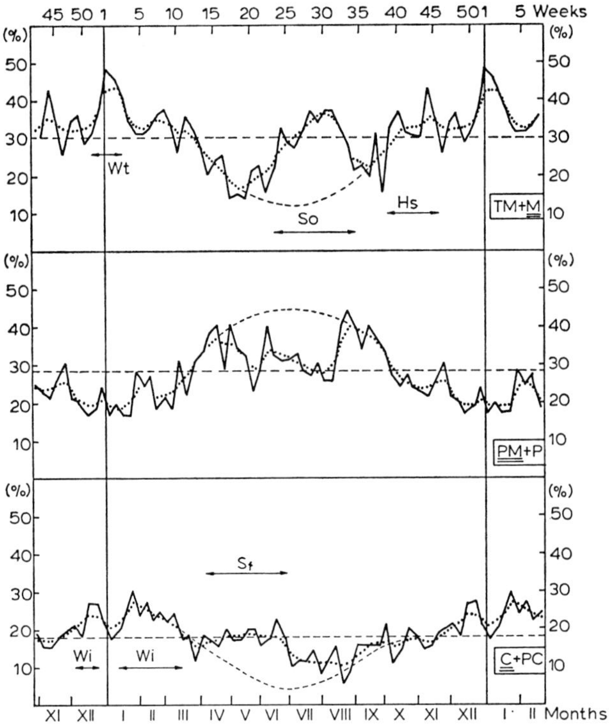

Fig.106. Mean annual course of air mass frequency in northern Germany (Potsdam, 1901–1940). Top: (TM + M) Center: (PM + M). Bottom: (C + PC) (description of air masses cf Fig.105.) Dotted line = even course; broken line = hypothetical course; *So* = European summer monsoon with prevailing maritime air masses; *Hs* = late fall with prevailing maritime and tropical-maritime air masses; *Wi* = winter phase with frequent presence of continental air masses; *Wt* = early winter thaw period; *Sf* = early summer phase with occasional presence of continental air masses (mainly alternating with maritime polar air masses).

by the heat contrasts between ocean and continent and consequent weather. One should note the midsummer peak in the appearance of maritime (*M*) air masses (cf. upper part of Fig.106), a phenomenon known as "European summer monsoon" (indicated in Fig.106 as *So*). After a short interruption there is again an increase in frequency in the last September decade of maritime air masses, often with tropical characteristics. This phenomenon (*Hs*) may last into October and accompanies the mild Indian summer days. In the sense of a land monsoon the *Wi* phase acts as a precursor (cf. lower part of the figure) to a relative predominance of continental cold air masses. This continues through spring and early summer and is then often a cause for dry and warm weather in those seasons (*Sf*).

Two peaks are found in the annual course of the air mass groups *PM* and *P* (cf. middle part of Fig.106) which appear most frequently during the summer semester. The first one is found in April and May. The regular weather phenomena of April showers and sudden cold phases in May in the lowlands spring are part of the phenomenon of this air mass circle. At the beginning of June the so-called "sheep cold" (cold spell) is caused by *PM* air masses. In late summer there is another peak in this air mass group which often brings the first surface frost of the season. During the central summer the incoming flow of maritime air masses from middle latitudes dominates the naturally bound central European aerosol phenomena, whereby instead of meridional air transport (*PM* + *P*) mainly zonal air transport (*TM* + *M*) takes place.

To one unfamiliar with meteorology the picture of warm and cold air frequency (upper and middle curve in Fig.106) seems paradoxical. A well-known feature is the appearance of the ominous Christmas thaw (*Wt*) which can even reach high Alpine regions and may cause unpleasant circumstances for the holidays. In winter the mild oceanic air masses dominate over the comparatively cold air masses which progress towards central Europe from NW and N. It must be taken into consideration, however, that at the same time continental polar air and continental air from medium latitudes balance the maritime air masses.
Such time-conditioned frequency analyses of air masses and air mass groups elucidate the specific absorption capacity of these complex elements also for the foreign aerosols which are carried into the atmosphere. They deal not only with stages of fresh air masses but also with those of aged air masses including transition phases.

Thermo-hygric environment as an aid for site evaluations based on aerosol, for the example of Alpine regions

While Figs.105 and 106 reflect aerosol conditions of a general and regional–climatological type, Fig.107 gives a climatological sketch of the vertical structure of aerosol's site-conditioned behaviour.
In the upper section of the figure the seasonal behaviour of relative humidity *f* is illustrated for an extra-Alpine lowland station (Basel at 318 m) and an Alpine valley site (Davos at 1,592 m), whereby the different behaviour at these two reference stations as well as that of Davos and Weissfluhjoch above Davos (2,660 m) is used. Interesting perspectives result for climatological aerosol characterizations in the Alpine and lowland regions.
The greatest differences in humidity relations of the lowlands and the valley site of the central Alps extend temporally from late summer till late winter (end of August till end of February). This period which is not identical with the calendar winter semester comprises the period of most frequent haze and fog formation in the lowlands and a corresponding development with peaks between the end of September and the beginning of January. In contrast, the smallest *f* differences in the example Basel–Davos are found between the middle of March and the middle of August. However, an almost reversed behaviour of *f* is encountered in altitude ranges between the high vally site and the crest site (Weissfluhjoch–Davos). The highest *f* differences in this case are observed in late winter and spring, the smallest in December at winter solstice.
In the case of the altitude range Weissfluhjoch–Davos between 1,600 and 2,600 m the difference maximum of *f* fixed from March till May is based on the circumstance that at that time the vertical air temperature gradient has its highest annual values. At the same time the temperature and wind speed rises at crest sites causes the snow to evaporate very quickly. During April the latter causes the reduction of snow height, a process completed at crest sites only during July. After that the *f* difference is rapidly reduced. In late fall and early winter (October till December) the *f* difference between high valley and mountain crest rapidly approaches a minimum. The higher *f* amounts in the valley cause a negative difference value because of great frequency of anticyclonic inversions. Special interest centers on the seasonal isopleth illustration given by LAUSCHER (1949) (Fig.108) of relative humidity in Eastern Alpine areas for altitude ranges from 0 to 3,000 m. The same situation is obvious as described for the individual example of

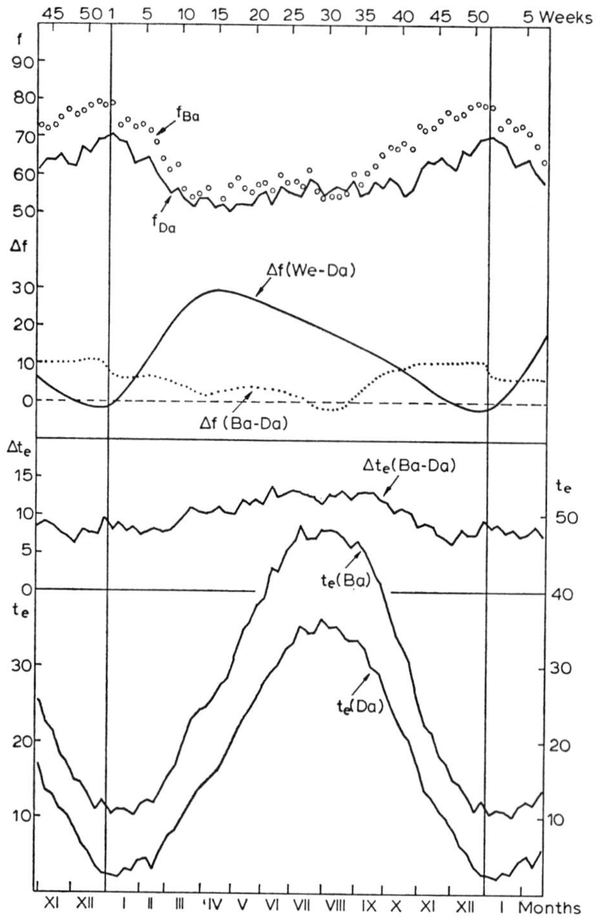

Fig.107. Upper part: mean seasonal behaviour of relative humidity f (%) in the lowlands (Basel, 318 m) and in Alpine regions (Davos, 1,592 m, and Weissfluhjoch above Davos, 2,667 m), and the pertinent value differences (latter equalized). *Ba* = Basel, *Da* = Davos, *We* = Weissfluhjoch.
Lower part: mean seasonal behaviour of equivalent temperature t_e (°C) in the lowlands (Basel) and in Alpine regions (Davos), based on mean values of midday observation point from a 10-year period.

Weissfluhjoch–Davos, namely, that at altitudes above 2,600 m compared with the lowlands there can also occur a humidity inversion in Alpine valleys. The isopleth illustration shows furthermore that at altitudes between 1,500 and 2,000 m there is a range in which only little changes in annual variation of f are observed. This is an analogous case to the vertical distribution of relative sunshine duration in the Alps. CONRAD (1939) provided the proof that in the Alpine altitude range between 1,800 and 2,500 m there are only very small annual amplitudes of relative sunshine duration.

Supplementary to the graph (f) in the upper section, the lower part of Fig.107 (also referring to Basel and Davos) illustrates the seasonal heat content of the air by means of the complex element equivalent temperature t_e. It reveals, in connection with relative humidity, the potential influence of the seasonal changes on the state of the aerosol. A remarkable fact is that precisely during the summer semester with the lowest differences in relative humidity at both reference points there are considerable differences in the air's heat content and hence also in the water vapour supply. This last fact is of great meteorobiological importance and also affects site evaluation.

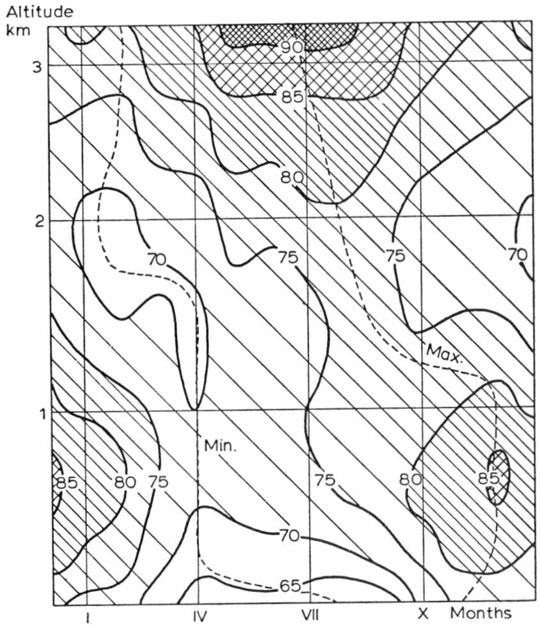

Fig.108. Mean annual course of relative humidity f (%) in the Alpine atmosphere of the eastern Alpine countries for altitudes up to 3,000 m (maxima and minima emphasized by broken curves), period 1938–1947 (LAUSCHER, 1949).

STEINHAUSER (1936) contributed to this specialized subject (Fig.109) by comparing the values from a 25-year observation period by frequency statistics of relative humidity f (example 14h00) on the Sonnblick (3,106 m) and at the base station Salzburg (430 m). In winter at crest sites a broad spread of the f groups may be observed, while summer conditions show a concentration in high f ranges. The cause in the first case is the relatively frequent anticyclonic situation with subsiding air, but in the second case much condensation occurs at these levels in Alpine regions. At the base station Salzburg the seasonal variation of the f groupings is reversed compared with the altitude regions. A strong tendency towards f values below 50% is particularly characteristic.

The characteristics of relative humidity f are not alone responsible in the important range of haze aerosol formation. Temperature and water vapour supply are of equal importance (as illustrated by Table XLVII). The table presents an example of the percentage distribution of air temperature t and vapour pressure e for the sultry t_e group 45°–49°C (central European acclimatization conditions). For the case of air temperature it is particularly interesting to see that in the Alpine valley site the t_e groups 24°–28°C are relatively more represented than in the lowland climates to the north and south of the Alpine crest. In the lower regions the peak is found in the t range between 22° and 26°C. The reason for this is that in Alpine sites (valley) the air is heated up more than would be possible in the mostly hazy lowlands because of its relatively high transparency and back radiation from the slopes.

Regarding vapour pressure behaviour and water vapour supply the illustration shows that in the lowlands the highest percentages are observed at higher vapour pressure values than in the high mountain areas. In the mentioned example the peaks are at 12–16 Torr (lowland) and 8–12 Torr (mountains). At the upper level the e range of 8–10 Torr represents nearly 60% of the cases. With reference to thermal physiology this means that

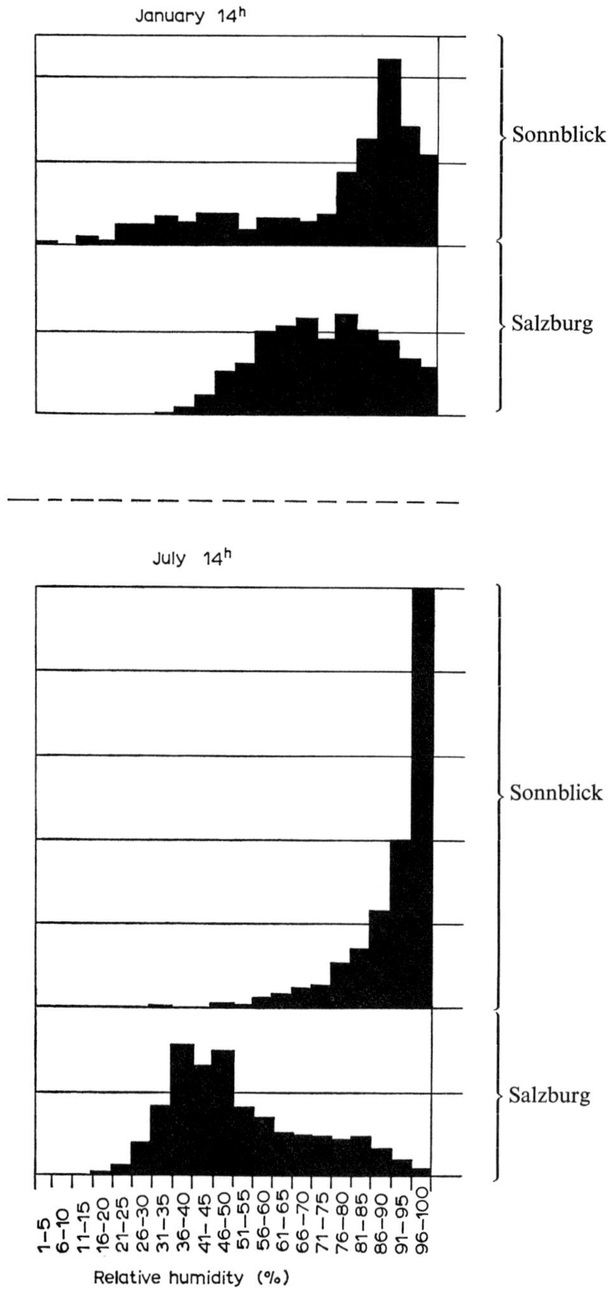

Fig.109. Mean frequency distribution of relative humidity in Salzburg (430 m) and on the Hohen Sonnblick (3,106 m) in Austria. Based on mean values of midday observation point (January and July, resp.) from a 25-year period. (STEINHAUSER, 1936.)

in Alpine sites, under the mentioned conditions, there are ideal comfort conditions and even the incoming heat radiation causes no stress. In the lowlands, however, this state (t_e = 45°–49°C) creates thermal conditions which are close to, or even above, the sultriness limit.

Such frequency statistics are significant for the introduction of meteorobiological considerations into bioclimatological site evaluations.

TABLE XLVII

MEAN PERCENTUAL DISTRIBUTION OF AIR TEMPERATURE t (°C) AND VAPOUR PRESSURE e (Torr) IN THE LOWLANDS (BASEL) AND ALPINE REGIONS (DAVOS) FOR THE EQUIVALENT TEMPERATURE GROUP 45–49°C BASED ON MEAN VALUES OF MIDDAY OBSERVATION POINT FROM A 10-YEAR PERIOD

t =	18–20	20–22	22–24	24–26	26–28	28–30	30–32	32–34	34–36
Temperature distribution:									
Basel (318 m)	8.5	20.8	34.9	21.2	7.4	4.1	0.7	—	—
Davos (1,592 m)	12.6	12.5	24.3	28.1	9.4	3.1	—	—	—
Locarno (379 m)	6.6	20.7	39.7	21.8	7.5	1.7	—	—	—

e =	6–8	8–10	10–12	12–14	14–16	16–18	18–20
Vapour pressure distribution:							
Basel (318 m)	0.7	4.8	44.6	43.5	6.3	—	—
Davos (1,592 m)	9.4	59.4	31.3	—	—	—	—
Locarno (379 m)	—	5.3	45.7	45.7	3.6	—	—

Further aspects on the application of aerosol climatology based on thermo-hygric factors

The vertical behaviour of relative humidity in the Alpine region is important. This climatic feature was pointed out by LAUSCHER (1949) and GEIGER (1961). Table XLVIII presents the mean changes in climate conditions with altitude (eastern Alps) and the behaviour of relative humidity (column 16). It may be concluded that between 1,000 and 2,000 m f is noticeably reduced. FLACH (1969) made similar observations for the northern side of the Swiss Alps. Furthermore, Table XLVIII provides some orientation for other bioclimatologically important elements which could supplement vapour pressure and equivalent temperature. The table also contains valuable indications on vertical behaviour of heat radiation (global radiation $T = S + D$), on annual precipitation, frequency of snowfall and snow height as well as on soil characteristics. All these individual elements provide valuable information on aerosol influences.

Bioclimatological evaluations of mountain areas, in particular of Alpine regions, despite their numerous specific laws governing climatic structures there, show many fine shadings of a regional or local nature. These always necessitate a detailed site evaluation.

The use of relative humidity in climatological studies of aerosol behaviour necessitates refined observing methods. Agricultural meteorology has already made valuable contributions to this topic (WMO, Technical Note No.21, 1958). Further important information on relative humidity f has also been supplied by the First International Conference on Humidity and Moisture (Washington, May 1963).

Mean climato-physiological values and especially the meteorobiological state of a region or specific site is strongly affected by the average behaviour of aerosols. Therefore it is quite useful in bioclimatological analyses to include, besides the seasonal and time-of-day dependence of the saturation degree and water vapour supply, those factors which represent natural topographic elements affecting weather. This concerns chiefly humidity sources and sink areas. Humidity sources are: mountain flanks, frequently acting as windward sides; coastal areas, interior lakes and river areas; forests and their surroundings; sites with permanently damp soil (swamp and marshes, etc.).

Humidity sink areas are: deserts; mountain flanks frequently appearing as foehn zones; interior countrysides with predominantly dry soil.

TABLE XLVIII

MEAN VERTICAL BEHAVIOUR OF THE MAIN CLIMATE ELEMENTS IN THE LOWLANDS AND MOUNTAIN AREAS OF THE EASTERN ALPS
(After GEIGER, 1961)

Altitude (m)	Mean daily amounts of global radiation (cal. $cm^{-2}d^{-1}$)				Mean air temperature (°C)				Annual number of			
	cloudless		overcast		Jan.	July	year	annual fluctuation	summer days	frost-free days	frost change days	frost days
	June	Dec.	June	Dec.								
(1)	(2)	(3)	(4)	(5)	(6)	(7)	(8)	(9)	(10)	(11)	(12)	(13)
200	691	130	155	30	−1.4	19.5	9.0	20.9	48	272	67	93
400	708	136	168	32	−2.5	18.3	8.0	20.8	42	267	97	98
600	723	141	180	34	−3.5	17.1	7.1	20.6	37	250	78	115
800	735	146	192	36	−3.9	16.0	6.4	19.9	31	234	91	131
1,000	747	150	205	38	−3.9	14.8	5.7	18.7	15	226	86	139
1,200	759	154	220	40	−3.9	13.6	4.9	17.5	11	218	84	147
1,400	771	157	236	43	−4.1	12.4	4.0	16.5	7	211	81	154
1,600	782	160	253	47	−4.9	11.2	2.8	16.1	4	203	78	162
1,800	791	163	272	50	−6.1	9.9	1.6	16.0	2	190	76	175
2,000	799	166	293	54	−7.1	8.7	0.4	15.8	0	178	73	187
2,200	807	168	314	58	−8.2	7.2	−0.8	15.4	0	163	71	202
2,400	814	169	336	62	−9.2	5.9	−2.0	15.1	0	146	68	219
2,600	821	170	358	66	−10.3	4.6	−3.3	14.9	0	125	66	240
2,800	828	171	380	70	−11.3	3.2	−4.5	14.5	0	101	64	264
3,000	834	171	403	75	−12.4	1.8	−5.7	14.2	0	71	62	294

Altitude (m)	Annual number of days with		Relative air humidity (%)	Annual precipitation (mm)	Relative snow frequency (%)		Number of days with snowfall	Mean yield (cm d^{-1})	Sum of new snow heights (cm)	Maximum snow height	
	dry ground	snow cover			summer	winter				height (cm)	date of appearance
(1)	(14)	(15)	(16)	(17)	(18)	(19)	(20)	(21)	(22)	(23)	(24)
200	187	38	71	615	0	49	27	4.6	51	20	Jan. 18
400	173	55	74	750	0	61	32	5.2	116	31	Jan. 23
600	160	81	77	885	0	70	38	5.8	182	51	Jan. 28
800	147	109	78	1025	0	79	45	6.4	247	73	Feb. 3
1,000	133	127	76	1160	0	85	53	7.0	313	93	Feb. 11
1,200	120	138	74	1295	1	90	62	7.6	379	100	Feb. 14
1,400	107	152	73	1430	2	93	73	8.2	444	120	Feb. 21
1,600	93	169	73	1570	5	96	85	8.8	510	142	Mar. 3
1,800	80	189	74	1700	10	97	98	9.4	575	168	Mar. 14
2,000	67	212	74	1835	16	98	113	10.0	641	199	Mar. 26
2,200	53	239	75	1970	24	99	128	—	707	242	Apr. 8
2,400	40	270	78	—	34	100	143	—	—	296	Apr. 20
2,600	27	301	80	—	44	100	158	—	—	366	May 3
2,800	13	332	82	—	55	100	173	—	—	446	May 15
3,000	0	354	84	—	67	100	188	—	—	545	May 29

Meteorological–climatological analyses of the site-conditioned control factors of water vapour aerosol are also part of regional climatology and particularly of microclimatology and of an element aerosol climatology. GEIGER's (1961) survey demonstrates the variety of aspects of water vapour supply and disappearance.

Bioclimatology oriented towards water vapour aerosol is greatly aided by use of the results from topogeographic studies (TROLL, 1967).
Causal relations of climatological situations must also take into account the effects of "climate modification" (FLOHN, 1973). In this sense a true bioclimatological example is a climate changed by "urbanization" (LANDSBERG, 1973).
Widespread bioclimatological site evaluation is fully justified even outside of purely climatological studies by the general requirements of urban planning, by the guidelines of environmental protection striven for and by the expansion of tourism. This aspect of bioclimatology is particularly worthy of promotion. One must not, however, stop at the common evaluation methods but try and create a detailed and complex evaluation system. Fundamental knowledge of meteoro-biological phenomena and their climato-pathological interpretation is essential. This objective, however, can only be pursued on an international basis, such as the MAB programmes.

Acknowledgement

I owe thanks to Miss Helen Staeubli for her help in translating this chapter into English, a difficult task in view of the subject matter.

References

ADOLPH, E. F., 1947. *Physiology of Man in the Desert.* Interscience, New York, N.Y.,

ADOLPH, E. F. and MOLNAR, G. W., 1946. In: *Am. J. Physiol.*, 146: 507.

AICHINGER, F. and MUELLER, I. L., 1953. In: H. A. GOTTRON und W. SCHÖNFELD, *Dermatologie und Venerologie*, Bd. I, 2. Thieme, Stuttgart, 1962.

ALBRECHT, E. and ALBRECHT, H., 1967. Metabolismus unter O_2-Mangel im Höhenklima. *Pflügers Arch. Ges. Physiol.*, 293: 1–8.

AMBROSETTI, FL., 1965. Die Ueberwärmungsgrösse auf die Alpensüdseite. *Carinthia*, II, 24.S.-Heft: 145–146.

AMELUNG, W., 1952. Einführung in die Balneologie und medizinische Klimatologie. In: H. VOGT und W. AMELUNG (Herausgeber), *Bäder- und Klimaheilkunde*, 2.Aufl. Springer, Berlin.

AMELUNG, W., 1962. Wesen der Klimabehandlung (Verfahren und Dosierung). In: W. AMELUNG und A. EVERS (Herausgeber), *Handbuch der Bäder- und Klimaheilkunde*. Schattauer, Stuttgart, pp. 701–717.

AMELUNG, W., JUNGMANN, H. und SCHULTZE, E.-G., 1962. Klimakuren. In: W. AMELUNG und A. EVERS (Herausgeber), *Handbuch der Bäder- und Klimaheilkunde*. Schattauer, Stuttgart, pp.718–729.

ASCHOFF, J., 1948. Hundert Jahre Homoiothermie. *Naturwissenschaften*, 35: 235.

ASCHOFF, J. und WEVER, R., 1958. Kern und Schale im Wärmehaushalt des Menschen. *Naturwissenschaften*, 45: 477.

BALTRUSCH, M., 1974. Dreidimensionale Analyse des CO_2-Konzentrationsfeldes über einer Flächenquelle. *Arch. Meteorol. Geophys. Bioklimatol., Ser. B*, 22: 73–108.

BAUMGARTNER, A., 1971. Wald als Austauschfaktor in der Grenzschicht Erde/Atmosphäre. *Forstwiss. Zentralbl.*, 90: 174–182.

BAUR, F. und PHILIPPS, H., 1935. Ausstrahlung, Gegenstrahlung und meridionaler Wärmetransport bei normaler Solarkonstante. *Gerlands Beitr. Geophys.*, 45, 82.

BECKER, F., 1972. Bioklimatische Reizstufen für eine Raumbeurteilung zur Erholung, mit einer Karte "Die bioklimatischen Zonen in der BRD". *Veröff. Akad. Raumforsch. Landesplanung*, 76(3): 45–61.

BENER, P., 1963. Der Einfluss der Bewölkung auf die Himmelsstrahlung. *Arch. Meteorol. Geophys. Bioklimatol., Ser. B*, 12: 442–357.

BENER, P., 1964. Tages- und Jahresgang der spektralen Intensität der ultravioletten Global- und Himmelsstrahlung bei wolkenfreiem Himmel in Davos (1,590 m NN). *Strahlentherapie*, 123: 306–316.

BENER, P. 1964. Investigation of the influence of clouds on ultraviolet sky radiation. *Contr. AF 61(052)-618. Tech. Note, No. 3 (Davos).*

BERG, H., 1961. Richtlinien für eine an der Biometeorologie des Menschen ausgerichtete Klimaklassifikation. *Arch. Meteorol. Geophys. Bioklimatol., Ser. B,* 10: 1–10.

BERNHARDT, O., 1933. Heliotherapie der Tuberkulose. *Strahlentherapie,* 48, 60 pp.

BIDER, M. und VERZÁR, F., 1957. Mehrjährige Registrierungen der Zahl der Kondensationskerne in St. Moritz. *Geofis. Pura Appl.,* 36: 110–117.

BIDLOT, R. and LEDENT, P., 1947. Que savons-nous des limites de température humainement supportables? *Commun. Inst. Hyg. Mines,* 28: 14.

BRADTKE, F. und LIESE, W., 1952. *Hilfsbuch für Raum- und aussenklimatische Messungen,* 2nd ed. Springer, Berlin, 108 pp.

BRUENER, H., 1959. Arbeitsmöglichkeiten unter Tage bei erschwerten klimatischen Bedingungen. *Int. Z. Angew. Physiol. einschl. Arbeitsphysiol.,* 18: 31–61.

BRUENER, H., 1961. Luftdruck und Luftzusammensetzung. In: G. Lehmann (Herausgeber), *Handbuch der gesamten Arbeitsmedizin, Vol. I. Arbeitsphysiologie.* Urban und Schwarzenberg, München, pp.654–679.

BUETTNER, K., 1938. Physikalische Bioklimatologie. In: *Probleme der kosmischen Physik,* Bd. 18. Akad. Verlagsges., Leipzig, 155 pp.

BUETTNER, K. J. K., 1951. Physical aspects of human bioclimatology. In: *Compendium of Meteorology.* Boston, pp.1112–1125.

BULLRICH, K., 1948. Die Leuchtdichte des Himmels und die Globalbeleuchtungsstärke während der Dämmerung und in der Nacht. *Ber. Dtsch. Wetterd.,* 4, 27 pp.

BULLRICH, K., 1964. Scattered radiation in the atmosphere and the natural aerosol. *Adv. Geophys.,* 10: 99–260.

BURTON, A. C. and EDHOLM, D. G., 1955. *Man in a Cold Environment.* London.

CARLSON, L. D. and BUETTNER, K., 1957. Thermal stress and physiological strain. *Fed. Proc.,* 16: 609–613.

CASTENS, G., 1925. Ueber Tropenklimatologie, Tropenhygiene und den Lettow-Feldzug. *Ann. Hydrol.,* 53: 177.

COBLENTZ, W. W. and STAIR, J., 1934. In: *J. Res. Nat. Bur. Stand.,* p.13.

CONRAD, V., 1939. Die Komponenten der Jahresschwankung der Sonnenscheindauer. *Helv. Phys. Acta,* 12: 38.

CORDES, H., 1968. Ueber das Klima von Mexiko. *Meteorol. Rundsch.,* 21: 1–2.

COURT, A., 1948. Wind chill. *Bull. Am. Meteorol. Soc.,* 29: 487–493.

COURVOISIER, P., 1951. Ueber das Eindringen von Schwankungen der meteorologischen Elemente in Gebäude. Zur Theorie der Wetterfühligkeit. *Arch. Meteorol. Geophys. Bioklimatol., Ser. B,* 2: 161–173.

COURVOISIER, P., 1957. Luftdruckschwankungen und Wetterfühligkeit. *Arch. Geophys. Bioklimatol., B,* 1: 115.

COURVOISIER, P. und WIERZEJEWSKI, H., 1954. Das Kugelpyranometer Bellani. *Arch. Meteorol. Geophys. Bioklimatol., Ser. B,* 5: 413.

CURRY, M., 1946. *Bioklimatik,* 2 Bände. Riederau (Ammersee), Amer. Bioclimatol. Res. Inst.

DALRYMPLE, P. C., 1966. A physical climatology of the Antarctic Plateau. In: M. J. RUBIN (Editor), *Studies in Antarctic Meteorology.* 9: 195–231.

DAMMANN, W., 1951. Klima und Energieentfaltung der Völker. *Med.-Meteorol. Hefte,* 6: 38–48.

DAMMANN, W., 1964. Die Schwüle als Klimafaktor. *Ber. Deutsch. Landeskd.,* 32: 100–114.

DAMMANN, W., 1968. Physiologische Klimakarte. *Stat. Atlas Oeffent. Gesundheitswes. BRD,* 3: 23–25.

DAMMANN, W., 1969. Klimagliederung in der Bundesrepublik. *Dtsch. Bäderkalender,* 1969, pp.50–56.

DAVIES, R., 1969. Aerobiology and the relief of asthma in an Alpine valley. *Acta Allergol.,* 24: 377–395.

DE QUERVAIN, A., 1912. Merkwürdiger Himmelsanblick im Innern Grönlands im Sommer 1912. *Meteorol. Z.,* 29: 587.

DE RUDDER, B., 1938. *Grundriss einer Meteorobiologie des Menschen,* 2.Aufl. Springer, Berlin, 234 pp.

DE RUDDER, B., 1952. *Grundriss einer Meteorobiologie des Menschen,* 3rd ed. Springer, Berlin, 303 pp.

DE RUDDER, B., 1960. Wetter, Jahreszeit und Klima als pathogenetische Faktoren. In: *Handbuch der Allgemeine Pathologie, Umwelt I, Strahlung und Wetter.* Springer, Berlin, pp.370–390.

DIETERICHS, H., 1958. Dauer und Häufigkeit schwüler Stunden in San Salvador (C.A.). *Arch. Meteorol. Geophys. Bioklimatol., Ser. B,* 8: 369–377.

DIRMHIRN, I., 1964. *Das Strahlungsfeld im Lebensraum.* Akad. Verl. Ges., Frankfurt/Main, 426 pp.

Dorno, C., 1911. *Studie über Licht und Luft des Hochgebirges.* Vieweg, Braunschweig, 153 pp.

Dorno, C., 1912. Ueber den Einfluss der gegenwärtigen atmosphärisch-optischen Störung auf die Strahlungsintensitäten der Sonne und des Himmels, sowie auf die luftelektrischen Elemente. *Meteorol. Z.* 29: 580–584.

Dorno, C., 1919. Himmelhelligkeit, Himmelpolarisation und Sonnenintensität in Davos 1911 bis 1918. *Veröffentl. Preuss. Meteorol. Inst., Abh. VI,* 303 pp.

Dubief, J., 1971. Die Sahara, eine Klima-Wüste. In: H. Schiffers (Herausgeber), *Die Sahara und ihre Randgebiete,* I. Weltforum-Verlag München, pp.227–348.

Du Bois, E. F., 1951. *Western J. Surg.,* 59: 476. Cited by H. Hensel, Mensch und warmblütige Tiere. In: H. Precht, J. Christophersen und H. Hensel, *Temperatur und Leben.* Springer, Berlin, 1955.

Edgar, J. L. and Paneth, F. A., 1941. The separation of ozone from other gases. *J. Chem. Soc.,* 511.

Eiden, R. and Eschelbach, G., 1973. Das atmosphärische Aerosol und seine Bedeutung für den Energiehaushalt der Atmosphäre. *Z. Geophys.,* 39: 189–228.

Ellenberg, H., 1972. Belastung und Belastbarkeit von Oekosystemen. *Tagungsber. Ges. Oekologie,* 19–26.

Flach, E., 1939. Ergebnisse von Kern- und Staubuntersuchungen im westsächsischen Mittelgebirge. In: H. Burckhardt und H. Flohn (Herausgeber), *Die atmosphärischen Kondensationskerne.* Heft 3 der Abh. a.d. Gebiet der Bäder- und Klimaheilkunde, Springer, Berlin, pp.101–112.

Flach, E., 1952. Ueber kontinuierlich durchgeführte Kernzahlbestimmungen, ihre meteorologische und bioklimatische Bedeutung. *Ber. Dtsch. Wetterd., US-Zone,* 38: 258–264.

Flach, E., 1952. Ueber ortsfeste und bewegliche Messungen mit dem Scholz'schen Kernzähler und dem Zeiss'schen Freiluftkonimeter. Ein Beitrag zur bioklimatischen Geländeaufnahme. *Z. Meteorol.,* 6: 97–112.

Flach, E., 1954. Wetter und Klima in ihren Beziehungen zum menschlichen Organismus. In: K. Walther (Herausgeber), *Lehrbuch der Hygiene.* VEB-Verlag, Berlin, pp.243–323.

Flach, E., 1954. Zum Problem der bioklimatischen Kurortklassifikation. *Abh. Gebiet Phys. Ther.,* 1: 113–119.

Flach, E., 1956. Das thermische und lufthygienische Klima in einer Mittelgebirgshöhle und deren unmittelbaren Umgebung nach gleichzeitigen Messungen. Unpublished manuscript.

Flach, E., 1957. Grundbegriffe und Grundtatsachen der Bioklimatologie. In: F. Baur, *Linkes Meteorologisches Taschenbuch* (Neue Ausgabe), III, pp.178–271.

Flach, E., 1960. Bioklimatische Grundlagen des Sports und der Sportmedizin. In: A. Arnold (Herausgeber), *Lehrbuch der Sportmedizin,* 2nd ed. Ambrosius Barth, Leipzig, pp.480–520.

Flach, E., 1963. Grundzüge einer spezifischen Bewölkungsklimatologie. *Arch. Meteorol. Geophys. Bioklimatol., Ser. B.,* 12: 357–403.

Flach, E., 1964. Die Bedeutung der Abkühlungsgrösse für die Bioklimatologie. *Z. Angew. Bäder-Klimaheilkd.,* 11: 463–475.

Flach, E., 1965. Klimatologische Untersuchung über die geographische Verteilung der Globalstrahlung und der diffusen Himmelsstrahlung. *Arch. Meteorol. Geophys. Bioklimatol., Ser. B,* 14: 161–183.

Flach, E., 1968. Zum Strahlungsklima der ariden Zonen. Unpublished manuscript.

Flach, E., 1968. Vergleichende bioklimatologische Untersuchungen zum Verhalten der Sonnenscheindauer und der kurzwelligen Wärmestrahlung im Hochgebirge und Flachland. *Z. Angew. Bäder- und Klimaheilkd.* 15: 11–35.

Flach, E., 1969. Zum jahreszeitlich differenten Verhalten von Lufttemperatur und Luftfeuchtigkeit im Hochgebirge und dessen bioklimatologischer Bedeutung. *Arch. Phys. Ther.,* 21: 487–495.

Flach, E., 1972. Zur Klimatologie der Aenderungen der Beleuchtungsstärke und deren klimaphysiologischen Bedeutung. Vortrag. St. Moritz.

Flach, E., 1972. Zur klimatologischen Infrastruktur der Aequivalenttemperatur und deren bioklimatologischer Bedeutung. Manuscript.

Flach, E., 1973. Die bioklimatischen Eigenschaften des Hochgebirgsklimas. Vortrag Aerztl. Fortbildungskurs in Bad Nauheim. Manuscript.

Flach, E. and Moerikofer, W., 1967. *Supplemental Results of the Climatology of Cooling Power: New Stations of Medium Altitude and of Arctic Regions.* European Research Office US Army Contract DA-91-591-EUC-3637. Final Scientific Report, IV.

Fleisch, A. und Von Muralt, A., 1944 & 1948. Klimaphysiologische Untersuchungen in der Schweiz, 1 und 2. *Suppl. Helv. Physiol. Acta.*

Fleischer, R., 1955/56. Die Ultrarot-Strahlungsströme aus Registrierungen des Strahlungsbilanzmessers. *Ann. Meteorol.,* 7: 87–95.

Fleischer, R. und Gräfe, K., 1953/54. Die Kälteperiode vom 30.1. bis 10.2.54. im Spiegel der Ultrarotstrahlung. *Ann. Meteorol.,* 6: 220–231.

FLOHN, H., 1954. *Witterung und Klima in Mitteleuropa. Forschungen zur Deutschen Landeskunde*, Bd. 75. Hirzel, Zürich, 214 pp.

FLOHN, H., 1961. Man's activity as a factor in climatic change. *Ann. N. Y. Acad. Sci.*, 95: 271–281.

FLOHN, H., 1970. Beiträge zur Meteorologie des Himalaya. *Khumbu Himal.*, 7: 25–47.

FLOHN, H., 1973. Natürliche und anthropogene Klimamodifikationen. *Annal. Meteorol., N.F.*, 6: 59–66.

FOITZIK, L. und HINZPETER, H., 1958. *Sonnenstrahlung und Lufttrübung*. Akad. Verlagsges., Leipzig, 309 pp.

FORSTER, R. E., FERRIS, B. G. and DAY, R., 1946. The relationship between heat exchange and blood flow in the hand at various ambient temperatures. *Am. J. Physiol.*, 146: 600.

FORTAK, H., 1971. Meteorologie. In: W. VON BRAUN (Herausgeber), *Wissen der Gegenwart*. C. Habel, Berlin, 287 pp.

FORTAK, H., 1973. Physikalische Probleme der Luftverschmutzung. *Annal. Meteorol., N.F.*, 6: 35–46.

FUHRMAN, F. A., 1963. Modification of the action of drugs by heat. *Unesco Arid Zone Res.*, pp.223–265.

GAERTNER, W. und GOEPFERT, H., 1964. Topographische Untersuchungen über die Strahlungseigenschaften der lebenden Haut. *Pflügers Arch. Ges. Phys.*, 280: 224–235.

GALINDO, I. G. and MUHLIA, A., 1970. Contribution to the turbidity problem in Mexico City. *Arch. Meteorol. Geophys. Bioklimatol., Ser. B*, 18: 53–82.

GEIGER, R., 1961. *Das Klima der bodennahen Luftschicht*, 4th ed. Vieweg, Braunschweig, 646 pp.

GEIGY Documenta, 1954. *Wissenschaftliche Tabellen* 6th ed. R. Geigy, Basel, (Ed. K. Diem).

GEORGII, H. W., 1954. Untersuchungen über den Luftaustausch zwischen Wohnräumen und Aussenluft. *Arch. Meteorol. Geophys. Bioklimatol., Ser. B*, 5: 191–214.

GEORGII, H. W., 1956. Messungen der Scheindiffusion von Aerosolen in Bodennähe bei windschwachen Wetterlagen. *Z. Aerosolforsch.-Ther.*, 5: 303–313.

GEORGII, H. W. und GRAVENHORST, G., 1972. Untersuchungen zur Konstitution des Aerosols über dem atlantischen Ozean. *Meteorol. Rundsch.*, 25: 180–181.

GEORGII, H. W. und SCHAEFER, H. J., 1968. Ueber die Einwirkung meteorologischer Parameter auf die Immissionskonzentration luftverunreinigender Komponenten. *Arch. Tech. Mess.*, 1968, pp.257–262, 723–730.

GEORGII, H.-W. und SCHAEFER, K., 1967. SO_2-Konzentration bei einer Inversionslage in der Grosstadt. *Arch. Tech. Messen.*

GIAJA, J., 1938. *L'Homéothermie*. Paris.

GLASER, E. M., 1963. Circulatory adjustments in the arid zone. *Unesco Arid Zone Res.*, pp.131–147.

GLASSER, M. and GREENBERG, L., 1971. Air pollution, mortality and weather. *Arch. Environ. Health*, 22: 334–343.

GOETZ, F. W. P., 1954. *Klima und Wetter in Arosa*. Huber, Frauenfeld, 148 pp.

GRANDJEAN, E., 1967. *Physiologische Arbeitsgestaltung (Leitfaden der Ergonomie)*, 2nd ed. Ott, München, 268 pp.

GRANDJEAN, E., 1973. Epidemiologie der Luftverunreinigungen. *Naturwiss. Rundsch.*, 26: 323–329.

GROSSE-BROCKHOFF, F., 1969. *Pathologische Physiologie* 2nd ed. Springer, Berlin, 797 pp.

HALDANE, J. S., 1905. The influence of high air temperatures. *J. Camb.*, 5: 494.

HARDY, J. D., 1972. Peripheral inputs to the central regulator for body temperature. In: S. ITOH, K. OGATA and H. YOSHIMURA, (Editors), *Advances in Climatic Physiology*. Springer, Berlin, Part I, pp.3–21.

HARDY, J. D. and DU BOIS, E. F., 1938. Basal metabolism, radiation, convection and vaporization at temperatures of 22° to 35°C. *C.J. Nutr.*, 15: 477.

HARTMANN, H. and VON MURALT, A., 1934. Pulsfrequenz und Höhenanpassung *Helv. Acta Physiol.*, 1: 38.

HAUSSER, K. W. und VAHLE, W., 1922. In: Strahlentherapie, 13: 41.

HEBERDEN, W., 1826. An account of the heat of July 1825. *Phil. Trans*, Pt. II.

HELLPACH, W., 1950. *Geopsyche*, 6th ed. Stuttgart.

HENSCHKE, U. und SCHULZE, R., 1939. Ueber Pigmentierung durch langwelliges Ultraviolett. *Strahlentherapie*, 64: 14–42.

HENSEL, H., 1955. Mensch und warmblütige Tiere. In: H. PRECHT, J. CHRISTOPHERSEN and H. HENSEL (Herausgeber), *Temperatur und Leben*. Springer, Berlin, pp.329–465.

HENSEL, H., 1966. *Allgemeine Sinnesphysiologie, Hautsinne, Geschmack, Geruch*. Springer, Berlin, 345 pp.

HENSEL, H., BRUECK, K. and RATHS, P., 1973. Homeothermic organisms. In: H. PRECHT, J. CHRISTOPHERSEN, H. HENSEL and W. LARCHER (Herausgeber), *Temperature and Life*. Springer, Berlin, pp.503–763.

HENSEL, H., PRECHT, H., CHRISTOPHERSEN, J. and LARCHER, W., 1973. *Temperature and Life*. Springer, Berlin, 779 pp.

HERPERTZ, E., ISRAEL, H. und VERZÁR, F., 1957. Vergleich luftelektrischer Messungen mit der Kondensationskernzahl auf Jungfraujoch (3578 m). *Geofys. Pura Appl.*, 36: 218–232.
HERRMANN, H., 1959. *Die Schwüle—eine vergleichende Untersuchung*. Dissert., Köln.
HILDEBRANDT, G., 1962. Biologische Rhythmen und ihre Bedeutung für die Bäder- und Klimaheilkunde. In: W. AMELUNG und A. EVERS (Herausgeber), *Handbuch der Bäder- und Klimaheilkunde*. Schattauer, Stuttgart, pp.730–785.
HILL, L., 1923. *The Katathermometer in Studies of Body Heat and Efficiency*. His Majesty's Stationary Office, London.
HINZPETER, A., 1971. *Physik als Hilfswissenschaft, 5. Optik*. Vanderhock und Rupprecht, Göttingen, 110 pp.
HITZLER, J. und LAUSCHER, F., 1970. Die Bestrahlung des Menschen durch die Sonne in allen Zonen der Erde. *Wetter Leben*, 22: 231–244.
HODGE, P. W. and LAUTAINEN, N., 1973. Transparenz und Trübung der Atmosphäre. *Umschau*, 73: 325–331.
HOEGL, O., 1957. Einteilung der schweizerischen Klimakurorte. *Arbeitsgem. Klimafragen, Bull. Eidg. Gesundheitsamt Bern*.
HOESCHELE, K., 1970. Ein Modell zur Bestimmung des Einflusses der klimatischen Bedingungen auf den Wärmehaushalt und das thermische Befinden des Menschen. *Arch. Meteorol. Geophys. Bioklimatol., Ser. B*, 18: 83–99.
HOGAN, A. W., MOHNEN, V. A. and SCHAEFER, V. J., 1973. Comments on "Oceanic Aerosol deduced from Measurements of the Electrical Conductivity of the Atmosphere". *J. Atmos. Sci.*, 30: 1455–1460.
HOINKES, H., 1961. *Antarktischer Alltag*. CIBA-Symposium, 9: 271–282.
HOINKES, H. C., 1961. Studies of solar radiation and albedo in the Antarctic. *Arch. Meteorol. Geophys. Bioklimatol., Ser. B*, 11: 281–291.
HOINKES, H. C., 1961. Studies of solar radiation in the Antarctic. *Arch. Meteorol. Geophys. Bioklimatol., Ser. B*, 10: 175–181.
HOINKES, H. C., 1971. Studies of solar radiation and albedo in the Antarctic (Little America V and South Pole 1957/58). *Arch. Meteorol. Geophys. Bioklimatol., Ser. B*, 10: 175–181.
HOLLWICH, F., 1966. Augenlicht und vegetative Funktionen. *Nova Acta Leopoldina, Abh. Dtsche Akad. Naturforsch. Leopoldina, Halle, N.F.*, 31: 189–217.
HOPPE, U., 1973. Neue hautaffine Lichtsubstanzen. *J. Soc. Cosmet. Chem.*, 24: 317–330.
HOUGHTON, F. C. and YAGLOU, C. P., 1924. *Trans. ASHVE*, 30, 103. In: H. E. LANDSBERG, WMO, Techn. Note No. 123, 1972.
HUBER, B., 1952. Ueber die vertikale Reichweite vegetationsbedingter Tagesschwankungen im CO_2-Gehalt der Atmosphäre. *Forstwiss. Z.*, 71: 372–380.
HUNTINGTON, E., 1945. *Civilisation and Climate*. New Haven.
ISRAEL, H., 1957/61. *Atmosphärische Elektrizität*, I und II. Akad. Verlagsges., Leipzig.
ISRAEL, H., 1962. Luftelektrizität und Radioaktivität. In: W. AMELUNG und A. EVERS (Herausgeber), *Handbuch der Bäder- und Klimaheilkunde*. Schattauer, Stuttgart, pp. 578–593.
JACOBI, W., 1965. Die natürliche Strahleneinwirkung auf den Atemtrakt. *Biophysik*, 2: 282–300.
JACOBI, W., 1966. *Ausscheidung und Verteilung von Aerosolen im Atemtrakt*. Hahn-Meitner-Institut für Kernforschung, Abt. Strahlenphysik. Vortrag Symposium über Spurenstoffe der Luft in St. Moritz, Juni 1966. Birkhäuser, Basel, pp.60–73.
JACOBS, M. B., 1960. *The Chemical Analysis of Air Pollutants*, Vol. X. Interscience, New York, N.Y., 430 pp.
JANSEN, G., 1961. Einwirkungen des Lärms auf den Menschen. In: G. Lehmann (Herausgeber), *Handbuch der gesamte Arbeitsmedizin*, Bd. I. *Arbeitsphysiologie*. Urban und Schwarzenberg, München, pp.680–701.
JAUREGUI, E., and SATO, C., 1967. Wet-bulb temperature and discomfort index areal distribution in Mexico. *Int. J. Biometeorol.*, 11: 21–28.
JESSEL, U., 1970. Probleme der medizinischen Meeresluftforschung. *Christiana Albertina*, 9: 36–43 (Kieler Univ.-Schrift).
JONES, F. N., 1958. Scales of subjective intensity for odors of diverse chemical nature. *Am. J. Psych.*, 71: 305–310.
JUNGE, CHR., 1952. Die Konstitution des atmosphärischen Aerosols. *Beih. Annal. Meteorol.*, 5, 55 pp.
JUNGE, CHR., 1963. *Air Chemistry and Radioactivity*. Acad. Press, New York, N.Y.
JUNGE, CHR., 1971. Der Stoffkreislauf der Atmosphäre-Probleme und neuere Ergebnisse der luftchemischen Forschung. *Jahrb. Max-Planck-Ges. Fördg. Wiss.*, pp.149–181.

JUNGE, CHR., 1974. The requirements of the World Meteorological Organisation for measurements of atmospheric pollutants. *WMO Spec. Environ. Rep., No.3, Geneva*, pp.1–13.

JUNGMANN, H., 1962. *Das Klima in der Therapie innerer Krankheiten.* Barth, München, 132 pp.

JUSATZ, H. J., 1967. *Regional Studies in Geographical Medicine. Geomedical Monograph Series.* Springer, Berlin (Heidelberger Akad. Wiss).

KAWAMURA, R., 1965. Distribution of discomfort index in Japan in summer season. *J. Meteorol. Res.*, 17: 460–466.

KAYSER, K., 1972. Ergebnisse der Registrierung der Schwefeldioxyd-Immission in Westerland/Sylt mit dem Ultragas 3-Gerät von Wösthoff. *Arch. Meteorol. Geophys. Bioklimatol., Ser. B*, 20: 79–87.

KEIDEL, W. D., 1971. *Sinnesphysiologie, I. Allgemeine Sinnesphysiologie, visuelles System.* Springer, Berlin, 229 pp.

KEMENY E. and HALLIDAY, E. C., 1972. The influence of yearly weather variations on smoke and sulphur dioxide pollution in Pretoria. *Arch. Meteorol. Geophys. Bioklimatol., Ser. B*, 20: 49–78.

KING, E., 1955. Ein empirisches Schwülemass. *Medizin-Meteorol. Hefte*, 10: 5.

KLEEMANN, A., 1969. Einsatzwert eines Schiffes aus arbeitsmechanischer Sicht. *Wehrmed. Wehrpharm.* 7: 134–149.

KLEIN, E., SEITZ, E. D. und MEYER, H. A. E., 1955. Ergebnisse und Fortschritte auf dem Gebiet der Anwendung der ultravioletten und infraroten Strahlung in der Medizin. *Ergeb. Phys. -Diätet. Ther.*, 5: 252–366.

KLOEPPER, R., 1972. Einführung zur Thematik "Landschaftbewertung für die Erholung". *Forschungsber. Akad. Raumforsch. Landesplanung*, Vol. 76, 3. *Raum und Fremdenverkehr.* Janecke, Hannover, pp.1–8.

KNEPPLE, R., 1948. Ueber die Ursachen der Schwüle. *Z. Meteorol.*, 2: 366.

KNEPPLE, R., 1956. Die biologischen Wirkungen der infraroten Eigenstrahlung (Gegenstrahlung) der Atmosphäre auf den Menschen. Die tonischen Wirkungen des Wetters. *Z. Angew. Meteorol.*, 2: 275.

KNOCH, K. und SCHULZE, A., 1952. *Methoden der Klimaklassifikation.* Perthes, Gotha, 78 pp.

KNOCH, K., 1962. Problematik und Probleme der Kurortsklimaforschung als Grundlage der Klimatherapie. *Mitt. Deutsch. Wetterd.*, 30 (4), 62 pp.

KNOCHE, W., 1905. Ueber die räumliche und zeitliche Verteilung des Wärmegehaltes der unteren Luftschicht. *Arch. Dtsch. Seewarte*, II.

KÖPPEN, W. und GEIGER, R., 1936. *Handbuch der Klimatologie.* Borntraeger, Berlin.

KRAMMER, M., 1970. Untersuchungen der atmosphärischen Trübung in Basel und ihre Abhängigkeit von den Wetterlagen. *Arch. Meteorol. Geophys. Bioklimatol., Ser. B*, 18: 53–82.

KREBS, A., 1968. Strahlenbiologie. *Verständliche Wiss.*, 95, Springer, Berlin, 127 pp.

KREUTZ, W., 1941. Kohlensäuregehalt der unteren Luftschichten in Abhängigkeit von Witterungsfaktoren. *Angew. Bot.*, 3.

KUHN, M., 1972. Global pollution in Antarctic air documented by solar radiation depletion. *Antarct. J. USA*, II: 35–37.

KUHN, M., 1973. Natural illumination of the Antarctic plateau. *Arch. Meteorol. Geophys. Bioklimatol., Ser. B*, 21: 55–56.

KUHNKE, W., 1962. Wärme, Schwüle und Todesfälle. *Z. Angew. Bäder- Klimaheilkd.*, 9: 509–519.

LADELL, W. S. S., 1951. *J. Physiol.*, 115: 296 (cited from H. Hensel).

LANCASTER, A., 1898. De la manière d'utiliser les observations hygrométriques. *Congr. Intern. d'Hydrol., 5eme, Liège.*

LANDSBERG, H., 1938. Atmospheric Condensation Nuclei. *Ergeb. Kosm. Physik*, III, pp.155–259. (Akad. Verlagsges., Leipzig.)

LANDSBERG, H. E., 1954. Bioclimatology of housing. *Meteorol. Monogr.*, 2: 81–98.

LANDSBERG, H. E., 1963. Global distribution of solar and sky radiation. In: H. E. LANDSBERG, H. LIPPMANN, K. H. PAFFEN and C. TROLL, *World Maps of Climatology.* Springer, Berlin, 28 pp.

LANDSBERG, H., 1964. Controlled climate (outdoor and indoor). *Med. Climatol.*, pp.663–701.

LANDSBERG, H. E., 1964. Die mittlere Wasserdampfverteilung auf der Erde. *Meteorol. Rundsch.*, 17: 102–103.

LANDSBERG, H. E., 1969. *Weather and Health.* Anchor Books, Doubleday, New York, N.Y., 148 pp.

LANDSBERG, H. E., 1972. The assessment of human bioclimates. A limited review of physical parameters. *WMO, Tech. Note, Geneva*, 123, 36 pp.

LANDSBERG, H. E., 1973. Climate of the urban biosphere. *Biometrics*, 5: 71–83.

LANDSBERG, H. E. and VAN MIEGHEM, M., 1959. Proceeding of symposium of atmospherics diffusion and air pollution held at Oxford. *Adv. Geophys.*, 6, 470 pp. (Acad. Press, New York).

LAUSCHER, F., 1949. Normalwerte der Relativen Feuchtigkeit in Oesterreich. *Wetter Leben*, 1: 289–297.

LAUSCHER, F. und SCHWABL, W., 1934. Untersuchungen über die Helligkeit im Wald und am Waldrand. *Bioklimatol. Beibl. Meteorol. Z.*, 1: 60–65.

LEE, D. H. K., 1958. Proprioclimates of man and domestic animals. *Climat. Rev. Res. UNESCO, Paris*, pp.102–125.

LEE, D. H. K., 1963. Physiology of the arid zones. *UNESCO, Arid Zone Res. XIII, Environmental Physiology and Psychology in Arid Conditions*, pp.15–73.

LEE, D. H. K., 1969. Proposals for a new definition of biometeorology. In: S. W. TROMP, *Report on the 5th International Biometeorological Congress, Montreux, 1969.*

LEHMANN, K., 1936. Mikroklimatische Untersuchungen der Abkühlungsgrösse in einem Waldgebiet. *Veröff. Geophys. Inst. Leipzig*, 7, Heft 4.

LEISTNER, W., 1964. Die praktische Bedeutung eines geeigneten Schwülemasses. *Z. Phys. Ther.*, 16: 67–76.

LINKE, F., 1922. Das Prött-Theorem. *Meteorol. Z.*, 39: 267–272.

LINKE, F., 1924. Ergebnisse von Messungen der Sonnenstrahlung und Lufttrübung über dem Atlantischen Ozean und in Argentinien. *Meteorol. Z.*, 41: 42–46.

LINKE, F., 1935. Die physikalisch-meteorologischen Grundlagen der medizinischen Klimatologie. *Verh. Dtsch. Ges. Inn. Med., 47. Kongr., Wiesbaden.*

LINKE, F., 1938. Bedeutung und Berechnung der Aequivalenttemperatur. *Meteorol. Z.*, 55: 345–350.

LOTMAR, R., 1972. Die Ultraviolettstrahlung und ihre biologisch-medizinische Bedeutung. *Naturwiss. Rundsch.*, 25: 89–99.

LUEDI, W. und VARESCHI, V., 1935. Die Verbreitung und das Blühen, sowie der Pollenniederschlag der Heufieberpflanzen im Hochtal von Davos. *Ber. Geobot. Inst. Rübel, Zürich.*

LUFT, U. C., 1941. Die Höhenanpassung. *Erg. Physiol.*, 44: 256.

MACFARLANE, W. V., 1963. Endocrine functions in hot environments. *Unesco Arid Zone Res.*, pp. 153–211.

MACHTA, L., 1972. Mauna Loa and global trends in air quality. *Bull. Am. Meteorol. Soc.*, 53: 402–420.

MACPHERSON, R. K., 1962. The assessment of the thermal environment. *Brit. Ind. Med.*, 19: 151–164.

MANI, A. and CHAKO, O., 1963. Measurements of solar radiation and atmospheric turbidity with the ÅNGSTROEM pyrheliometer at Poona and Delhi during IGY. *Indian J. Meteorol. Geophys.*, 14: 3.

MARTINI, E., 1955. *Wege der Seuchen*, 3.Aufl. Enke, Stuttgart, 203 pp.

MCCONNELL, W. J. and SPIEGELMANN, M., 1940. Reactions of 745 checks to summer air conditioning. *Heat. Piping Air Cond.*, 12: 317.

MEYER, H. A. E. und SEITZ, E. O., 1949. *Ultraviolette Strahlen.* De Gruyter, Berlin, 390 pp.

MIESCHER, G., 1960. Biologie und Pathologie des sichtbaren Lichts, des Ultravioletts und des Infrarots. In: *Handbuch der allgemeine Pathologie, Bd. 10, Umwelt I.* Springer, Berlin, pp.288–330.

MISSENARD, A., 1949. *Klima und Lebensrhythmus.* Deutsch. Ausgabe, Meisenheim.

MOELLER, F., 1965. On the backscattering of global radiation by the sky. *Tellus*, 17: 350–355.

MOELLER, F. und QUENZEL, H., 1972. Ueber die Wechselwirkung zwischen Bodenreflexionsvermögen und Himmelshelligkeit. *Gerlands Beitr. Geophys.*, 81: 407–413.

MOERIKOFER, W., 1944. *Die Heilklimate der Schweiz.* Arch. Fachkurse Fremdenverkehr, Gesundh. Faktor, 22 pp.

MOHRING, D., 1971. *Touristikmedizin.* Thieme, Stuttgart, 231 pp.

MUELLER, E. A., 1961. Der Schutz gegen ungünstige Klimawirkungen. In: *Handbuch der Ges. Arbeitsmedizin, I. Arbeitsphysiologie.* Urban und Schwarzenberg, München, pp.605–632.

MUELLER, E. A. und WENZEL, H. G., 1961. Die Beurteilung des Arbeitsklimas. In: *Handbuch der Ges. Arbeitsmedizin*, I. *Arbeitsphysiologie.* Urban und Schwarzenberg, München, pp.588–604.

NEUROTH, G., 1948. Die Hauttemperatur im Dienst der Wärmeregulation. *Pflügers Arch. Ges. Physiol.*, 250: 396–413.

NEUWIRTH, R., 1966. Die Nullwertwetterlage vom Aerosolstandpunkt aus. *Z. Angew. Bäder- Klimaheilkd.*, 13: 692–699.

NEUWIRTH, R., 1973. Einfluss der Staubbelastung an Kurorten. *An. Meteorol.*, N.F., 6: 161–164.

NYBERG, A., 1972. The effect of local and distant sources of sulphur on the precipitation contents of sulphur in Scandinavia. *Geophysica*, 12: 33–42.

OLSCHOWY, G., 1971. *Belastete Landschaft—gefährdete Umwelt.* Samml. "Das wissenschaftliche Taschenbuch," Abt. Naturwiss., Goldmann, München, 346 pp.

OPITZ, E., 1941. Ueber akute Hypoxie. *Erg. Physiol.*, 44: 315.

PEARSON, J. E. and JONES, G. E., 1965. Soil concentrations of "emanation radium^{226}" and the emanation of radon^{222} from soils and plants. *Tellus*, 18: 655–662.

PFLEIDERER, H., 1938. Die bioklimatische Bedeutung der Strahlen. In: H. WOLTERECK (Herausgeber), *Die Welt der Strahlen.* Quelle und Meyer, Leipzig.

PFLEIDERER, H., 1961. Ueber den heutigen Stand der Möglichkeiten zur Klimacharakterisierung bezüglich der thermischen Verhältnisse. *Arch. Phys. Ther.*, 13: 101–107.

PFLEIDERER, H. und BÜTTNER, K., 1935. *Die physiologischen und physikalischen Grundlagen der Hautthermometrie*. Barth, Leipzig, 52 pp.

PFLEIDERER, H. und BUETTNER, K., 1940. Bioklimatologie. In: H. VOGT (Herausgeber), *Lehrbuch der Bäder- und Klimaheilkunde*, II. Teil, Abschn. V, pp.609–949, Springer, Berlin.

PFOTZER, G., 1966. Die hohe Atmosphäre nach Messungen mit Raketen und Satelliten. *VDI-Zschr.*, 108: 1134–1140.

PORTMANN, A., 1963. *Licht und Leben*. Reinhardt, Basel, 51 pp.

PROETT, C. H., 1919. *Pröttmeter*. Selbstverlag, Rheydt, 24 pp.

PUESCHEL, R. F., ELLIS, H. T., MACHTA, L., COTTON, G., FLOWERS, E. C. and PETERSON, J. C., 1972. *Normal Incidence of Solar Radiation Trends on Mauna Loa, Hawaii*. Paper IUGG Symposium, Sendai, 1972.

QUENTZEL, H., 1967. Optische Bestimmung der Kontinuum-Absorption maritimer Luftmassen im Spektralbereich der Sonnenstrahlung. *Meteorol.-Forschungsergeb.*, *B*, 1: 36–40.

RANKE, O. F., 1953. Die optische Simultanschwelle. *Z. Biol.*, 105: 224–231.

READER, J. D. and WHYTE, H. M., 1951. Tissue temperature gradient. *J. Appl. Physiol.*, 4: 396.

REITER, R., 1960. *Meteorobiologie und Elektrizität der Atmosphäre*. Akad. Verlagsges., Leipzig, 424 pp.

REITER, R., 1968. Die erweiterte Fernübertragungsanlage zur Registrierung aerologischer Daten von Seilbahngondeln—Aerosolstudien an Inversionen. *Meteorol. Rundsch.*, 21: 73–81.

REITER, R., CARNUTH, W. and SLADKOVIC, R., 1972. Ultraviolettstrahlung in alpinen Höhenlagen. *Wetter Leben*, 24: 231–247.

REUTER, H. und CEHAK, K., 1966. Zur Luftverunreinigung durch turbulente Diffusion. *Arch. Meteorol. Geophys. Bioklimatol., Ser. 1A*, 15: 192–204.

ROBERTSON, D. F., 1972. *Geographic Pathology of Skin Cancer*. Diss. Univ. Queensland, 131 pp.

ROBITZSCH, M., 1951. Eine einfache Auflösungsmethode für die Psychrometerformel. *Z. Meteorol.*, 5: 143–147.

ROBITZSCH, M., 1931. Beiträge zur Behandlung klimatologischer Fragen auf physiologischer Grundlage. *Ann. Hydrogr. Marit. Meteorol.*, 59: 73.

RODENWALDT, E., 1968. Leon Battista Alberti—ein Hygieniker der Renaissance. In: H. J. JUSATZ (Herausgeber), *Sitz. Ber. Heidelb. Akad. Wiss. Math. Naturw. Kl., 4.Abh.*, 104 pp.

RODENWALDT, E. und JUSATZ, H. J., 1952. *Weltseuchenatlas*. Falk, Hamburg.

ROLLIER, A., 1951. *Die Heliotherapie*. München, Berlin.

RUBNER, M., 1896. Atmosphärische Feuchtigkeit und Wasserdampfabgabe. *Arch. Hyg.*, 11, 135 pp.

RUGE, H., 1932. Das Verhalten der Lufttemperatur und Luftfeuchtigkeit auf einem modernen Kreuzer in den Tropen. Ein Beitrag zur praktischen Brauchbarkeit von Schwülekurven. *Veröff. Marine-Sanitätswesen, Berlin*, Heft 22.

SARGENT II, FREDERICK, 1963. Tropical neurasthenia: giant or windmill? *Unesco, Arid Zone Res.*, pp.273–309.

SAUBERER, F., 1954. Zur Abschätzung der Gegenstrahlung in den Ostalpen. *Wetter Leben*, 6: 53–56.

SAUBERER, F. und RUTTNER, F., 1941. *Die Strahlungsverhältnisse der Binnengewässer*. Akad. Verlagsges., Leipzig, 240 pp.

SCHARLAU, K., 1943. Die Schwüle als messbare Grösse. *Bioklimatol. Beibl. Meteorol. Z.*, 10: 19.

SCHARLAU, K., 1952. Die Schwülezonen der Erde. *Ber. Dtsch. Wetterd. (US-Zone)*, 42: 246–249.

SCHMIDT-KESSEN, W., 1965. Hat Heliotherapie noch Berechtigung? *Aerztl. Praxis*, 17: 1254–1258.

SCHMIDT-KESSEN, W., 1965. Klimatherapie des bekleideten Patienten. *Z. Angew. Bäder- Klimaheilkd.*, 12: 255–263.

SCHMIDT-KESSEN, W., 1973. Bioklimatologie. In: W. WACHSMUTH (Herausgeber), *Aertzliche Problematik des Urlaubs*. Springer, Berlin, pp.86–97.

SCHMIDT-KESSEN, W. and PLEHN, I., 1965. Die Erythropoese beim Aufenthalt im Mittelgebirge. *Mediz. Welt*, pp.102–105.

SCHNELLE, K.-W., 1962. Bäder- und Kurorte. In: W. AMELUNG und A. EVERS (Herausgeber), *Handbuch der Bäder- und Klimaheilkunde*. Schattauer, Stuttgart, pp.1033–1056.

SCHOENBEIN, C. F., 1840. In: *Bericht über die Verhandlg. d. Naturforsch. Gesellschaft in Basel*, IV, pp.58, 66.

SCHOENHOLZER, G., 1968. Sportliche Leistungsfähigkeit in der Höhe. In: I. VON DESCHWANDEN, K. SCHRAM und J. C. THAMS (Herausgeber), *Der Mensch im Klima der Alpen*. Huber, Bern. (Wiss. Abh. d. Kongresses.)

SCHRAM, K. and THAMS, J. C., 1967. Die kurzwellige Strahlung von Sonne und Himmel auf verschieden orientierte und geneigte Flächen. *Arch. Meteorol. Geophys. Bioklimatol., Ser. B*, 15: 99–126.

SCHROEER, E., 1949. Theorie der Entstehung, Zersetzung und Verteilung des atmosphärischen Ozons. *Ber. Dtsch. Wetterd.*, (US-Zone), 11: 13–23.

SCHUEPP, W., 1949. Die Bestimmung der Komponenten der atmosphärischen Trübung aus Aktinometermessungen. *Arch. Meteorol. Geophys. Bioklimatol., Ser. B*, 1: 257–346.

SCHULTZE, E. G., 1962. Einfluss des Meeresküstenklimas. In: W. AMELUNG und A. EVERS (Herausgeber), *Handbuch der Bäder- und Klimaheilkunde.* Schattauer, Stuttgart, pp.683–700.

SCHULTZE, E.-G., 1973. *Meeresheilkunde.* Urban und Schwarzenberg, München.

SCHULZE, R., 1970. *Strahlenklima der Erde. Wiss. Forschungsber.*, 72. Steinkopff, Darmstadt, 217 pp.

SCHWEIZERISCHE VEREINIGUNG DER KLIMAKURORTE, 1961. *Das kleine Klimabuch der Schweiz.* 116 pp.

SIEDENTOPF, H. und REGER, E., 1944. Die Beleuchtung durch die Sonne. *Meteorol. Z.*, 61: 144.

SIPLE, P. A. and PASSEL, C. F., 1945. Measurements of dry atmospheric cooling in subfreezing temperatures. *Proc. Am. Phil. Soc.*, 89(1): 177–199.

SLOAN, R., SHAW, J. H. and WILLIAMS, D., 1955. Infrared emission spectrum of the atmosphere. *J. Opt. Soc. Am.*, 45: 45.

STEINHAUSER, F., 1936. Ueber die Häufigkeitsverteilung der relativen Feuchtigkeit im Hochgebirge und der Niederung. *Meteorol. Z.*, 53: 223–226.

STEINHAUSER, F., 1960. Statistische Untersuchung der Inversionen im Luftraum über Wien. *Arch. Meteorol. Geophys. Bioklimatol., Ser. A*, 11: 427–457.

STEINHAUSER, F., 1960. Ergebnisse von Registrierungen des Ozongehaltes der Luft in Wien. *Arch. Meteorol. Geophys. Bioklimatol., Ser. A*, 11: 368–382.

STEINHAUSER, F., 1963. Der Tagesgang der Luftverschmutzung in Wien. *Arch. Meteorol. Geophys. Bioklimatol., Ser. B*, 12: 109–123.

STEINHAUSER, F., 1966. Die Aenderungen der Radioaktivität der Luft. *Wetter Leben*, 18: 45–54.

STEINHAUSER, F., 1970. Untersuchungen über die SO_2-Ablagerungen und den SO_2-Gehalt der Luft in Wien. *Arch. Meteorol. Geophys. Bioklimatol., Ser. B*, 18:383–395.

STEINHAUSER, F., 1974. Strontium-90 Ablagerungen aus dem Niederschlag im Raume von Nordafrika bis zur Arktis. *Arch. Meteorol. Geophys. Bioklimatol., Ser. B*, 22: 55–72.

STEVENS, S. S., 1957. On the psychophysical law. *Psych. Rev.*, 64: 153–181

STEVENS, S. S., 1959. On the validity of the loudness scale. *J. Acoust. Soc. Am.*, 31: 995–1003.

STIX, E., 1971. Vorkommen von Pollen und Sporen in der Luft. *Mitt. Dtsch. Forsch. Gem.*, VI: 46–73.

STONE, R. G., 1943. The practical evaluation and interpretation of the cooling power in bioclimatology. *Bull. Am. Meteorol. Soc.*, 24: 295, 327.

SUESSENBERGER, E., 1973. *Jahresbericht 1972 des Deutschen Wetterdienstes.* Selbstverlag, Offenbach/ Main.

SZAVA-KOVATS, J., 1938. Die Verteilung der relativen Luftfeuchtigkeit auf der Erde. *Ann. Hydrogr. Marit. Meteorol.*, 66: 373–378.

TENNENBAUM, J., SOHAR, E., ADAR, R. and GILAT, T., 1961. The physiological significance of the cumulative discomfort index. *Harefuah, J. Med. Ass. Israel*, 60: 315–319.

TERJUNG, W. H., 1967. The geographical application of some selected physio-climatic indices to Africa. *Int. J. Biometeorol.*, 11: 5–19.

TEUB, B., 1970. *Die Herkunft des Luftjods an der Nordsee.* Diss. Heidelberg, 35 pp.

THAMS, J. C., 1961. Der Einfluss der Bewölkungsmenge und -art auf die Grösse der diffusen Himmelsstrahlung. *Geofys. Pura Appl.*, 48: 181.

THAMS, J. C., 1962. Zum Problem der Messung der Abkühlungsgrösse in warmen Klimaten. *Arch. Meteorol. Geophys. Bioklimatol., Ser. B*, 11: 292–300.

THAMS, J. C. und WIERZEJEWSKI, H., 1963. Die Grösse der diffusen Zirkumglobalstrahlung. *Arch. Meteorol. Geophys. Bioklimatol., Ser. B*, 12: 47–63.

THAUER, R., 1939. Der Mechanismus der Wärmeregulation. *Ergebn. Physiol.*, 41: 607.

THAUER, R. und ZOELLNER, G., 1953. Der insensible Gewichtsverlust als Funktion der Umweltbedingungen. Seine Abhängigkeit von dem Wasserdampfdruck der Luft im indifferenten Temperaturbereich. *Pflügers Arch. Ges. Physiol.*, 258: 58.

THILENIUS, R. und DORNO, C., 1925. Das Davoser Frigorimeter (ein Instrument zur Dauerregistrierung der physiologischen Abkühlungsgrösse). *Meteorol. Z.*, 42: 57.

THOM, E. C., 1957. A new concept for colling degree days. *Air Cond. Heat Vent.*, 54: 73–80.

TOPERCZER, M., 1931. Die Horizontalkomponente der Strahlung. *Gerlands Beitr. Geophys.*, 26: 98–110.

TRENDELENBURG, W., 1961. Der Gesichtssinn, Grundzüge der physiologischen Optik. In: *Lehrbuch der Physiologie*, 2. Aufl. Springer, Berlin, 440 pp.

TROLL, C., 1967. Die geographische Landschaft und ihre Erforschung. In: *Zum Gegenstand und zur Methode der Geographie*. pp.417–463.

TROLL, C., 1969. Die räumliche und zeitliche Verteilung der Schwüle und ihre graphische Darstellung (mit besonderer Berücksichtigung Afrikas). *Erdkunde*, 23: 183–192.

TROMP, S. W., 1969. *Report on the 5th International Biometeorological Congress, Montreux, 1969.*

TRONNIER, H., 1972. Die Wirkungsweise der therapeutischen Anwendung von Licht. In: B. LÜDERITZ (Herausgeber), *Zur Wirkungsweise unspezifischer Heilverfahren.* Hippokrates Verlag, Stuttgart, pp.27–36.

TURNER, D. B., 1964. A diffusion model for an urban area. *J. Appl. Meteorol.*, 3: 83.

UNESCO, 1958. *The Problems of The Arid Zone. Proceedings of the Paris Symposium.*

UNESCO-FAO, 1963. *Bioclimatic Map of the Mediterranean Zone.* Unesco-FAO, Rome, XXI, 58 pp.

UNZ, F., 1969. Die Konzentration des Aerosols in Troposphäre und Stratosphäre aus Messungen der Polarisation der Himmelsstrahlung im Zenit. *Beitr. Phys. Atmosph.*, 42: 1–35.

URBACH, F., 1969. *The Biologic Effects of Ultraviolet Radiation.* Pergamon, New York, N.Y., 704 pp.

VALKO, P., 1962. Untersuchungen über die vertikale Trübungsschichtung der Atmosphäre. *Arch. Meteorol. Geophys. Bioklimatol., Ser. B*, 11: 143–210.

VALKO, P., 1963. Ueber das Verhalten des atmosphärischen Dunstes am Alpensüdfuss. *Arch. Meteorol. Geophys. Bioklimatol., Ser. B*, 12: 458–474.

VALKO, P., 1971. Das kurzwellige Strahlungsfeld der Atmosphäre—Richtwerte für Ingenieure und Architekten, 4. *Heizung Lüftung*, 38: 121–126.

VERNEKAR, A. D., 1975. A calculation of normal temperature at the earth's surface. *J. Atmos. Sci.*, 3: 2067–2081.

VERNON, H. M., 1930. The measurement of radiant heat in relation to human comfort. *J. Phys. Lond.*, 70: 15.

VERNON, H. M. and WARNER, C. G., 1932. The influence of the humidity of the air on capacity for work at high temperatures. *J. Hyg. Camb.*, 32: 431.

VINCENT, I., 1890. Nouvelles recherches sur la température climatologique. *Ciel et Terre*, 10.

VON BETZOLD, W., 1894. In: *Z. Luftschiff. Phys. Atmos.*, Bd. 13 p.1ff.

VON MURALT, A., 1954. Krankheiten durch verminderten Luftdruck und Sauerstoffmangel. In: *Handbuch der inneren Medizin*, Bd. IV, 2. Springer, Berlin.

VON MURALT, A., 1968. Der Beitrag zur Höhenphysiologie zur allgemeinen Bioklimatologie. In: J. VON DESCHWANDEN, K. SCHRAM und J. C. THAMS (Herausgeber), *Der Mensch im Klima der Alpen*, Huber, Bern. (Wiss. Abh. d. wiss. Kongresses).

WACHSMUTH, W., 1973. *Aertzliche Problematik des Urlaubs.* Springer, Berlin, 230 pp.

WANNER, H. U. und GILGEN, A., 1966. Die hygienische Bedeutung des Ozons. *Arch. Hyg.*, 150, pp.62, 68.

WEBB, C. G., 1960. Thermal discomfort in an equatorial climate. *J. Inst. Heat Vent. Eng.*

WEIDEMANN, H., ROSKAMM, H., SAMEK, L. und REINDELL, H., 1968. Ueber das Problem der Höhenanpassung unter besonderer Berücksichtigung des Olympischen Spiele 1968 in Mexico City. *Meteorol. Rundsch.*, 21: 152–155.

WENZEL, H. G., 1956/59. Unveröffentlichte Versuche am Max-Planck-Institut für Arbeitsphysiologie in Dortmund.

WENZEL, H. G., 1961. Die Wirkung des Klimas auf den arbeitenden Menschen. In: G. LEHMANN (Herausgeber) *Arbeitsphysiologie, Handbuch der Ges. Arbeitsmedizin*, Band I .Urban und Schwarzenberg, München, pp.554–588.

WEZLER, K., 1950. Der Einfluss klimatischer Faktoren auf den Menschen. *Schriftenr. Dtsch. Bäderverband*, 5: 163–197.

WEZLER, K. and NEUROTH, G., 1949. In: *Z. Ges. Exp. Med.*, 115: 127 (cited from H. Hensel).

WEZLER, K. and THAUER, R., 1948. Erträglichkeitsgrenze für wechselnde Raumtemperatur und -feuchte. *Pflügers Arch. Ges. Phys.*, 250:192–199.

WIERZEJEWSKI, H., 1963. Die Messprinzipien von Frigorigraph und Frigorimeter und ihre Eignung für die Beurteilung der Beanspruchung des menschlichen Organismus durch die thermischen Umweltbedingungen. *Interne Ber. Phys.-Meteorol. Observ. Davos*, No. 217.

WIESNER, J. und AMELUNG, W., 1962. Unspezifische und allergische Erkrankungen der Atemwege. In: W. AMELUNG und A. EVERS (Herausgeber), *Handbuch der Bäder- und Klimaheilkunde.* Schattauer, Stuttgart, pp.866–880.

WILSON, O., 1967. Objective evaluation of windchill index by record of frostbite in the Antarctic. *Int. J. Biometeorol.*, 11: 29, 32.

WINSLOW, C. E. A., HERRINGTON, L. P. and GAGGE, A. P. 1937. Physiological reactions of the human body to varying environmental temperatures. *Am. J. Physiol.*. 120: 1.

WMO, 1958. Measurement of evaporation, humidity in the biosphere and soil moisture. *WMO Tech. Note, Geneva*, No. 21.

WMO, 1974. Urban climates. *WMO Tech. Note, Geneva*, No. 108.

WOERNER, H., 1958. Der Jahresgang der Tageshelligkeit und seine Beeinflussung durch die Witterungsverhältnisse. *Arch. Meteorol. Geophys. Bioklimatol., Ser. B*, 8: 202–214.

YAGLOU, C. P., 1927. Temperature, humidity and air movement in industries. The effective temperature index. *J. Ind. Hyg.*, 9: 297, 351.

YAGLOU, C. P. and MILLER, W. E., 1924. In: *Trans. Am. Soc. Heating Vent. Eng.*, 30: 339.

ZENKER, H., 1971. Schwüle im Mittelgebirge. *Z. Physio-Ther.*, 23: 17–20.

Chapter 2

Agricultural Climatology

AUGUSTINE Y. M. YAO

Introduction

Man, in this space age, is still deeply dependent on climate for the adequacy of his food supplies. The production of food depends on understanding the interrelationship of the total environment which includes climate, soils, genetic material of plant and animal, insects and disease, and man himself. Of these important environmental variables, weather or the longer phase "climate" is the only uncontrollable variable which needs special attention.

Understanding the physical principles of climatic parameters such as radiation, temperature, precipitation, and others, together with the understanding of the life processes of plants and animals in relation to climatic variables, is of primary importance in the struggle to supply food. The collective ability to optimize food production through appropriate land use and to minimize the weather hazard is the next important step in the struggle with nature and should not be overlooked. Yield and productivity analysis is the final goal of an agricultural climatological study. By understanding the relationships between weather and crop yield through statistical models or other methods, the agricultural climatologist may be able to predict crop yield in advance. He may also provide information about the climatic potential of an area and how it can be used to maximize its potential.

Climatic factors affecting agriculture

Radiation[1]

Radiant energy from the sun is the prime source of energy for terrestrial life. Practically all of the energy for all physical and biological processes occurring on earth arrives in the form of solar radiation. It is the ultimate cause of all changes and motion of the atmosphere and the single most important control of climate. It is a meteorological element of the highest importance.

---

[1] The subject of radiation with particular emphasis on general principles and global aspects, and the subject of net incoming radiant energy at the surface and the energy balance near the ground have been discussed thoroughly in Vol. I and Vol. II, respectively, of the *World Survey of Climatology*. To avoid repetition, these subjects have been discussed only briefly in this chapter.

Incoming radiation

At the outer limit of the earth's atmosphere, measured on a surface normal to the incident radiation and at the earth's mean distance from the sun, the energy value is 0.136 Watt cm^{-2} (1.95 cal. $cm^{-2}min^{-1}$ or Langley(ly)min^{-1}). This is known as the "solar constant" (JOHNSON, 1954). The solar constant is not entirely constant. The variations are not yet well known, but a large portion lies in the ultraviolet part of the spectrum (RENSE, 1961).

In passing through the earth's atmosphere, solar energy is depleted by molecules and particles. As the remaining energy reaches the earth's surface, part of it is further reduced by reflection, generally referred to as "surface albedo", defined as the ratio of the amount of radiation reflected to the incident amount, commonly expressed as a percentage. Sometimes only the visible portion of the spectrum is considered; sometimes the totality of wavelength in the solar spectrum. MONTEITH (1959) suggested that albedo be used exclusively for the visible light while the term "reflection coefficient" be used for total short-wave radiation. The reflection coefficient changes with the spectral quality of the radiation, and depends upon the angles of incidence of the radiation and nature of the surface. KUHN and SUOMI (1958) measured the reflectances from aircraft as shown in Table I. BLAD and BAKER (1972) measured the reflected incoming solar radiation continuously over a soybean crop for three growing seasons at St. Paul, Minnesota. They reported that the average daily albedo ranged from 0.24 to 0.27 with complete soybean cover. The albedo of moist soil was approximately 10.5% with no cover and increased at the rate of about 1.3% for each 10% increase in crop cover.

TABLE I

SURFACE REFLECTANCE MEASURED FROM AIRCRAFT
(From KUHN and SUOMI, 1958)

Surfaces	Reflectance
Dry grass	0.17–0.21
Wet	0.20
Green field	0.09–0.13
Small grain	0.21–0.25
Sand hill (outcropping)	0.21
Large river	0.04

The earth's surface receives, in addition to direct sunlight, diffuse skylight caused by scattering in the atmosphere. The intensity of diffuse radiation depends on solar elevation, height, cloudiness and atmospheric turbidity. In higher latitudes it may be an important factor.

Outgoing radiation

The radiation absorbed by the earth's surface is re-emitted as long-wave radiation to space. The intensity of the terrestrial radiation may be expressed by the following equation:

$$R = \varepsilon \sigma T^4$$

where R is the terrestrial outgoing radiation, ε the emissivity of the earth's surface, σ the Stefan-Boltzman constant ($7.92 \cdot 10^{-11}$ly° $K^{-4}min^{-1}$), and T is the absolute temperature in Kelvin. This "black body radiation" is the maximum amount of radiant energy which can be emitted by a body at a given temperature. A good absorber is also a good radiator at the same temperature. The surface of the earth can be considered a black body. GEIGER (1965) and VAN WIJK (1963) have given the emissivity of soils from 0.95 to 0.98, while BROOKS (1959) shows an emissivity of plant surface of 0.90 to 0.95. However, the emissivity varies with the wavelength.

Water vapor, carbon dioxide and ozone in the atmosphere control the exchange of radiation between the earth's surface and sky. These gases absorb much of the outgoing radation from the earth's surface and reradiate a portion back to the ground, producing the "greenhouse effect", which effectively prevents excessive cooling at night. Fig.1 shows that water vapor absorbs strongly in the vicinity of 6.0 and beyond 22 μ, while carbon dioxide captures energy in the 14–16 μ region. The atmospheric windows through which the terrestrial radiation escapes freely are located in the vicinity of 5.0 and again between 8 and 12.0 μ.

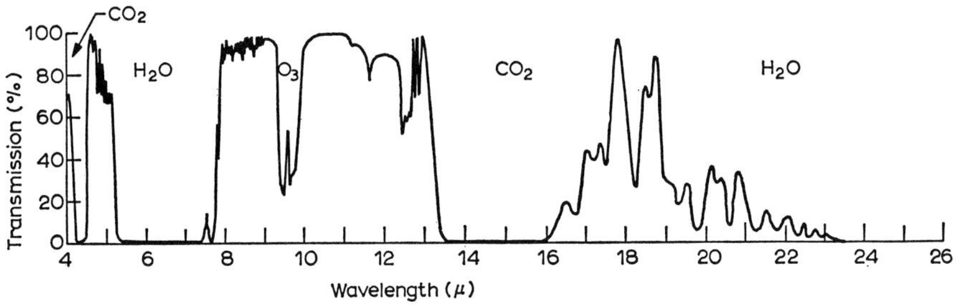

Fig.1. The transmission of the earth's atmosphere of long-wave thermal radiation as a function of the wavelength. (After GATES, 1965.)

Net radiation

The net radiation is the difference between total upward and downward radiation flux. It is the energy available at the earth's surface. It is expressed by the following equation:

$$R_n = R_S\downarrow - R_S\uparrow + R_L\downarrow - R_L\uparrow$$

where R_n is the net radiation, $R_S\downarrow$ the incoming short-wave solar radiation (direct and diffuse radiation), $R_S\uparrow$ the short-wave radiation reflected from the surface (or albedo), $R_L\downarrow$ the incoming long-wave radiation which is reradiated principally by water vapor and carbon dioxide, and $R_L\uparrow$ the upward long-wave radiation emitted by the surface according to its temperature.

The energy balance

The net radiation may be partitioned into the following components:

$$R_n = A + LE + S + P + M$$

where A is the flux of sensible heat (heat transferred by convection), LE, the flux of latent heat (heat used in evaporation or condensation), S, the ground heat storage (heat flow into or out of the ground by conduction), and P and M are small amounts of energy utilized by plant for photosynthesis and other exchanges such as heat storage in the crop if the ground is covered by vegetation.

GEIGER (1957) presented a schematic diagram from the outer limits of atmosphere to the earth's surface showing heat exchange at noon for a summer day in Germany (Fig.2). The various processes involving radiation balance are clearly shown by this diagram. The width of the arrows in the figure give an idea of the relative amounts of the transferred heat totals.

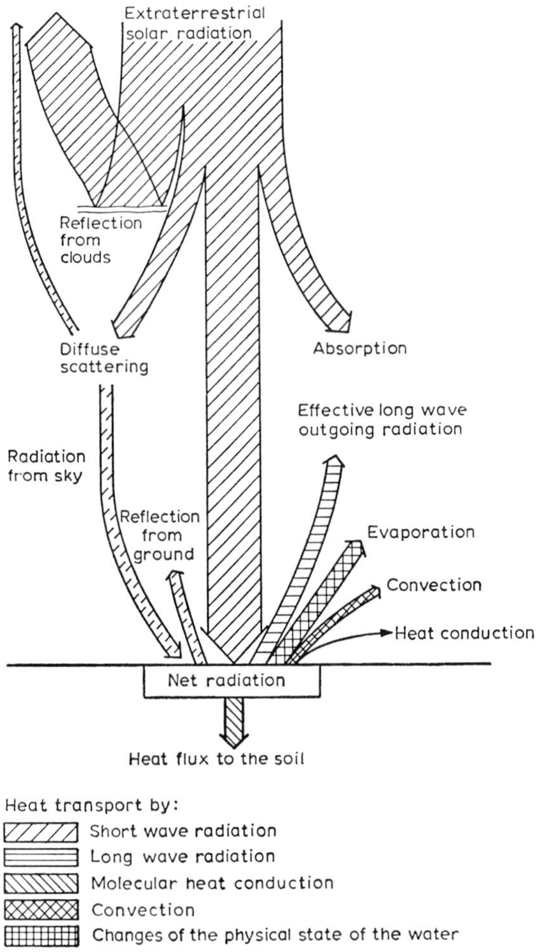

Fig.2. Heat exchange at noon for a summer day. (After GEIGER, 1957.)

Spectral distribution of solar radiation

The spectral distribution of the sunlight (diıect) at the top of the earth's atmosphere and at the earth's surface is shown in Fig.3. Most of the ultraviolet radiation from the sun is absorbed by the oxygen high in the atmosphere, and by ozone at a height of 25 km. CHANG (1968) noted that for a solar altitude of 30°, the energy in the ultraviolet

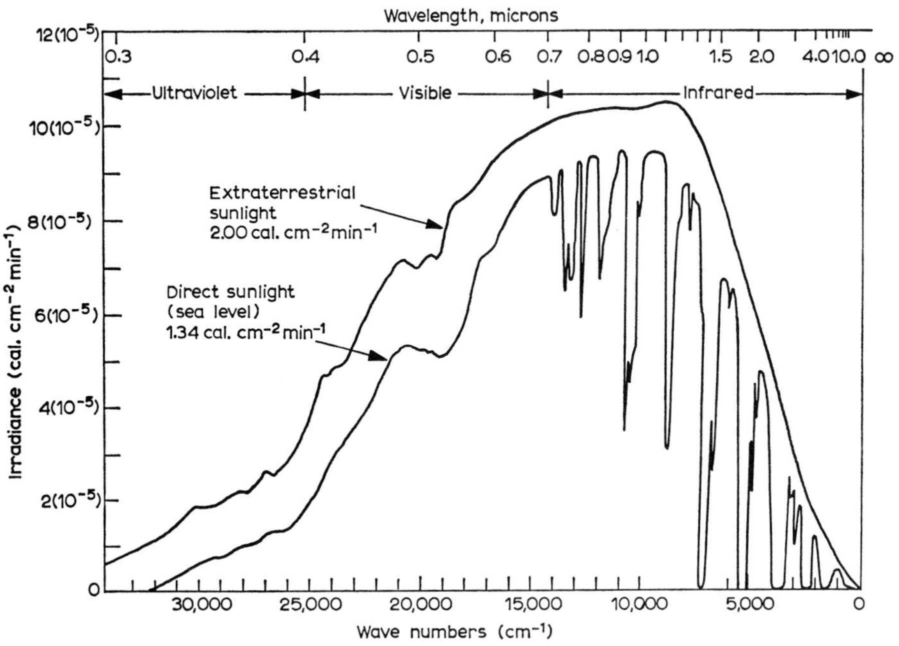

Fig.3. Spectral distribution of extraterrestrial solar radiation and of solar radiation at sea level for a clear day. Each curve represents the energy incident on a horizontal surface. (After GATES, 1965.)

below the wavelength of 0.4 μ comprises about 9% of the total incident energy, while the energy in the visible region comprises 41% and that in the infrared, beyond 0.72 μ, contains about 50%. The spectral distribution of sunlight changes with altitude and with cloudiness. Ultraviolet and infrared radiation is reduced much more on a cloudy day than on a sunny day. At high altitudes ultraviolet radiation is enriched.

Based upon the physiological response of plants to solar radiation, The Dutch Committee on Plant Irradiation (WASSINK, 1953) has recommended the following bands of solar spectrum:

1st band: greater than 1.0 μ. "No specific effects of this radiation are known. It is acceptable that this radiation, as far as it is absorbed by the plant, is transformed into heat without interference of biochemical processes."

2nd band: 1.0–0.70 μ. "This is the region of specific elongating effect on plants. Although the spectral region of elongating effects does not coincide precisely with the limits of this band, one may provisionally accept that the radiant flux in this band is an adequate measure of the elongating activity of the radiation."

3rd band: 0.70–0.61 μ. "This is almost the spectral region of the strongest absorption of chlorophyll and of the strongest photosynthetic activity in the red region. In many cases it also shows the strongest photoperiodic activity."

4th band: 0.61–0.51 μ. "This is a spectral region of low photosynthetic effectiveness in the green and of weak formative activity."

5th band: 0.51–0.40 μ. "This is virtually the region of strong chlorophyll absorption and absorption by yellow pigments. It is also a region of strong photosynthetic activity in the blue-violet and of strong formative effects."

6th band: 0.40–0.315 μ (ultra-violet A). Produces fluorescence in plants.

7th band: 0.315–0.280 μ (ultra-violet B) is detrimental to plants. Practically no solar radiation wavelengths shorter than 0.29 μ reaches the earth's surface.
8th band: Wavelength shorter than 0.28 μ (ultra-violet C). This radiation has strong germicidal action.

Radiation and light distribution within the plant canopy

The radiation distribution within the plant canopy is affected by plant leaf area, density and height, leaf arrangement, the sun's angle, inclination and transmissivity of the leaves.
The light distribution in the canopy may be expressed by Beer's law:

$$I = I_0 \, e^{-kF}$$

where I is the light intensity at a given height within the plant canopy; I_0 is the light intensity at the top of the plant canopy; e is the base of natural logarithm; F is leaf area index; k is the extinction coefficient, determined primarily by leaf inclination and arrangement and secondarily by leaf transmissivity.
MONTEITH (1965) presented another equation dealing with light distribution within the plant canopy:

$$I = [S + (1 - S)\tau]^F I_0$$

where S is the fraction of light passing through a unit leaf layer without interception; τ is the leaf transmission coefficient; and F is the leaf area index.
Both equations indicate that leaf area is an important factor while the transmissivity of the leaf may be only of secondary importance. The rest of the factors are either included in k of the first equation or in S of the second equation.

Leaf area

Leaf area index (*LAI*) was developed by WATSON (1947) and is defined as the leaf area subtended per unit area of land. There are many ways of estimating the leaf area index, some of them involving tedious work, others subject to significant error. Other methods may be quite accurate, but the cost of required instruments is high. Generally the leaf area index for most of crop plants ranges from 2 to 6. In extreme cases it may be doubled, examples including sugar cane and rice. EIK and HANWAY (1966) measured the leaf area index of corn in Iowa, and found that yields of corn grain tended to be linearly related to the leaf area index at silking time. The linear relationship did not continue beyond the leaf area index value of 3.3.

Leaf arrangement

The leaf arrangement also plays an important role in the light interception within the plant canopy. Nicheprovich (1962—cited by CHANG, 1968), considered the ideal arrangement of leaves is that the lowest 13% of the leaves lay at angles between 0° and 30° to the horizontal, the adjacent 37% of the leaves lay at 30° to 60°, and the upper 50% of the leaves lay at 60° to 90°. Incident radiation intercepted by plants varies with the angle of sun, the row spacing, and the orientation of the rows. SHAW and WEBER (1967) simulating various degrees of lodging of soybean leaves and branches, found significant changes in light interception and yield. For the three years of observations, maximum light penetration occurred with a moderate amount of lodging. Generally, the largest part of the light interception occurred in the outer 15–30 cm of the plant canopy. Greater light penetration resulting in a greater amount of the plant canopy having light intensity above 150 ft.-c* generally resulted in greater yield. STEVENSON and SHAW (1971) compared upright orientation against naturally exposed leaves. They found the leaf resistance to water vapor diffusion for upright leaves was less than that of naturally exposed leaves on 8 of the 9 experimental days and averaged 1.38 sec/cm less. Similarly, leaf temperatures were less for upright leaves. On overcast days there was no difference between treatment for both resistance and temperature. PENDLETON et al. (1968) conducted a field investigation at Urbana, Illinois, concerning the relationship of leaf angle and canopy shape to grain yield and apparent photosynthesis of corn. They found that varieties with erected leaves produced 40% more grain than those counterparts with normal type leaves. Apparent photosynthesis measurements on individual corn leaves show that the relative efficiency of CO_2 fixation/unit of incoming sunlight steadily increases as the leaf angle decreases.

Light distribution in canopy

Many experiments have been conducted during the past few decades to measure the light distribution within the plant canopy for plants and trees. Each experiment may present evidence for one or a combination of factors that affect the penetration of light within the plant community. As early as 1925, Ångström reported the classical observation on the light distribution with respect to height within the canopy of meadow grass. His experiment clearly indicated the exponential relationship between light intensity and height in the canopy. Experiments by Sauberer (1938, cited by GEIGER, 1965) showed the difference between the vertically leaved, sparsely populated barley and densely populated rye and the broad-leaved clover. Clover intercepted more light than rye while barley had the least light interception.

DENMEAD et al. (1962) measured the net radiation profile within a mature crop canopy. They found marked differences between clear and cloudy days. Fig.4 is the average distribution of net radiation in the crop canopy for clear and cloudy days after maximum leaf area development. It showed that upper leaves intercepted much higher energy on clear days while the distribution of energy in the profile is fairly even on cloudy days.

YAO and SHAW (1964) conducted experiments to measure the effects of plant populations and planting patterns of corn on net radiation. There were two populations, 34,594 (single planting) and 69,188 (double planting) plants per hectare; three row spacings: 53 cm, 81 cm, and 107 cm; and two row directions: N–S and E–W. Net radiation was measured at 1 m above the crop surface $R_n(T)$, and 15 cm above the soil surface $R_n(G)$. The average ratio of $R_n(G)/R_n(T)$ for double plantings was generally lower than

* Foot-candle: the illuminance of light source of one candle at a distance of 1 ft. (30.5 cm) equals to 1 lumen incident per 10.76 m².

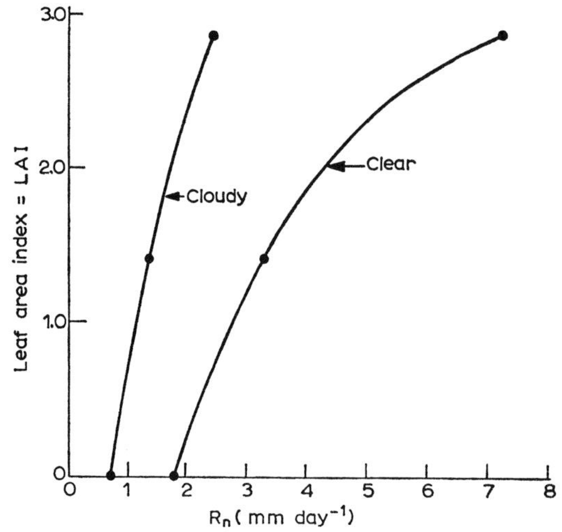

Fig.4. Average distribution of net radiation in the crop canopy for clear and cloudy days after maximum leaf area development. LAI is cumulated from the ground. (After DENMEAD et al., 1962.)

for single planting, because the higher population produced a greater leaf area, allowing less incident energy to reach the soil. With the same population the closer the row spacing the lower was the ratio, because the closer spacing resulted in a more uniform leaf arrangement and less penetration of radiation.

SAKAMOTO and SHAW (1967b) conducted a study to determine the pattern of light interception and distribution in a field soybean community. They found light interception occurred primarily at the periphery of the canopy. When the open space between rows closed or when it nearly closed, interception was primarily at the top of the canopy, (see Fig.5). LUXMOORE et al. (1971) measured the solar radiation in a soybean canopy. In the 310–2750 nm wave band, from 65 to 85% of the radiation received at a plane in

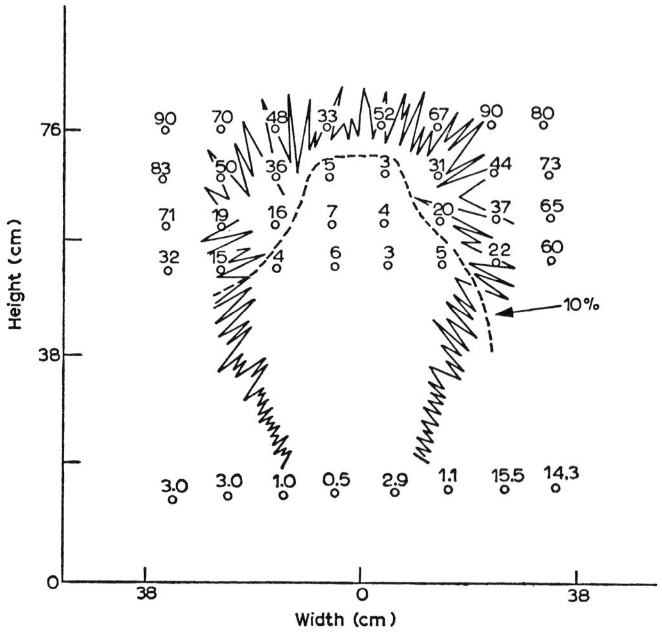

Fig.5.Cross-section of percent light interception pattern of a 36-inch (91.4 cm) spacing soybean community at stage 5 in 1964. Plants were permitted to lodge slight. (After SAKAMOTO and SHAW, 1967b.)

the crop came from the upper hemisphere and the remainder was reflected to the under surface. PENDLETON and HAMMOND (1969) investigated the relative photosynthetic potential (*RPP*) for grain yield of various leaf canopy levels of corn. They reported that the *RPP* of various leaf canopy levels was twice as high in leaves in the top third of the canopy as in the middle leaves, and 5 times as high as in leaves in the bottom third. There was also a linear decrease in *RPP* as defoliation treatments were applied at weekly intervals following tasseling.

Photosynthesis and climate

Photosynthesis is the process by which certain carbohydrates are formed by chlorophyll in cells from CO_2 and water with energy furnished by light. Photosynthesis occurs only in chlorophyllous plants and chiefly in the leaves.
The process of photosynthesis may be illustrated by the following chemical equation:

$$6CO_2 + 6H_2O \xrightarrow[\text{chlorophyllous cells}]{\text{radiant energy}} C_6H_{12}O_6 + 6O_2$$

It is essentially an oxidation-reduction reaction between CO_2 and water. Water is oxidized, CO_2 is reduced, and hydrogen atoms are transferred from water to CO_2. Light in the visible region is used as energy for this process. GAASTRA (1962) has summarized the following important processes for photosynthesis.
(*1*) A diffusion process—movement of CO_2 from external air near the leaf toward the reaction center in the chloroplast.
(*2*) A photochemical process—conversion of light energy into chemical energy which can be used for the reduction of CO_2 to carbohydrates. This process is influenced by light only.
(*3*) Biochemical process—energy produced by light conversion is used for the reduction —oxidation reaction between water and CO_2.
This process is strongly affected by temperature.

Carbon dioxide

Most of the higher land plants use CO_2 for photosynthesis. CO_2 in the air constitutes, by volume, on the average only about 0.03%. This small amount plays an extremely important role in the biosphere. Green plants are continuously removing CO_2 from the atmosphere, which is replenished from respiration and decomposition of plants, animals, and micro-organisms, volcanic activity, and combustion of organic fuel. The ocean is an important reservoir of CO_2.
The CO_2 fluctuates with low concentrations occurring generally in a growing crop at mid-day when the photosynthetic rate is high and lack of turbulent mixing of the air. Increasing the concentration of CO_2 in the air has a continuously favorable effect on photosynthesis when light is not limiting. THOMAS and HILL (1949), GAASTRA (1959), MOSS et al. (1961), and others have all indicated that with increased CO_2 in the air photosynthesis also increases. EGLI et al. (1970) experimented with three varieties of soybeans at Urbana, Illinois, with different CO_2 concentration. Their results showed close agreement with earlier researches. Fig.6 shows that increasing CO_2 concentration

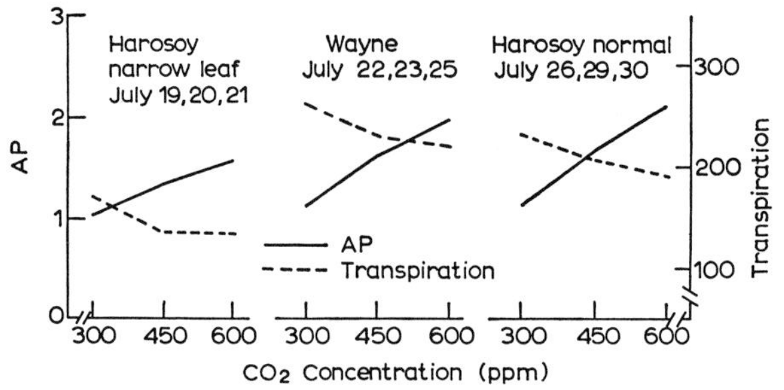

Fig.6. Mean daily apparent photosynthesis (mg CO_2 per m^2 grd. area per 15 min) and transpiration (ml H_2O per m^2 grd.area per 30 min) at three CO_2 levels. The linear effect of CO_2 on AP and transpiration was significant at $\alpha = 0.01$. (After EGLI et al., 1970.)

from 300 to 600 ppm results in an average increase, across varieties, of 72% in the mean daily apparent photosynthesis. However, a very high level of CO_2 may have a detrimental effect upon plant growth (THOMAS and HILL, 1949). Potato plants artificially exposed to ten times the atmospheric concentration of CO_2 during daylight deteriorated within less than 2 weeks.

Light intensity

Only a portion of the narrow band of the short wave, in the visible region of the solar spectrum, is absorbed by plant leaves for photosynthesis. The ability of plant leaves to absorb the visible spectrum of light varies considerably depending upon the kind of leaf and the intensity of the light. The rate of photosynthesis of most plant leaves increases nearly linearly with light intensity to some point where the leaf becomes light saturated. Photosynthetic rate becomes independent of the light intensity beyond this point. Many experiments have indicated that light saturation occurs in the vicinity of 2000–3000 ft.-c for many species of plants. However, the saturation light intensity varies considerably with plants (Table II). Fig.7 shows the light response curves for four species which represent three categories in which most of the common plants can be placed.

TABLE II

SATURATION LIGHT INTENSITY FOR PLANTS

Plant	ft.-c	Source
Sun species	2,500	BÖHNING and BURNSIDE (1956)
Shade species	1,000	BÖHNING and BURNSIDE (1956)
Sugar beets	4,400	THOMAS and HILL (1949)
Wheat	5,300	THOMAS and HILL (1949)
Corn	2,500–3,000	VERDIUM and LOOMIS (1944)
Soybeans (initial flowering)	6,000–6,400	SAKAMOTO and SHAW (1967a)
Soybeans (pod-formation and filling)	5,500	SAKAMOTO and SHAW (1967a)

Temperature

Air temperature also plays an active role in photosynthesis at saturation light intensity and normal CO_2 concentration. Fig.8 shows the rate of photosynthesis as a function of

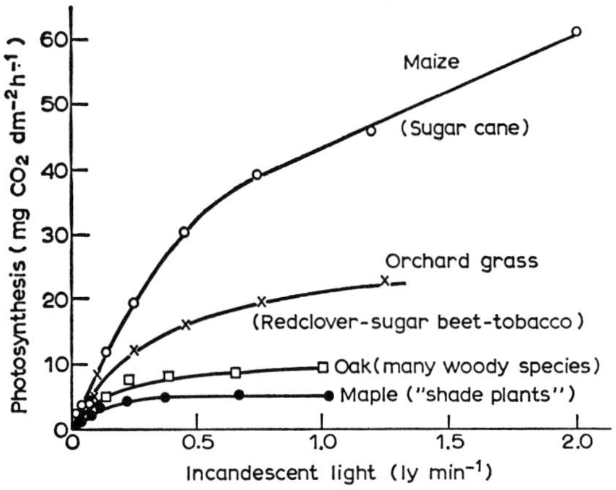

Fig.7. Effect of light intensity (in cal. $cm^{-2}min^{-1}$, or ly min^{-1}) upon the photosynthesis of 4 groups of plants. (Hesketh and Moss, 1963, cited by Moss, 1965.)

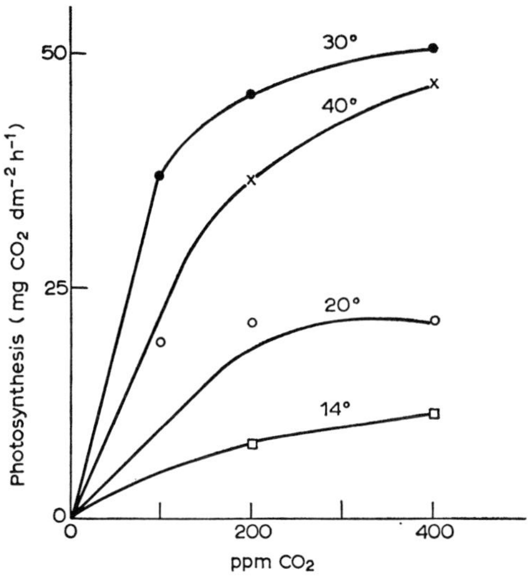

Fig.8. Relation between photosynthesis of maize and CO_2 concentration at different temperatures. (After Moss, 1965.)

temperature and CO_2 concentration. At a low temperature of 14°C, the activity of maize is very low. Activity increases as the temperature is raised to 30°C, but lowers again at 40°C. Fig.9 shows the relationship between temperature and photosynthetic rate for potato, tomato, and cucumber leaves. The photosynthetic rate increases with temperature, reaches a maximum and then decreases sharply at higher temperatures. The minimum temperature necessary for assimilation lies somewhere between 0° and 5°C, and photosynthesis ceases between 45° and 50°C. The optimum photosynthetic rate is in the neighborhood of 30°–35°C.

Dry matter accumulation depends upon both photosynthesis and respiration. The interaction of these two processes may be expressed in the following form:

$$N_p = P - RSP$$

where N_p is the net photosynthesis or net assimilation of dry matter. P is the photosynthetic rate while RSP is the rate of respiration. Photosynthesis occurs in the leaves

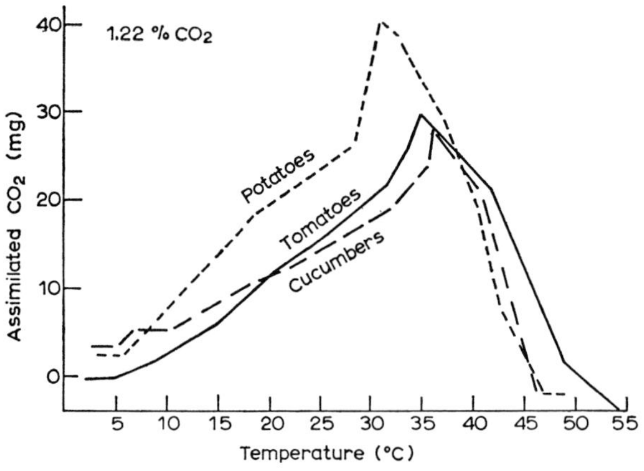

Fig.9. Dependence on temperature of photosynthesis in the leaves of potatoes, tomatoes and cucumber: *1* = potato leaves; *2* = tomato leaves; 3 = cucumber. (After VENTSKEVICH, 1961.)

when light is available, while respiration is a continuous process day and night throughout the plant, provided temperature is not limiting.

Respiration rates increase with temperature up to a maximum; this maximum temperature is different with different plants. THOMAS and HILL (1937) measured the respiration rates of alfalfa with the temperature range of 0°–22°C. They found that the relationship between temperature and respiration can be expressed by the lineaı regression equation:

$$Y = 0.533 + 0.078X$$
$$r = 0.927$$

where Y is CO_2 respired per 64 min (grams), X is temperature (degrees Celsius), and r is the regression coefficient.

Fig.10 shows the relationship between photosynthesis, respiration, and temperature. Net photosynthesis at a given time is the total assimilation minus the area respiration.

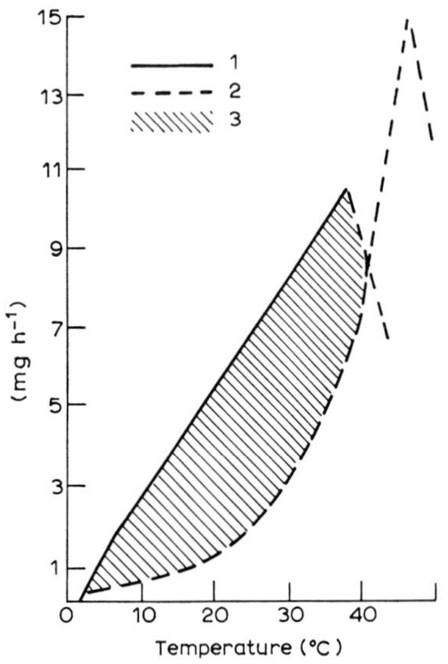

Fig.10. The relation between assimilation and respiration and the temperature of the outer environment. *1* = assimilation; *2* = respiration; *3* = organic matter increment. (After VENTSKEVICH, 1961.)

Water

Although less than 1 % of the water absorbed by plants is used in photosynthesis, experiments show that reducing the water content of leaves usually also decreases the rate of photosynthesis. Moisture stress in leaves causes the stomata to close, thus limiting the entrance of CO_2. Moss (1964) examined the photosynthetic rate of a tobacco leaf by illuminating the leaf from above by 5000 ft.-c and either 0 or 1000 ft.-c from below. His result showed that when light was from above only, the rate of photosynthesis was constant at 25 mg $dm^{-2}h^{-1}$. However, when a relative low illumination of 1000 ft.-c was added to the bottom of the leaf, the lower stomata partially closed and the rate of photosynthesis dropped to 14 mg $dm^{-2}h^{-1}$. This closuie could be prevented by circulating moist air around the leaf, thus maintaining the higher rate of photosynthesis.

Turbulence

In still air, CO_2 reaches the leaf by diffusion, a relatively slow process. There are local fluctuations in CO_2 concentration within the plant canopy. Fig.11 shows idealized profiles of CO_2 concentration in the air in an actively photosynthesizing corn field for different periods of a sunny day. CO_2 concentration within the corn canopy is very low during calm daytime periods when the photosynthetic rate is high. CO_2 near the leaves is depleted, while at some distance away from the leaf the concentration remains about 0.03 % (normal concentration in the atmosphere).

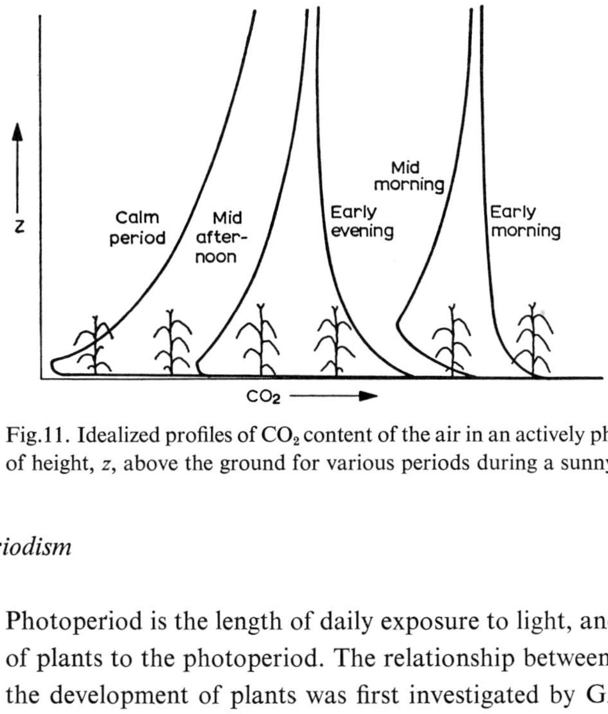

Fig.11. Idealized profiles of CO_2 content of the air in an actively photosynthesizing corn field as a function of height, *z*, above the ground for various periods during a sunny day .(After LEMON, 1960.)

Photoperiodism

Photoperiod is the length of daily exposure to light, and photoperiodism is the response of plants to the photoperiod. The relationship between the length of day and night and the development of plants was first investigated by GARNER and ALLARD (1920, 1923, 1930). The experiments were conducted in 1920 at Washington, D.C. Maryland. Mammoth tobacco grown outdoors always continued in a purely vegetative condition. It flowered profusely when grown under short days in the greenhouse in the winter.

Photoperiodic classification

GARNER (1933) classified plants into three groups. These are long-day (plants flowering only under daylengths longer than 14 h), short-day (plants flowering with less than 10 h of daylength), and day-neutral (plants can form their flowers under any period of daylength). An additional group was suggested by ALLARD (1938) known as intermediate-day (plants flower at a daylength of 12–14 h but are inhibited in reproduction by daylength either above or below this duration of daylength). Long-day plants include small grains, timothy, bromegrass, orchardgrass, and sweet clover; while soybean, millet, sweet potato, lespedeza, hemp, strawberry and certain varieties of tobacco belong to the short-day group. The day-neutral plants are crops such as cotton, tomato, buckwheat, annual bluegrass, carrot, celery and many kinds of beans.

Geographical distribution of plants

Generally, plants which originated in the tropics require relatively short days and long nights for flowering, and plants native to the high latitudes are long-day and short-night types. Through many generations of natural selection, species, varieties, and biotypes have developed responses to photoperiods by which they flower and fruit in definite seasons of the year. This difference in the comparative lengths of days and nights, as related to latitude, is an important factor in the natural distribution of plants.

There are also wide variations of photoperiodic response within plants species. In a given environment genotypes adapted to the daylength of that area may soon predominate. OLMSTEAD (1944) grew 12 strains of side-oat gramma *Bauteloua curtipendula* in Illinois, collected over a wide range of geographical areas from Texas to North Dakota. His results showed that these strains exhibited differences, particularly with respect to photoperiodic adaptation.

Light and temperature effect

HAMNER and BONNER (1938) and HAMNER (1944) conducted experiments to show that the photoperiodic stimulus was brought about during both the light and dark periods. They concluded that the most critical factor was the absolute length of the dark period for both long- and short-day plants. Instead of the long-day and short-day classification of photoperiodism, BORTHWICK et al. (1950) suggested a photoperiodic classification based upon the length of the dark period. A long-day plant is really a short-night plant and vice versa. His suggestion was based on the fact that a break in the middle of a 10- or 12-h dark period prevents flowering in the Biloxi soybean or cocklebut, while a similar break in the middle of a 12-h dark period promotes flowering in Wintex winter barley and other long-day plants.

Other than light, temperature is probably the most important factor affecting photoperiodic response. Many plants do not respond to critical photoperiod conditions provided their thermal requirements are met. Robert and Struckmeyer (1939, cited by WILSIE and SHAW, 1954) observed in a greenhouse at Madison, Wisconsin, that red clover grew most rapidly and flowered best with long days at cool temperatures. Sweet clover responded best to long days at warm temperatures. Alfalfa grew and flowered satisfactorily under long, warm days but did not set seed if the nights were too warm. SANDHU and HODGES (1971) reported that a combination of high intensity of light, a 16-h photoperiod and a temperature of 22.5°C produced more flowers and seed than other treatment combinations of chickpea (*Cier arietinum*); their treatment combinations included 3 temperatures: 15°, 22.5° and 30°C; 2 light intensities: 16,136 and 28,063 lux and 3 photoperiods, 8, 12 and 16 h.

Photoperiodic induction

It is interesting to note that some species can alter their daylength requirements in different stages of development. A long-day plant for floral initiation can be a short-day plant for fruiting and vice versa. Strawberry and spinach are the examples. The photoperiodic response can also be altered by artificial arrangement of light exposure. For instance, STOUGHTON and VINCE (1954) found that Mexican sunflower (*Tithonia*) requires exposure to short days of approximately 10 hours for 2 to 3 weeks to enable it to form initial flower buds, but after initiation, flowering continues regardless of day length in all subsequent shoots. Thus, *Tithonia* plants exposed to short days in the seedling stage will subsequently flower throughout the summer. Untreated plants, however, will not flower until autumn when nights are longer. KASPERBAUER (1969, 1970) gave a similar report on floral induction of Burley tobacco. Young seedlings of tobacco were started and grown to the 5–6 leaf stage under long, warm days, which favored growth, followed by controlled exposure of plants to short, cool days, which favored floral induction, for 2 to 3 weeks. They were then returned to long, warm days after floral induction and initiation to hasten growth and development of the flower. This effect of changing daylength and temperature in different stages of development is known as photoperiodic induction.

Temperature

Air temperature

Temperature is one of the primary factors affecting plant growth and its geographical distribution. Tropical plants have problems completing their life cycle in a cold environment, while cold region plants find it very difficult to adapt to the warm climate.

Cardinal temperature

There are limits of temperature beyond which plants cannot survive. These are the lethal temperatures. There is another set of temperature limits beyond which plant growth ceases. Between these limits there exists a point or a narrow range of temperature with which plant growth is most favored. These maximum, optimum, and minimum temperatures are generally referred to as "cardinal temperatures". The cardinal temperature can be defined within a narrow range. The cardinal temperature varies with plant species, and within different stages of growth for a given plant. For instance, the minimum temperature requirement for cotton during the stage of sprouting is 14°–15°C, while during the stage of fruiting it is 15°–20°C. The approximate range of cardinal temperatures varies with plant. Generally for cool-season crops such as wheat, oat, barley, rye etc., the cardinal (minimum, optimum, and maximum) temperatures are 0°–5°C, 25°–31°C, and 31°–37°C, resp. For warm-season crops such as melon, sorghum, rice, soybean, and corn, the cardinal temperatures are 15°–18°C, 31°–37°C and 44°–50°C.

Vernalization

LYSENKO (1925) discovered that a cold treatment of the germinated seeds before sowing or of seedlings before transplanting can hasten the time of flowering and result in an early harvest. This process of cold treatment is known as "vernalization", which can be applied to the cold-climate plants, and constitutes an important contribution in plant science.

Growing season

The growing season is defined as the length of the interval in days between the last killing frost in the spring and the first in the fall. It would be better defined as the length of the interval during which plants can grow. The length of the growing season varies from year to year. In general, the growing season increases in length from polar to equatorial latitudes, and decreases in length with increasing altitude. The temperature at which frost damage occurs depends on the resistance to cold of the vegetation. It would be more logical to use the occurrence of certain specific minimum temperatures such as 10°C, 5°C, 2°C, 1°C, and 0°C. The use of climatological data to determine probability dates for spring and fall frosts is a valuable asset to the farmer.

The growing degree day

The growing degree day may also be referred to as a temperature summation or heat unit system. It is an arithmetic accumulation of daily mean temperature above a certain threshold value, a simple means of relating plant growth, development, and maturation to air temperature. The essence of the growing degree day concept is that plants generally take the same number of heat units to pass through a specified portion of a developmental sequence. Early workers generally assumed that there was one base temperature throughout the life of the plant and that the plant response to temperature is linear over the entire temperature range, and that day and night temperatures are of equal importance to plant growth. Many investigators have since sought to overcome this weakness.

BROWN (1960) grew soybeans under controlled conditions and found that the average rate of development showed a curvilinear relationship to temperature. The threshold or base temperature occurred near 10°C and the optimum near 30°C (see Fig.12). BROWN (1964) further developed "heat units" for corn production in Ontario. Both maximum and minimum temperatures were taken into consideration. During daytime hours 10°C is usually considered to be the lowest temperature at which corn will grow. It grows fastest at 30°C as long as a plentiful supply of soil moisture is available. Above 30°C the growth rate drops off. This relationship is illustrated by the curve and equation shown in Fig.13. At night 4.4°C is assumed to be the lowest temperature at which corn will grow.

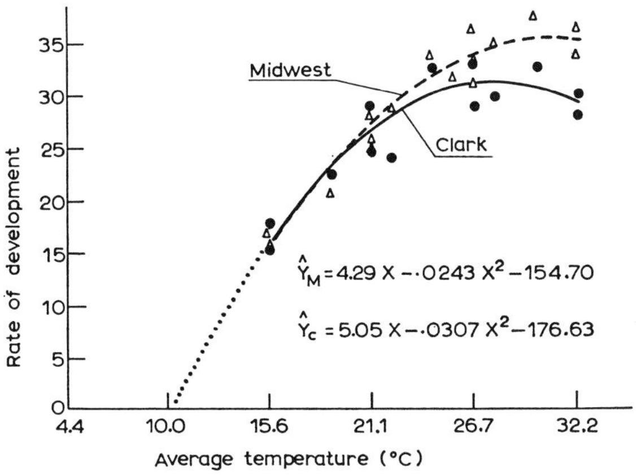

Fig.12. Relationship between temperature and rate of development of two soybean varieties grown under controlled conditions. (After BROWN, 1960.)

Temperatures are seldom too high for growth during the dark hours, thus the night-time relationship is a straight line as shown in Fig.13. The daily corn heat unit is the average of the maximum and minimum heat units. NEWMAN et al. (1967) introduced accumulation of radiation units in lieu of heat units and concluded that net radiant heat loads on vegetative surfaces in Langleys per unit time are a more sensitive climatic index of orange fruit maturity cycles than sensible air temperature. But LOWRY and VEHTS (1968), using the method proposed by Newman et al., found they are less well related to actual development than the traditional heat units.

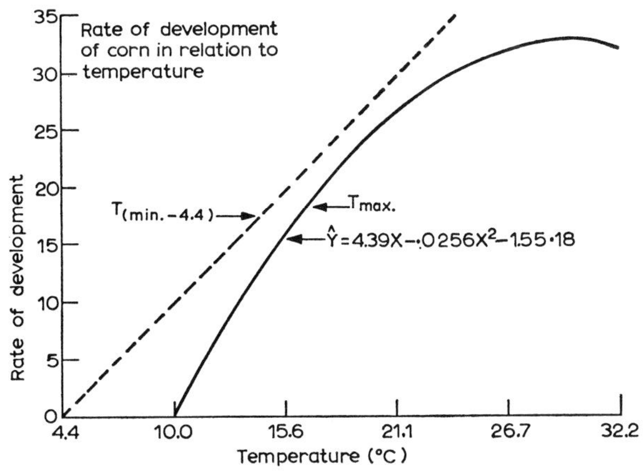

Fig.13. Relationships between daily rate of development of corn and night-time (minimum) and day-time (maximum) temperatures. (After BROWN, 1964.)

CAPRIO (1971) reported another method of rating plant growth, the solar-thermal unit (STU). This method takes into consideration not only the thermal unit (temperature), but also the incoming radiation. The equation of the STU is:

$$STU = \sum_{T_e}^{BB} Rad\,[\tfrac{1}{2}(T_{max} - T_{min}) - X]$$

where Rad = global radiation (0.3—4 μm) in ly per day; max = daily maximum temperature; min = daily minimum temperature; X = base of temperature, the original

equation use 31 (it could be another base with another crop); $T_e = \frac{1}{2}(\max - \min) - X =$ effective temperature; BB = time plant begins to bloom.
Caprio concluded that the STU concept provides the best model available for estimating the rate of development for many plant species in Montana.
The base temperature for growing degree days varies with plant. Generally for cool-season plants such as peas, lettuce, and small grains commonly 4.4°C is used while for warm-season crops such as corn, and beans 10°C is often used as the base temperature. For still other crops sensitive to low temperatures such as tomatoes or oranges, a base temperature between 13°–15.5°C is used.
Comparative studies among different methods of computing degree days have been many; GILMORE and ROGERS (1958) studied fifteen different methods of calculating heat units and found that corrections for low and high temperatures could improve the commonly used method of computing degree days.

Thermoperiodicity

It is common to many plant species that the greater growth rate and stem elongation occurs during the night. These plants are likely to have different optimum temperatures during both day and night. Experiments in the greenhouse by WENT (1944, 1945, 1950 and 1956) have shown that optimal growth for tomato plant was achieved by the combination of a 26.5°C temperature during the day and 17.5°C during the night. Went indicated that the optimal night temperature is 17°C with a possible deviation of 1°–2°C due to varietal difference. However, the process of sugar translocation has a different optimum temperature, with the requirement decreasing with increasing night temperatures. Night temperature is also important for many other plants. Rubber formation of the rubber plant proceeded at a fast rate at night temperatures between 5° and 10°C and was very slight above 15°C (BONNER, 1944). Valencia peanuts grow best with relatively high day and night temperature (JACOBS, 1951). Recently, HAROON et al. (1972) grew flue-cured tobacco (*Nicotiana tabacum* L.), under controlled conditions. They found a day/night temperature of 26°/22°C resulted in the best overall response with respect to total rate and uniformity of rate of seedling establishment.

Soil temperature

Soil temperature is a primary control of plant growth and development. In many instances, it has greater ecological significance to plant life than air temperature. Soil temperature first affects the germination of seed and later influences the root development and the growth of the entire plant.

Plant growth and soil temperature

Numerous experiments have been conducted to study the relationship between soil temperature and plant growth and development. THARP (1960) reported that the minimum soil temperature for both germination and early seedling growth of Upland varieties of cotton averaged about 15.6°C, the maximum about 38.9°C and the optimum near 34°C. HADDOCK (1964) recommended an average temperature of 21°C or warmer

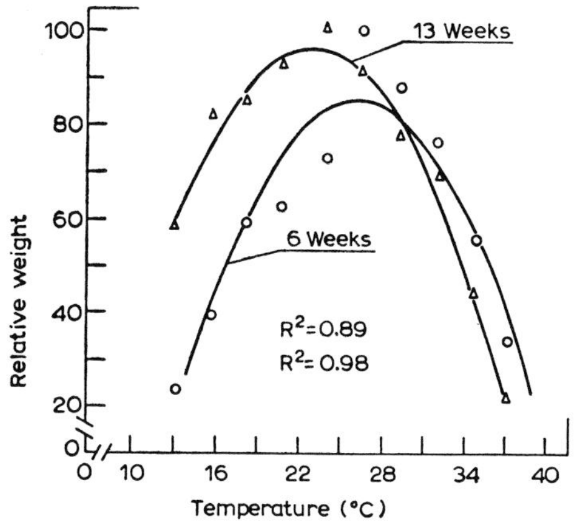

Fig.14. Relative dry matter (sugar beet tops) 6 and 13 weeks after emergence. Relative weights of 100 correspond to 12.0 and 210 g per six plants for 6 and 13 weeks after emergence, respectively. (After RADKE and BAUER, 1969.)

at the 2-inch depth for optimum cotton seed germination. RADKE and BAUER (1969) grew sugar beets (*Beta vulgaris* var. *saccharifera*) in a greenhouse with ten different constant root temperatures. They found that optimum root temperature for sugar beet emergence occurred in the range of 25°–35°C. The optimum root temperature for dry matter production in sugar beet tops progressively decreased from 26.3°C at 6 weeks to 23°C at 13 weeks following emergence. Conversely, high sucrose yields were obtained between 18° and 32°C with rapid decrease outside of this range (see Figs.14 and 15). WANJURA et al. (1970) presented an empirical soil temperature model for predicting initial emergence time of cotton planted at 5.1 cm. The correlation coefficient between predicted and actual emergence was 0.99. They concluded that the close agreement between predicted and actual emergence times indicates that seed level temperature is the chief determinant of emergence time.

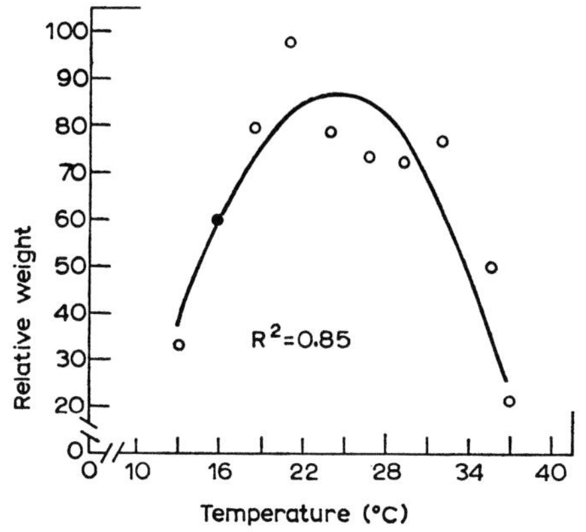

Fig.15. Sugar beet roots grown at the indicated temperatures (13 weeks after emergence). (After RADKE and BAUER, 1969.)

Environmental effects on soil temperature

The physical, chemical, and biological processes in the soil all are strongly affected by temperature. These processes in turn affect plant growth and development. BOUYOUCOS (1915) studied the effect of temperature on movement of water vapor and capillary moisture in soil. He found that the field capacity of mineral soils decreased as temperature increased. The chemical weathering of the parent material of soil, the rate of organic matter decomposition, and the mineralization of organic forms of nitrogen all increase with increasing temperature. The rate of growth and multiplication and activity of soil micro-organisms is influenced by soil temperature.

Physical properties

There are two independent parameters which control the thermal behavior of soils. These are "thermal conductivity" and "thermal diffusivity".

Thermal conductivity is generally defined as the quantity of heat which flows through a unit area of unit thickness in unit time under a unit temperature gradient. The thermal conductivity of soil depends upon its composition and its water and air content. Porous soils have much lower heat conductivity. However, when this soil is wetted, the heat conductivity is increased. Soil texture and organic matter content of soil also affect heat conductivity.

Thermal capacity is the amount of heat required to raise the temperature of 1 g (by weight C_p) or 1 cm^3 (by volume C_v) of some material by 1°C. The thermal capacity of a soil depends upon its mineral composition, texture, moisture, and air content. Water has a much higher thermal capacity than mineral soil. Therefore, the heat capacity of soil varies greatly according to its moisture content. Dry peat has a lower heat capacity than sandy and clay soils; however, in the wet state the reverse is true, because the peat's greater porosity can hold more water.

Thermal diffusivity is the quotient of thermal conductivity and thermal capacity used to measure temperature changes of any given substances. The thermal diffusivity of a given soil largely depends on its moisture. In general, it rises with increasing moisture content, reaches a maximum, and then decreases. The thermal diffusivity is usually low in the surface soil layer, especially during the summer when evaporation is high and surface of the soil is dry. Organic matter has a low value of thermal diffusivity, because of its low density. When the soil is compacted the value of thermal diffusivity is increased. Most of the common soils have a range of thermal diffusivity from 0.01–0.001 $cm^2\ sec^{-1}$.

Factors affecting soil temperature

The external factors affecting soil temperature include meteorological elements, such as solar radiation, air temperature, rainfall, humidity, winds, etc. Among these elements solar radiation (direct and indirect) is by far the most important element. (For the heat balance between the ground and the air, see p.192 above.) The intrinsic factors are discussed below.

Soils are classified according to particle size from coarse to fine, as sand, loam and clay. The porosity and the water holding capacity of soils increase as the soil changes from sand to loam to clay. The heat conductivity of soils is in the order sand > loam > clay > peat (VON SCHWARZ 1879; WAGNER 1883; KERSTEN 1949). However, SMITH and BYERS (1938) found that, with the possible exception of organic matter content, heat conductivity varies little among soils. The thermal capacity of soils is higher for clay than for sandy soils.

It was often observed that sandy soil warms up faster than fine textured loam and clay soils in the spring, while in the fall the reverse is usually true. These changes in temperature during spring and fall are largely attributable to the fact that sandy soil has a lower thermal capacity than loam and clay soils. The difference in moisture content among these soils also plays a role. The cooling effect of evaporation slows the warming of clay soils in the spring.
The daily maximum soil temperatures usually occur in the order sand > loam > clay especially in the upper layer of the soil. Fig.16 shows that, because of the small thermal capacity, sand has higher maximum temperatures than loam while loam is warmer than clay. Peat has a higher temperature than clay presumably due to its dark color.

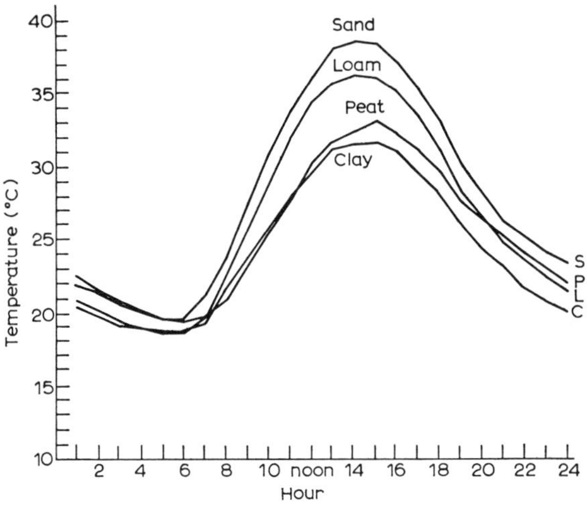

Fig.16. Daily course of soil temperature at 5 cm depth on clear summer days at Sapporo, Japan. (After CHANG, 1968.)

Soil temperature is also affected by its moisture content. Thermal conductivity and capacity change with the soil moisture content. Water percolated through the soil profile carries heat with it. Evaporation of water from the soil creates chilling effects. The effect of color on soil temperature has been studied extensively by WOLLNY (1878). He found dark soils are warmer during the warm season, and have a wider diurnal range of soil temperature than light-colored soils. The effect of slope on soil temperature is an effect of angle of incidence of insolation. The effect of exposure is generally small in low latitudes, while in middle and high latitudes of the Northern Hemisphere, southern slopes receive more insolation per unit area than northern slopes. A detailed discussion on topoclimate has been presented by R. Geiger in Volume 2 (Ch.3) of this series.

Daily and seasonal changes in soil temperature

The daily and seasonal changes of soil temperature are due largely to concurrent changes in solar and terrestrial radiation. The heat cycle, in the soil profile, is delayed and weakened with depth. The time lag of the maximum and the minimum of the heat cycle of homogeneous soil can be expressed as:

$$t_2 - t_1 = \frac{z_2 - z_1 T}{2\pi} \sqrt{\frac{\pi \delta C_s}{Tk}}$$

where t_2 and t_1 are the time required to reach the maximum or minimum temperature at the depth of z_2 and z_1, respectively, T is the oscillation period of the heat cycle, δ is the density of the soil, C_s is the specific heat, and k is the heat conductivity. The temperature range in the soil profile at any given point (ΔT) can be computed by the following equation:

$$\Delta T_z = \Delta T_1 \, e^{(z_1 - z_2)} \sqrt{\frac{\pi \delta C_s}{Tk}}$$

where $k/\delta C_s$ is the thermal diffusivity and k is the thermal conductivity and δC_s is the thermal capacity. Fig.17 shows the hourly average air and soil temperature at the indicated level for September 1954, Argonne, Illinois, while Fig.18 shows the annual progression of soil temperature based on 3-year (1954–56) average for each month at the same site. These figures clearly indicate the daily march and annual cycle of soil and air temperature and the time lag for maximum and minimum air and soil temperature.

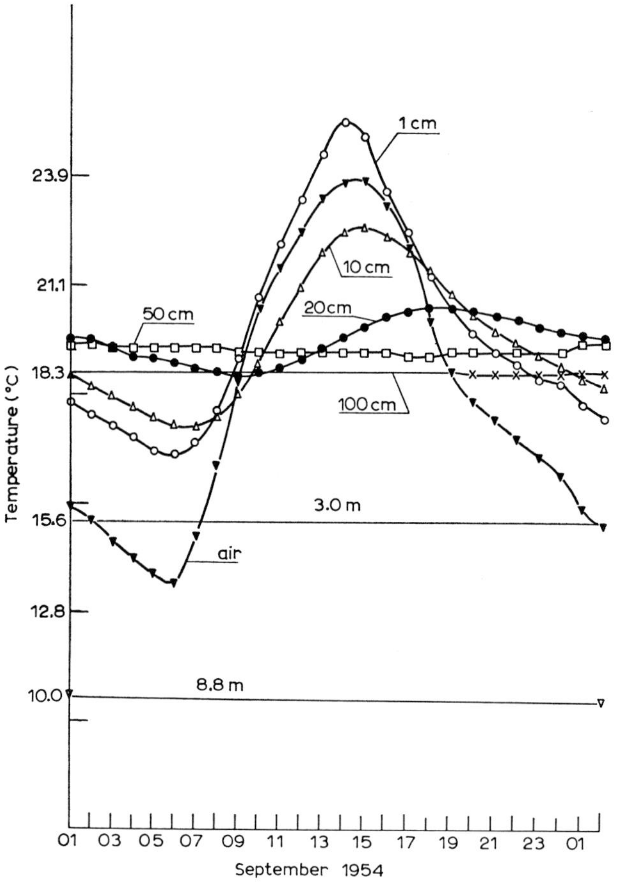

Fig.17. Hourly average air and soil temperature at indicated levels. (After CARSON, 1961.)

Leaf and canopy temperature

Leaf temperature

The energy balance of a leaf may be written as follows:

$$R_n = LE + A + P$$

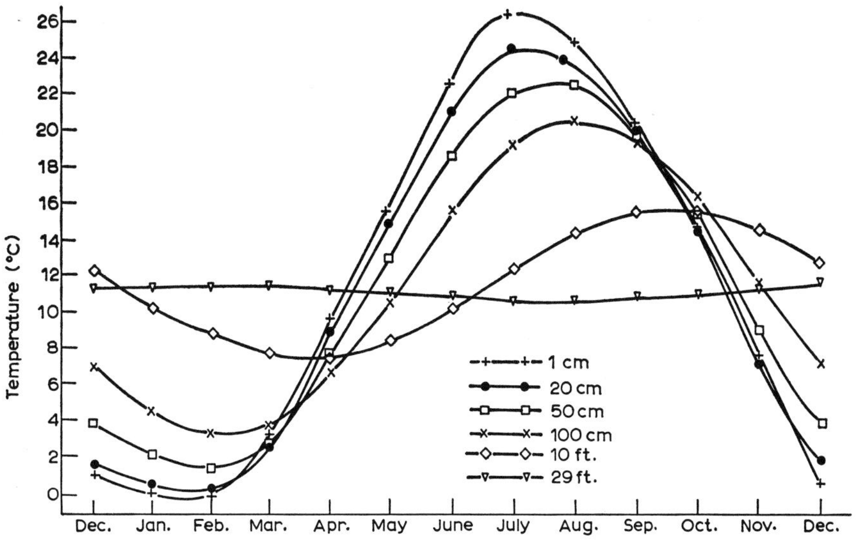

Fig.18. Annual progression of soil temperatures based on 3-year averages for each month. (After CARSON, 1961.)

where R_n is the net radiation; LE = transpiration; A = sensible heat flux, and P, a very small amount of energy gained from photosynthesis or lost by respiration, which is negligible and usually omitted.

Leaf temperature is influenced by many environmental variables acting upon it singly or in combination. Solar radiation, soil moisture, wind speed, vapor pressure, and ambient temperature are important variables affecting leaf temperature. WAGGONER and SHAW (1952) observed leaf temperatures of potato and tomato plants. A leaf perpendicular to insolation was 3.2°C warmer than a leaf parallel to insolation. Shaded leaves on all parts of the plant had nearly equal temperatures, while leaves exposed to insolation at the top of the plant and near the soil line were 7.8° and 12.0°C warmer, respectively, than the shaded leaves.

Leaves exposed to sunlight have higher temperature than the surrounding air (WAGGONER and SHAW, 1952), especially for thick leaves which retain more absorbed solar energy. However, there are also cases where the air temperature exceeds leaf temperature during the day. LINACRE (1964), using 42 different observations, established a highly significant linear relationship between air and leaf temperature of 0.90 (see Fig.19). Linacre concluded that the linear regression line crosses the line of equality at about 35°C and that above the temperature of equality there is a tendency for normal leaves exposed to bright sunshine to be below air temperature. Below 35°C the leaves tend to be relatively warmer than air temperature. PRIESTLEY (1966), studying the same problem, concluded that maximum temperatures observed over any wet surface including plants should be about 34°C. Leaves whose surface temperature exceeds 34°C suffer from water deficit. GATES et al. (1968) found that the temperature of small leaves of many desert plants is within 3°C of the air temperature, while the *Opuntia*, in the same locale, has a temperature of 10°–16°C above that of the air. Hence, leaf size is important to leaf temperature, and leaves 1 by 1 cm or less will remain close to air temperature. Leaves larger than 1 by 1 cm may have temperatures very much above air temperature, particularly if the leaves' internal resistance to moisture is large. It would seem that there

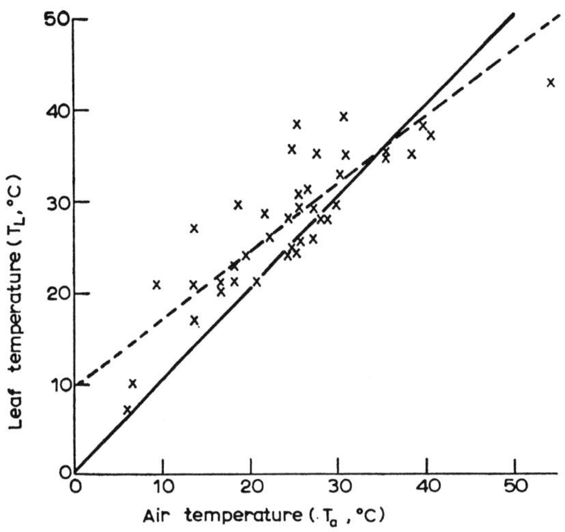

Fig.19. Relationship between representative published data on leaf and air temperatures. (After LINACRE, 1964.)

may be a physiological advantage for plants in arid or semiarid regions to have small leaves. Apparently, *Opuntia*, having large blades and using little water, has evolved a protein structure which is stable at high temperature (see Fig.20). CARLSON et al. (1972) studied the relationship between the temperature and moisture content of soybean leaves

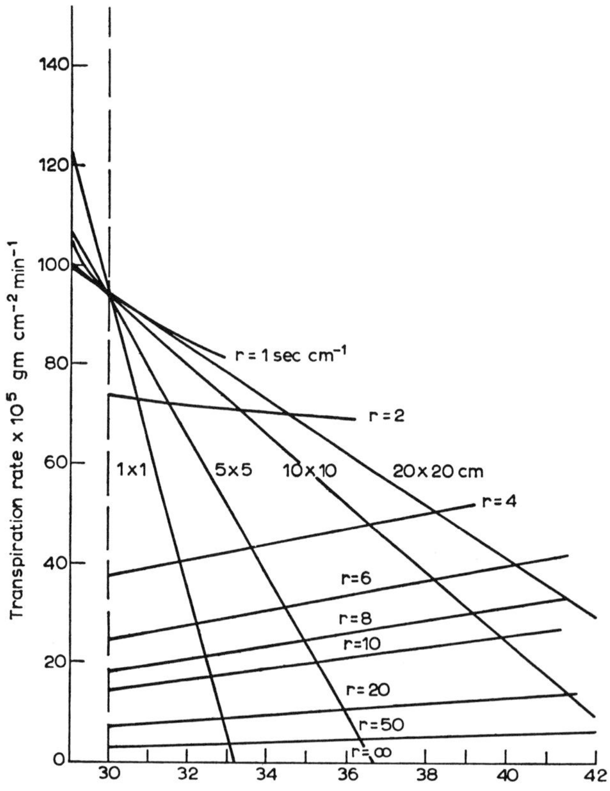

Fig.20. Transpiration rate versus leaf temperature as a function of leaf size and internal diffusion resistance at 1.2 cal. $cm^{-2}min^{-1}$ of absorbed radiation, for an air temperature of 30°C, a relative humidity of 20%, and a wind speed of 100 cm sec^{-1}. (After GATES et al., 1968.)

and air temperature. These temperatures were highly correlated with relative leaf water content (see Fig.21). Vapor pressure deficit and air temperature were also affecting leaf temperature. Leaf temperature increased with decreasing values of both relative leaf water content and vapor pressure deficit, as shown also in earlier work (GATES, 1964; TANNER, 1963; WIEGAND and NAMKEN, 1966; VAN BAVEL and EHRLER, 1968). Wind speed may change the vapor pressure gradient and therefore affect leaf temperature.

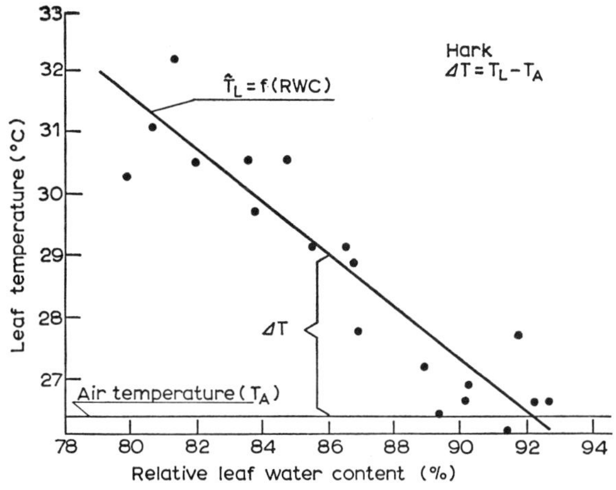

Fig.21. Measured leaf temperature (T_L) plotted vs. relative leaf water content for Hark soybeans on day 6. (After CARLSON et al., 1972.)

During the night, leaf temperatures are generally about the same as the surrounding air. But on calm and clear nights the leaf temperature can be much lower than the air temperature. Tomato leaves 1 cm high, had a temperature 1.8°C lower than the surrounding air (SHAW, 1954). On calm clear nights, the leaf temperature was 3.0°–4.5°C cooler than the temperature in the standard shelter; however, the difference decreased to 2°C for an overcast night. NOFFSINGER (1961) measured the leaf temperature for both papaya and pineapple plants. His results agree with Shaw's findings.

Canopy temperature

Many of the agro-climatological investigations of plant growth and temperature relationships used the air temperature recorded in the standard thermometer shelter. It is obvious that the shelter temperature differs from that of the canopy. NOFFSINGER (1961) indicated that the shelter air temperature at plant level does provide a good measure of plant temperature. It is therefore the measurement of canopy temperature that is important. RAHN and BROWN (1971) studied the relationship between the maximum and minimum temperature in a corn canopy and air temperature in the standard thermometer shelter. They found that canopy temperatures can be estimated by knowing the shelter temperature, the daily insolation, and the number of days since 0.1 inch (2.5 mm) or more of rainfall.

Moisture

Precipitation

Precipitation falls either in liquid (rainfall) or in solid form (snow, hail, sleet, etc.). It is the principal source of the soil moisture reserves, which in turn are the only source of the water required by plants. The importance of precipitation to agriculture is obvious. Precipitation is characterized by the amount of fall per unit time and is measured by the thickness of the layer of water which would have formed if the precipitation were prevented from runoff, seepage, and evapotranspiration. The total amount as well as the intensity of precipitation are both important. Heavy rainfall not only produces much runoff, but also can result in extensive soil erosion and, in the extreme case, land or mud slides. Light rain, infiltrating into the soil, is stored as soil moisture which can readily be used by plants. The form of precipitation is also important to agriculture. Generally, crops are most benefited by liquid precipitation. However, at high latitudes, snowfall contributes to an appreciable amount of moisture to plant growth when it melts in the spring. The time and frequency distribution of precipitation is of paramount importance to agriculture. Details on the frequency distribution are given by Essenwanger (Vol.1 of this Series).

Soil moisture

Soil moisture is directly related to plant growth. Most agricultural plants absorb their water from the soil. It is therefore important that the behavior of soil moisture should be understood.

Physical properties of soil moisture

BRIGGS (1897) classified soil water into "hygroscopic water", which is absorbed from an atmosphere of water vapor as a result of attractive forces on the surface of the particles; "gravitation water", which is the water that drains out of a soil by the gravitational force; and "capillary water", which is held by surface tension forces as a continuous film around the particles and in the capillary spaces. Capillary water can be removed by air-drying or plant absorption, but resists movement out of the soil by force of gravity. Capillary water is free water in that it can move from the soil into plant roots or move within the soil. The amount of water in the soil, at the time the plant can no longer absorb moisture from the soil and wilting occurs, is expressed in percent of the oven-dry weight of the soil and is commonly known as wilting point. "Field capacity" is the moisture condition of the soil when downward movement of capillary water into dry soil has virtually ceased. For a well-drained soil this condition occurred usually after 2–3 days of rain. BRIGGS and MCLANE (1907) developed the method to measure the field capacity by centrifuging a wet sample of soil at 1,000 times gravity. The amount of water in the soil determined by this method is known as moisture equivalent, which is approximately the same moisture content as field capacity. "Available water" is the water between the field capacity and the wilting point.

Energy relationship of soil water

Capillary potential. BUCKINGHAM (1907) introduced the idea that the flow of water through the soil could be compared to the flow of heat through a metal bar. The driving force was visualized as the difference in attraction for water between two portions of the soil that are not equally moist. Buckingham suggested the term capillary potential, which measures the attraction of the soil at any given point for water. It is the work required to pull a unit mass of water away from a unit mass of water against capillary forces in a column of soil, from a free water surface to a given point above this surface.
GARNER (1920a, b) showed that the capillary potential is a linear function of the reciprocal of the moisture content at least over a considerable range, which can be written as:

$$\psi = (e/H + b)$$

where e and b are constants, and H is the moisture content. RICHARD (1928) experimented with different soils, as explained by BAVER (1956) and found that a coarse-textured soil exhibits a high negative potential at a low moisture content. Finer-textured soils contain much more water at the same potential. Richard also concluded that moisture content, size of particles, and state of packing affect the value of the capillary potential (see Fig.22).

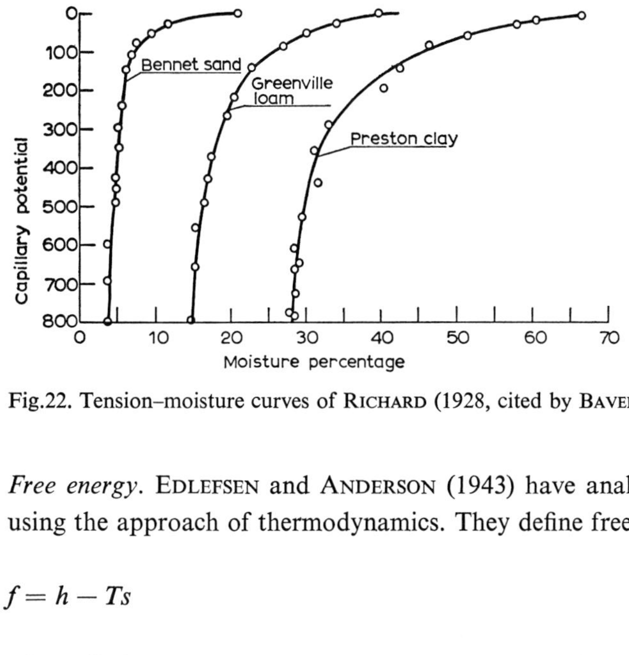

Fig.22. Tension–moisture curves of RICHARD (1928, cited by BAVER, 1956).

Free energy. EDLEFSEN and ANDERSON (1943) have analyzed soil-moisture phenomena using the approach of thermodynamics. They define free energy as:

$$f = h - Ts$$

where f is free energy, h is heat content, T is the absolute temperature and s is the entropy of the system. Any change in the free energy of a system Δf from one value to another represents the maximum useful work Δwm that can be obtained from the process when it occurs:

$$-\Delta f = \Delta wm$$

The free energy in saturated soil is zero. In unsaturated soil it is negative. Since conditions of constant temperature and pressure prevail under most instances where there are changes in soil moisture, the free energy function is considered a more appropriate function than the potential approach (BAVER, 1956).

Soil moisture measurement

The measurements of soil moisture may be classified into three different approaches: the direct measurement of soil moisture content (e.g., by sampling), the measurement of soil moisture tension, and the indirect measurement of soil moisture, e.g., by the use of neutron scattering devices. The best known instrument is the tensiometer using a porous clay cell and manometer. Its use was described by BAVER (1956).

The gravimetric method is direct and simple. A soil sample core is taken, and the moisture content determined by the difference in weight of a moist and oven-dry sample. A more recent method is the neutron scattering method, which is based upon the premise that hydrogen is more effective in slowing down fast neutrons than any other elements and that most hydrogen nuclei present in the soil are found in water molecules. A source of fast neutrons and a detector are housed in a probe which is lowered into a soil access pipe. Fast neutrons, radiated into the soil, are absorbed by hydrogen and then scattered back into the detector as slow neutrons. The count of slow neutrons is a measure of soil moisture content in the soil.

Soil moisture estimation and water balance

Knowing the rate of *PE* and soil moisture holding capacity, the water balance in the soil can be estimated with the available climatological data. THORNTHWAITE (1953) presented a simple bookkeeping method for scheduling time and rate of irrigation. The computations are simple, involve precipitation *PE*, *ET*, soil moisture storage, surplus, and deficit. Evapotranspiration is subtracted from the soil moisture storage while the precipitation is added. Precipitation plus the existing soil moisture in excess of the soil moisture storage capacity is considered as surplus (runoff and deep percolation), while moisture deficit is the amounts of moisture which have to be added to the available soil moisture in order to bring it to the level of potential rate of evapotranspiration.

SHAW (1962, 1963, 1964) developed a method of estimating soil moisture changes under corn, oats, and meadow. Moisture survey data from Iowa in the spring were used as a starting point. Factors used in this technique are soil moisture characteristics, precipitation, runoff, evaporation, degree of soil moisture stress, and active root zone depth at different stages of plant development. During the early part of the growing season 0.25 mm per day is used as the evapotranspiration from the top soil layer. Thereafter, adjusted open pan evaporation, which varies according to the stage of development and moisture stress, is used as potential crop evapotranspiration. Runoff from the field was estimated by means of an antecedent precipitation index which was developed by BUSS and SHAW (1960). The observed and the predicted soil moisture were reported highly correlated throughout the growing season. Correlation coefficients were 0.92 for corn and 0.97 for meadow.

BAIER and ROBERTSON (1966) presented a new technique for the estimation of daily soil moisture on a zone-by-zone basis which is known as versatile soil moisture budget (*VB*), using standard meteorological data. This new method as according to Baier and Robertson is modified from Holmes and Robertson's method known as modulated budget (HOLMES and ROBERTSON, 1959b). Modifications include simultaneous moisture withdrawals from several soil layers of varying moisture contents, using different types of soil dryness curves, and adjustments for the effect of different atmospheric demands on the *ET/PE* ratio. The new technique (*VB*) can be written in the following form:

$$ET_i = \sum_{j=1}^{n} \left[k_j \frac{S'_j(i-1)}{S_j} Z_j \, PE_i \, e^{-w(PE_i - \overline{PE})} \right]$$

Where ET_i = actual evapotranspiration for day i ending at the morning observation of day $i+1$; k_j = coefficient accounting for soil and plant characteristics in the jth zone; $S'_j(i-1)$ = available soil moisture in the jth zone at the end of day $i-1$, that is at the morning observation of day i; S'_j = capacity of available water in the jth zone; Z_j = adjustment factor for different types of soil dryness curves; PE_i = potential evapotranspiration for day i; w = adjustment factor accounting for effects of varying PE rates on ET/PE ratio; $\overline{PE}$ = average PE for month or season.
The coefficients k express the amount of water in percent of PE extracted by plant roots from different zones during the growing season. Baier and Robertson reported that comparison between daily soil moisture reading from Colman blocks and estimates from the modulated budget and from the *VB* showed the feasibility of estimating daily soil moisture from standard meteorological data. The estimates of the *VB* were superior to those from the modulated budget on a zone-by-zone basis from a statistical analysis.

Water use and plant growth

The function of water in plant growth can be summarized as follows. (*1*) Water is used as a solvent in which gases, minerals, and other nutrients move from cell to cell. (*2*) Water is the major constituent of physiologically active plant tissue. (*3*) Water is a reagent in photosynthesis which was discussed in the section on photosynthesis and the hydrolytic process, and also is an essential element for the maintenance of turgidity. (*4*) Water is needed also for transpiration.

Evapotranspiration

About 70% of the water which falls on the continental U.S. is lost not by runoff or deep seepage, but rather through direct evaporation from soil by transpiration through plants. Though it is a rough estimate, the relative importance of evapotranspiration to agriculture is clearly indicated. Evapotranspiration includes evaporation from all surfaces and transpiration through plants, while potential evapotranspiration, as defined by PENMAN (1956) is "the amount of water transpired in unit time by a short green crop completely shading the ground, of uniform height and never short of water".

Soil moisture and evapotranspiration

When soil moisture is maintained at the field capacity, water loss from the soil is maintained at the potential rate. The potential evapotranspiration is primarily determined by meteorological factors. As soil moisture is depleted below field capacity, the actual water losses lag behind the potential rate.

VEIHMEYER and HENRICKSON (1955) have discussed that evapotranspiration is essential for equal availability of soil water from field capacity to the wilting percentage. Their conclusion was supported by lysimeter observations of perennial rye grass (VEIHMEYER et al., 1960). THORNTHWAITE and MATHER (1955) presented a linear relationship between soil moisture content and evapotranspiration, based on the O'Neill, Nebraska, experiment. PIERCE (1958) presented still another curve, which was essentially exponential in nature. DENMEAD and SHAW (1962) reported from Iowa that the shape of the curve between evapotranspiration and soil moisture content was determined primarily by the potential transpiration rate, which in turn was affected by solar radiation (see Fig.23).

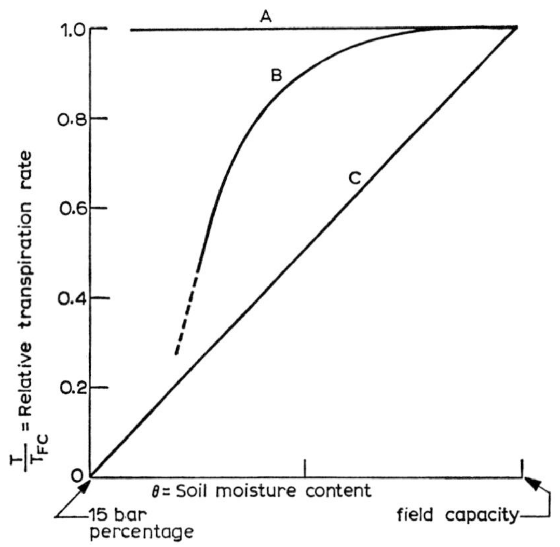

Fig.23. Various proposals for the relationship between relative transpiration rate and soil moisture content. For further explanation see text. (After DENMEAD and SHAW, 1962.)

Denmead and Shaw used three types of weather conditions: clear (*A*), partly cloudy (*B*) and overcast (*C*) to represent high, moderate, and low potential evapotranspiration rates. They found that when potential transpiration is high, the actual transpiration rate decreased as soil moisture content decreased and the decrease in transpiration rate occurred at a very low moisture tension. On the contrary, when the actual transpiration rate is low (overcast) the transpiration rate did not decline until the soil moisture tension was near the wilting percentage. From this experiment, Denmead and Shaw concluded that Veihmeyer and Henrickson's curve represented low potential transpiration, while Thornthwaite-Mather's curve was obtained at a high potential transpiration rate. Pierce's curve was obtained from a weighing lysimeter over several weeks time. During this period environmental condition would be expected to vary widely. His curve agreed well with those obtained under usual weather conditions of a moderate potential transpiration (partly cloudy sky).

Estimation of evapotranspiration

There are many different methods of estimating evaporation including both theoretical and empirical approaches. The theoretical study of evaporation is approached in two quite different ways, as reported by SUTTON (1953). The first method involves the mechanism of vapor removed by diffusion, and is primarily applicable for determining the local rate of evaporation. The second method considers the energy balance and is particularly applicable to large areas. ROSENBERG et al. (1968) presented a detailed discussion on the development, testing comparison, and applicability of the modern techniques of estimating evaporation.

There have been a large number of attempts to estimate evaporation based on various principles. Many of these are based on DALTON's (1802) discovery that evaporation from a free water surface is governed by the saturation vapor pressure e_s at a free water surface and the prevailing vapor pressure of the air above the water e_a. It became soon clear that the wind's removing vapor from the surface also played a great role, a number of empirical relations, all of the form $E_0 = (a+b\bar{u})(e_s - e_a)$ were developed with the constants a and b determined by local conditions (CARPENTER, 1891; ROHWER, 1931). Most widely used has been the one developed by PENMAN (1948), who used the wind speed at two meters u_2:

$$E_0 = 0.4(1+0.17u_2)(e_0 - e_a)$$

where e_0 is the saturation vapor pressure at the temperature of the surface.

Aerodynamic methods (see also E. L. Deacon, Vol.2, Ch.2, of this series) were first proposed by THORNTHWAITE and HOLZMAN (1939):

$$E = \delta k^2 \frac{(q_1 - q_2)(u_1 - u_2)}{[\ln(z_2/z_1)]^2}$$

where δ is air density; k is von Kármán's constant ($\sim$0.4); u_1, u_2, q_1, q_2 are measured wind speeds and specific humidity at height z_1 and z_2 above the surface.

PASQUILL (1950) modified this equation by subtracting a displacement height d from the two height values in the denominator of the equation. This is a height at which the wind vanishes. Other methods include those of PRIESTLEY (1955, 1958), DEACON and SWINBANK (1958), WEBB (1960), WILLIAMS (1961), and PRUITT (1963). ROSENBERG et al. (1968) presented a critique of these equations and concluded: "The original equations have undergone many refinements; the majority of these directed toward stability corrections which appear necessary for reliable estimates over most surfaces... Before widespread use of the techniques can be made additional refinements in measuring techniques and theory are required. The stringent requirements for fetch which appear necessary for proper experimentation with aerodynamic methods tend to reduce the practicality of these methods and their applicability to the topography and geometry typical of agricultural fields in many regions".

SWINBANK (1951) proposed an eddy correlation method to evaluate the turbulent flux of heat or vapor by using fast-response sensors. His equation is:

$$E = \overline{(\delta' w')q'}$$

where E is the upward vapor flux, δ', w' and q' are the averaged simultaneous departures from a temporal mean of air density, the vertical wind speed, and the specific humidity. Many different measuring instruments have been developed in recent years for measuring the eddy and its characteristics, including TAYLOR and DYER (1958), DYER and PRUITT (1962)—the evapotrone—and TAYLOR and WEBB (1955)—the machine analyzer. DYER et al. (1967) have reported on the development of the fluxatrone for measuring sensible heat flux above the ground. The eddy correlation method to estimate the vertical flux of heat and vapor is theoretically sound, but the difficulties of the instrumentation restrict its use.

The energy balance at the earth's surface (see p.191) may be written as:

$$R_n = S + A + LE + P + M$$

The last two terms are very small and can be omitted:

$$R_n = S + A + LE$$

Of these three factors, the term S (heat flux to the soil) has less effect than the other two terms, especially under a dense vegetative cover. DECKER (1959) reported in Missouri that for a moist soil in a corn field when the soil is incompletely covered, heat flux into the soil is about 15%. When the corn is fully developed, the heat flux into the soil decreased to 4%.

BOWEN (1926) recognized that soil heat flux S was only a small fraction of the net radiation, when soil moisture was not limiting. He proposed the Bowen ratio, a method of partitioning the energy used in evaporation and in heating the air:

$$B = A/E = \gamma(K_H)/(K_V)\,(T_s - T_a)\,(e_s - e_a)$$

where B is the Bowen ratio; γ is the psychrometric constant (0.65 for degree C and mbar, 0.27 for degree F and mm Hg); and K_H and K_V are the eddy diffusivities of heat and water vapor, respectively; T_s is the temperature of the surface; T_a is air temperature; e_s is vapor pressure of the surface; and e_a is the vapor pressure of the air.

The Bowen ratio is negative when heat is transferred from the air to soil and positive when the heat flow is in the reverse direction. In the computation of the Bowen ratio the equality of $K_H = K_V$ has also been assumed, but is doubtful. The Bowen ratio is, however, sufficiently accurate for practical use in computing evaporation. SUOMI and TANNER (1958) computed evaporation by using:

$$LE = R_n - S/1 + B$$

They found that error in LE is considerably smaller than any error in B if the assumption of identity between K_H and K_V is in error. FRITSCHEN (1965) has developed instrumentation for direct determination of the Bowen ratio over crop surfaces that was later improved by SARGEANT and TANNER (1967). He has also reported that errors of ET estimation were found to be on the order of 5% as compared with precision weighing lysimeters in Tempe, Arizona.

PENMAN (1948, 1952, 1956) proposed an equation which combines the aerodynamic and heat budget approaches. This equation is:

$$E = \left(\frac{\Delta}{\gamma} H + E_a\right) / \left(\frac{\Delta}{\gamma} + x\right)$$

where E = evaporation; E_a = an expression for the drying power of the air, involving wind speed and saturation deficit equals to $0.35(e_d - e_a)(1 + u_2/100)$; Δ = the slope of the saturation vapor pressure curve at mean air temperature obtained form standard table; γ = the constant of the wet- and dry-bulb psychrometric equation. The ratio of Δ/γ is dimensionless, and is effectively a weighing factor in assessing the relative effects of energy supply and ventilation on evaporation;

$$H_0 = (1 - r)R_1 - R_B \text{ mm/day} = 0.95R_A(0.18 + 0.55n/N) - \sigma T^4(0.56 - 0.09\sqrt{e_a})$$
$$(0.10 + 0.90n/N)$$

r = reflection coefficient; R_A = the theoretical maximum solar radiation that could reach the site in the absence of the earth's atmosphere; n/N = the ratio of actual to possible hours of bright sunshine; σT^4 = the theoretical black body radiation at mean air temperature; e_d = the mean vapor pressure of the atmosphere; u = the wind speed at a height of 2 m; e_a = the saturation vapor pressure at mean air temperature.

PENMAN (1963) explained that for open water $E = E_0$ and the quantity H is that appropriate to open water H_0, with $r = 5\%$ and $x = 1$. For a green crop $E = E_T$, and H has the value H_T with $r = 25\%$ and in the form given in 1952, x and $1/SD$ where S is a factor dependent on stomatal geometry ($S < 1$), and D is a day-length factor ($D < 1$) i.e. $x > 1$. BUSINGER (1956) discussing this equation suggests that it is better in practice to set $x = 1$ for this case.

Penman's equation has a much sounder physical basis than those of Thornthwaite and many other empirical formulae. The main difficulty in the use of Penman's equation for computing evapotranspiration lies in that it requires knowledge of vapor pressures, sunshine duration, net radiation, wind speed, and mean temperature. These parameters are measured regularly only at a few locations in the world. In addition, the conversion of E_0 to E_T is sometimes difficult.

Experiments to verify Penman's computed evapotranspiration values against lysimeter data, pond and lake evaporation include those of PENMAN (1949, 1952, 1956, 1963), VAN BAVEL and WILSON (1952), GERBER and DECKER (1961), WANG and WANG (1962). The comparison between Penman's equation and other empirical equations will be discussed below.

In humid or subhumid regions where available soil moisture is high, a high correlation of E_T and net radiation is generally obtained. Therefore, evapotranspiration in these areas can be determined by measuring net radiation. TANNER (1957) reported that E_T measured from a lysimeter was highly correlated with net radiation over an alfalfa-brome hay field. GRAHAM and KING (1961) found in Ontario, Canada a highly significant correlation between R_n and lysimeter data over a corn field. TANNER and LEMON (1962) indicated that when soil moisture is available with a substantial crop coverage most of the net radiation energy is used in the evapotranspiration.

In arid regions, in the presence of advected energy, the potential evapotranspiration may substantially exceed the net radiation. HUDSON (1965) measured evapotranspiration of lucerne at a rate of 2.5 cm a day. Two thirds of this transpiration was accounted for by advected energy. In the arid regions one often finds isolated agricultural lands which are generally irrigated. Solar radiation absorbed by the vast arid areas is mostly converted into sensible heat, which ultimately raises the temperature of the air mass. Solar radiation absorbed by those agricultural lands is largely converted into latent heat which tends to lower the temperature of the surface near air. Thus, in these agricultural land areas, the evapotranspiration is higher, and the air temperature is lower than the surrounding arid land. This vertical transfer of energy from the air above the crop in the agricultural field is referred to as oasis effect. In a small field, and for short period of duration the effect on microclimate can be large, however, in large oases and in the mean condition the effect is generally small. Table III shows the mean monthly differences in relative humidity and temperature between the large Amu Darya Oasis and the adjacent desert. It can be seen from this table that the differences of mean monthly relative humidity and temperature between those two areas can not be considered as large.

TABLE III

RELATIVE HUMIDITY AND TEMPERATURE DIFFERENCES BETWEEN THE LARGE AMU DARYA OASIS AND ADJACENT DESERT AT 2 M ABOVE THE GROUND
(From ALISSOW et al., 1956)

Hour	Mar.	Apr.	May	June	July	Aug.	Sep.	Oct.	Nov.
	Relative humidity (%):								
7:00 A.M.	11	4	7	16	22	23	21	19	10
1:00 P.M.	5	2	5	7	8	8	8	4	1
9:00 P.M.	3	3	7	16	21	23	13	16	1
	Temperature (°C):								
Mean monthly	−1.2	−0.6	−1.1	−2.2	−3.1	−2.8	−2.3	−1.7	−0.8
1:00 P.M.	−0.3	−0.2	−0.7	−1.5	−2.0	−2.1	−1.4	−1.3	−0.3

Most of the empirical methods of evapotranspiration computation rely on a high correlation of one or more meteorological parameters (usualy air temperature) with evaporation or evapotranspiration. Therefore, these techniques have geographical limitations. THORNTHWAITE (1948) proposed to estimate potential evapotranspiration by:

$$PE = 1.6(10T/I)^a$$

where PE is the potential evapotranspiration for a 30 day month; T is the mean air temperature, °C; I is a heat index, which is the sum of 12 monthly indices i given by:

$$i = (T/5)1.514$$

and a is a cubic function of I.
Temperature is the only parameter in computing potential evapotranspiration with a day length adjustment to correct the relationship for season and latitude. CHANG (1968) stated "the Thornthwaite formula works well in the temperate continental climate of North America, where the formula was derived and where temperature and radiation

are strongly correlated. This has been substantiated by MATHER (1954) in New Jersey, BAKER (1958) in Minnesota, DECKER (1962) in Missouri, BURMAN and PARTRIDGE (1962) in Wyoming. However, in other parts of the world the Thornthwaite approach has been less successful".

Shortcomings of the method include evaporation lags behind the annual maximum heat during the late spring and is out of phase in the fall; the incorrect assumption that evapotranspiration will cease when the mean temperature is below 0°C; significant errors when applied to short time periods with excessive variation in mean air temperature.

BLANEY and MORIN (1942) proposed an empirical equation by correlating pan evaporation with monthly mean temperature, relative humidity, percentage of total yearly daylight hours for each individual month, later modified by BLANEY and CRIDDLE (1950):

$$U = K_s F = \Sigma K_m f$$

where U = consumptive use in inches during the period of interest; K_s = seasonal or growing period consumptive use coefficient; K_m = monthly consumptive use coefficient; f = monthly consumptive use factor = $tP/100$; t = monthly mean temperature; P = monthly percent of total annual daylight; F = sum of the monthly factors (f) for the season (sum of the products of mean monthly temperature (t) in degrees °F and monthly percentage of annual daytime hours (P).

This equation has been used extensively, particularly in the western United States. Because of the development of the original equation in the semiarid region of New Mexico and Texas, this method works well in the semiarid region of the western U.S.A. For example, WILLIAMS (1954), SCHLEUSENER et al. (1961), and SOMNERHOLDER (1962) used the Blaney-Criddle equation to compute the water need for corn and pasture in Nebraska. TOMLINSON (1953) computed water need for native hay in Wyoming. A close agreement was found between actual water need and the computed consumptive use of water by using the Blaney-Criddle method in those areas.

There are many other similar empirical equations such as those of LOWRY and JOHNSON (1942), HAUDE (1952), MAKKINK (1957), and TURC (1961). Each may use one or more meteorological parameters.

Many comparative studies among various empirical methods for computing evapotranspiration can be found in the literature. VAN BAVEL and WILSON (1952) compared the Thornthwaite, Blaney-Criddle, and Penman methods with pan evaporation. They concluded that a fair agreement exists among these methods. HALKIAS et al. (1955) compared the Thornthwaite, Penman, and Blaney-Criddle methods against atmometer evaporation and gravimetric soil moisture measurements and concluded that only soil moisture samples yield reliable data on extraction. YAO and SHAW (1956) tested the same three methods with pond and pan evaporation in central Iowa. They reported that better results were obtained from Penman's method than from the other two, especially in short periods of less than 7 days. STANHILL (1961) tested 8 methods of computed potential evapotranspiration data against lysimeter data under arid conditions in Israel. He concluded that the Penman equation gave the best results for monthly and weekly periods. MCGUINESS and BORDNE (1972) tested as many as fourteen different methods plus lake and pan evaporation against lysimeter evaporation in Coshocton, Ohio for the period 1948–1965. They reported that fourteen methods of computing *PE* daily values were segregated into groups depending upon the climatic inputs required. In the group using temperature only, the Blaney-Criddle method gave the closest fit to the standard lysimeter data.

Lysimetry

A lysimeter is a tank buried in the ground to measure water loss through evapotranspiration and percolation. The early lysimeters were mostly very simple. BRIGGS and SHANTZ (1916) grew plants in large weighable pots to measure diurnal patterns of transpiration. KIESSELBACK (1916) grew plants in large cans of soil which were watered from reservoirs connected to perforated coils in the soil. Changes in system weight were determined with a block and tackle device. VEIHMEYER (1927) grew plants in a tank of soil; the quantity of water transpired or evaporated was determined by derrick and suspension scale weighings.

Only in recent years have sophisticated weighing lysimeters been designed in different parts of the world. Some of these lysimeters have an accuracy on the order of a hundredth of a millimeter over a period shorter than one hour. For instance, VAN BAVEL and MEYER (1962) reported that a lysimeter in Tempe, Arizona can measure a weight of change within 10 g. Fig.24 is one of the standard lysimeters which was installed at the University of California, Davis, California. It has a high sensitivity of $\pm$ 0.029 mm evapotranspiration.

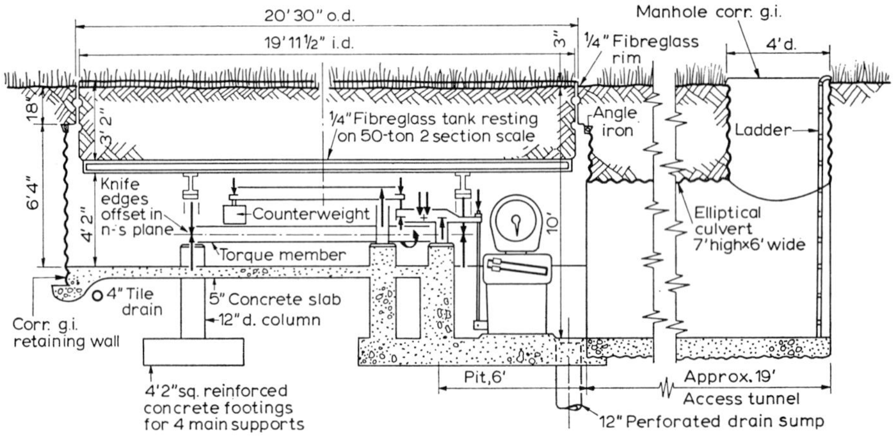

Fig.24. Weighing lysimeter. (After PRUITT and ANGUS, 1960.)

The lysimeter is considered to be the most direct, accurate and reliable means at present time available for determining true evaporation. However, a proper design taking into consideration those important factors affecting the accuracy of the lysimeters is extremely important. PELTON (1961) has discussed the design and the practice of weighing and floating lysimeters and potential evapotranspirometers. He considered the following to be of major importance in proper design of lysimeters: (*a*) a large area/rim ratio; (*b*) uniformity of soil mass to surroundings; (*c*) forced drainage or very deep construction; and (*d*) similarity to the surface surrounding.

A comprehensive report on the constructions and comparisons of different types of modern lysimeters was included in the World Meteorological Organization publication by GANGOPADHYAYA et al. (1966).

Water requirement for plant growth

Optimum water requirement

Evapotranspiration from plants can, in many cases, reach the potential rate. However, this high rate of water use does not necessarily indicate that it is the most efficient evapotranspiration rate. DENMEAD and SHAW (1962) and DALE and SHAW (1965a) have suggested the moisture stress day concept in which moisture stress days are defined as days on which the potential evapotranspiration is expressed as the percent of available soil moisture necessary to prevent the plant from losing turgor. Fig.25 shows that the turgor loss point of corn is a function of soil moisture supply and the atmospheric moisture demand. This curve in Fig.25 can be regarded as representing the optimum evapotranspiration for corn. This optimum is different for various plants, although for most of the field crops such as corn and soybeans, Dale's curve may be a close estimate for the optimum water requirement. For tropical plants a 1:1 ratio between actual and potential evapotranspiration may be needed. CHANG et al. (1963) studying water use and sugar cane yield in Hawaii found the actual yield approaches the potential evapotranspiration.

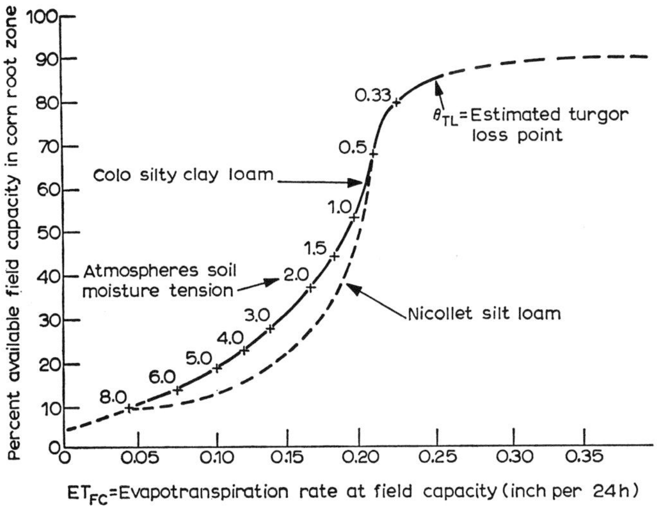

Fig.25. Estimated percent available field capacity in the corn root zone at the turgor loss point, θ_{TL}, as a function of the evapotranspiration at field capacity, ET_{FC}. The solid curve is from DENMEAD and SHAW (1962) for Colo silty clay loam, and the dashed curve is adjusted to 5 mm aggregate Nicollet silt loam by means of soil moisture tension curve. (After DALE and SHAW 1965a.)

Irrigation requirement

Many experiments have been conducted to determine the critical time that water should be applied. SIMONIS (1947) has studied the effect of water on yield of crimson clover with two levels of soil moisture, 80 and 40% of field capacity. His results showed that there was a significant yield difference between these two levels of soil moisture. The dry matter production was 128.7 mg for the 80% and 83.6 mg for the 40% levels of soil moisture content. SCHWANKE (1963) indicated for corn soil moisture levels below 60% as stress day, and found high correlation of stress days with plot yield. LIGON and BENOIT (1966) studied the effect of four levels of soil moisture treatments 100, 75, 50 and 25% of field capacity on tobacco plants. Sufficient water was not added to the soil until all soils reached their specific moisture levels. Their results showed that total dry leaf weight was significantly reduced for treatments in which soil moisture was allowed to reach the 50% field capacity.

Dew

Dew is water condensed onto plant and other objects near the ground from the water vapor of the surrounding clear air. The temperatures of these objects have fallen below the dew point of the surface air due to radiative cooling. The presence of dew can affect transpiration, leaf temperature, and disease development.

The relative importance of water contributed by dew was reported by BURRAGE (1972) who measured dew on wheat in Ashforn, England, in June–July, 1963. On the ten occasions measured, dew lasted from 4 to 14 h with quantities deposited ranging from 0.02 to 0.33 mm/night. In Germany, HOFMANN (1958) gave the upper limit of dewfall as 0.3 to 0.4 mm/night. MONTEITH (1963) measured the maximum dewfall on crops at Rothamstead, England: spring wheat 0.26, sugar beet 0.47 and grass 0.20 mm/night. FRITSCHEN (1972) measured dew on a Douglas fir tree using a weighing lysimeter in the state of Washington during a week of clear days in May 1972. The weight changes of the lysimeter indicate a considerable amount of dewfall on one day; the dewfall was equivalent of 0.87 mm of water while the total evapotranspiration during the days was 5.13 mm of water, indicating that the dewfall amounted to about 17% of the water balance. TULLER and CHILTON (1973) studied the role of dew in the moisture balance of a summer-dry climate. Dew amounted to 12–14% of normal monthly rainfall in mid-summer. In the unusually dry year 1970 it reached 154% of the rainfall in August.

Dew is generally not a significant contributor of moisture to plant growth. However, in arid regions it plays a dual role in meeting plant water requirements. It usually delays the morning rise in leaf temperature, thereby reducing the rate of evapotranspiration. WAGGONER et al. (1969) reported simulated dewfall on corn plants. Their results show that the rate of evaporation from the wetted crop (simulated dewfall) was at least twice that from the unwetted plant. Thus, transpiration losses are reduced when dew is evaporating. However, dewfall does not save an equivalent amount of water loss from a crop surface.

Dew may have an adverse effect on plants by providing the necessary medium for the development of many plant diseases. The late blight of potato requires dew-covered leaves for disease development. Southern corn leaf blight, which devastated the U. S. Corn Belt in 1970, requires a minimum of 8 h of free water for a viable spore to germinate and penetrate the leaf surface.

Fog

Fog differs from a cloud only in that the base of fog is at the earth's surface while clouds are above the surface.
In some instances the amount of moisture contributed by fog to plant growth can be significant. WENT (1955) noticed on the coast of southern California, immediately adjoining the ocean, a zone one-half to a few miles wide where tomatoes, peppers, beans and other vegetables can be grown in summer. These crops developed well even without irrigation, although no rain falls during the growing season from May to October. The soil in the region may contain enough moisture for the last few months. The source can hardly be anything but dew or coastal fog. In some mountainous areas fog interception by trees can be an important moisture source.
The physics of fog formation and its classifications are given by E. L. Deacon in Vol.2, Ch.2, of this series.

Humidity

Atmospheric water vapor content may be expressed in several ways, such as absolute humidity, specific humidity, mixing ratio or relative humidity. The measure commonly used is the relative humidity (R.H.), a dimensionless ratio between the actual mixing ratio and the saturation mixing ratio at the ambient air temperature. Air humidity has both beneficial and adverse effects in agriculture. The moisture contribution to plant growth is small, although many plants can directly absorb moisture from an unsaturated air of high humidity. HAINES (1952) and SLATYER (1958) report on such cases.
Air humidity has also adverse effects on plant growth, by promoting plant pathogens. WALLIN (1967) lists temperature and air humidity as the most important climatic factors affecting plant diseases. Relative humidities favorable for the growth and spreading of pathogens for different host plants and trees range from 85 to 100%. According to BOURKE (1955) the severity of plant disease is related to the number of hours R.H. exceeded 90% at any given temperature.

Wind

Wind affects evapotranspiration, CO_2 intake by plant for photosynthesis, mechanical damage, transport of pollen, insects as well as diseases and soil erosion.
Wind effects on photosynthesis and CO_2 concentration have been discussed in the section on photosynthesis (see pp.201).

Wind effect on evapotranspiration

DALTON (1802) already suggested that wind is a factor for evaporation. Since then, methods for estimating evaporation and evapotranspiration have incorporated a wind factor. The effect of wind on transpiration or evapotranspiration varies with plant species, the roughness of the surface, and other weather elements. Usually transpiration increases with increased wind speed up to some maximum, then levels off (STALFELT, 1932). But slight decreases at high wind speed have also been observed (HESSE, 1954).

In arid regions dry, hot winds sweeping over the plant surface, can result in rapid water losses.

Wind damage to plants

Strong winds may be harmful to plant growth. Plants suffer from hurricanes and typhoons not only by the torrential rainfall but also in many cases by the destructive winds. Leaves damaged by wind have a reduced capacity for photosynthesis and translocation as reported by HARTT (1963). To protect crops from destructive wind and to reduce evapotranspiration, shelter belts have been effectively used in recent years. This subject will be discussed later.

Transportation of pollen, disease, and insects by wind

Wind carries pollen pathogens, and insects from one area to another. These pollen, pathogens, and insects are generally harmful to plant and agriculture. The crop-killing locusts are an example of a pest which repeatedly in history has caused famine in Asia and Africa. HURST (1964) studied the migration of moths of Britain. In his analysis, both surface and upper air wind flow were used as important factors affecting the moth's movement. LOMAS and GAT (1967) reported on the effect of wind-borne salt on citrus production in Israel. They pointed out that salt concentration in the plantation followed closely the distribution of wind.

Wind profile near the ground

The wind profile near the ground has been discussed by E. L. Deacon in Vol.2, Ch.2, of this series. Deacon's discussion is for a smooth surface and applies to short crops. The structure of the wind profile over a tall crop is different from that over a short crop. LEMON (1963) presented wind profiles in and above a corn canopy. Wind profiles change abruptly at a height of about 220 cm, slightly below the canopy. Above this height, the logarithmic relationship seems to hold, while below it the wind speed is greatly reduced. The wind profile equation for a tall crop is modified to:

$$u = \frac{1}{k} \sqrt{\frac{\tau}{\delta}} \ln \left(\frac{z-d}{z_0}\right)$$

where u is the wind velocity at the height of z, k is von Kármán's constant, τ is the shearing stress, δ is air density, and d is the zero plane displacement, which is roughly on the order of the depth of the layer of air trapped among the plants, z_0 is the roughness of the surface; z_0 and d are geometric constants for the surface. The relationship along z_0, d and wind speed u is shown in Fig.26.

Generally roughness (z_0) increases with the height of the vegetation. Fig.27 shows the relationship between vegetation height and roughness parameter.

The value of the zero plane displacement (d) is a function of plant height, plant density and the mechanical properties of the plant. Plants standing erectly in the wind generally have a higher value of d than if they bend and weave with the wind. Considerable varia-

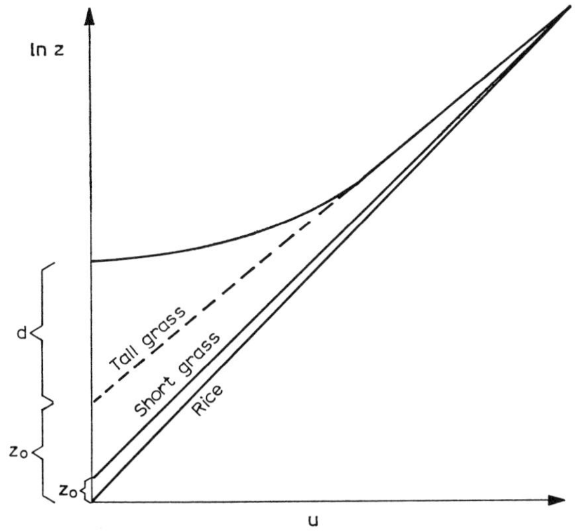

Fig.26. Schematic diagram showing relationship between wind speed and logarithm of height. (LEMON, 1963, cited by CHANG, 1968.)

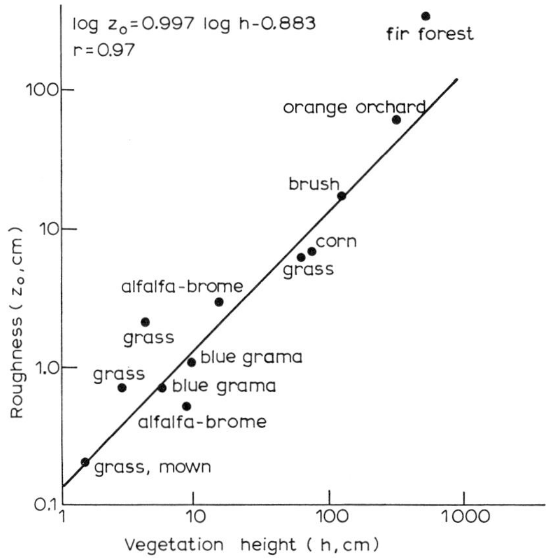

Fig.27. Graph for estimating roughness length from vegetation height. (After Kung and Lettau, cited by CHANG, 1968.)

tion of both z and d for tall crops which bend and weave with the wind have been observed by RIDER (1954) for oats, by LEMON (1960) for corn, by PENMAN and LONG (1960) for rice and wheat, and by INOUE (1963) for rice.

Agricultural climate

There are three subdivisions in this section, namely (*1*) crop, (*2*) forest, and (*3*) animal climate. Only a few of the more important agricultural crops are discussed in detail. These are: food for humans—rice (warm season) and wheat (cold season); food for animals—corn; fiber crop—cotton; oleaginous crop—soybean; vegetable crop—tomato; forage crop—alfalfa; and fruit crop—apple.

Crop climate

Rice

Rice (*Oryza sativa* L.) is one of the leading food crops, being the principal item in the diet of nearly one-half of the world's population. Rice is generally considered to be a tropical crop, but, it is also important in subtropical and temperate regions. FAO* (1963) reported that the continent of Asia including China, Japan, Korea, India, Pakistan, and the tropical region proceduces nine-tenths of the world's output of rice and that rice is steadily growing in importance in Africa. There has also been a considerable extension of rice cultivation in the Western Hemisphere, where the United States is second to Brazil in rice production.

Climate of rice

The climate favorable for rice is characterized by high temperature and abundance of available moisture. It is also very sensitive to the intensity and duration of sunshine. The variability of rice yield is sensitive to climatic variability.

Photoperiod

Rice is generally considered a short-day plant. Short daylength decreases the rice plant growth period. This type of reaction to photoperiod is basically similar in all varieties. However, the sensitivity of rice plant to photoperiod varies among varieties. In the tropical region the change in daylength is small. As the daylength increases the number of days from sowing to flowering of rice (in the Philippines) also increases. In Japan, MATSUO (1955) reported that paddy fields in the region west of Kanto are devoted mainly to the rice varieties of high photoperiodic response, while low photoperiodic response rice varieties are distributed among Hokkaido, Northeast, and Hokuriku. AZMI and ORMROD (1971) reported that rate of net CO_2 assimilation was consistently higher for four rice varieties in the U.S.A. (Bluebonnet, Caloro, Dokri, and Kangni cultivars) grown under short daylength than those grown under long (14 h) daylength.

Moisture

Water is the single most important factor affecting rice production. Rice fields are generally covered with water when the plants are from 15–20 cm high and remain submerged under 9–15 cm of water until the crop is nearly mature. Upland rice, which is relatively unimportant, is grown without flooding. Most rice production requires continuous flooding, hence high rainfall during the growing season or abundant irrigation water is essential for rice growth and development. The onset of the monsoon in tropical Asia usually determines the planting time.

The seasonal water requirement for rice cultivation varies from 750–1,500 mm. LOURENCE and PRUITT (1971) studied the water use of rice (*Oryza sativa* L.), using lysimeter data at

* Food Agriculture Organization, United Nations.

Davis, California. For a May 1 to October 1 period, a normal seasonal use of water by rice was 920 mm, a value only 3.5% greater than the 10-year mean of 889 mm of water used by fescue (*Festuca eliator* L.) during the same period.

Irrigation

Much of the rice crop is grown under irrigation. Table IV is the world rice acreage under irrigation. Data in this table were compiled from FAO 1963 World Rice Economy Report.

TABLE IV

WORLD RICE ACREAGE UNDER IRRIGATION

Countries	Percentage of rice acreage under irrigation
Japan	95%
Ceylon, Indonesia, Korea, Malaya, Taiwan (China)	50% or more
India	33.3%
Burma, Cambodia, Pakistan, Philippines, Thailand	16.7–25%
Australia, Egypt, European countries, U.S.A.	100%
Africa (excluding Egypt)	negligible
Central and South America	<50%

Temperature

Temperature is also a limiting factor in rice cultivation, but usually critical only in the temperate region. There minimum, rather than maximum, temperatures are important for rice plant growth, because the rice plant is fairly tolerant to high temperature. In the tropics, rice can be planted any time, provided moisture is not limited. In the temperate region, such as Japan, moisture is generally not the limiting factor during the growing season, but temperature often creates problems. Cold weather during the late spring may delay planting, while cold summer weather can seriously affect rice growth and final yield. INOUE et al. (1965) reported that the critical temperature for transplanting rice in Japan is 15°C for daytime mean temperature, but the growing of rice is possible until mean temperatures drop to 10°C in autumn. GRIST (1959) gave a range of average temperature required throughout the life cycle of the rice plant from 20° to 37.8°C. But VAN ROYEN (1954) suggested a minimum temperature for germination from 10° to 12°C for non tropical and 15.6°–20°C for tropical varieties and an optimal temperature between 32.2° and 37.8°C. Table V shows the experimental results from IRRI* (1970).
Optimum temperature was based on 100% germination while maximum and minimum temperatures were based on at least 10% germination.
The blooming or opening of the spikelets depends primarily on temperature, moisture, and light intensity. RAMIAH and RAO (1953) indicate that 28.9°–31.1°C appears to be the most favorable temperature while van Royen suggests the minimum temperature for flowering is 22.2°C.

* The International Rice Research Institute, Los Banos, Laguna, Philippines.

TABLE V

OPTIMUM, MAXIMUM AND MINIMUM TEMPERATURE FOR GERMINATION 3 AND 6 DAYS AFTER SOAKING

Variety	Temperature (°C)					
	3-day			6-day		
	opt.	max.	min.	opt.	max.	min.
IR5	26–30	33	19	19–33	40	16
IR20	26–30	40	19	19–33	40	16
IR8	26–30	33	19	19–33	40	16
IR22	26–30	40	26	26–33	40	19
Fugisata 5	28–33	33	26	26–33	33	19
Kula	26–30	40	26	19–40	40	19

Night temperature

The fact that low night temperature benefits rice growth is probably due to low rate of respiration, which in turn results in a higher rate of net assimilation. ABE and WADA (1957) reported from the warmer districts of Japan that high night temperature (at least 2°C higher than natural) promoted the Akiochi phenomenon (physiological disease—early death of lower leaves and root injury) and decreased the grain yield. However, their experiment also indicated that higher temperature during the heading stage resulted in superior yield. MATSUSHIMA and TSUNODA (1958) reported a marked effect on ripening in the period from the reduction-division stage in yellow-ripe stage. In most cases favorable effects were brought about by the daily range of temperature of 10° or 15°C, but the optimum ranges for ripening varied with day temperature and growth stage. Their experiment clearly showed that high temperatures in nighttime were much more harmful for ripening of rice grains than those in daytime.

Water temperature

Water temperature is also important to rice plant growth, in many cases even more so than air temperature. In general, standing water temperature is higher than ambient air temperature. INOUE et al. (1965), citing ENOMOTO (1937) and others to the effect that the rice plants growing in water with temperatures below 19°C, averaged over the period from transplanting to heading, never bear fruit even though the air temperature is sufficiently high. Rice plants need water warmer than 23°C to ripen perfectly. During the growing season, irrigation water is always colder than the standing water in the rice field. Cold irrigation causes chilling injuries to rice plants. MATSUO (1957) indicated that for rice plant growth a water temperature of 40°–43°C is the maximum and 13°C is the minimum. A suggested optimum temperature for seedling growth is 23°C. CHAPMAN and PETERSON (1962) found that water temperature between 25° and 30°C was most favorable for rice seedling establishment. Emergence of the shoot from water was most rapid at 30°C, but root development was favored at lower temperature. HERATH and ORMROD (1965) observed that generally shoot lengths and weights, root lengths and weights, leaf lamina and sheath lengths, and stomata number were greater at higher water temperatures. Stomata were larger and utilization of seed reserves was slower at lower temperatures.

Light intensity

CHANDLER (1963) found a linear relationship between yields of rice and the percent of possible sunshine. Cold and rainy weather reduces insolation and decreases photosynthesis and is therefore detrimental to rice growth. Rice yields are much higher in mid-latitude regions (Japan, Italy, Australia, China, and the United States) than in the tropics. Many environmental and generic factors contribute to these differences. However, differences in quantity of light in these areas are thought to be important yield factors. MURATA (1964), in the study of rice productivity, found a positive linear correlation between the yield and the mean daily insolation over a period from August to September. As solar radiation increases yield also increases in northern Japan. The low correlation in the southern part of Japan was interpreted as resulting from higher air temperature in southern Japan. ADACHI and INOUE (1970) observed that the flowering response of rice plants decreases with decreasing light intensity, and that the critical light intensity for flower initiation in the tested varieties is in the neighborhood of 900 lx under continuous light. STANSEL (1975) has also reported that the level of sunlight had a variable influence on each stage of development and therefore, on yield. He showed that the most critical sunlight-requiring period for grain yield is between the panicle differentiation and medium dough stage.

Wind

Light wind is considered beneficial to rice plant growth, because it results in CO_2 concentration for photosynthesis near the plant canopy. However, strong winds, such as during a typhoon, often damage rice plants and reduce yield. The rice growing regions in Asia are often threatened by typhoons, especially during the period after heading and at pollination time.

Wheat

Wheat is grown extensively throughout the world, because it can adapt to a wide range of climate. The most extensive wheat growing areas in the world are in the north and south temperate zones between 30°–55°N and 25°–40°S, such as the Great Plains of the U.S.A., the Prairie Provinces of Canada, southern U.S.S.R., northern China, Australia, Argentina, India and southern European countries which are the world wheat producing centers.

Wheat is an annual grass belonging to *T. monococcum*, and varieties are classified as winter, intermediate, and spring (referring to their inherent habits of growth under different environments). Winter wheat is generally sown in the fall, because when it is sown in the spring it usually remains prostrate on the ground throughout the growing season, producing no seed. Spring wheat is sown in the spring, but it can be sown in the fall and successfully completes its life cycle in a mild climate, as is the case in the southern Great Plains of the U.S.A. The difference between spring and intermediate wheat varieties is that the intermediate wheat varieties can not be sown in late spring in order to complete their life cycle whereas the spring wheat varieties can.

Climate of wheat

The most favorable climatic conditions for growing wheat were described by NUTTONSON (1955): a cool, moderately moist, growing season during which the basal leaves become well developed and tillering proceeds freely, merging gradually into a warm, bright, and preferably dry harvest period. KLAGES (1942) found that most of the wheat growing centers are located in the B, C, and D climates of the Köppen classification. Generally wheat is not extensively cultivated in regions having a growing season of less than 90 days nor in regions having annual precipitation less than 230 mm. Most of the important wheat regions of the world have average annual precipitation of less than 750 mm. High rainfall, especially if accompanied by moderate or high temperature, is generally unfavorable to wheat production, because it intensifies disease and insect attacks. The climates favorable for winter and spring wheat are different. Winter wheat requires a chilling period during the early growth to promote its later development, while spring wheat does not require a winter hardening period.

Photoperiod

Photoperiodically, wheat is classified as a long-day plant. There are great variations of required daylength between varieties. Some wheat varieties are able to complete their life cycle successfully in a wide range of photoperiod conditions. For instance, NUTTONSON (1955) reported that Marquis has a latitudinal range from at least 19°–64°N, the daylength of 11.6–20.3 h during the period of emergence to heading. Table VI shows that as the daylength decreased the growing period of all varieties increased. However, the daylength increases of from 12 to 24 h only moderately affected the northern varieties. The table seems to suggest that the geographical origin of the spring wheat varieties may have a certain imprint on the photoperiod reaction of wheat plants.

TABLE VI

GROWING PERIODS OF WHEAT (in days) UNDER VARIOUS DAYLENGTHS
(After Fedoseyeva (1939, cited by NUTTONSON, 1955)

Variety and origin	10-h day		Normal day		24-h day	
	without fertilizer	with NPK complete fertilizer	without fertilizer	with NPK complete fertilizer	without fertilizer	with NPK complete fertilizer
Novinka (northern Sweden)	75	75	58	58	55	55
Aurora (northern Sweden)	80	80	66	68	57	57
Hordeiforme 189 (Krasnyy Kut. U.S.S.R.)	93	91	70	68	70	70
Cicerella (Italy)	95	95	70	70	70	70

Temperature

Temperature is an important factor for germination and emergence. In Russia, ULANOVA (1975) reported that the optimum temperature for seed germination is 14°–20°C. Data from controlled growth chambers indicated that spring wheat grew best at a temperature

TABLE VII

AIR TEMPERATURE AND DAYS REQUIRED FOR WHEAT EMERGENCE
(After VENTSKEVICH, 1961)

Mean daily air temperature (°C)	6	8	10	12	14	16	18
No. of days between sowing and sprouting	25	17	12	10	8	7	6

of 14°–24°C. The average emergence was 43% lower at 14°C than at 24°C. According to VENTSKEVICH (1961) emergence of wheat will occur within a certain number of days after sowing depending on the temperature if soil moisture is not limited (Table VII). The higher the temperature after sowing the shorter the period required for emergence. The minimum, maximum, and optimum temperatures for effective germination vary with other environmental conditions (such as soil moisture, soil compactness, and the depth of planting) and with plant species, varieties, and the condition of the seed. However, Table VIII can be considered to present approximately cardinal temperatures for germination of wheat.

TABLE VIII

AVERAGE GERMINATION TEMPERATURE (°C) FOR WHEAT

	Minimum	Optimum	Maximum
Germination A	0–5	25–31	31–43
Germination B	3.5–5.5	20–25	35

A, from HALL, 1945.
B, from LOMAS, 1971.

During the winter, ULANOVA (1975) reported that the main causes of damage or death of winter wheat are freezing, soaking damage to plants by ice crusts and a number of other causes (erosion, heaving, and drying of the crops); sometimes damage and death of winter wheat in winter stem not from one factor, but from the joint action of two or more unfavorable factors.

During the vegetative period, temperature is the main factor influencing the duration of the vegetative period and of the growth phase of winter wheat. Table IX shows the in-

TABLE IX

TEMPERATURE EFFECTS ON DURATION OF INDIVIDUAL PHENOLOGICAL STAGE OF WHEAT
(After NUTTONSON, 1955)

Variety	Emergence to booting		Difference	Booting to headed		Difference	Total difference
	10°–12°C	20°–25°C		10°–12°C	20°–25°C		
Hordeiforme 10	28	20	8	54	33	21	29
Melianopus 69	28	20	8	45	28	17	25
Lutescens 62	29	19	10	52	24	28	38
Cesium 111	25	19	6	63	35	28	34
Milturum 321	30	21	9	64	30	34	43
Lutescens 25652 (Africa)	30	20	10	47	20	27	37
Borzum (Norway)	25	23	2	62	28	34	36

fluence of temperature on the duration of individual stages. It demonstrated that the higher the temperature (within limits) the shorter the duration of individual phenological stage, especially during the booting-to-headed phase when the shortened time resulting from higher temperature is noted. This appears to indicate that wheat growing during the booting-to-headed stage demands higher temperature. The optimum temperature for heading as reported in Russian data from a variety of regions for heading of spring wheat is 21.1°C. Temperatures above 22.2°C seem to slow down the beginning of heading. The temperature requirement during the ripening stage is also important. Palmova (1935, cited by NUTTONSON, 1955) reported that the northern wheat varieties ripen best at or above 12.2°C while the southern wheat varieties do not ripen in this temperature range. Spring wheat is generally subject to a considerable frost hazard prior to maturity, while a late spring frost corresponding with the flowering and early stages of kernel development constitutes a hazard in the production of winter wheat. LEONARD and MARTIN (1949) reported that wheat loss from low temperature (winter killing) is nearly as great as the loss brought about by all wheat diseases combined. Winter wheat may survive even as low a temperature as −40°C provided it had a chance to harden prior to the cold weather and is protected by a sufficient snow cover. They further pointed out that winter wheat in hardened condition, but without snow protection, may withstand as low a temperature as −31.7°C; that spring wheat during the early stages of its growth may withstand temperatures as low as −26.1°C. A temperature of −2.2° to −1.1°C during the heading stage of wheat is conducive to sterility and grain development is checked if a light frost occurs prior to the ripening of the crop. Wheat may be grown under rather high temperature conditions, provided that the period of high temperature does not coincide with a period of high atmospheric humidity. However, overly high temperatures during the earliest heading stage lead, according to NESTOROVA's (1939) observation, to a decrease in the number of grains in the spike and to lower yield.

Moisture

Wheat is not grown extensively in the humid regions due primarily to the prevalence of diseases and leaching of nutrients from the soil. However, in the spring and winter wheat growing region, lack of moisture is probably the major factor resulting in abandonment of wheat acreage. Winter wheat depends on favorable moisture conditions during the fall months for its best development. This is essential to the proper establishment of the plants prior to the advent of the period of dormancy enforced by low temperature during the winter months. Spring wheat needs moisture in late spring and summer. During the germination period, given a particular temperature, the higher soil moisture content shortens the period required for germination. Fig.28 shows the daily water use by winter wheat from late filling to ripening stage. Water use in early May was 1.00, 1.25, 1.50 mm/day for the three nitrogen treatments. Daily water use increases continuously and reaches a maximum during the heading-flowering stage (4.00, 4.50, and 5.25 mm/day) and decreases gradually thereafter. HANKS et al. (1969) grew winter wheat in a lysimeter in the central Great Plains. The results show, as Fig. 29 indicates, that the yield of winter wheat is highly correlated with evapotranspiration (*ET*). An approximately linear relation exists between *ET* and yield from the beginning of the measurements until maturity was reached in 1967. The wheat did not mature in 1966, because of severe hail damage.

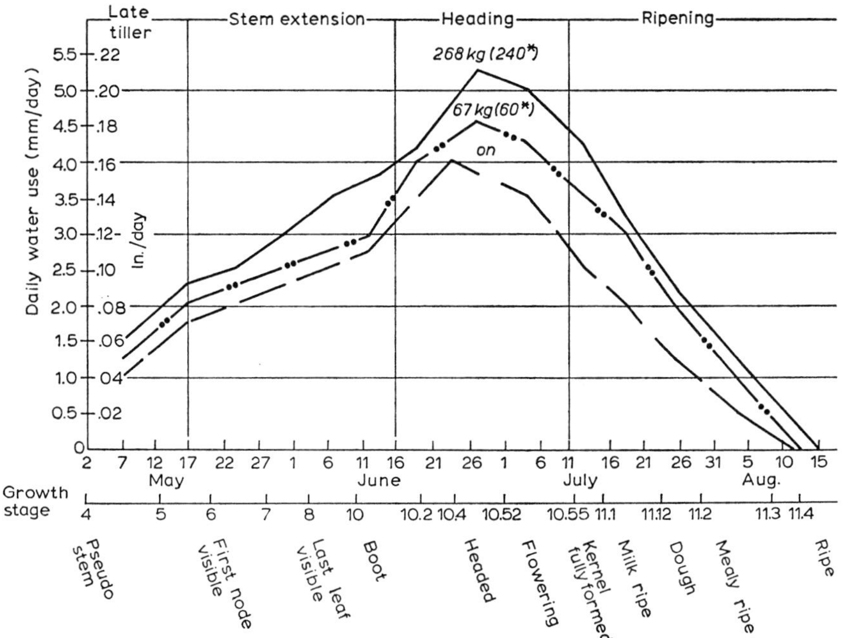

Fig.28. Daily water use by dryland winter wheat grown under favorable climatic conditions with 0, 67, and 268 kg ha^{-1} N rates, 1968. (After BROWN, 1971.)

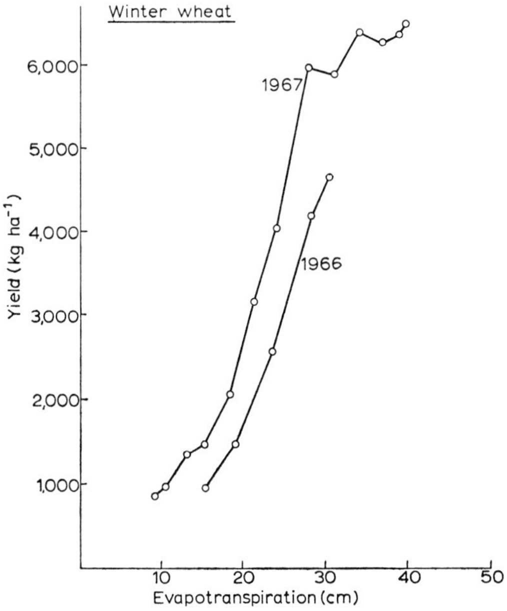

Fig.29. Cumulative evapotranspiration–cumulative yield relation for winter wheat. (After HANKS et al., 1969.)

Soil moisture stress at any stage of growth decreased grain yield (DAY and INTALAP, 1970). Moisture stress at any stage of growth lowers wheat pastry flour quality (DAY and BARMORE, 1971).

Light intensity

Light is also important to wheat growth. Cloudy days during the growing season result in less photosynthesis and less net assimilation, and consequently less grain yield. The relationship between light intensity and plant growth and the final yields was studied by PENDLETON and WEIBEL (1965). They concluded that shading greatly reduced wheat grain yields. Light is very critical during the heading stage and even slight restrictions for short periods result in reduced yields. CAMPBELL et al. (1969), in a study of the influence of solar radiation in the growth and yield of Chinook wheat, found that mean leaf area ratio decreased and mean net assimilation rate and relative growth rate increased linearly with increase in the logarithm of light intensity. In a lysimeter shading study, they found that days to heading and the percentage of crude protein content were negatively related to light intensity, but all the other characters were positively related.

Corn

Corn (*Zea mays*) is a warm season crop, originated from the remote Andropogonaceous ancester in the lowland of South America. It is widely cultivated in the world from Canada and Russia (near 58°N) in the Northern Hemisphere to southern Australia (40°S) in the Southern Hemisphere. The major producing center is the Corn Belt of the U.S.A., with nearly half of the world production. Next in rank is Argentina, followed by China, Brazil, Romania, Yugoslavia, the U.S.S.R., Italy, India, and many other countries.

Climate of corn

Corn is a heat-loving plant. It is not resistant to frost. A relatively mild frost will damage it at all stages of development. Despite its heat requirement corn also cannot withstand high temperature. The climatic requirements for corn can best be illustrated by the Tables X and XI.

TABLE X

IDEAL CORN CLIMATE (TEMPERATURE AND PRECIPITATION IN CENTRAL OF THE CORN BELT)
(From WALLACE and BRESSMAN, 1949)

Month	Mean temperature (°C)	Precipitation (mm)
May	18.3 (warmer than ave.)	89
June	21.7	89
July	22.8 (cooler than ave.)	114
August	22.8	114
Sept.	warmer and drier than ave.	
October	same as September	

Table XI shows the average weather conditions of two years with highest average state yield in Iowa 1948 (3.75 t/ha) and 1952 (3.88 t/ha). It can be seen that the weather of the two highest years of corn yield in Iowa was close to the ideal corn climate.

Photoperiod

Photoperiodically corn is in the short-day or in the day-neutral class. The response of the corn plant to changes in daylength is not so pronounced as for some other plants. The

TABLE XI

MONTHLY TEMPERATURE (°C) AND PRECIPITATION (mm) FOR 1948 AND 1952, IOWA
(After SHAW, 1955)

Month	1948		1952	
	temp.	precip.	temp.	precip.
May	15.0	58.4	15.6	96.5
June	20.0	86.4	23.3	137.2
July	23.9	104.1	23.9	96.5
August	23.3	63.5	21.1	121.9
September	warmer and drier than normal		very dry	

period from emergence to flowering is reduced by short days and increased by long days. The flowering of early-maturing northern varieties adapted to long summer days is hastened when these are grown nearer the Equator, while the growing season of southern varieties is lengthened when they are moved north. However, through many generations of natural selection the strains of corn grown in different latitudes have adapted to the length of day in the locality in which they are grown.

Temperature

Temperature governs corn germination. As temperature rises, the time required for emergence decreases (Table XII). ALESSI and POWER (1971) conducted field and growth room experiments to evaluate the effects of seed depth and soil temperature on corn germination and emergence. Increasing soil temperature from 13.3° to 26.7°C reduced the time for 80% emergence. Temperature had a much greater effect than seed depth.

TABLE XII

TEMPERATURE AND CORN EMERGENCE DAYS
(After WALLACE and BRESSMAN, 1949)

Temperature (°C)	Emergence days
10.0–12.8	18–20
15.6–18.3	8–10
>21.1	5–6
<12.5	susceptible to root rot

During the vegetative growth period (from emergence to tasseling) young corn plants are relatively resistant to cold weather with a temperature of −1.1°C or perhaps a little lower generally killing exposed above-ground parts (HANNA, 1924). Chilling and high temperature could also result in damage to the plant resulting in yield reduction during this early stage of growth. In general, the younger the plant the greater the injury. Six weeks old plants chilled at a temperature 0.05°–5.0°C for different lengths of time recovered and produced seed if less than 25% of the area was injured soon after chilling but those injured more than 50% seldom recovered (SHAW, 1955).

Temperature previous to the silking time is very important in determining the time of tasseling. After tasseling starts, heat no longer plays such an important part. Hot weather has less to do with hastening ripening as it does with causing rapid growth before tasseling SHAW and THOM (1951) found that in a hot dry year the rate of silking was much slower than in a normal year. The period prior to silking is very sensitive to weather. The period from silking to maximum dry weight is relatively independent of weather variations.

Moisture

The importance of water in corn production was noted by SHAW et al. (1958) with water use in the early season averaging nearly 2.5 mm/day. Similar results were also reported by ROBINS and RHOADES (1958). From early June to mid-August in the Corn Belt the water use is about 4.5 mm/day. Peak use may be for short intervals up to 10 mm/day, mostly for evapotranspiration with little runoff or percolation. Water use gradually decreases after mid-August with a maturing plant cover and lower solar energy. HARROLD and DREIBELBIS (1951) reported that the amount of water transpired and evaporated in the production of corn crops, computed on an acre basis, ranged from 442 to 625 mm for three different years. Depletion of soil moisture by corn was the greatest in July. The corn frequently suffered from lack of water in August and September.

The rapidity of germination is also related to the soil moisture content. Germination increased at 80% soil moisture. At 10% there was no germination, whereas at 100% saturation or above germination was retarded or prevented owing to lack of oxygen. Thus, at planting time preseasonal moisture stored in the soil is a primary factor in determining success or failure of corn production, especially in regions where water supply is generally deficient.

The period from tasseling to kernel formation is critical and has the highest water need (HOLT and VAN DAREN, 1961). Moisture stress during the silking stage reduced yield about twice as much as when similar amounts of stress occurred during the vegetative period or during ear development (DENMEAD and SHAW, 1960). Severe stress for one or two days during the tasseling or pollinating period can reduce grain yield more than 20% (ROBINS and DOMINGO, 1953). During the periods when water use exceeds supply moisture stress hurts corn the most.

Other factors such as light intensity, winds, relative humidity and hailstorms in the Corn Belt all result in comparatively small effects on corn growth and production.

Cotton

Cotton is a plant of the warm regions of the world, but it is also adapted to the temperate regions, it belongs to the genus of *Gossypium* and is long-lived tropical and subtropical perennial. The latitudinal distribution of cotton in the American continent is roughly between 37°N and 32°S, while in the old world the northern limit is near the 47° parallel (Ukraine) and the southern limit is near 30°S in Australia. The major cotton producing centers include the Cotton Belt of the U.S.A., the cotton growing areas of India and Pakistan, the Yangtze Valley and the North China Plain of China, the U.S.S.R., Egypt, Brazil, Peru and other minor areas.

Climate of cotton

Cotton is a heat-loving plant. It cannot withstand frost. A growing season of 180–200 days is required. Climate rather than soil is the chief factor in determining where cotton can be grown. DAINGERFIELD (1929) gave the best climate for cotton as: "Other things being equal, the ideal year for cotton would be one in which there was good soil moisture storage during the preceding winter, which should be sufficiently cold to destroy the hibernating pests. This is followed by an early spring of moderate rainfall, promoting planting and cultivation, with a moderately dry, hot summer, with abundant sunshine, but not really droughty and not subject to sharp reversals in rainfall or temperature, thus favoring care and growth of crop and holding down weevil infestation. Finally, a fairly dry bright autumn and late frost to remove all of the cotton leaves from the field without deterioration or loss."

Photoperiod

Photoperiodically, cotton is day-neutral. Thus, daylength does not appreciably affect cotton growing. However, LEWIS and RICHMOND (1960) indicated that cotton has numerous characteristics, such as plant height, square formation, and flowering, that are affected by photoperiodic variance. Photoperiod is the primary factor in the control of fruiting and some stocks do not initiate flowers unless the daylength is below 12 h, regardless of the number of nodes the plants had developed (HUTCHINSON, 1959).

Temperature

Cotton seed germinates poorly at temperatures lower than 21.1°C, and the seedlings grow more slowly and less vigorously at these low temperatures than those which germinate at 21.1°–29.4°C. The minimum soil temperature for both germination and early seedling growth of upland varieties averages about 15.6°C, the maximum about 38.9°C and the optimum near 33.9°C (THARP, 1960). VENTSKEVICH (1961) indicated for Russia that emergence of cotton occurs within 12–15 days at the temperature of about 15°C whereas at 20°C it appears at about 6–8 days. HOLEKAMP, et al. (1960) reported that soil temperature can be used as a guide to timely cotton planting. They found that soil temperatures at 20 cm depth that average 15.6°C or above for the 10 days preceding planting are more likely to give good cotton stand. They also found that planting according to the specified soil temperature resulted in a higher lint yield and earlier maturity. There is a sharp decrease in emergence time as seed level temperature increases from a range slight above 10°C to little over 20°C. According to WANJURA et al. (1967), 103 h of seed level temperature above 18°C are required for initial emergence of the cotton variety (Blightmaster). Unfavorable soil moisture, soil texture and variation in planting depth could affect the emergence index.

During the active growing stage and the period of bloom low temperature is very detrimental. Generally 15.6°C is the threshold temperature. High temperatures aıe favorable for development of cotton, with optima at 25°–30°C. Temperatures of 15°–20°C retard growth and reduce fruiting. Temperatures of 30°–40°C and above have, however, detrimental effects on the rate of growth, reducing the percentage of fiber and lowering its quality. Cold nights and hot days favor the cotton aphid or cotton louse, which frequently does serious damage to young plants. GIPSON and JOHAM (1969) found that decreased night temperature resulted in lower oil and nitrogen content of the seed.

Moisture

Growth of cotton depends on soil moisture but humid weather is unfavorable. Studies (JENSEN, 1971) of the effects of soil water tension on emergence and vigor of cotton show little influence of changes in soil water tension between 1/3 and 3 bar. Emergence at 4 bar tension was significantly less than at low tension with the number of emerging seedlings dwindling as the tension increased from 4.0 to 7.5 bar and a sharp decrease at a tension of 8 bar. No emergence occurred at tensions greater than 12 bar. During the active growing stage in the Cotton Belt of the U.S.A., May and June are critical months when heavy rainfall (especially if accompanied by low temperature) is very detrimental. This promotes the development of diseases and insect pests. Shedding is closely related to the water supply of the plant. GERALD and NAMKEN (1966) reported a linear relationship between available moisture and yield on a fine-textured soil. At the stage of setting and maturing of bolls large amounts of water are needed. Otherwise, shedding is likely to occur. However, heavy continuous rain also results in shedding. As the cotton matures and the bolls begin to open, rainy weather retards maturity, interrupts picking, and discolors or damages the exposed fiber.

Air humidity

Air humidity particularly during the harvest strongly influences cotton moisture content. WOODRUFF et al. (1967) conducted experiments to study the influence of relative humidity, temperature, and light intensity during the boll-opening period on cottonseed quality. Seed quality was reduced as relative humidity increased above 60% during a 21-day exposure period. Relative humidities to 80% and above cause a rapid increase in free fatty acid in seed oil.

Light intensity

Sunshine also influences cotton growth, especially when the plants are in bloom. Much sunshine with abundant warmth and low humidity are beneficial to plant growth and hold down weevil ravages. The moisture content in the cotton is influenced by the amount of energy received from the sun. Bottom defoliation, which permits better penetration of sunshine, has a beneficial influence (RILEY, 1962). It reduces boll rot loss by reducing high humidities in the canopy and enhances ventilation (RANNEY et al., 1971).

Wind

High winds cause mechanical damage and in sandy soil areas cause abrasive injury to cotton plants. ARMBURST (1968) found that an increase in the amount of soil striking a young cotton plant decreased the plant growth rate and yield, delayed first bloom, and increased the number of plants lost.

Soybeans

The soybean (*Glucine max*, L.) is an annual legume. The plants are used for forage, pasturage, and soil-improvement, and the beans provide food and oil.
Soybeans are widely cultivated in different parts of the world with successful production dependent on climate. There are two major production centers in the world—northeastern China (Sung-Liao Plain) and the Corn Belt of the U.S.A. Brazil has increased its soybean acreage considerably in recent years. Other soybean producers include Japan, Korea, India, Philippines, Thailand in Asia, and Germany, England, France and the U.S.S.R. in Europe.

Climate of soybean

Soybeans thus cover a wide range of climates. Maturity periods vary from 75–200 days and a medium resistance to frost exists. Light frost of −2°C is not dangerous to the plant. It is quite possible to grow soybeans successfully in any region in which there is a 5-month period of growth with a total heat accumulation of 2,400°C from May to September and an annual precipitation of 300 mm (RIEDE, 1938). Soybeans grow in much the same climate as corn.

Photoperiod

Soybeans are sensitive to daylength, and soybean flowering and maturity are dependent on it. Many varieties are incapable of flowering unless there are 10 or more hours of darkness daily. All varieties flower more quickly with daily dark periods of 14–16 h than with shorter dark periods. Only in a narrow belt of latitude is the length of the daily dark period favorable for a full season crop. Changes in natural photoperiod affect the maturity of soybeans primarily prior to flowering. Rates of development in subsequent stages are also influenced by photoperiod, but the intervals from seed set or end of flowering to maturity are similar for all soybean varieties regardless of maturity (BORTHWICK and PARKER, 1939; PARKER and BORTHWICK, 1951).

Temperature

Germination is the most critical stage for soybeans, abnormally cold, excess moisture, or prolonged drought at this time are likely to cause injury. Soybeans like other beans, germinate at a temperature of 8°–10°C. Higher temperatures within limits shorten the germination period. DELAUCHE (1953) found maximum germination in the shortest time with a constant temperature of 30°C, but Inoue (1953, cited by HOWELL, 1963) gives as optimum range of germination temperature 33.9°–36.1°C, a minimum range of 2.2°–3.9°C and a maximum range of 42.2°–43.9°C. Temperature and moisture both affect the rate of percentage of germination (PHILLIPS, 1968).
Blooming dates are also affected by temperature (GARNER and ALLARD, 1930). Sustained summer temperatures below 23.9°–25.0°C will ordinarily delay blooming. A decrease of 0.6°C will cause a delay of 2–3 days. Interannual differences in flowering date of a given soybean variety planted on a particular date is due chiefly to differences in temperature.

The minimum temperature for most growth processes, for all practical purposes, is about 10°C. Floral induction is greatly inhibited at 10°C or lower (PARKER and BORTHWICK, 1943).
During the pod-filling period temperature correlates with the seed quality of soybeans. HOWELL and CARTTER (1953, 1958) conducted a controlled experiment by growing soybean at 29.4°, 25.0° and 21.1°C during the pod-filling stage. The final seed oil analyses were 22.3, 20.8, and 19.5%, respectively.
Even brief periods of high temperature early in the season have adverse effects (HOWELL, 1956) and reduce the rate of node formation and the rate of growth of internodes. Seed quality is adversely affected by high temperatures during the seed development (GREEN, 1961), and it influences storage of oil in the seed.

Moisture

HUNTER and ERICKSON (1952) found that a moisture content of about 50% was required for germination of soybean seed. Short periods of moisture change (slightly dry or wet) do not affect plant growth during the subsequent period of growth. Moisture stress affects yield with a maximum reduction when plants were stressed in periods *5*, *6*, and *7* (see Fig.30), during the bean filling stage, and the last week of the pod-development period (LAING, 1965). Soil water suction and root temperature affects transpiration of soybeans (*Glycine max*, L.). With increasing root temperature, transpiration increases slowly at first and then rapidly until an optimum temperature is reached, followed by a decrease at high temperature. Rates of transpiration decreased with increasing soil water suction.
The summer optimum relative humidity during fruiting varies from 70 to 75% while the minimum varies from 50 to 65% (Rudenko, 1950, cited by VENTSKEVICH, 1961).

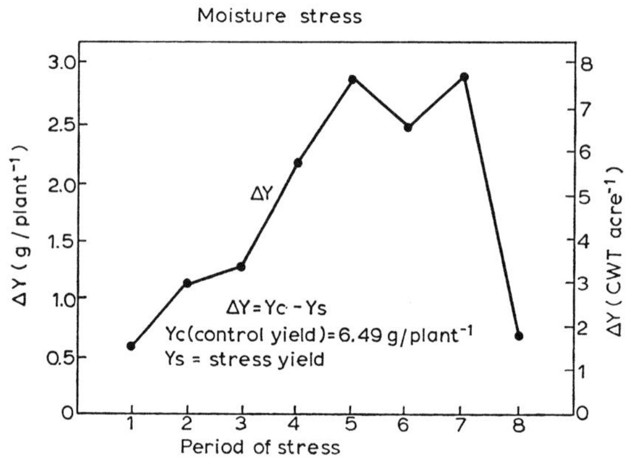

Fig.30. Change in soybean yields due to moisture stress applied at selected periods of growth. (After LAING, 1965.)

Light intensity

Light during the flowering and pod-development period affects the number of pods. This number depends upon the vigor of the plant during the time of blooming. If the plants are shaded during this period, the proportion of pods that abort will be high. Shading is significant in relation to natural variation in light intensity during intervals of cloudiness, especially at critical periods in plant development. Various degrees of lodging significantly change light interception and hence, highest yield and oil content of seeds result from a dense plant canopy.

Tomato

The parent species of the garden varieties of tomato are two tender perennials, *Lycopersicon esculentum* and *L. pimpinellifolium*. When cultivated, they are treated as annuals and are divided into two main varieties, "earlies" and "main crops", differing mainly in the time of maturity. Although tomatoes grow well in warm weather, they are not classified as tropical or subtropical plants. They are grown in many parts of the world: in the Americas they grow from Canada to Mexico; in China, tomatoes are allegedly planted outdoors in the latitude of 46°N (near Harbin, a frost free period of only 5 months) and in Lhasa (on the Tibetan Plateau, elevation of 3,658 m).

Tomato climate

Sunshine and warm weather favor tomato growth as a tender annual and as one of the first plants to be damaged by fall frosts. Even slight frost harms the tender foliage. It is a rapid-growing, short-lived plant which bears under favorable conditions the first fruit in 90–120 days from seeding.

Photoperiod

Recent research indicates that most of the species of tomato flower earlier under light of 11–13 h than under longer light hours. It is a day-neutral plant.
The age at which flower buds first appear is correlated with daylength (COOPER, 1961). The condition of equal light and dark may also be important in that a daylength of less than 12 h has a greater effect for a given change in light duration than a daylength of more than 12 h. Also the duration of the period from germination to the start of fruiting was related to daylength, and the rate of fruit volume production during this period of rapid increase was related to daylength (COOPER, 1963). This relation was curvilinear (see Fig.31).

Temperature

During its life process the tomato plant requires a long period of warmth. The cardinal threshold temperatures of tomato are around 15°C minimum, 33°C maximum and 22°–24°C optimum. The lethal threshold temperature is about −0.5° to −0.8°C.
Chilling the young tomato plant generally improves plant growth and final yield. A short period at an average temperature of 10°C during the early growth of tomato plant increased the number of flowers on the first inflorescence. The yield of these plants whose vegetative apex was removed after the first inflorescence was approximately 25% greater than that of similar unchilled controls (HURD and COOPER, 1970). During the period beginning at cotyledon expansion and ending approximately nine days later, tomato

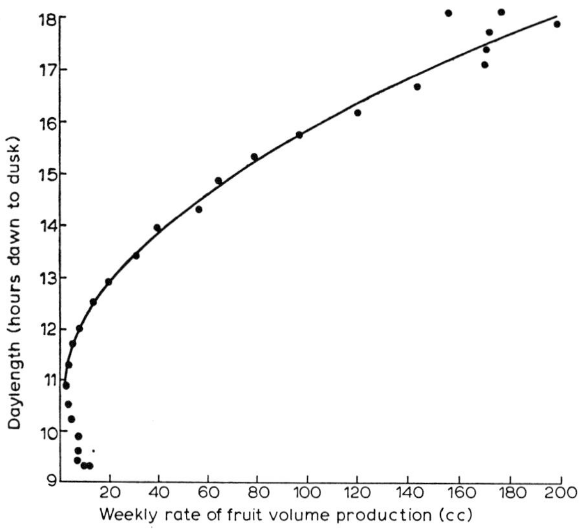

Fig.31. Relation between rate of fruit volume production and daylength during period of increasing fruit volume production. (After COOPER, 1963.)

seedlings can be vernalized (CALVERT, 1957). On completion of the vernalization period the seedling becomes sensitive for the number of flowers in the first inflorescence. This period extends to the 12th and 15th day after cotyledon expansion; low temperature increases and high temperature decreases the number of flowers.
The tomato plant has a good resistance to high temperature. In many cases it can endure temperatures as high as 40°C for a short duration of time.
Table XIII shows the number of days required for germination at different temperatures. The lowest temperature for germination is 11°C. However, at this temperature the rate of germination is low. As temperature rises, the period of days required for germination decreases and the rate of germination increases. But both temperature and water regimes influenced the percentage of seedling emergence (ABDELHAFEEZ and VERKERK, 1969). The 24°C-temperature and wet conditions showed the earliest and best emergence, followed by 18°C. Seeds at 9°C failed to emerge even after 42 days. The young seedlings of tomato plants are susceptible to cold temperature especially when the root system is not established after transplanting. Low temperature during this period induces flower abortion and reduces the final yield.

TABLE XIII

TEMPERATURE AFFECTING SEED GERMINATION
(After Kotowski, cited by CHEN and HSU, 1957)

Germinating temperature (°C)	4	8	11	18	25	30
Germination beginning (days)	–	–	22	7	5	4
Germination completed (days)	–	–	33	14	10	8
Rate of germination (%)	–	–	75	97	96	94

During the flowering period, temperature is critical. Mean maximum temperature higher than 26°–27°C and mean minimum temperature lower than 18°–19°C can result in serious flower abortion. During this period night temperature is also important. Optimal night temperatures during the flowering period are about 15°–18°C. PHATAK et al. (1966) found that plant temperatures determine the position as to node number, of the first inflorescence, whereas root temperatures influence the number of flowers in the first inflorescence. Top temperatures of 10°–12.8°C significantly reduced the number of nodes below the first inflorescence as compared to 15.6°–18.3°C or 18.3°–21.1°C. Conversely at 10°–12.8°C root temperature the number of flowers was significantly increased as compared to 15.6°–18.3°C or 18.3°–21.1°C.

WANG (1963) collected data from 20 sources and constructed a comprehensive picture which shows the thermal response of tomato growth at different stages (Fig.32).

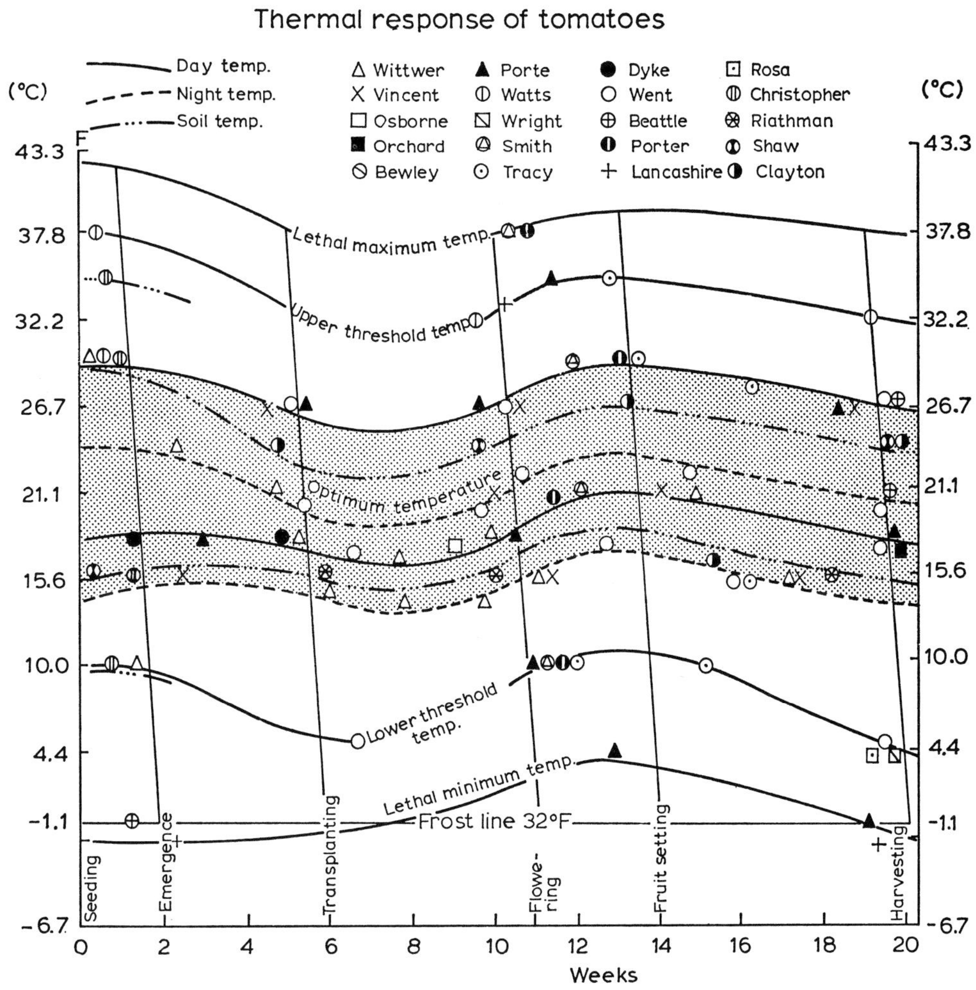

Fig.32. Thermal response of tomatoes. (After WANG, 1963.)

Moisture

Tomatoes do not require a large water supply. Excess soil moisture has adverse effects on plant growth. Generally the optimum soil moisture is in the neighborhood of 80% of field capacity. However, during the fruiting stage tomatoes need a higher soil moisture. Table XIV shows the relationship between soil moisture and air temperature and the yield of tomatoes. Highest yields were with 86% of soil moisture and at a temperature

TABLE XIV

SOIL MOISTURE, TEMPERATURE AND EFFECT ON YIELD
(After FOSTER and TADMAN, 1940)

Soil moisture (%)	Temp. (°C)	Yield (g)	Ave. yield (g)
59	23.3	1,092	
	21.1	1,078	
	15.6	1,064	1,064
72	23.3	1,378	
	21.1	1,360	
	15.6	1,250	1,329
86	23.3	1,520	
	21.1	1,528	
	15.6	1,393	1,480

21.1°C; and the lowest yields were with 59% of soil moisture and at a temperature 15.6°C. In general, tomatoes grow best in a relatively dry climate, with ample sunlight and with application of irrigation water, especially during the fruiting period.
High air humidity is also important especially when temperature is high. Such a condition encourages development of insect and plant diseases. The optimum air humidity is in the neighborhood of 45–55%. Very low air humidities often result in flower abortion.

Light intensity

Tomato growth requires large amounts of light. Shaded and poorly lighted fields produce low yields. Light intensity affects the nutritional value of the fruit. Different light intensity has different effects on different species of tomato. Generally plants grown under weak light resulted in a taller plant, but the diameter of the stem and the total weight of the plant are less than those grown under strong light. Shading the tomato plant delays fruiting and allows a higher rate of stem extension to be attained before fruiting begins (COOPER, 1961).

Alfalfa

Alfalfa (*Medicago sutiva*, L.) is a perennial leguminous forage crop. It does best in a relatively dry climate where water is available for irrigation. However, the production of alfalfa has been increased rapidly in humid and temperate areas of the world in recent years.
Alfalfa is grown over a wide range of temperature and moisture conditions. In north America it is an important crop from the southern valleys of Arizona and California to the prairie provinces of Canada. The world distribution of alfalfa includes the drier portion of India, central Asia, throughout the Balkan area and in the southern part of the U.S.S.R., in the Mediterranean and adjoining area, and in Argentina, Chile, Peru, and Mexico.

Climate of alfalfa

Though alfalfa is grown over a wide range of temperature and moisture conditions, high temperature, expecially combined with high atmospheric humidity, has an adverse effect on alfalfa growth. In the southwestern U.S.A., it generally experiences a marked decline in growth and yield during the summer months when the temperature is high (commonly called "summer slump").

Photoperiod

Most alfalfa cultivars are long-day plants. However, different cultivars have different photoperiodic responses. NITTLER and KENNY (1964) found that alfalfa flowered most profusely when exposed to continuous light. Profuseness of flowering decreased under 20-h photoperiods and was least under two 10-h photoperiods with a 2-h dark period between each light period. The change from vegetative to reproductive stage of development requires repeated exposure to appropriate photoperiodic cycles. Photoperiodic responses in alfalfa are enhanced by an increase in light intensity. With the same day-length, high intensity resulted in an earlier flowering. BULA and MASSEGALE (1972) indicated that when studying the photoperiodic requirement of alfalfa it is important to consider the influence of temperature during the induction and initiation period.

Temperature

Although different varieties have significantly different rates of seed germination, higher temperature (within limits) generally increases the rate of germination and emergence. The caıdinal temperatures for seedling emergence are 10°, 25°, and 35°C. UENO and SMITH (1970) grew three cultivars of alfalfa (*Medicago sativa*, L.) for 35 days from seed in three temperature regimes: hot (H) 32°C day/27°C night; warm (W) 27°/21°C; and cool (C) 21°/15°C. Plant maturity was delayed as temperature decreased. Shoot weight, heights, and plant parts for each cultivar generally were greatest in the W regime; followed in order by H and C regimes. Temperature has also a marked influence on floral intitiation. First-flower stage or growth "Vernal" alfalfa was reached in 21 days in the warm temperature regime of 32°/24°C, but not until 37 days in the cool regime of 18°/10°C. Herbage yields, however, were considerably higher in the cool than in the warm regime. Temperatures during the period from forage removal to full bloom have the greatest influence on seed quality and quantity. Greatest reduction of flowers per raceme, number of pods, and seeds per pod occurred when maximum and minimum temperatures were highest (DOTZENKO et al., 1967).

Moisture

Alfalfa is considered a drought-resistant plant, yet available soil moisture greatly affects seed germination and the growth of alfalfa seedlings and fruiting growth. Moisture supply during these stages is important. However, excess moisture also adversely affects the growth, because excess moisture reduces soil aeration and results in shallow root systems and small crowns. Alfalfa grown at high soil moisture stress (both in growth chamber and field) yielded less dry matter than at low soil moisture stress (VOUGH and MARTIN, 1971). Low temperatures and low soil moisture stress cause a higher yield of a higher-quality forage than high temperatures and high soil moisture stress. Alfalfa in Nebraska used 11.4 cm water per metric ton of 88% dry matter hay (DAIGGER et al., 1970). Water is used more efficiently in early May and June than in July and August. Average water use per day was 4.1 mm for the first harvest, 5.6 mm for the second harvest, and 5.9 mm for the third harvest.

Light intensity

Alfalfa growth is dependent on light intensity. BULA et al. (1959), using light intensities up to 32,000 lx, observed that total plant-dry-weight accumulation was essentially proportional to the intensity of light. Alfalfa seedlings are not tolerant to low light intensities (MATCHES et al., 1962) which reduce plant dry weight. Light intensity also affects the number of stems per plants and has a marked effect on leaf morphology and quality of the plant tissues. Dry matter accumulation and cellulose content of two alfalfa varieties increase with advancing maturity and light intensity, but crude protein and in-vitro digestibility decrease (GARZA et al., 1965). Soluble carbohydrates are greatest under high light intensity wity alternating day and night temperature of 30°C and 15°C, respectively.

Apple

The apple is the most important temperate zone fruit tree. Regions best suited to apples on the North-American continent include the Pacific coast, the Great Lakes, the Mississippi Valley. Apples are grown in most of the European countries, northern China, Japan, and Korea, and many other regions of the world. All common apple sorts are modifications of *Pyrus malus*. Most apple varieties are self-unfruitful and need pollen from another compatible variety if good crops of fruit are to be produced. Generally, apples can be classified as winter and summer varieties. The former requires 125–150 frost–free days for its fruit to mature while the latter requires 150–185 such days.

Climate of apple—Temperature

Temperature limits the cultivation of apples. Yet apples are the least heatdemanding fruit. There are three stages annually when temperature plays a role. Concerning winter injury, apple trees have been able to resist severe cold during the winter. In Russia the northern varieties are able to withstand cold as low as −45° to −50°C, but the southern varieties can be damaged by temperatures below −30°C (VENTSKEVICH, 1961). The root system of an apple tree is not very winter-hardy and in the absence of a snow cover suffers damage at temperatures of −12° to −15°C. The damage caused by frost is most serious during the spring when the leaves begin to turn green. A cold spring is the least desirable climate for the bud stage of apples. A chilling temperature during the earlier stage of growth can hasten the growth of the apple tree (ABBOTT, 1962). Apple trees may be forced into growth early in the year after a period of four weeks of cold treatment followed by a suitable photoperiod in a heated glasshouse (MACNEILL, 1969).

The temperature is extremely important for the flowering stage of apple. In Russia, temperatures of −5.0 to −8.0°C at flowering time cause total damage and destroy the crop. Flowering of apples is related to temperature. In Norway an average temperature of about 13.5°C from May to September gives superior taste to the fruit (LJONES, 1970). During pollination, temperature is critical if too low in daytime.
The tree itself is sensitive to wide diurnal temperature variation especially in early spring. On clear days at the end of winter when the bark is subjected to strong insolation during the day and low temperatures at night, tissues may become torn and frost cracks and blisters form on the bark.

Moisture

After flowering, evapotranspiration and hence water requirement are high. Low soil moisture results in the abortion of ovaries and decrease in the size of fruit. However, excessive soil moisture also adversely affects apple trees and their root systems. Growth and crop responses to irrigation treatments A, B, and C, consisting of 25.4, 50.8, 76.2 mm water applied at calculated soil moisture deficits of 25.4, 76.2 and 1.27 mm, respectively, were studied (GOODE and INGRAM, 1971). The trees in treatment D were not irrigated. Regimes A, B, and C yielded 42, 37, and 25%, respectively, more than the unwatered trees. The amount of available soil moisture strongly affects radial trunk growth (TAERUM, 1964).
The frost resistance of apple blossoms is affected by soil moisture level. Under conditions of relatively severe drought, the water content of blossoms is reduced; the blossom supercooled to a greater extent, less ice was formed in the tissues, and frost damage was reduced (MODLIBAWSKA, 1961). But small changes in the water content of the soil at near field capacity have no effect on susceptibility of apple blossoms to frost, but relatively severe drought increases frost resistance.
High air humidity, expecially on a cloudy day, inhibits the flight of insects and prevents complete pollination of the flower.

Light intensity

During the period of flowering and before fruit maturity sunshine is important. The degree of exposure to sunlight of Red Delicious and McIntosh apples during the growing season affected several fruit characteristics (HEINICKE, 1969). Color development was directly related to exposure. The best color was in fruit exposed to more than 70% of possible full sunlight (FS), adequate color was in 70–40% FS, and inadequate color when exposed to less than 40% FS. Fruit exposed to less than 50% FS was of small size. The soluble solid content of both varieties was related to light exposure. Well-exposed McIntosh were less firm than those from heavily shaded areas.

Forest climate

Forest climates differ from the climate of other plant communities by the marked layerlike structure of the canopy: crown space, trunk space, and forest floor. Each of these layers has its own climatic characteristics. Climatic stratifications in the forest canopy can be clearly recognized, while in the relatively low plant communities profile characteristics within a canopy are often undiscernible.

Radiation

Radiation distribution in the forest canopy has been extensively studied. A general picture (Fig.33) was presented by GEIGER (1965) in describing the results of Trapp's photometric measurements in a 120–150 year old red beech stand, 31 m tall. These experiments compare the light distribution inside the canopy as the percentage of the outside of the canopy. The total light intensity is lower on cloudy days, but its rate of decrease is slower than under the sunny day, because a higher percentage of diffuse sky radiation is present on a cloudy day. The fraction of diffuse sky radiation transmitted by forest or a crop each day is not much affected by weather, for a uniform crop shows little spatial variation (ANDERSON, 1970). The fraction of direct solar radiation transmitted by a forest varies with solar altitude, canopy structure, and cloud duration. The age as well as the productivity of a forest also affects the penetration of light. DENMEAD (1969) measured the radiation balance in a pine forest and in a wheat field as shown in Table XV. In this table, R_S is daily total solar radiation; R, net radiation; H, sensible heat transfer; LE, latent heat transfer; G, soil heat flux; and P, energy equivalent of downward CO_2 transfer. There is a large difference in energy absorption and dissipation

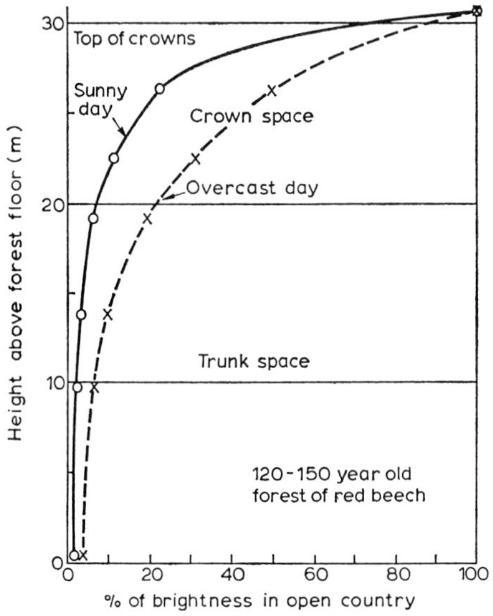

Fig.33. Decrease of light in a stand of red beech with dense foliage. (After Trapp, cited by GEIGER, 1965.)

TABLE XV

MEAN SOLAR RADIATION AND COMPONENTS OF THE ENERGY BALANCE IN A PINE FOREST AND IN WHEAT FIELD (After DENMEAD, 1969)

Community	No. of days	R_S	R	H	LE	G	P
Wheat	7	695	378	110	234	26	8
Pine forest	4	608	461	98	329	19	15

between plant communities growing in the same environment. Differences in energy absorption are caused by contrast in canopy development and reflectivity of the communities. Net radiation was higher in the forest canopy, and more latent heat dissipation contributed to that result.

During the day, radiation received in the canopy is smaller and decreases with decreasing height, while during the night stronger outgoing longwave radiation occurred outside than inside the canopy. The bell shape of the daily radiation distribution in the open flattens off inside the forest canopy. On the forest floor the diurnal variation of radiation is small as shown in Fig.34. Excellent reviews of radiation penetration in the forest canopy are: MILLER, 1955, 1959, 1965; VEZINA, 1961, 1963; VON WILHELMI, 1962; ANDERSON, 1964; REIFSNYDER and LULL, 1965.

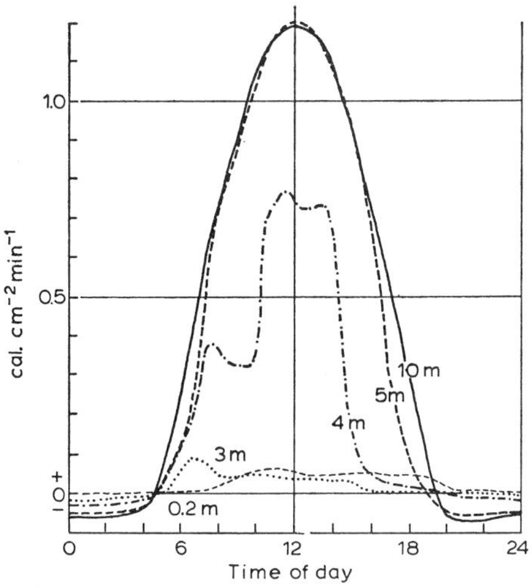

Fig.34. Smoothed radiation variation at five different heights in a young fir plantation during a dry period in midsummer. (After Baumgartner, cited by GEIGER, 1965).

Temperature and humidity

Air temperature in a forest canopy is mainly under the influence of the upper active surface, which during the day absorbs the solar energy and reradiates the heat at night to the air layer above the canopy. Energy exchange by the forest floor is small. The temperatures of the forest and its soil during the day are lower and at night higher than that of an open area. Before sunrise, temperature is at its lowest in the radiating crowns and warmest on the stand floor. At sunrise the temperature of the air layer above the crowns begins to increase as a result of the first level rays penetrating and warming up the tree top space, this process continues rapidly. As the sun climbs higher, the crown space warms up more rapidly and finally becomes the warmest layer for the rest of the daytime. About noon, a state of equilibrium is reached. Temperature decreases from the maximum in the crown area upward into the free atmosphere and down into the trunk area. The temperature decrease in the evening is in contrast to the increase in the morning.

The temperature distribution in the forest canopy varies with the structure of the canopy, the density and the age of the stand, and other environmental factors.

The annual temperature range in the forest is generally lower than in the open. The mean monthly temperature is lower in the forest than in the open for every month of the year, but the difference is highei in summer than in winter.

The air humidity in the forest canopy is also affected by these factors which influence temperature distribution in the canopy. Table XVI portrays the humidity distribution with height inside and outside the canopy. The daily average humidity is highest near the forest floor and lowest at and above the top of the forest.

TABLE XVI

TEMPERATURE AND RELATIVE HUMIDITY IN A DENSE FIR PLANTATION
(After GEIGER, 1965)

Height of measurement (m)	Temperature (°C)		Average water-vapor pressure (mm-Hg)	Relative humidity (%)		
	daily average	daily fluctuation		daily average	daily fluctuation	average for over-cast days
10.0, above the forest	22.3	16.4	11.9	63	58	76
5.0, in tree-top area	21.6	19.4	11.2	63	62	80
3.0, in crown area	21.1	19.0	12.2	70	62	84
2.5, in trunk area	20.8	18.4	11.7	69	60	86
1.5, in area of dead branches	19.6	16.5	11.5	71	60	87
0.2, at forest floor	18.3*	14.0*	12.5	79	45*	90

The diurnal change of humidity is highest in the morning, decreasing with time to reach its minimum in the evening hours. Air humidity is generally higher in the canopy than in the open.

Precipitation

Precipitation falling on a forest is partly intercepted by the crown (the needles, leaves, and twigs), partly occurs as trunk flow, and the rest as throughfall. Intercepted rainfall may be evaporated or may be accumulated to form large drops and finally reaches the ground, this depending on the intensity and duration of precipitation. Precipitation measured on the forest floor varies greatly. The annual rainfall amount lost from leaf and stem interception by evaporation is commonly between 20 and 40% in conifers and between 10 and 20% in hard woods (ZINKE, 1967). Generally, the forest floor with a closed canopy receives less precipitation than one with an open canopy. Different types of trees have different precipitation distribution on the ground. The physical properties of rain, such as raindrop size, are very important to its fall in the forest canopy. Fig. 35 shows the spectral distribution of drops of three different canopies. It shows that more than 50% of the weight of rain comes from tiny droplets with diameters less than 1 mm. It also shows that the total percentage weight of all drops increases as the drop size increases. The extremely large size of rain drops near the end of the distribution is the result of the accumulation of rain on leaves in the canopy. The intensity of the rain also affects the amount of water received at the forest floor. Table XVII shows clearly that heavy rain has a higher percentage of rainfall reception at the forest floor. Generally the amount of reception on the forest floor is directly proportional to the amount of rainfall. Younger stands allow more rain through to the forest floor (GEIGER, 1965).

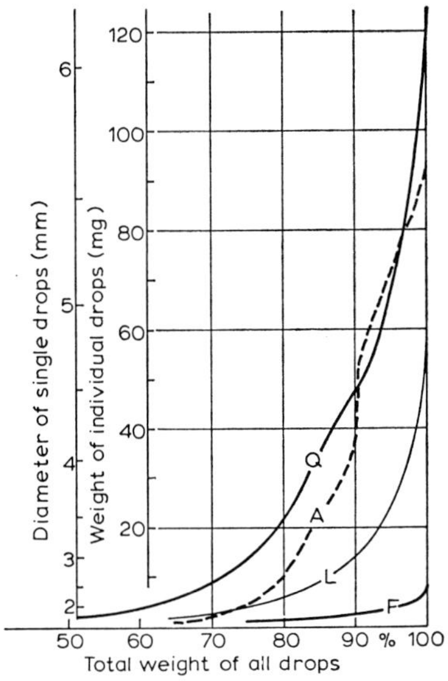

Fig.35. Spectral distribution of drops in oak (*Q*), Silver fir (*A*), and larch (*L*) wood compared with the open (*F*). (After GEIGER, 1965.)

TABLE XVII

AMOUNT OF RAIN REACHING THE FOREST FLOOR AS A PERCENTAGE OF OUTSIDE THE CANOPY WITH DIFFERENT INTENSITY

(After MOLGA, 1962)

Type of forest	Rain intensity less than 5 mm/d	15–20 mm/d
Pine (65 yr. old)	52	75
Spruce (65 yr. old)	29	69
Beech (90 yr. old)	62	87

Dew and snow

Dew deposit in a forest stand is insignificant. However, (as discussed in the section on dew), Fritschen has measured the dewfall on Douglas fir tree amounted to about 17% of the total water balance.

Snow reaches the forest floor more readily than rain. 90% of the total falling snow can be expected to penetrate the canopy and reach the floor. During the cold season evaporation is slow and accumulation of snow on the forest crowns is much less than that which finally reaches the ground. Snow cover in a forest stays longer than in open fields, because of reduction in radiation. This favors soil moisture accumulation in the forest.

Wind

Basically the wind distribution with height in and above forest canopy is not different from that presented in the section on wind. The term *d* (crop height) in the wind equation is, of course, much greater for forest than for other vegetation. However, the air movement in the trunk area of the forest may be less restricted. Wind profiles measured at three different wind speeds in a thin fir stand are shown in Fig.36 (GEIGER, 1965). Wind distribution in the forest is influenced by canopy structures (such as foliage difference), density of stands, thickets, and growth of saplings in the forest.

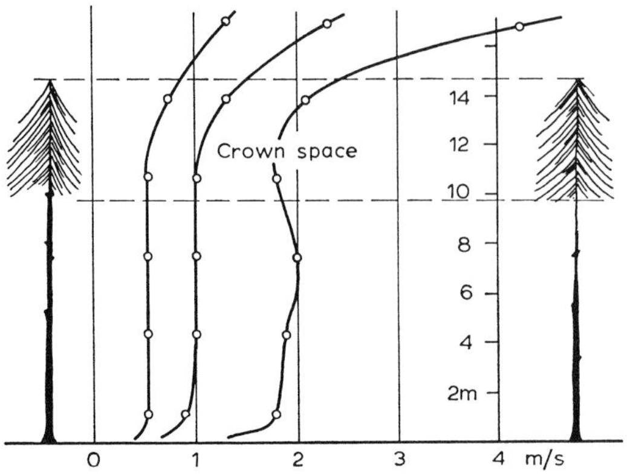

Fig.36. Wind profiles in a stand of pine of three ranges of wind speed. (After GEIGER, 1965.)

Mathematical models

The improvement and availability of the instruments permits measurements of micro-meteorological events inside the forest canopy, such as radiation fluxes, and the turbulent diffusive nature of wind in and above the forest canopy. Such data combined with an understanding of the physical processes of the micro-environment, make it possible to develop mathematical models to predict the physiological development, the photo-synthetic rate of plants and trees, and micro-environmental changes. WAGGONER (1969) pioneered in development of such a model, an electrical analog simulator of net photo-synthesis in a single leaf by corporating the effects of light, temperature, and CO_2 diffusion, and photochemical ability. REIFSNYDER et al. (1971) reported a model that identifies four components (direct, and indirect radiation penetrating holes in the canopy, and radiation reflected off vegetation) of the solar radiation appearing beneath tree canopies. Direct beam radiation appearing below the canopy follows an exponential-extinction law in the pine forest and a constant-ratio law in a hardwood forest. RUTTER et al. (1972) developed a predictive model of interception loss of rainfall in pine canopies that agreed satisfactorily with observed monthly interception losses in a stand of Corsican pine for a period of 18 months. BOHREN and THORUD (1973) presented two simple theoretical models to describe the effect of forest cover on radiation transfer to a snow-pack. These models partially explain melting rates which appear to be higher near the boles of trees than at more distant points.

Animal climate

The elements of the environment which directly and indirectly influence the performance of animals can best be explained by the "wagon wheel" diagram of BONSMA (1958) and modified by MCDOWELL (1967). It depicts the complexity of the problem. McDowell regards (Fig.37) man as the axle and the animal as the hub. The lubricant is manage-

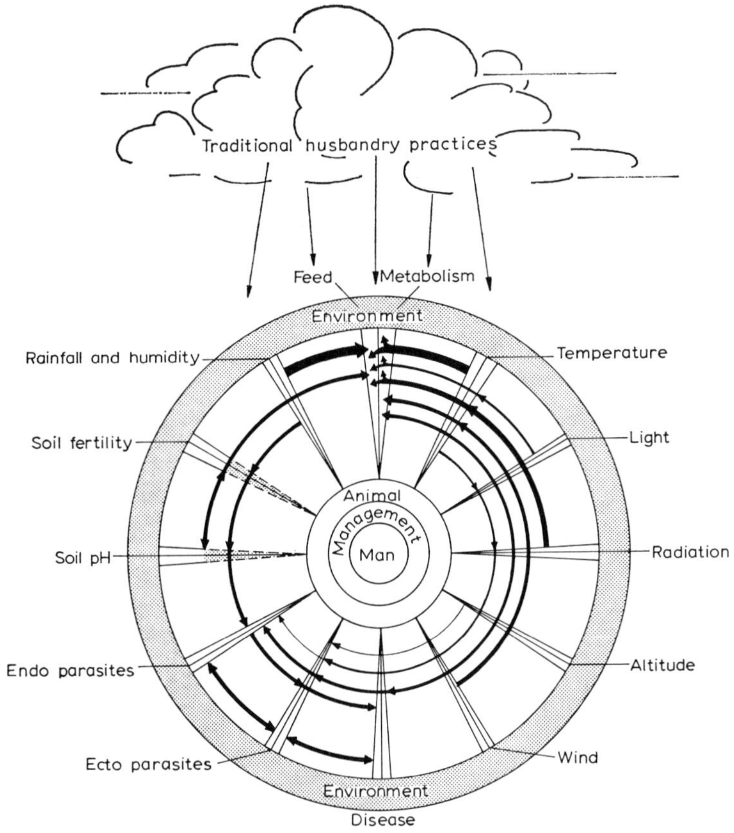

Fig.37. Elements of the environment which directly or indirectly influence the performance of animals. (After McDowell, 1967.)

ment, which facilitates rotation of the hub around the axle. The running surface of the wheel represents the total environment, and each spoke denotes a particular environmental factor (largely climatic elements). The concentric arrow represents interaction between the various components. If any particular element acts so drastically on the animal that it succumbs or degenerates, the leverage of the particular spoke will be so great that the hub will be broken and the delicate balance between the environment and the animal will be upset. Animals generally have, within limits, the ability to adjust physiologically to the extremes of the climatic environments. The adaptability of animals varies with the species. With few exceptions, all the present types of farm animals in the U.S.A. originated on other continents (Rhoad, 1941). The same applies to many countries.

Temperature

Temperature is the most important climatic factor affecting the performance of animals. It governs the physiological functions maintaining normal body temperature under diverse weather conditions. For most classes of livestock, there are optimal climatic conditions under which they will develop and produce best within the limits of their inherent capacity. Thus, the comfort zone for European breeds of dairy cattle is in the range between −1.1° and 10°C while for the Indian breeds it is 10°–26.7°C. The ideal temperature for most domestic species is generally within the range of 4°–24°C (Johnson, 1967). All of these species can normally survive in climatic regions where mean temperatures are greater or less than these, but they must adjust physiologically. As a result there are losses in productivity. The temperature affects size of eggs from pullets in different latitudes (WARREN, 1939). Larger eggs were produced in the northern than in the southern latitudes by pullets of the same breed, and summer eggs were smaller than winter eggs from the same birds (see Fig.38). Calorimetric studies with mature swine indicate that the critical temperature is approximately 21°C (CLAPSTICK and WOOD, 1922). Hens and hogs have a lower optimum production temperature (in the vicinity of 15°C) than cows and chicks (in the vicinity of 26°C). Cows have a wider optimum temperature zone than other animals; they adapt better to cold temperature than other animals. Low temperature responses of Holstein cows, in a loose-housing-type barn, showed that milk production does not drop markedly until the ambient temperature reached −3.9°C, but becomes very pronounced below −12.2°C (MCDONALD and BELL, 1958). HAHN and MCQUIGG's (1970) measurements agree with their study.

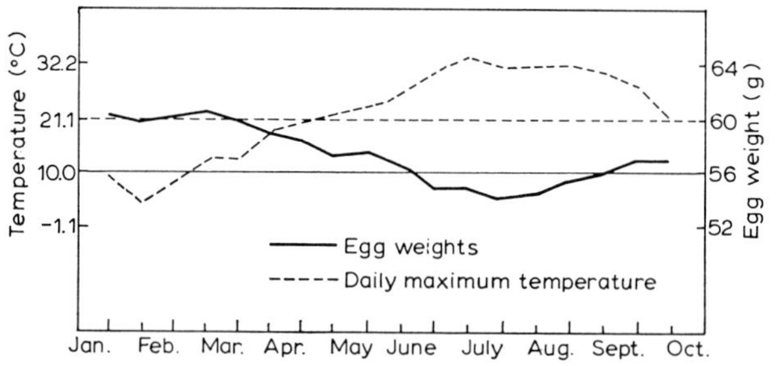

Fig.38. Mean maximum air temperature and mean egg weights at semimonthly intervals for 20 white Leghorn hens (second year) at Manhattan, Kansas, showing the influences of temperature on annual egg size curves, 1923–24. (After RHOAD, 1941.)

Radiation

RHOAD (1938) found that when cattle are moved from the shade and exposed to strong sunlight on a summer day, their respiration rate and body temperature rise, indicating increased difficulty in disposing of body heat. Less time is spent grazing in an open pasture on a calm summer day than on an overcast day. Fig.39 is a schematic diagram showing the radiation heat loads (ly h^{-1} at the animal's surface) received by shaded and unshaded animal on a typical cloudless summer day in El Centro, California. Shade may reduce the radiant heat load on an animal, it can not be assumed that shade will

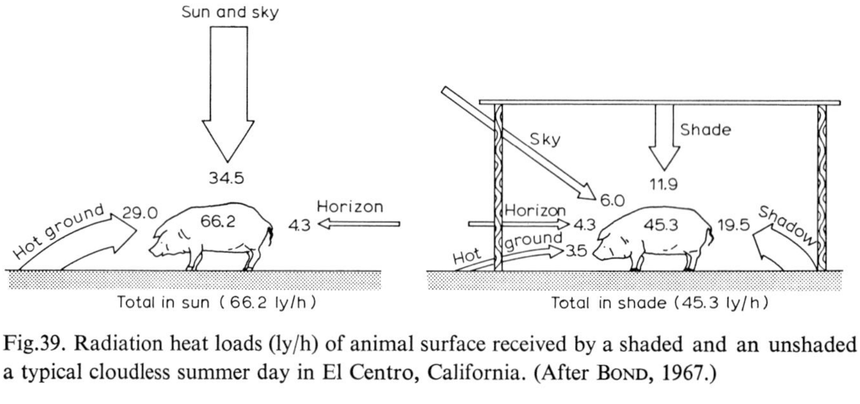

Fig.39. Radiation heat loads (ly/h) of animal surface received by a shaded and an unshaded animal on a typical cloudless summer day in El Centro, California. (After BOND, 1967.)

always improve the gain of all livestock in hot climate (BOND, 1967). For example, the performance of Hereford cattle at Davis during the summers of 1964 and 1965 was not improved by shade (GARRETT et al., 1966).

Air movement

Experiments to relate air movement to animal production have led to confusing results. Some report that air movement in a hot environment increases animal production. Others found no difference between forced air movement and calm air. BOND et al. (1965) made five field production tests with swine at three locations in California to study the effect of increased air movement. They found that increased air speed did not promote any increase in animal production during any of the test periods. However, BOND et al. (1957) also found when fans were added to a wood canopy to increase average air flow over the animal from 0.3 to 1.65 m s^{-1} during a summer feeding period, seven Herefords averaging 300 kg initially gained an average of 1 kg day^{-1}, compared to 0.6 kg day^{-1} for an unfanned group. Air temperature and radiation load were the same in the two wood pens, hence differences in animal gain were interpreted as a result only of difference in air flow.

Humidity

BERRY et al. (1964) presented the effect of humidity as a functional relationship between weather events and livestock production response:

$$M.Dec. = -2.37 - 1.736NL + 0.0247NL \times THI$$

where *M.Dec.* is the decline of milk production; *NL* is the normal level of production of 450 g day^{-1}; and *THI* is the daily mean-temperature humidity index. In this relation temperature and humidity are the main factors affecting milk production. Fig.40 shows that the combination of high humidities and temperatures is unfavorable for animal production. Environmental influences on animal physiological responses and productivity were reported by SCHOLANDER et al. (1950), STEWART et al. (1951), HART (1957), HEITMAN

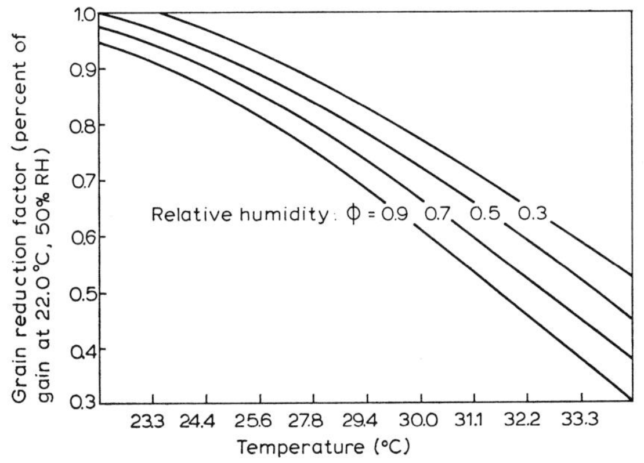

Fig.40. Ratio of weight gain for 67.5 kg pigs at a given relative humidity and temperature to that at 22.2°C and 50% relative humidity. (After Morrison et al., 1969, cited by HAHN and MCQUIGG, 1970.)

et al. (1959), BLAXTER (1962), HEITMAN (1962), INGRAM and WHITTOW (1962) (ambient temperature effects); POMEROY (1953), HOLUB (1957), ALEXANDER and MCCANCE (1958), (temperature effect on newborn animals); GARRETT et al. (1960, 1966), MCCORMICK et al. (1963) (shading effects). There are also numerous reviews of the same subject, HANCOCK (1954), BRODY (1956), HESS and BAILEY (1961), (temperature response of dairy cattle); MACFARLANE (1963), CARLSON (1964), LIND (1964) (temperature regulations); HENSCHEL (1964) (climatic chamber study).

Animal breeding and housing

The lack of adaptability of certain types of animals to tropical climatic conditions, as evidenced by discomfort, low production, and frequently degeneration in type, can best be overcome by selective breeding (RHOAD, 1941). Artificial shelters (animal houses) with climatic controls are becoming more common in many parts of world. The design, ventilation, and the pattern of the air flow in animal housing and also the quantitative relationship between environment comfort and animal production from housed pigs is found in papers by SMITH (1964a,b,c).

Meteorological hazard and weather modification

The protection of plants from meteorological hazards or from unfavorable weather conditions is a weather modification technology. Weather modification involves small-scale microclimatic alteration as well as large-scale climatic effects. In general, the agricultural applications of weather modification are mostly in terms of small scale microclimatic alterations, with some mesoscale projects. Large-scale climatic modification has mainly indirect effects on agriculture. The discussion in this section is concentrated on climatic modification which directly affects agriculture.

Radiation and heat

Modification of radiation and heat near the ground

Surface albedo: soil color and moisture determine the heat absorption of soils. HANK and VÁSÁHELYI (1954) found that coal dust covering the soil surface increases soil temperature and thereby lengthens the growing season. Dusting the tobacco leaves with white powder in summer reduces transpiration. *Reflecting light*: reflected light contributes a large amount of energy to the vicinity of plants and affects various microclimatic factors such as soil and air temperature. In Alaska, tomato plant flowering is hastened by aluminum foil which reflects the light to the plant and increases the soil and air temperature. *Mulch* is a common practice in agriculture. It modifies the temperature near the ground and conserves soil moisture. Different materials are used for ground mulch, such as straw, paper polyethylene, and others. The difference in soil temperature when covered with various colors and polyethylene are shown in Table XVIII. The black film consistently increased the mean temperature, largely through increasing the minimum temperature and decreasing the loss and gain of energy to and from the sky and

TABLE XVIII

THE DIURNAL MEAN AND RANGE OF TEMPERATURES 3 CM BELOW FILMS EXPRESSED AS THE DIFFERENCE FROM A BARE SOIL
(After WAGGONER et al., 1960)

Date	Sky	Mean less mean beneath bare			Range less range beneath bare		
		black	translucent	aluminum	black	translucent	aluminum
Sept. 11	Partly cloudy	5.0°C	8.7	2.1	−3.9°C	5.0	−8.2
12	Clear	2.8	7.2	1.2	−3.7	6.1	−10.8
27	Overcast	0.2	0.6	−0.5	−1.5	−0.8	−2.0
Oct. 1	Overcast	0.6	2.0	−0.2	−0.9	−0.4	−2.4
2	Clear	1.2	5.7	1.2	−4.9	5.4	−7.5
3	Mostly cloudy	2.0	5.4	1.9	−3.5	2.5	−6.2
4	Partly cloudy	1.2	3.9	0.8	−2.6	0.2	−4.5
5	Partly cloudy	1.0	2.8	1.4	−1.3	2.0	−2.7
6	Clear	1.8	4.2	1.8	−4.3	1.2	−6.0
11	Clear	2.2	7.2	1.5	−6.1	4.1	−10.4
12	Clear	1.6	6.6	−0.3	−6.5	3.9	−11.1
13	Overcast	2.8	4.5	2.5	−0.9	−0.6	−3.0

air. The translucent film was even effective in increasing the mean temperature; on most days it brought this about by increasing the gain of energy, but it brought it about on overcast days by decreasing the loss. The aluminum film generally increased the mean temperature slightly. The effect of shading on plant growth has been discussed earlier in this chapter and, therefore, will not be repeated here.

Frost weather modification

The ability of plants to resist frost temperature varies with species. For instance, spring wheat has the ability to resist temperatures as low as −9° to −10°C, −1° to −2°C, −2° to −3°C, whereas corn can only resist temperatures as low as −2° to −3°C, −1° to −2°C, −2° to −3°C, respectively, during the stages of germination, flowering and fruiting (VENTSKEVITCH, 1961). Tomatoes are even more susceptible to low temperature. For the same plant, the frost killing temperature may vary widely with the manner of the temperature change, time of year, physiological development stage of the plant, and many other environmental conditions.

There are radiative and advective frosts. One results from radiational cooling which occurs usually on clear calm nights. A temperature inversion is created near the ground, and as a result the temperature near the ground is decreased to the dangerous low. The advective frost results from moving air already with temperature below freezing, or by cold air drainage. It is not bound to time of day. A combination of these two types can also occur, that makes protection even harder.

Frost protection attempts to raise the plant temperature a few degrees higher than the critical temperature for a few to several hours. The expense prevents raising the temperature very high or covering a large area for a long interval.

Frost protection can be passive or active. Passive methods include agricultural practices such as clean cultivation, maintenance of soil moisture, wrapping plants with insulating material and enclosing the basal part of the plant, proper selection of location (avoiding locations where there is danger of frost), choice of growing season, and breeding of cold-resistant varieties. Most of these passive methods involve no environmental modification. All these passive methods have met with varying degrees of success. They are precautionary in nature and should always be practiced to minimize frost danger. The active methods of frost protection are many. The most commonly used methods are: heaters, wind machines, sprinkling, protein foam, and weather forecast.

Heaters

Heaters or small fires for frost protection use various fuels (including wood, coal, coke, oil and other burning materials). The heat released counteracts the heat loss by radiation and advection and keeps a tolerable temperature near the ground. Heaters are most effective on a night with a strong temperature inversion. Many small fires are more effective than a few large fires, but highly radiant heaters are more effective under windy, weak inversion conditions. The oil consumption of the heater as reported by the WORLD METEOROLOGICAL ORGANIZATION (1963) is 132–264 litres per hour per hectare, depending to some extent upon meteorological conditions and to some extent upon the number, size, and nature of the fires used. The disadvantage of heaters arises from the soot, smog, and other combustion particles that fall on the plant surface. There have been successful uses of electric heating cables in the soil that are nonpolluting.

Wind machines

Wind machines mix the bottom layer of air and destroy a temperature inversion. This method offers less protection than heaters especially during windy and cold night conditions. However, the machines are clean, they require a minimum of labor, they are inexpensive to operate, and they utilize the heat stored in the atmosphere. Despite being less efficient in frost protection, wind machines are widely used in the U.S.A. California alone had more than 28,000 wind machines (BROOKS et al., 1952).

Sprinkling

The sprinkling method of frost protection makes use of a physical property of water—latent heat of fusion. When a film of water is spread over a leaf or other surface which is cooling by radiation or otherwise below freezing, the latent heat of fusion is released on the plant. As long as a film of water is maintained, the temperature of the plant parts thus protected will stay at the freezing point. Sprinkling water particles suspended in the air effectively retard the flow of outgoing radiation, especially if they are atomized and create an artificial fog. The freezing process on the leaf or other protected surface will result in building up a layer of ice which must be supported by the plant. BILANSKI et al. (1954) listed the disadvantages: bacterial action in the soil was delayed, soil was compacted, soil became excessively wet and reduced soil air content, and soil fertility leached. The advantages of this system are low operating cost, very complete protection under proper conditions and no air pollution.

Protein foam

A nontoxic, easy to wash protein-based foam was used as protective insulation of plants against frost damage. The foam permitted survival of the lower parts of tomato and coleus plants after an overnight frost that reached −6.6°C. Temperature differences between the control plot and the foam-covered plot were as high as 9.0°C. However, the weak physical strength of foam limits its use. The cost is rather high, and application rates are slow (SIMINOVITCH et al., 1967).

There are many more other methods being used for frost protection. For instance, in Coachella Valley, California, kraft paper shields are used commercially to protect off-season vegetable crops including tomatoes, peppers, and squash. Kraft paper is attached to arrowed stems on the northside of the east–west rows leaning over the plants. During the day, the shields deflect radiation to the plant and soil, while during the night, the shields prevent radiation loss to the sky. Artificial clouds of various kinds have also been used to modify the radiation balance with varying degrees of success. SCHNELLE (1963) presented a comprehensive monograph on frost protection in two volumes. The author tried to exhaust all special aspects of frost protection in the light of the recent knowledge. Volume I of the monograph covers the meteorological and biological bases of frost prevention while Volume II deals with the relevant practices.

Weather forecasting

Frost warnings are important for an active frost protection program. An accurate weather forecast of frost may not only save the crop from frost attack, but also save protection costs. Farmers who operate their heaters on the basis of the forecast can delay lighting the heaters for several hours on some nights, which means a saving of money.

Frost forecasts are often more reliable than the general forecasts and take into consideration the nature of crop, the critical states of plant growth, the topographical difference, and other physical as well as synoptic effects. Frost forecasts require a high degree of precision. A 0.5°–1.0°C difference in temperature during the frost night can mean the difference between no damage and severe crop loss.

Moisture

Cloud modification

Schaefer (1946, cited by GILMAN et al., 1965) discovered that small amounts of dry ice introduced into a chamber of supercooled cloud resulted in the formation of millions of ice crystals. If such seeding is done in nature in supercooled water clouds the crystals will grow following the Bergeron-Findeisen process, at the expense of the droplet. In warm clouds the large drops increase their size by coalescing with smaller droplets. Since the micro-physical structures of these two types of cloud are fundamentally different, the methods of cloud seeding are also different. In the cold cloud the principle of modification is to trigger the Bergeron-Findeisen mechanism, encouraging the development of ice crystals with artificial nuclei while the seeding process with warm clouds is intro-

duction into the cloud of large hygroscopic particles such as sodium chloride or liquid water droplets which initiate the coalescence mechanism.
The effects of cloud seeding have remained elusive and controversial (HESS, 1974). Even after 30 years of experiments there is no firmly established technology to extract precipitation from clouds. It remains an experimental field. For farmers the sad climatic fact is that there usually are no modifiable clouds in major drought situations.

Hail suppression

Cumulonimbus clouds are often referred to as hail-producing clouds, because of strong persistent updrafts and large concentrations of supercooled water droplets favorable for hail growth. The present postulate of hail suppression is to add freezing nuclei so as to produce smaller ice particles and thus to promote the growth of hailstones of smaller size than nature would produce. Any experiments on hailstone suppression follow this prescription. There is another method of hail suppression which is similar to rainfall suppression. This involves slowing coalescence and accretion process by introducing great numbers of condensation nuclei into the storm updraft, thus reducing the average drop size and narrowing the drop size spectrum (WEICKMANN 1953, 1964; LUDLAM, 1958). The material used in hail suppression is mostly silver iodide. Russian investigators introduce the silver iodide particles by rocket into the cloud volume where the first radar returns appear (SULAKVELIDZE et al., 1967). The hail suppression research in the Caucasus and Transcaucasus area, on the basis of empirical evidence, has convinced the Russian experimenters that their technique is highly successful in hail suppression (BATTAN, 1965). However, the experimental design does not allow for adequate statistical evaluation.
Although much experimentation has gone on elsewhere, the results are not unambiguous. A carefully designed experiment in northeastern Colorado carried on by the National Center for Atmospheric Research in cooperation with several universities, used airplanes to seed clouds in potentially hail-producing storms. The results showed apparently decreases in hail in some sorties and a possible increase in others (ATLAS, 1977). Presently, full understanding is still lacking, and the techniques remain experimental.

Drought alleviation

Drought means different things to different persons. A universal definition of drought is difficult to find. PALMER (1965) listed seven different ways of defining drought, while HOUNAM, who in 1971 summarized drought definitions based on meteorological, hydrological, soil moisture, and crop parameters, provided an even larger collection of definitions. However, to the agricultural meteorologist drought may be defined as lack of available soil moisture to support plant growth, causing temporary or permanent wilt and resulting in crop yield reduction.
Drought develops when precipitation lacks over an extended period of time. Other factors such as high temperature, intense solar radiation, hot dry wind also contribute to the severity of drought.

Modification of drought weather

Most large-scale persistent droughts are the result of anomalies of planetary circulation. To break a large-scale drought, according to GILMAN et al. (1965) one must change the whole hemispheric wind field, a task outside human capabilities. Localized and transient dry conditions can be alleviated by moisture conservation practices such as surface mulch, shelter belts, snow fences, and suppression of evaporation and transpiration. The most effective drought control in the farming area is irrigation.

Drought indices and irrigation

Drought indices are a convenient tool for indicating severity of drought and for planning of irrigation. There are many procedures which express plant water relationship which can be used as drought indices useful to express irrigation requirements. These include soil moisture accounting procedures for wheat (MACK and FERGUSON, 1968), the moisture stress day for corn (DENMEAD and SHAW, 1962; DALE and SHAW, 1965a), and others including formulations by VAN BAVEL (1953, 1959), HOLMES and ROBERTSON (1959b, 1963), BAIER and ROBERTSON (1965, 1967, 1970) and YAO (1969).

Palmer drought index

PALMER (1965) developed a general index which measures the relative wetness or dryness of each successive time interval in a specific locality or region. Briefly, the method requires a climatological analysis of a long record in order to derive 5 parameters which define certain moisture characteristics of the climate of the area of interest. α is the coefficient of evapotranspiration ($\overline{ET}/\overline{PE}$), β the coefficient of recharge ($\overline{R}/\overline{PR}$), γ the coefficient of runoff ($\overline{RO}/\overline{PRO}$), and δ the coefficient of loss ($\overline{L}/\overline{PL}$), where P in each of the coefficients represents the potential values. The constant K is an empirically derived weighting factor:

$$K = \overline{PE} + \overline{R}/\overline{P} + \overline{L}$$

where R is net gain in soil moisture during a given period of time and L is net loss of soil moisture during the same period of time. The computed precipitation ($\hat{P}$) is the amount of precipitation expected during a particular time period to sustain the evapotranspiration, runoff, and moisture storage that is accepted as "normal" for the climate, having taken account of antecedent moisture conditions. The equation is:

$$\hat{P} = \alpha PE + \beta PR + \gamma PRO - \delta PL$$

where the potential values are those that apply to the particular time in question.

The final drought index (X) depends on the sequence of Z values. These were combined by the empirical equation:

$$X_i = X_{i-1} + (Z/3.0) - 0.103 X_{i-1}$$

$$Z = Kd$$

where d is the difference between actual and computed precipitation, and X_i takes values in general range from around +6 (very much wetter than normal) to −6 (extreme severe drought) and with near zero approach normality.
These index values are realistic and, in many cases, accurately reflect the actual situation. The index is mathematically simple, but the computation is tedious by hand. The method is better suited for climatological analysis than for operational use. However, during periods when a major drought is developing and spreading, it affords a useful means for routinely assessing the areal distribution of various degrees of drought severity.

Crop moisture index (CMI)

PALMER (1968) also developed the crop moisture index, a general information service type of index which imparts the direct crop and moisture relationship. Equations for the computation of CMI are:

1. $DE = (ET - CET) / \alpha^{\frac{1}{2}}$
2. $Y'_i = .67\,Y'_{i-1} + 1.8\,DE$
3. $-Y = -Y'$
4. $+Y = M(+Y')$
5. $M = (SP + SS + SU) / 2(SP + PR)$
6. $G_i = G_{i-1} - H + MR + RO$
7.

G_{i-1}	H
0	0
<.50	G_{i-1}
.50 to 1.00	.50
>1.00	$.5\,G_{i-1}$

8. $C = Y + G$

where DE = relative evapotranspiration anomaly for a given period of time say a week; ET = computed actual evapotranspiration for the week; CET = cafec evapotranspiration for the week = $\alpha \times PE$; α = coefficient of evapotranspiration = $\overline{ET/PE}$; Y' = 1st approximation to Y; i and $i-1$ refer to values at end of this week and at end of last week; Y = index of evapotranspiration deficit; M = average percent of field capacity during the week; $SP = S'$ = inches or millimeters of water in soil at start of week; $SS + SU = S$ = inches or millimeters of water in soil at end of week; $SP + PR$= water storage capability, a constant, AWC; G = index of excessive moisture, never <0; H = a "return to normal" factor, a function of G_{i-1}; R = computed recharge (both layers) this week; RO = computed runoff this week; C = crop moisture index; PE = potential evapotranspiration (Thornthwaite); cafec = climatically appropriate for existing condition.
The CMI involves only one coefficient $\alpha = ET/PE$, which is a weighting factor. The final CMI covers the entire range of crop moisture conditions from >3 (excessive wet) to <−4 (ruinous drought). HOUNAM (1971) concluded that evidence to date indicates that the CMI values describe the time and space variation of crop moisture conditions fairly realistically most of the time. Of course, the interpretation of the effect of a given

moisture condition on crops depends primarily on the kind of crop and its stage of development.

Probability of drought

Drought is caused mainly by prolonged insufficient rainfall. BARGER and THOM (1949) defined drought in terms of rainfall deficit. Any period of n weeks during the corn-growing season in Iowa, where $1 \leqslant n \leqslant 16$, which has a rainfall total smaller than the minimum amount determined to be just sufficient for that duration experiences a rainfall deficit. The largest deficiency for the season, regardless of the length of period involved, is taken to be the measure of drought intensity and resulting effect on corn yield for that year. Thom and Barger also found that the frequency distribution of weekly precipitation is similar to that of the γ distribution, which is discussed by Essenwanger in Volume 1 of this Series.
Barger and Thom's work did not take into consideration the soil moisture, which in many cases is very important, especially during the early growing season.
YAO (1969) developed, therefore, the R-index which is defined as:

$$R = ET/PE$$

where $0 \leqslant R \leqslant 1$ and ET is defined as:

$$ET = S_\Delta + P$$

where $S_\Delta \infty S_0$; and where the assumption is made that no runoff occurs until S_0 reaches field capacity (S_0 = soil moisture storage in antecedent month or week; S_Δ = soil moisture loss as evapotranspiration).
The R-index is mainly a function of the atmospheric energy which causes evaporation from the soil and plant surface and precipitation which replenishes the soil moisture. These two interacting factors govern the magnitude of R and constitute a moisture supply and demand relationship.
In the study of the frequency distribution of the R-index, Yao found that the β distribution can be used as a suitable probability model for the variable of R. The β distribution can be written as:

$$I_x(p,q) = 1/\beta(p,q) \int_0^x t^{p-1}(1-t)^{q-1}\mathrm{d}t$$

where:

$$\beta(p,q) = \int_0^1 t^{p-1}(1-t)^{q-1}\mathrm{d}t = \Gamma(p)\Gamma(q)/\Gamma(p+q)$$

p,q are two positive parameters and x has a range from 0 to 1. This distribution function represents a two-parameter family of distributions.
The parameters p and q of the β distribution are computed by the first two moment coefficients and as given by PEARSON (1934).

$$\mu_1'(\mu_1' - \mu_2') / \mu_2' - \mu_1'^2$$

$$(1 - \mu_1')(\mu_1' - \mu_2') / \mu_2' - \mu_1'^2$$

YAO (1972) using the *R*-index to study the historical droughts in the U.S.A. computed the probability occurrence of each of the major droughts: 1934, 1936, 1954, and 1956.

Wind

Windbreaks

The reduction in wind speed caused by the presence of a wind barrier has a real influence on the microclimate of the surrounding area. BATES (1911) classified the shelter influence into two distinct effects, mainly: competitive effect—those effects near barriers arising directly from interference of the barrier with the radiation, rainfall and other plant physiological effects; and windbreak effect—the reduction in wind speed beyond barriers and the consequent modification of other micrometeorological factors. In this section we are dealing mainly with these latter effects.

JENSEN (1954) and CABORN (1957) have reviewed experiences and experiments with windbreaks. Two later, excellent reviews have appeared (VAN DER LINDE, 1962; VAN EIMERN, 1964). More recent work is by ROSENBERG (1966), MARSHALL (1967), BROWN and ROSENBERG (1971), and BROWN (1972).

Wind

Modifying the microclimate can greatly influence the growth of plants in the protected area. The extent of protection is a function of height, density and width of the windbreak, the distance between windbreaks, and their orientation to the prevailing wind.

The permeability or the density of the shelter is probably the most decisive factor in determining wind speed reduction. Fig.41 shows how the different permeabilities affect the reduction of wind with distance from the barrier. Dense barriers have the greatest wind reduction zone immediately behind the barrier. Wind speed recovers sharply after the minimum is reached. At the 20% wind reduction level, very dense barrier extended at the distance about $9 \times H$ from the barrier, while for 50% permeability (medium) the distance is almost doubled (*H* designates the height of the windbreak). Similar results were found by WALKER (1951), MATJAKIN (1952), ROSENBERG (1966), BROWN and ROSENBERG (1971). Well-designed windbreaks are typically 50% porous and have a spacing of 10 times the height differential between windbreaks and the protected crop (BROWN, 1972).

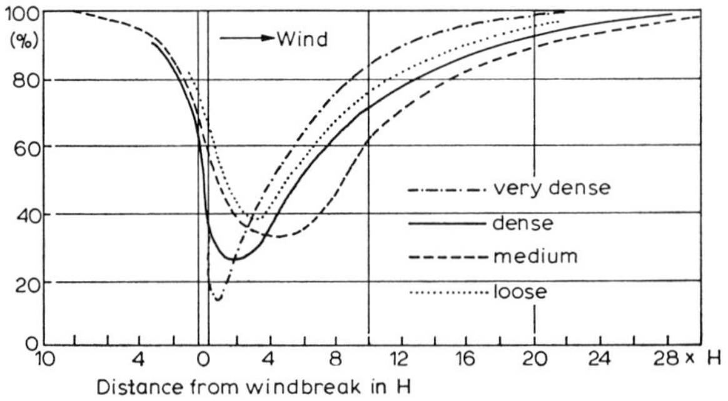

Fig.41. The wind speed reduction by different shelterbelts (Naegeli, 1946, cited by VAN EIMERN, 1964).

Naegeli (cited by Van Eimern, 1964) showed that narrow belts with smooth vertical walls offer sufficient protection. Growing vegetation is also used for shelter belts. The width of these types of shelter varies with climate and other factors.
The roughness of the ground surface determines the vertical increase in wind speed near the ground and hence is a measure for the degree of wind reduction at and near the shelterbelt. The roughness term as related to wind speed near the ground has been discussed earlier.

Air temperature

The difference in air temperature between sheltered and unsheltered area varies according to time of the day, season, weather, turbulent mixing, the local air circulation, soil moisture (Gol'cberg, Matjakin, van Der Linde-Woudenberg: see Van Eimern, 1964; Rosenberg, 1966). Generally night air temperatuie differs only slightly (lower) but during the day temperature is higher (usually less than 2°C) in the protected area than in the open (Van Eimern et al., 1964; Aslyng, 1958; Woodruff et al., 1959; Jaworski, 1961; Rosenberg, 1966). Based on theoretical considerations, Van Wijk and Hidding (1965) showed that the average daily amplitude near the ground could be 1°C greater at a distance of up to 30 × *H*, 2°C greater up to 10 × *H*.

Soil temperature

The change of soil temperature resulting from wind reduction is limited and uneven, depending on the ground covering. Jensen (1954) in Denmark, Steubing (Van Eimern, 1964) in Germany, and other Western-European workers have consistently noted increased soil temperature in sheltered zones and a proportional increase in soil temperature with increasing wind protection. Burnacki (cited by Van Eimern, 1964) found that in the Kamannaja steppe of the U.S.S.R. bare soil was warmer in the sheltered zone. After a wheat crop became established, the opposite held true. Generally highest soil temperature occurred in the afternoon and the lowest in the morning (Casperson, 1957; Kaiser, 1960).

Moisture conservation

The redistribution and conservation of water, especially in arid or subarid climate, is a beneficial effect of shelterbelts. The vapor pressure gradient between the leaves and the air is lower in the protected area than in the open (Van Eimern, 1964; Rosenberg, 1966). This reduces evaporative demand. In some environments in Canada, moisture conservation in the protected area was not pronounced (Staple and Lehane, 1955). According to Tomato (1956), in the leeward open field evaporation increased with the distance from the shelterbelt. The decrease in evaporation on the leeward side was about 40, 60, and 80% of open field evaporation at the points 1 × *H*, 5 × *H*, and 10 × *H*, respectively. The influence of the windbreak on evaporation was perceptible for a distance of 20–25 × *H* on the leeward side.
Shelterbelts also affect, in a secondary way, other climatic factors such as air humidity (Van Eimern et al., 1954; Stoeckler, 1962; Matjakin, 1963); precipitation and dew

(STEUBING, 1954; GEIGER, 1965); and carbon dioxide (RÜESCH, 1955; ROSENBERG, 1966, 1967; BROWN and ROSENBERG, 1971).

Crop yield

The response of plant growth to the microclimatic alterations of shelterbelts varies with crop species, varieties, stage of growth, and canopy structure of the crop. VAN EIMERN (1964) reviewed the comprehensive work on the effect of wind protection on agricultural yield, particularly in Russia and Denmark, done by Smith, Naegeli, and Andersen. In addition, VAN EIMERN (1964) also listed many other workers who indicated that yield was increased in the protected area. In the U.S.A., a survey showed that 274 of the 331 farmers interviewed estimated crop yield increases in one or more of three years (FEBER et al., 1955). Only two estimated decreases. FOSKETT (1955), BAGLEY and GOWEN (1960) and FELCH (1964) also reported increases in yield in the sheltered areas under irrigation conditions for different crops, while LEONTIEVSKY (1934), STAPLE and LEHANE (1955), and PELTON and EARL (1959) noted increases in yield under natural conditions in the protected area. In general, in the dry years yield increase is higher than in normal or wet years.

The presence of a windbreak decreases the vertical turbulent transport in the air above the protected crop (BROWN and ROSENBERG, 1971). Decreased turbulent transport causes an increase in the water vapor pressure in the air above the protected crop (BROWN, 1972). The temperatures of the air and crop surface are also increased, but the vapor pressure gradient between the leaves and the air is lower in the protected area than in the open. This reduces evaporative demand in the protected area. Hence, evapotranspiration is lower during these periods, the stomata remain open wider and longer between irrigation and rainfall. The increased stomatal aperture allows for greater CO_2 flux and thus more photosynthesis.

BROWN (1972) has summarized the effect of barriers on micrometeorology and other indicated factors, in a model which is a modified scheme of MARSHALL (1967). It is shown in Fig.42.

Weather and plant diseases

Plant diseases constitute a chief hazard for crop yields. STAKMAN and HARRAR (1957) conservatively estimate that 20% of the world production was lost yearly to diseases, pests and direct weather damage.

Plant disease is the result of complex interactions between many environmental and physiological factors. Plant disease epiphytotics occur only if three factors prevail simultaneously: susceptible host, pathogen or pest, and favorable environmental conditions. The interaction between agent and host is the key relationship, a struggle between the virulence of the agent and the resistance of the host (BOURKE, 1968). The physical environment, principally weather, may tip the balance of the struggle one way or another by its influence on the life cycles of both agent and host.

Temperature and moisture control or affect the reaction of the pathogen and host, with air movement the close third factor (especially for insect infestation). Other weather

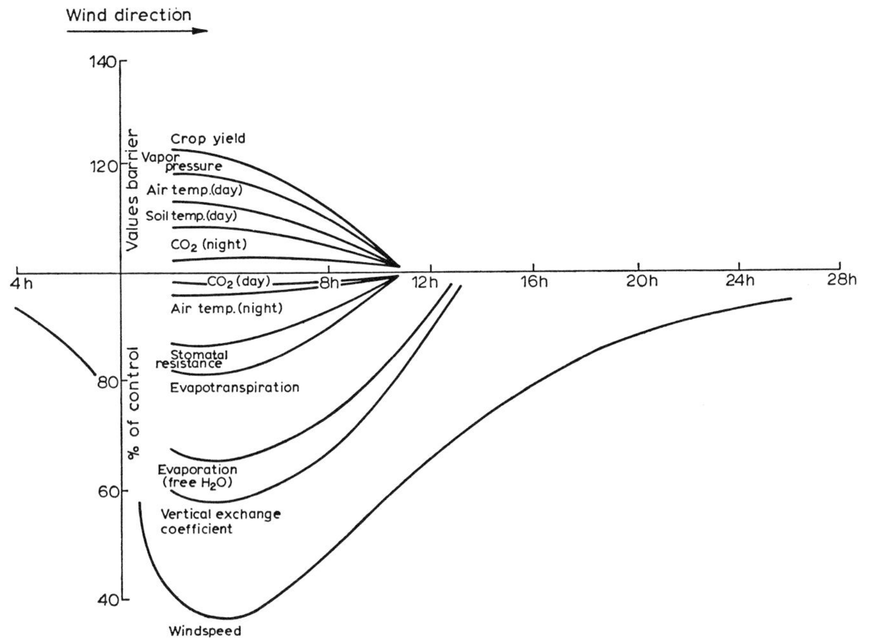

Fig.42. Summary diagram of the effect of barriers on micrometeorological and other indicated factors. (After MARSHALL, 1967; BROWN, 1972.)

factors, including rainfall, dew, soil moisture, soil temperature, and light intensity, may affect pathogen development. Under favorable weather conditions, the severity of a disease outbreak is determined by the other two factors, host and pathogen.

Weather and pathogen

Some diseases can survive over a wide range of climatic conditions, some others only a narrow range. Only a few of the more important diseases can be discussed here:

Potato and tomato late blights cause an estimated 4% of annual loss even with the best controls. The disease of the potatoes caused by the fungus *Phytophthora infestans* is intimately associated with meteorological factors. Climate was considered in relation to late blight epiphytotics by the Dutch in 1926 (VAN EVERDINGEN, 1926) and in England by WILTSHIRE (1931). In the U.S.A., CROSIER (1934) found that the temperature of 18°–25°C and a relative humidity of nearly 100% over a period of at least 6 h, or a temperature of 12°–15°C and a relative humidity of 100% over 12 h or more are necessary for the formation of sporangia.

Wheat rusts (*Puccinia* spp.) are world-wide diseases causing serious economic yearly losses of this important food crop. The major wheat rusts are the black (stem), the brown (leaf), and the yellow (stripe) rusts. HOGG et al. (1969) have summarized the results of research on the physiological processes and meteorological factors. The optimum temperatures for the development of rusts are slightly different for these three rusts. LANGE et al. (1958) found that notable infection was favored by a period of 7 h of darkness with dew at 24°C plus a 3-h period of dew at 29.4°C with light intensity exceeding $5.5 \cdot 10^3$ lx for black rust. Leaf rust develops most rapidly in the field when mean temperature is between 15° and 22°C. For yellow rust, greenhouse experiments indicated that 20°–25°C stimulates the rate of germination (Straib, in HOGG et al., 1969).

High relative humidity, greater than 95%, also favors germination especially with high sporulation temperature. Very low temperatures may stimulate the germination rate of some varieties.

For corn leaf blight (*Helminthisporium maydis*) critical factors in the development are temperature and moisture. For a viable spore to germinate and penetrate the leaf surface, it must be in contact with free water, either from rain, dew, or irrigation. The time requirement, however, is determined by the temperature. At an optimum temperature of 26.7°–29.4°C, only 6–8 h is needed for germination and penetration, but at 18.3°C, 12–18 h are required. The outbreak of southern corn leaf blight (SCLB) in the U.S.A. in 1970 and 1971 points up the sensitivity of disease and weather. For several years the SCLB pathogen was confined to the southern states in the U.S.A. because of the cooler temperature of the north (FELCH, 1973). The new Race T, first appearing in 1970, differed from the original Race O in two significant ways. It was severely pathogenic on corn containing the Texas male sterile cytoplasm, which gave it the label Race T. Its effect on nonsterile or normal cytoplasm varieties was nil, acting much like the original Race O. More importantly, the new Race T was able to survive and perform well under much cooler temperatures than the old Race O. By the end of the summer of 1970 it had spread through most of the eastern part of the U.S.A. and on into Canada, resulting in heavy yield reduction.

Other plant diseases such as boll rots in cotton, powdery mildew on lettuce, flax rust, sugar beet leaf-spot fungus, white pine blister rust, and epiphytotics of apple scab are all affected by weather conditions.

The weather and the host

Plants of different varieties have generally different degrees of disease resistance. Breeding of more disease-resistant plants is an important program for pest control. Besides the physical resistance of the plant there are many other properties of the host which may be important to epidemiology. The density of the crop and the extent of its cultivation affect humidity, light penetration, and air and soil temperature, which in turn affect plant disease. The use of recommended levels of plant population will usually provide adequate help for reducing an environment favorable for disease development. Soil temperature as well as soil moisture are both important. Crop rotation is also a way to combat disease. Cold soil retards seed germination and weakens the seedling, thus making the young plant more susceptible to infection. Excessively wet soil provides favorable conditions for disease development.

Forecasting disease development

A disease forecast permits timing of control measures. Such timing is very important. Chemicals, such as fungicides or insecticides, require accurate weather information (both present and for the future). Too early or too late application of chemicals not only results in ineffective disease control, but also causes needless expense and loss of time and also introduces undesirable chemicals into the environment.

The basic methods of predicting disease activities include an empirical procedure, in which correlations between plant disease and pest surveys in a particular area and the corresponding weather factors are used. Beaumont's criterion is an example. Beaumont's rule states that a "critical period", which generally precedes the first outbreak of potato blight by 7–21 days, is defined by the following weather conditions: temperature not less than 10°C and relative humidity not below 75% for a period of at least 48 h. Another method is to determine the relationship between environment and various stages of the life cycle in the laboratory and formulate a computer model mimicking the life cycle of the organism and predicting the rate at which it is developing. A simulator of southern corn leaf blight, named EPIMAY, is an example of a model for predicting disease activities (WAGGONER et al., 1972). The simulator has three components: weather observations, the reactions of pathogen to weather, and the computer program or instructions for presenting the current weather to the simulated pathogen. With these three parts of the simulator the computer calculates whether the epidemic is likely to rise, fall, or remain level.

Agricultural land utilization

Agricultural land properly utilized, based on its natural state will yield optimum returns. There are many environmental, social, and ecological factors affecting crop production and land use, but climatic factors are by far the most important variables affecting plant growth. With a given soil, the degree of land utilization can be estimated by the climatic potential.

LANDSBERG (1968b) has made the point that very few nations have the land resources and climatic conditions assuring their complete food independence. For the sake of self-sufficiency, some countries are now committed to a land use program which is neither especially suitable for the particular climate nor in the best interest of maximizing world food output. Misuse of agricultural land not only wastes time, energy, and resources, but in some cases causes economic disaster. The East Africa Groundnut Scheme is an example of how a miscalculation on land use can be very costly. Because of a critical shortage of oils and fats in Great Britain during and shortly after World War II, a project for the mechanized production of groundnuts in East and Central Africa was initiated. The project involved large capital investments of approximately £ 24,000,000. Operations began in 1948, followed in 1949 by a serious but not unusual drought causing the collapse of the project. YAO (1973) has conducted a climatic investigation of this scheme. The groundnut (peanut) plant is unique in that it bears its fruit 50–100 mm below the soil surface. Dry topsoil is detrimental to fruit development mainly because it reduces the uptake of soluble calcium by the fruit. Thus any study of soil moisture requirements for groundnut production must consider the availability of both top and subsoil moisture. Table XIX compares the frequency of occurrence of various durations of dry topsoil at Dodoma (near the groundnut farm area) and Norfolk and Tifton on the East Coastal Plain of the U.S.A. The significant feature of the table is that neither Tifton nor Norfolk, where production has been successful, show any year with a dry topsoil for more than 15 consecutive days. Meanwhile, in central Tanzania, if the last two class intervals are combined the table shows that on the average, the expected groundnut crop will suffer severe to total damage in 3–4 years out of every 10. The groundnut uses moisture from a soil profile as deep as 1.3–1.6 m. Table XX

TABLE XIX

FREQUENCY OF OCCURRENCE OF VARIOUS DURATIONS OF DRY TOPSOIL FOR DODOMA (1936–65), NORFOLK AND TIFTON (1940–69)*[1]

(After YAO, 1973)

Duration (days)	Number of occurrence*[2]			Frequency (%)		
	Dodoma	Norfolk	Tifton	Dodoma	Norfolk	Tifton
6–10	18	16	27	60	53	90
11–15	19	4	4	63	13	13
16–20	7	0	0	23	0	0
21–25	6	0	0	20	0	0
>26	5	0	0	17	0	0

*[1] Mid-January–mid-April at Dodoma, mid-May–mid-August at Norfolk and Tifton.
*[2] Includes no more than one occurrence in each duration class each year.

TABLE XX

PROBABILITY OF INADEQUATE ($R < 0.60$) AND OPTIMUM $R > 0.90$) ROOT ZONE MOISTURE EACH MONTH DURING THE GROWING SEASON

(After YAO, 1973)

		$R < 0.60$					$R > 0.90$					Freq. of occur. (%) 1 or more times $R < 0.60$ in 5-month prd.
		Dec.	Jan.	Feb.	Mar.	Apr.	Dec.	Jan.	Feb.	Mar.	Apr.	
Dodoma	1936–65	0.46	0.37	0.25	0.24	0.50	0.28	0.26	0.46	0.46	0.22	77
Kongwa	1954–66	0.42	0.34	0.30	0.26	0.48	0.26	0.44	0.54	0.41	0.26	76
Average		0.44	0.35	0.28	0.25	0.49	0.27	0.35	0.50	0.44	0.24	76
		Apr.	May	Jun.	Jul.	Aug.	Apr.	May	Jun.	Jul.	Aug.	
Norfolk	1940–69	0.01	0.01	0.05	0.08	0.04	0.99	0.77	0.60	0.70	0.67	10
Tifton	1940–69	0.05	0.05	0.09	0.13	0.09	0.90	0.73	0.61	0.63	0.54	10
Average		0.03	0.03	0.07	0.11	0.07	0.94	0.75	0.60	0.66	0.60	10

shows that in central Tanzania irrigation (R-index < 0.60) is needed in December and April in almost half of the years. In January, February, and March, irrigation is necessary only about one quarter to one third of the time. By comparison, the data for the southeast U.S.A. show the irrigation needs peak in July, but will be required only about 1 year in 10. Table XX shows in the wettest month in Tanzania, February, the optimum R-index > 0.90 is reached only about 50% of the time. In the eastern U.S.A. the optimum water requirement during the 5-month period ranges from 60 to 94%. Table XX also shows that the probability of an R-index of < 0.60 occurring one or more times during the 5-month period at Dodoma (30-year) and Kongwa (13-year) is 76% whereas in the southeast U.S.A. the frequency of the same R-index is only 10%. The rainfall regime is therefore altogether unfavorable for groundnut production in the region. Without irrigation water the probability of success in growing groundnuts in that area is too small to justify such use of the land.

Climatic classification

In the planning of agricultural land use, one must understand the climatic regimes which influence the area. There are several systems of climatic classification. Most of them involve either rainfall or temperature or both. Each system has its merit, and each also has its limitation. Details are in Volume 1 of this series (Bailey, 1979).

System analysis of land use in agriculture

The analysis of land use for agriculture is complex. Climate and many other environmental factors need to be considered. Climatic classification can give partial solution. A system analysis of land use in agriculture is presented including many factors other than climate, with weather in a prominent role. Fig.43 is an illustration of the system. It consists of flow patterns for matter and energy and for information and shows the basic components of the system. According to Landsberg, "In a complete systems analysis, the weather factors and their influences on the system could be readily lifted out and analyzed as a subsystem. A considerable part of the basic work for quantitative analysis of a weather subsystem in agriculture has already been done or data for it are readily available. Many probabilistic expressions for climatic risks have been established. Multiple regression for the effects of various weather elements on specific crop plants are also available. For most major crop plants the optimal phenological conditions are well known. If all this information is incorporated into the system, it becomes possible to assess the influence of any given variable singly or in combination with others. Computer techniques are now available to handle these complex problems."

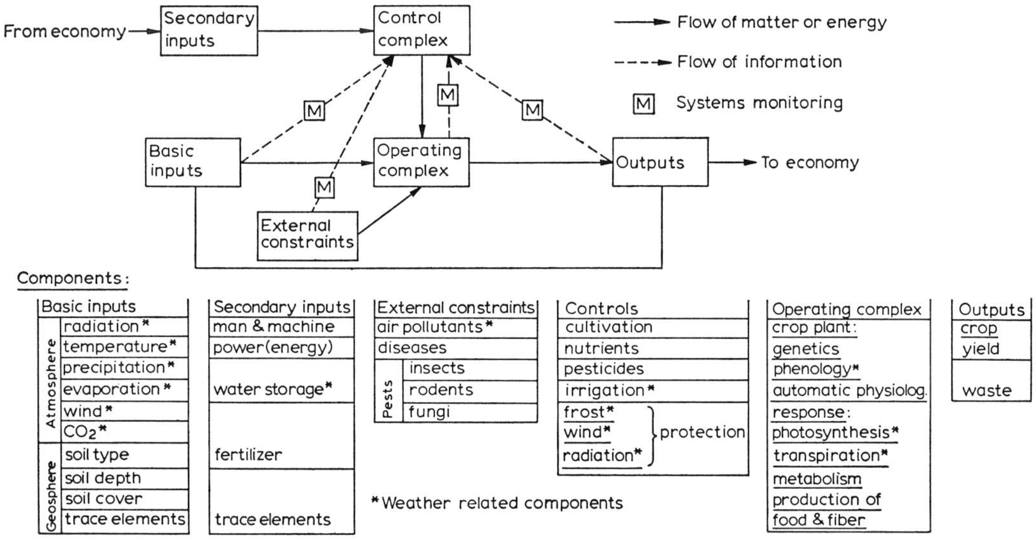

Fig.43. Scheme for a systems analysis of land use in agriculture. (LANDSBERG, 1968a.)

Land use in a small area mostly refers to the topographical differences affecting land use. Topo-climatology has been discussed thoroughly by R. Geiger in Vol. 2 of this Series. By knowing the climate of different topographic elements and by knowing the climatic requirement of the crops, it is not difficult to select a suitable geographical location for a particular crop. Thus, crops susceptible to low temperature should be planted on the slopes rather than the bottom land of a valley. Sunshine-loving plants have a better exposure on the southern slope rather than on the northern slope in the Northern Hemisphere.

Remote sensing

Remote sensing is a way of obtaining information about distant objects one cannot reach. The human eye is a good remote sensor in a narrow range of the electromagnetic spectrum (visible range) while the human ear can sense sound over a wide range of audio frequencies. Galileo's first telescope ushered in the era of remote sensing. In the 19th century, the second important instrument came into being—the camera. The combination of telescope and camera permitted recording objects remotely. The invention of aircraft permitted carrying instruments for remote survey of the planet. A new era began in 1960 with the launching of surveillance satellites. Equipped with sophisticated instruments, these spacecrafts orbit the earth and provide continuous views of the earth and its atmosphere.

Remote sensing and agricultural climatology

Remote sensing has been used to survey and to map the agricultural land during the past 3 or 4 decades in many parts of the world. But use of this information in agricultural meteorology is still in its infancy and in an experimental stage. Remote sensing techniques can be used to measure the reflectance, emittance, surface geometry and equivalent black body temperature of the earth's surface in the meaningful range of the electromagnetic radiation spectrum. It is entirely possible that in the future remote sensing may be used to estimate meteorological parameters which govern plant growth, including radiation and the water balance near the ground surface. Some recent research presented here indicates that the applicability of remote sensing to agricultural meteorology is promising for the future.

The vigor of vegetation can be determined photographically by recording foliage reflectance in the near infrared (COLWELL, 1971). The spongy mesophyll tissue of a healthy leaf (turgid) distended by water and full of air space is a very efficient reflector of any radiant energy and therefore also in the near infrared. When a plant's water relations are disturbed and it starts to lose vigor, the mesophyll collapses. As a result there may be great loss in the reflectance of near infrared energy by the leaves almost immediately after the damaging agent has struck a plant. These changes in reflectance can be detected photographically by a film sensitive to these near infrared wavelengths, which may be developed into indices for measuring plant disease, plant water stress, and irrigation requirements. VELLOSE (1971) reported using a remote sensing technique (color Ektachrome and IR Ektachrome aerial films) for estimating the diffuse reflectance of different degrees of defoliation in a coffee crop following a severe frost attack. Briefly, the recovery of the coffee crop production from a frost attack is related to the remaining live leaf area (other than that defoliated by frost). Vellose found a relationship between the diffuse reflectance and the leaf area index. From this, the recovery of coffee production can be predicted. BARTHOLIC et al. (1972) measured the irradiation, in the 8–14 μ wavelength interval, over an extensively instrumented agricultural area by using an aircraft-mounted thermal scanner. The area included soils differing in water and tillage condition, and replicated cotton plots with wide range of plant water stress. The observed irradiation corresponded to cotton plant canopy temperature differences up to 6°C between the most and the least water-stressed plots. The irradiance data from soil showed large differences as a function of time after tillage and irrigation. Hence thermal imagery is useful for delineating water-stressed and nonstressed fields, evaluating uniformity of irrigation, and evaluating surface soil water conditions.

Remote sensing from satellites and agricultural meteorology

Satellite observation systems for agricultural meteorology offer wide spatial coverage. The satellite observation systems can make observations continuously along the earth's surface. Although the analog record must be digitized for analysis creating a discontinuous yet useful record if the information content of one observation can be extrapolated to the next observation in space or time.

Multispectral cameras, electronic scanners, radars, and infrared devices are capable of providing data from which estimates of meteorological parameters can be made such as the surface temperature, snow cover, surface albedo, cloud amount, and atmospheric humidity etc. But remotely sensed surface data are not yet as accurate as ground observations, and in many cases the differences are significant. Take the radiometric estimate of surface temperature as an example. In the interpretation of surface temperature from radiometric data, the earth's surface and the atmosphere are assumed to radiate as a black body at the "window" channel radiation wavelength band 8–14 μ (WARK et al., 1962). In general, the lower the temperature of the observed surface the more accurate will be the assumption that it radiates as a black-body radiation (ATLAS et al., 1965). But in intervening atmosphere its water vapor absorption causes surface temperature to appear lower than the true surface temperature (FUJITA et al., 1968). The correct value of the water vapor mass absorption coefficient its distribution of the coefficient is not fully known so that in extreme cases the estimated temperature may be off by as much as 10°C (PLATT, 1972). Because of similar difficulties estimates of air moisture content within the first 300 m of the atmosphere, the region of most interest to agriculturalists, are not as yet attempted. However, in the future microwave radiometers may eventually enable surface layer soil moisture content to be estimated. Within the microwave range of radiation the intensity of emitted radiation is a function of both surface temperature and liquid water content. In addition, there is microwave attenuation in the atmosphere by water vapor and oxygen. Because the equivalent black-body temperature contrast between land and water (due to the large dielectric property differences) is about 100°K, water surfaces are easily identified. Liquid water in clouds over a land surface is not easily distinguishable from the land itself. The land surface emissivity varies significantly with soil moisture content. At a frequency of 13.4 GHZ the radiometer sees down to a depth of 5 cm and to 76 cm at frequencies as low as 1.4 GHZ (EDGERTON et al., 1968). The brightness temperature decreases with increasing soil moisture, but in snow the emissivity of the snow increases with increasing liquid water content. Precipitation over a water surface has a greater brightness temperature than the surrounding water surface, but over land precipitation may have a less brightness temperature than the land. Thus, the interpretation of microwave brightness temperature remains complex. Microwave radiometric data supplemented by IR radiometric and visual range data may help interpret images.

There are difficulties in interpretation of satellite-measured data such as surface temperature, moisture, and other meteorological parameters on and near the surface. These difficulties must be resolved before any real progress in agricultural meteorology can be made.

Recently, however, many empirical methods have been published to estimate precipitation, snow cover and depth, and soil moisture, using reflectance, brightness, size of clouds, or objects on or near the earth's surface of satellite imageries. These include FALLANSBEE (1975, 1976), SCOFIELD and OLIVER (1977) for precipitation; MCGINNIS et al. (1975), SCHNEIDER et al. (1976) for snow depth and extent, HIELMAN et al. (1977) for soil moisture, and many others.

Summing up all recent progresses, one can take an optimistic view on the applicability of remote sensing data by satellite to agricultural meteorology. The difficulties encountered today will soon be solved and agricultural meteorological problems relying on satellite information may become increasingly important as time progresses. This information will cover radiation, water balance of crops, evapotranspiration, drought and irrigation, plant disease prediction, and crop yield prediction.

Weather and crop yield and world distribution of productivity

The importance of weather, or the longer phase of weather, i.e. "climate", in influencing plant growth and final yield is unquestioned. Given that soil fertility is more or less constant, and varieties and other technological improvement change in a steady upward direction, year to year variations in yield are a function of fluctuations in weather.

Yield model

Investigations of climate and crop yield relationships have a long history. In earlier studies, most of the analyses were limited to correlations between monthly temperature or precipitation or both. SMITH (1904) correlated yield of corn in the Corn Belt of the U.S.A. with June, July, and August rainfall. In 1914 he employed the same method of correlation for temperature and precipitation. He found July precipitation is the most important factor affecting corn yield. WALLACE (1920) studied the problem by using simple and multiple correlation methods. July rainfall remained a dominant factor in Ohio but not in other parts of the Corn Belt. The problem of predicting corn yield from weather seemed relatively simple in the southern half of the Corn Belt, notably in Kansas, Missouri, and southern Illinois where drought and heat in June, July, and August are the chief influences, while in the northern half, prediction was difficult. FISHER (1924) developed a special statistical method for analyzing the effect of rainfall during the growing season on annual wheat yield. This was successfully used for studies of crop yields in regions where rainfall is limited. KINCER and MATTICE (1928) outlined a method of multiple correlations of weather data with crop yield. It was first applied to study the weather and spring wheat relationship in North Dakota. Subsequently, other crops were studied in different states using weekly, biweekly, and monthly weather variables. Results were satisfactory.

RUNGE and ODELL (1958) investigated the corn yield and weather relationship in Urbana,

Illinois. A modification of Fisher's method was used in the analysis by HENDRICKS and SCHOLL (1943). Open-pollinated corn yields were converted to hybrid corn yields by using simple linear regression. The upward trend of the corn yield was also removed by linear regression to avoid any spurious intercorrelations of trend and the various weather variables included in the analysis. Precipitation and maximum daily temperatures 50–74 before and 14–30 days after full tassel were used and the climatic data were treated such that the total length of maximum temperature and precipitation were divided into (*a*) 2-day, and (*b*) 8-day periods. Precipitation and maximum daily temperature for the aforesaid period explained up to 67% of the corn yield variability during the period 1903 through 1956. When the upward trend in yields was included in this analysis, approximately 75% of the corn yield variability was explained. RUNGE and ODELL (1960) showed that precipitation and maximum daily temperature from June 25 through September 20 explained 68% of the variation in soybean yields (from 1909 through 1957). Above-normal precipitation during July (period of major vegetative growth) and from mid-August to mid-September (grain-filling period) increased soybean yields, but abundant rainfall during other periods decreased yields. Maximum temperatures during July and August are generally too high for optimum soybean yields.

Weather and crop yield analyses involving multiple variables were possible in recent years with the advent of high-speed computers. THOMPSON (1962, 1963, 1969) studied corn, wheat, soybean, and sorghum in the Corn Belt and in the Great Plains of the U.S.A. Eight climatic variables (pre-season precipitation, and monthly precipitation and temperature during most of the growing season), their quadratic terms, and some of the interaction between temperature and precipitation were used. A linear trend for technological improvement such as fertilizer application was used in this analysis. In 1966 Thompson improved his trend estimate for corn by introducing two subjective technology variables. He found the most significant weather variable in the production of corn and soybeans to be July rainfall and August temperature. Higher than average rainfall in July and lower than average August temperature are desirable. For wheat Thompson found that the weather variables and the time trend factor accounted for 82% or more of the yield variation in each state (Kansas, Nebraska, Oklahoma, and North and South Dakota). The curvilinear regression equation involving a large number of weather variables was questioned by Dale and Shaw (cited by THOMPSON, 1963). But Thompson has argued that the analysis of variance, in conjunction with multiple regression, can avoid sometimes misleadingly high correlation coefficients.

Recently STEYAERT et al. (1978) discussed a new approach to crop yield modeling. They use principal components of large-scale atmospheric general circulation features, i.e., semipermanent features as predictors for wheat or rice yield. Long-term monthly sea level pressure or 700-mbar height data are the basic variables describing the general circulation for midlatitude and tropical regions. They also reported that the yield models have been run operationally for the period 1975–1977 for U.S.A., Canada, U.S.S.R., and Indian wheat and results compare closely to official final estimates.

Crop yield modelings in recent years include WILLIAMS (1973), SAKAMOTO and JENSEN (1975), POCHOP et al. (1975, WILLIAMS et al. (1975), SAKAMOTO (1976), SIROTENKO (1976), LEDUC (1976), SCHERER (1977), PITTER (1977), and MICHAELS and SCHERER (1977) for wheat; RUNGE and BENCI (1975), HAUN (1976), DALE (1977), HOISE (1977), and SHAW (1977) for corn; DAS (1971), and DAS and RAMACHANDRAN (1972) for rice.

Climatic indices are being used in weather and crop yield analysis. These indices are functions of different meteorological parameters. For an example the R-index given by YAO (1969) (actual over potential evapotranspiration ET/PE) involves temperature, radiation, wind and vapor pressure, precipitation, soil moisture content and soil water holding capacity. Many climatic indices are helpful for measuring plant growth and development. Climate acts upon plant growth as a unit. Some climatic parameters can affect plant growth, and such a response can be enhanced or tempered by other parameters. The indices represent an integration of all these effects. Statistically, the use of indices not only can eliminate the built-in effects of intercorrelation among climatic variables, but also conserves degrees of freedom.

OURY (1965) suggests the use of the composite "aridity" indices by de Martonne, and by Ångström. He presented the following three models:

$$Y = b + b_{\theta}\theta + b_{P}P + b_{T}T + e \qquad (1)$$

$$Y = b' + b'_{\theta}\theta + b_{M}(P/T + 10) + e' \qquad (2)$$

$$Y = b'' + b''_{\theta}\theta + b_{A}(P/1.07^{T}) + e'' \qquad (3)$$

where Y = yield per acre of the crop involved, dependent variable; θ = time trend; P = precipitation during selected period at selected location in millimeters; T = temperature, same period, same location in centigrade; b, b', b'' = constant for the respective equation; b_P, b_T = regression coefficients for the respective equation; b_M = regression coefficient of the de Martonne index M; b_A = regression coefficient of the Ångström index A; e, e', e'' residuals of the respective equations.

Oury tested eqs.2 (de Martonne's index) and 3 (Ångström's index) separately against eq.1. He found that both t and F tests were better for eqs.2 and 3 than for eq.1. However, he failed to provide a comparison between actual corn yield and the computed yield using eqs.2 and 3.

DALE and SHAW (1965b) presented a multiple curvilinear regression model to estimate the effect of moistuie stress and plant populations on corn yield at two fertility levels. Linear, quadratic and interaction terms were utilized in the equation. The climatic variable used in the equation was the total number of non-stress days for a period from 6 weeks before to 3 weeks after silking. The non-stress day is defined as any day on which the ET_{FC} (ET at field capacity) and available soil moisture combination fell below the turgor loss point (θ_{TL}). (See Fig.44, which is experimentally determined.) Dale and Shaw found that the weather variable was highly associated with the experimental plot corn yields at Ames, Iowa.

COOK (1971) used the Palmer crop moisture index as a tool to predict corn yield in Kentucky, U.S.A., by a simple linear regression model. The computed yield agreed very well with the actual yield.

Another crop–weather analysis model using a computer technique by BAIER (1973) is designed primarily to analyze the daily interacting effects of any three selected environmental variables on the response of an annual crop as reflected in the seasonal yield or yield components. The basic equation is as follows:

$$Y = \sum_{t=0}^{t=m} V_1 \times V_2 \times V_3 \qquad (1)$$

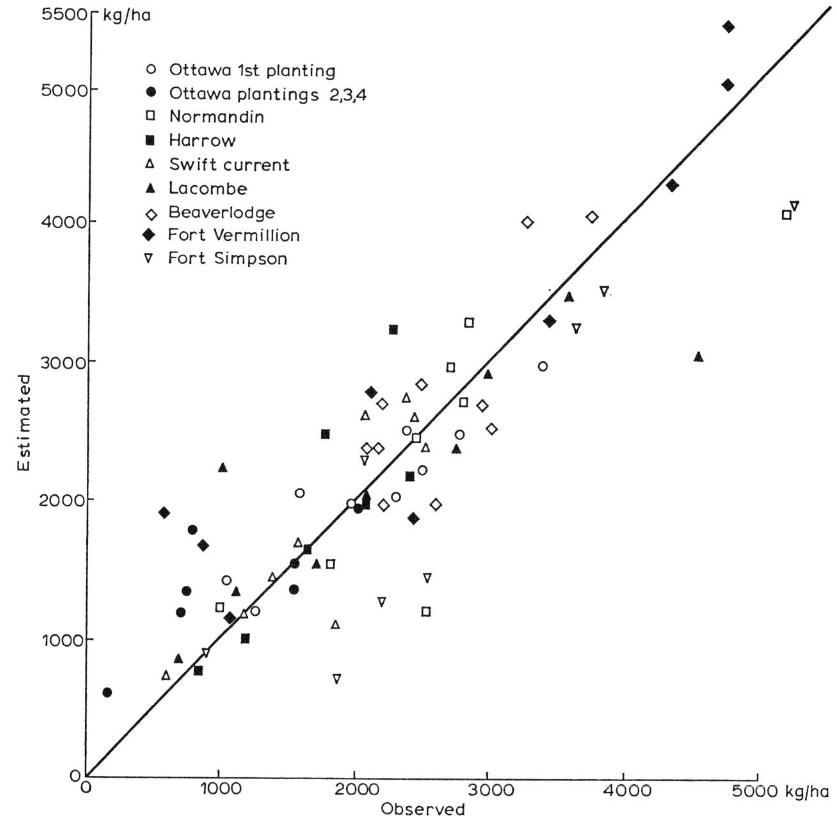

Fig.44. Scatter diagram of observed and estimated grain yields around 1:1 regression line. (After BAIER, 1973.)

where Y = "dependent" variable, such as the final observed crop yield or yield component at the end of a specified period. In the example of this paper, Y is the seasonal grain yield in kilograms per hectare $\sum_{t=0}^{t=m}$ = summation of daily V-values from biometeorological time $t = 0$ to $t = m$ according to the biometeorological time scale by ROBERTSON (1968), where 0 = planting (P), 1 = emergence (E), 2 = jointing (J), 3 = heading (H), 4 = soft dough (S), 5 = ripe (R). Intermediate values are expressed in decimals of the above time units. V_1, V_2, V_3 = functions of the selected independent variables. Each V-function is of the general form:

$$V = (u_1t + u_2t^2 + u_3t^3 + u_4t^4) + (u_5t + u_6t^2 + u_7t^3 + u_8t^4)X + (u_9t + u_{10}t^2 + u_{11}t^3 + u_{12}t^4)X^2 \quad (2)$$

where $u_1 < u_{12}$ are coefficients which are evaluated for each V in an iterative regression analysis and X represents in V_1, V_2, or V_3 any of the variables sected by the program controls.

Weather variables used in this model include both the standard meteorological observations and meteorological indices. Fig.44 is the scatter diagram of observed and estimated grain yield (using the proposed model). The individual observed and estimated yields are reasonably close to the 1:1 regression line and well distributed over the entire range (see also BAIER, 1977).

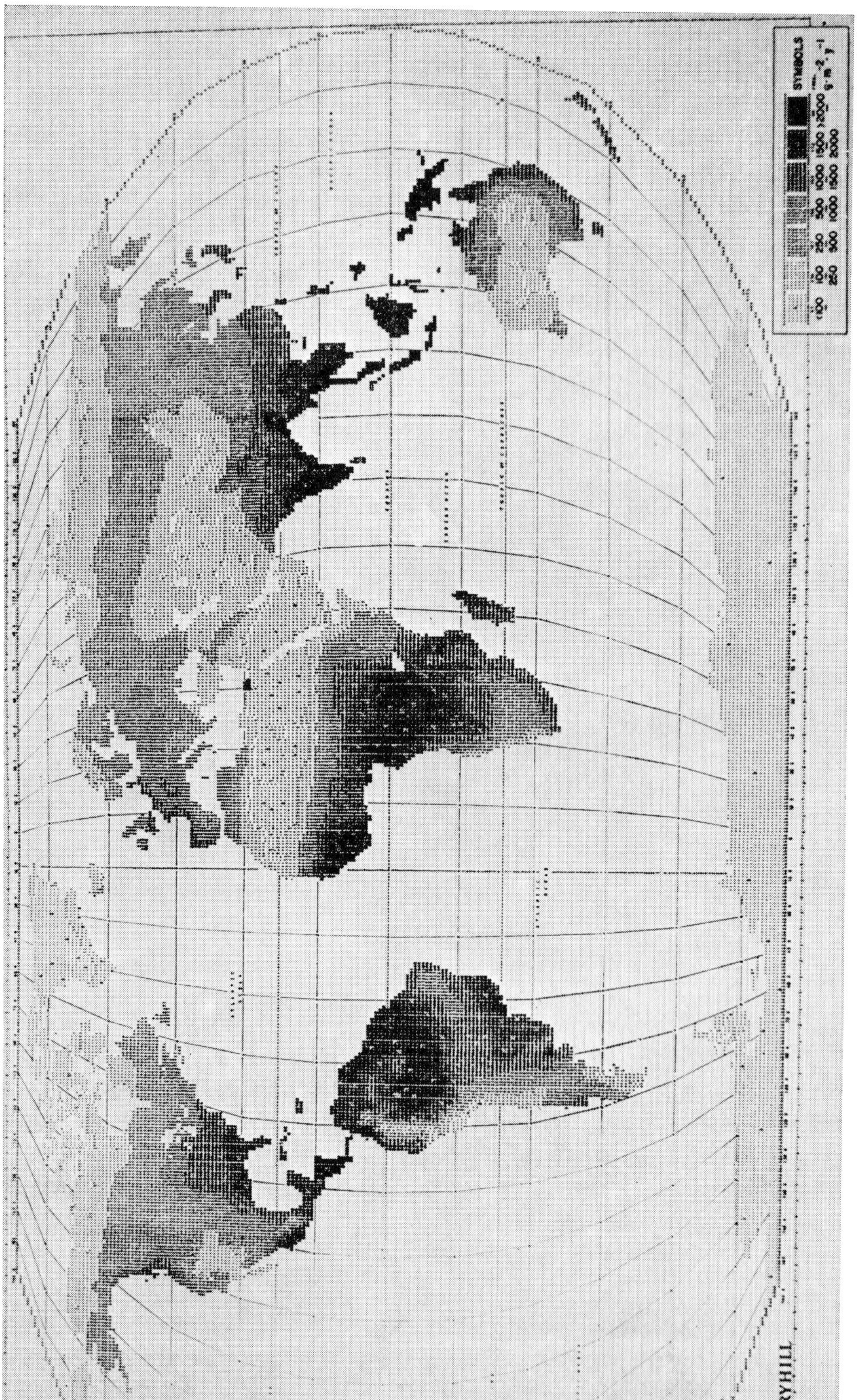

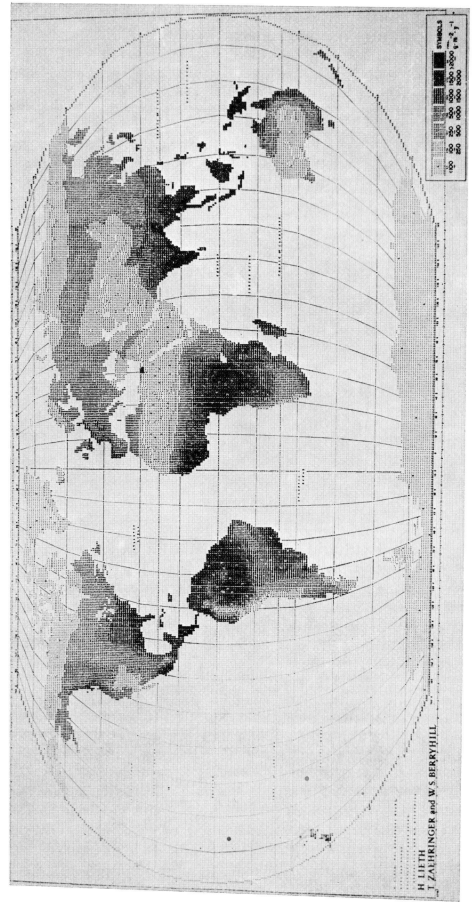

Fig.45. Primary productivity predicted from the precipitation and temperature averages of existing meteorological stations. "Miami" model. (After LIETH, 1972.)

World distribution of productivity

The prediction of the world productivity is very complex. It involves many environmental factors. However, the climatic variables are by far the predominant factor that affects productivity. Many recent investigations on climate and crop yield potential studies correlated yield and climatic parameters. LIETH (1972) cited the most recent work in Russia and the work under the International Biological Programme. He has presented a few models predicting the primary productivity of the world. The "Miami" model is one of them (see Fig.45). It was presented to the American Institute of Biological Science in Miami, Florida, October 1971. This model involves two basic regression models, one expresses the relationship between mean annual temperature and annual dry-matter production and the other expresses the relationship between annual average precipitation and annual dry-matter production. These two equations are:

$$Y = 3000/1 + e^{1.315-0.119x} \quad (1)$$

$$Y = 3000(1 - e^{-0.000664xI}) \quad (2)$$

where Y is the dry matter production, x is the mean annual temperature and xI is annual average precipitation. In the computation if temperature or precipitation predicted different productivity levels, then the lowest value was chosen.
Lieth concluded that more accurate maps can be constructed with improved productivity data from as many stations of the world as possible.

Acknowledgements

The author wishes to thank R. D. Felch, F. A. Godshall, J. D. McQuigg, J. J. Rahn, C. M. Sakamoto for their invaluable suggestions and editorial help. He is also deeply indebted to Dr. H. E. Landsberg of the University of Maryland, for his encouragement, comments, and discussion which broadened the author's view considerably. Finally, the author would like to thank the late Miss B. Van Stronder and Miss S. Souther for typing the manuscript.

References

ABBOTT, D. L., 1962. The effect of four controlled winter temperatures on the flowering and fruiting of the apple. *J. Hort. Sci.*, 37: 272–284.

ABDELHAFEEZ, A. T. and VERKERK, K., 1969. Effect of temperature and water regime on the emergence and yield of tomatoes. *Neth. J. Agric. Sci.*, 17: 50–69.

ABE, S. and WADA, M., 1957. Effect of night temperature at ear-developing stage on rice plant. *Proc. Crop Sci. Soc. Jpn.*, 26: 99–100.

ADACHI, K. and INOUE, J., 1970. Effects of photoperiod, light intensity and components of culture medium on flower initiation in highly thermosensitive paddy rice plants. *Proc. Crop Sci. Soc. Jpn.*, 41: 78–82.

ALESSI, J. and POWER, J. F., 1971. Corn emergence in relation to soil temperature. *Agron. J.*, 63: 717–719.

ALEXANDER, G. and MCCANCE, I., 1958. Temperature regulation in the newborn lamb, I. Changes in rectal temperature within the first six hours of life. *Aust. J. Agric. Res.*, 9: 339–347.

ALISSOW, B. P., DROSDOW, O. A. and RUBINSTEIN, E. S., 1956. *Lehrbuch der Klimatologie*. Veb Deutscher Verlag der Wissenschaften, Berlin (German translation of a Russian publication).

ALLARD, H. A., 1938. Complete or partial inhibition of flowering in certain plants when days are too short or too long. *J. Agric. Res.*, 57: 775–789.
ANDERSON, M. C., 1964. Light relations of terrestrial plant communities and their measurement. *Biol. Rev. Cambridge Phil. Soc.*, 39: 425–486.
ANDERSON, M. C., 1970. Interpreting the fraction of solar radiation available in forest. *Agric. Meteorol.*, 7: 19–28.
ARMBURST, D. V., 1968. Wind blown soil abrasive injury to cotton plants. *Agron. J.*, 60: 622–625.
ASLYING, H. C., 1958. Shelter and its effect on climate and water balance. *Oikos*, 9: 282–310.
ATLAS, D., 1977. The paradox of hail suppression. *Science*, 195: 139–145.
ATLAS, D., CUMINGHAM, R. M., DONALDSON JR., R. J., KANTOR, G. and NEWMAN, P., 1965. Some aspects of electromagnetic wave propagation. In: SHEA L. VALLEY (Editor), *Handbook of Geophysics.* Air Force Cambridge Res. Lab., pp.9.1–9.26.
AZMI, A. R. and ORMROD, D. P., 1971. Rate of net carbon dioxide assimilation in rice as influenced by growth temperature and photoperiod. *Agron. J.*, 63: 543–546.
BAGLEY, W. T. and GOWEN, F. A., 1960. Growth and fruiting of tomatoes and snap beans in the shelter area of a windbreak. *Proc. World For. Congr. ,5th, Seattle, Wash.*, pp.1–3.
BAIER, W., 1973. Crop–weather analysis model: Review and model development. *Appl. Meteorol.*, 12: 937–947.
BAIER, W., 1977. Crop–weather and their use in yield assessment. *WMO, Geneva, Tech. Note*, 151.
BAIER, W. and ROBERTSON, GEO. W., 1965. Estimating of latent evaporation from simple weather observation. *Can. J. Plant Sci.*, 45: 276–283.
BAIER, W. and ROBERTSON, GEO. W., 1966. A new versatile soil moisture budget. *Can. J. Plant Sci.*, 46: 299–315.
BAIER, W. and ROBERTSON, G. W., 1967. The performance of soil moisture estimates as compared with the direct use of climatological data for estimating crop yields. *Agric. Meteorol.*, 5: 17–31.
BAIER, W. and ROBERTSON, G. W., 1970. Climatic estimates of average and probable irrigation requirements and of seasonal drainage in Canada. *J. Hydrol.*, 10: 20–37.
BAKER, D. G., 1958. *A Comparison of Two Evapotranspiration Calculation Methods and the Application of One to Determine Some Climatic Differences Between Great Soil Groups of Minnesota.* Unpublished doctoral dissertation, Univ. of Minnesota.
BARGER, G. L. and THOM, H. C. S., 1949. Evaluation of drought hazard. *Agron. J.*, 41: 519–526.
BARTHOLIC, J. F., NAMKEN, L. N. and WIEGAND, C. L., 1972. Aerial thermal scanner to determine temperatures of soils and of crop canopies differing in water stress. *Agron. J.*, 64: 603–608.
BATES, C. C., 1911. Windbreaks, their influence and value. *U S. Dept. Agric. For. Serv. Bull.*, 86: 1–100.
BATTAN, L. J., 1965. A view of cloud physics and weather modification in the Soviet Union. *Bull. Am. Meteorol. Soc.*, 46: 309–316.
BAVER, L. D., 1956. *Soil Physics*, 3rd ed. Wiley, New York, N.Y.,
BERRY, I. L., SHANKLIN, M. D. and JOHNSON, H. D., 1964. Dairy shelter design based on milk production decline as affected by temperature and humidity. *Trans. Am. Soc. Agric. Eng.*, 7: 329–331.
BILANSKI, W., DAVIS, J. R. and KIDDER, E. H., 1954. Frost protection by sprinkler irrigation: a field survey. *Mich. Agric. Expt. Sta., Lansing, Q. Bull.*, 36: 357–363.
BLAD, B. L. and BAKER, D. G., 1972. Reflected radiation from a soybean crop. *Agron. J.*, 64: 277–280.
BLANEY, H. F. and CRIDDLE, W. D., 1950. Determining water requirements in irrigated areas from climatological data. *Soil Conserv. Serv., U.S. Dept. Agric. Wash., D. C., Tech. Publ.*, 96.
BLANEY, H. F. and MORIN, K. V., 1942. Evaporation and consumptive use of water empirical formulas. *Trans. Am. Geophys. Union*, 23: 76–83.
BLAXTER, K. L., 1962. *The Energy Metabolism of Ruminants.* Hutchinson, London.
BÖHNING, R. H. and BURNSIDE, C. A., 1956. The effect of light intensity on rate of seedling tomato plants variety E. S. 5. *Texas Agric. Expt. Sta., Coll. Station, Texas, Bull.*, 756.
BOHREN, C. F. and THORUD, D. B., 1973. Two theoretical models of radiation heat transfer between forest trees and snowpacks. *Agric. Meteorol.*, 11: 3–16.
BOND, T. E., 1967. Microclimate and livestock performance in hot climates. In: R. H. SHAW (Editor), *Ground Level Climatology.* Am. Assoc. Advan. Sci., Wash., D. C., pp.207–229.
BOND, T. E., KELLY, C. F. and ITTNER, N. R., 1957. Cooling beef cattle with fans. *Agric. Eng.*, 38: 308–309.
BOND, T. E., HEITMAN, H. and KELLY, C. F., 1965. Effects of increased air velocities on heat and moisture loss and growth of swine. *Trans. Am. Soc. Agron. Eng.*, 8: 167–174.
BONNER, J., 1944. Effects of temperature on rubber accumulation by the Guayule plant. *Bot. Gaz.*, 105: 233–243.
BONSMA, J. C., 1958. *Livestock Philosophy.* Univ. Pretoria, Johannesburg, Bull. No. 5.

Borthwick, H. A. and Parker, M. W., 1939. Photoperiodic responses of several varieties of soybeans. *Bot. Gaz.*, 101: 341–365.

Borthwick, H. A., Parker, M. W. and Hendricks, S. B., 1950. Recent developments in the control of flowering by photoperiod. *Am. Naturalist*, 84: 117–134.

Bourke, P. M. A., 1955. The forecasting from weather data of potato blight and other plant disease pest. *WMO No.42, Tech. Note*, 10: 1–49.

Bourke, P. M. A., 1968. Some principles and methods in plant disease forecasting. *Proc. Regional Training Seminar on Agro-meteorology, Wageningen*, pp.21–33.

Bouyoucos, G. J., 1915. Effect of temperature on movement of water vapor and capillary moisture in soil. *J. Agric. Res.*, 5: 141–172.

Bowen, E. G., 1952. A new method of stimulating convective clouds to produce rain and hail. *Q. J. R. Meteorol. Soc.*, 78: 37–45.

Bowen, I. S., 1926. The ratio of heat losses by conduction and by evaporation from any water surface. *Phys. Rev.*, 27: 779–787.

Briggs, L. J., 1897. The mechanics of soil moisture. *U.S. Dept. Agric. Bur. Soil, Bull.*, 10.

Briggs, L. J. and McLane, J. W., 1907. The moisture equivalent of soil. *Bur. Soils, U.S. Dept. Agric. Bull.*, 45.

Briggs, L. J. and Shantz, J. H., 1916a. Hourly transpiration rate on clear days as determined by cyclic environmental factor. *J. Agric. Res.*, 5: 583–650.

Briggs, L. J. and Shantz, J. H., 1916b. Daily transpiration during the normal growth period and its correlation with the weather. *J. Agric. Res.*, 7: 155–212.

Brody, S. 1956. Climatic physiology of cattle. *J. Dairy Sci.*, 39: 715–725.

Brooks, F. A., 1959. *An Introduction to Physical Microclimatology.* Univ. of California Press, Davis, Calif.

Brooks, F. A., Kelly, C. F., Rhoades, D. C. and Schultz, H. B., 1952. Heat transfer in citrus orchards using wind machines for frost protection. *Agric. Eng.*, 23: 74–78; 143–147.

Brown, D. M. 1960. Development-temperature relationships from controlled environment studies. *Agron. J.*, 52: 493–496.

Brown, D. M., 1964. *How to Determine Heat Units Available for Corn Production in Ontario.* Res. Foundation, Toronto, Ont.

Brown, K. W., 1972. Windbreak research—A review and outlook. *Abstr. Bull. Am. Meteorol. Soc.*, 53: 1129.

Brown, K. W. and Rosenberg, N. J., 1971. Turbulent transport and energy balance as affected by a windbreak in an irrigated sugar beet field. *Agron. J.*, 63: 351–355.

Brown, P. L., 1971. Water use and soil water depletion by dryland winter wheat as affected by nitrogen fertilization. *Agron. J.*, 63: 43–46.

Buckingham, E., 1907. Studies on the movement of soil moisture. *U.S. Dept. Agric. Bur. Bull.*, 38.

Bula, R. J. and Massegale, M. A., 1972. Environmental physiology. In: C. H. Hanson (Editor), *Alfalfa Science and Technology.* Am. Soc. Agron., Wisc., pp.167–184.

Bula, R. J., Rhykerd, C. L. and Langston, R. G., 1959. Growth response of alfalfa seedlings under various light regimes. *Argon. J.*, 51: 84–86.

Burman, R. D. and Partridge, J. R., 1962. Evapotranspiration of water by small grains, corn and beans in northwestern Wyoming. *Wyo. Agric. Expt. Sta. Circ.*, 174.

Burrage, S. W., 1972. Dew on wheat. *Agric. Meteorol.*, 10: 3–12.

Businger, J. A., 1956. Some remarks on Penman's equation for the evapotranspiration. *Neth. J. Agric. Sci.*, 4: 77–80.

Buss, S., and Shaw, R. H., 1960. *Prediction of Soil Moisture under Corn*, II. Final Rept. U.S. Weather Bur. Contr. Cwb-9560. Dept. of Agron., Iowa State Univ., Ames, Iowa.

Caborn, J. M., 1957. Shelterbelts and microclimate. *For. Comm., H. M. Stationery Off., Edinburgh, Bull.* 29: 1–35.

Calvert, B. A., 1957. Effect of the early environment on development of flowering in the tomato, 1. Temperature. *J. Hort. Sci.*, 32: 9–17.

Campbell, C. A., Pelton, W. L. and Nielson, K. F., 1969. Influence of solar radiation and soil moisture on growth and yield of Chinook wheat. *Can. J. Plant Sci.*, 49: 685–699.

Caprio, J. M., 1971. The solar-thermal unit theory in relation to plant development and potential evapotranspiration. *Mont. Agric. Expt. Sta., Mont. Sta. Univ., Bozeman, Circ.*, 251.

Carlson, L. D., 1964. Reaction of man to cold. In: S. Licht (Editor), *Medical Climatology.* Waverly, Baltimore, Md., 8: 196–228.

Carlson, R. E., Yarger, D. N. and Shaw, R. H., 1972. Environmental influence on the leaf temperature of two soybean varieties grown under controlled irrigation. Agron. J., 64: 224–229.

CARPENTER, L. G., 1891. The artesian wells of Colorado and their relation to irrigation. *Colo. Agric. Expt. Sta. Ann. Rept.*, 16: 1–28.

CARSON, J. E., 1961. *Soil Temperature and Weather Conditions.* Argonne National Laboratory, Argonne, Ill.

CASPERSON, G., 1957. Untersuchungen über den Einfluss von Windschutzanlagen auf den standörtlichen Wärmehaushalt (Investigations on the influnce of windbreaks on the local heat balance). *Angew. Meteorol.*, 2: 339–351.

CHANDLER, R. F., 1963. An analysis of factors affecting rice yield. *Int. Rice Comm., News Letter*, 41(4).

CHANG, J. H., 1968. *Climate and Agriculture.* Aldine, Chicago, Ill.

CHANG, J. H., CAMBELL, R. B. and ROBINSON, F. E., 1963. On the relationship between water and sugar cane yield in Hawaii. *Agron. J.*, 55: 450–453.

CHAPMAN, A. L. and PETERSON, M. L., 1962. The seedling establishment of rice under water in relation to temperature and dissolved oxygen. *Crop Sci.*, 2: 391–395.

CHEN, T. H. and HSU, C. M., 1957. *Study of Tomato.* Peiping Sci. Press., Peiping (in Chinese).

CLAPSTICK, J. W. and WOOD, T. B., 1922. The effect of change of temperature on the basal metabolism of swine. *J. Agric. Sci.*, 12: 257–268.

COLWELL, R. N., 1971. Applications of remote sensing in agriculture and forestry. *Earth Resour. Surv. System, Natl. Aeronaut. Space Admin., Wash., D.C.*, pp.67–90.

COOK, D., 1971. The Palmer crop moisture index as a tool to predict corn yield. *Conf. Agric. Meteorol., 10th, Columbia, Mo.* Am. Meteorol. Soc.

COOPER, A. J., 1961. Observations on flowering in glasshouse tomato plants considered in relation to light duration and plant age. *J. Hort. Sci.*, 36: 102–115.

COOPER, A. J., 1963. A consideration on the seasonal trends in the growth of the fruit of glasshouse tomato plants in relation to day-length. *J. Hort. Sci.*, 38: 95–108.

COOPER, A. J. and COOK, D., 1964. The effects of light intensity, day temperature and water supply on the fruit ripening disorder and yield of two varieties of tomato. *J. Hort. Sci.*, 39: 42–53.

CREASY, L. L., 1968. The role of low temperature in Anthocyamin synthesis in McIntosh apple. *Proc. Am. Soc. Hort. Sci.*, 93: 716–724.

CROSIER, W., 1934. Studies in the biology of *Phytophthora infectans* (Mont) *de Barry. Cornell Univ. Agric. Expt. Sta. Mem.*, 155.

DAIGGER, L. A., AXTHELM, L. S. and ASHBURN, C. L., 1970. Consumptive use of water by alfalfa in western Nebraska. *Agron. J.*, 62: 507–508.

DAINGERFIELD, L., 1929. Weather and cotton yield in Texas 1899–1929 inclusive. *Mon. Weather Rev.*, 57: 451–453.

DALE, R. F., 1977. *An Energy-Crop Growth Variable for Identifying Weather Effects upon Corn Growth and Yield.* Preprint from the 13th Agriculture and Forest Meteorology Conference, Purdue University, Indiana.

DALE, R. F. and SHAW, R. H., 1965a. The climatology of soil moisture, atmospheric evaporation demand, and resulting moisture stress days for corn at Ames, Iowa. *J. Appl. Meteorol.*, 4: 661–669.

DALE, R. F. and SHAW, R. H., 1965b. Effect on corn yields of moisture stress and stand at two fertility. levels. *Agron. J.*, 57: 475–479.

DALTON, J., 1802. Experiments and observations made to determine whether the quantity of rain and dew is equal to the quantity of water carried off by river and raised by evaporation, with an inquiry into the origin of springs. *Mem. Lit. Phil. Soc.*, 5: 346–372.

DAS, J. C., 1971. Forecasting the yield of principal crops in India on the basis of weather—paddy/rice. *Indian Meteorol. Dept. Poona, Sci. Rept.*, 120.

DAS, J. C., and RAMACHANDRAN, G., 1972. Forecasting the yield of Bajra on the basis of weather parameters of (Ahmadabad district). *Indian Meteorol. Dept. Poona, Meteorol. Monogr. Agrimet.*, 3.

DAVIS, D. A., 1954. Experiments on artificial stimulation of rain in East Africa. *Nature*, 174: 256–258.

DAY, A. D. and BARMORE, M. A., 1971. Effects of soil moisture stress on the yield and quality of flour from wheat. *Agron. J.*, 63: 115–116.

DAY, A. D. and INTALAP, S., 1970. Some effects of soil moisture stress on the growth of wheat (*Tritcum aestivum L. em Thell*). *Agron. J.*, 62: 27–29.

DEACON, E. L. and SWINBANK, W. C., 1958. Comparison between momentum and water vapor transfer. *Proc. Canberra Symp. Climatology and Microclimatology, Arid Zone Res., UNESCO*, pp.38–41.

DECKER, W. L., 1959. Variation in the net exchange of radiation from vegetation of different height. *J. Geophys. Res.*, 64: 1617–1619.

DECKER, W. L., 1962. Precision of estimates of evapotranspiration in Missouri climate. *Agron. J.*, 54: 529–531.

DELAUCHE, J. C., 1953. Influence of moisture and temperature levels on the germination of corn, soybeans and watermelons. *Proc. Assoc. Official Seed Analysts*, 43: 117–126.

DENMEAD, O. T., 1969. Comparative micrometeorology of a wheat field and a forest of *Pinus radiata. Agric. Meteorol.*, 6: 357–371.

DENMEAD, O. T. and SHAW, R. H., 1960. The effects of soil moisture stress at different stages of growth on the development and yield of corn. *Agron. J.*, 52: 272–274.

DENMEAD, O. T. and SHAW, R. H., 1962a. Availability of soil water to plants as affected by soil moisture content and meteorological conditions. *Agron. J.*, 54: 385–390.

DENMEAD, O. T. and SHAW, R. H., 1962b. Spatial distribution of net radiation in a corn field. *Agron. J.*, 54: 505–510.

DOTZENKO, A. D., COOPER, C. S., DOBRENZ, A. K., LAUDE, H. M., MASSENGALE, M. A. and FELLNER, K. C., 1967. Temperature stress on growth and seed characteristics of grasses and legumes. *Colo. Agric. Expt. Sta. Tech. Bull.*, 97.

DYER, A. J. and PRUITT, W. O., 1962. Eddy-flux measurements over a small irrigated area. *J. Appl. Meteorol.*, 1: 471–473.

DYER, A. J., HICKS, B. B. and KING, K. M., 1967. The fluxatron—a revised approach to the measurement of eddy fluxes in the lower atmosphere. *J. Appl. Meteorol.*, 6: 408–413.

EDGERTON, A. T., MANDL, R. M., POE, G. A., JENKINS, J. E., SOLTIS, F. and SAKAMOTO, S., 1968. *Passive Microwave Measurements of Snow, Soils, and Snow–Ice–Water System.* Aerojet-General Corp., El Monte, Calif. Contract No. NONR 4767 (00) NR. 387-033.

EDLEFSEN, N. E. and ANDERSON, B. C., 1943. Thermodynamics of soil moisture. *Hilgardia*, 15: 31–298.

EGLI, D. B., PENDLETON, J. W. and PETERS, D. B., 1970. Photosynthetic rate of three soybean communities as related to carbon dioxide levels and solar radiation. *Argon. J.*, 62: 411–414.

EIK, K. and HANWAY, J. J., 1966. Leaf area in relation to yield of corn grain. *Agron. J.*, 58: 16–18.

ENOMOTO, N., 1937. Effects of cold water irrigation on growth of rice plant. *Rept. Exp. Farm Kyoto Univ.*, 1: 3–16 (in Japanese).

FALLANSBEE, W. A. 1975. Estimation of average daily rainfall from satellite cloud photographs. *NESS, NOAA, Wash., D. C. Tech. Mem.*, NESS 44

FALLANSBEE, W. A., 1976. Estimation of daily precipitation over China and the U.S.S.R. using satellite imagery. *NESS, NOAA, Wash., D. C. Tech. Mem.*, NESS 81.

FEBER, A. E., FORD, A. L. and MCCRORY, S. A., 1955. Good windbreaks increase South Dakota crop yields. *South Dakota Sta. Coll. Agric. Expt. Sta. Circ.*, 118: 1–16.

FELCH, R. E., 1964. *Growth and Phenological Responses of Irrigated Dry Beans to Changes in Microclimate Induced by a Wind Barrier.* M.S. Thesis, University of Nebraska, Lincoln, Nebr.

FELCH, R. E., 1973. Weather and plant disease. *Crops Soil*, April–May, 1973.

FISHER, R. A., 1924. The influence of rainfall on the yield of wheat at Rothamsted. *Phil. Trans. R. Soc. Lond.*, B213: 89–142.

FOOD AND AGRICULTURE ORGANIZATION, 1963. The world rice economy, II. Trends and forces commodity. *FAO Rome, Bull. Ser.*, 36.

FOSKETT, R. L., 1955. Wind barriers increase vegetable yields. *South Dakota Farm Home Res.*, 4 (2): 26–49.

FOSTER, A. C. and TADMAN, E. C., 1940. Influence of environmental factors on the transpiration and growth of tomato plants. *J. Agric. Res.*, 61.

FRITSCHEN, L. J., 1965. Evapotranspiration rates of field crops determined by the Bowen ratio method. *Agron. J.*, 58: 339–342.

FRITSCHEN, L. J., 1972. Dewfall on a Douglas Fir. *Abstr. Bull. Am. Meteorol. Soc.*, 53: 1037.

FUJITA, T., BARALT, G. and TSUCHIYA, K., 1968. Aerial measurement of radiation temperatures over Mt. Fuji and Tokyo areas and their application to the determination of ground and water surface temperatures. *Satellite Meteorol. Res. Proj., Dept. Geophys. Sci., Univ. Chicago*, pp.1–32.

GAASTRA, P., 1959. Photosynthesis of crop plants as influenced by light, carbon dioxide, temperature and stomatal diffusion resistance. *Mededel. Landbouwhogesch. Wageningen*, 59: 1–68.

GAASTRA, P., 1962. Photosynthesis of leaves and field crops. *Neth. J. Agric. Sci.*, 10: 311–325.

GANGOPADHYAYA, M., HARBECK JR., G. E., NORDENSON, T. J., OMAR, M. H. and URYVAEV, V. A., 1966. Measurement and estimation of evaporation and evapotranspiration. *WMO, Geneva, Tech. Note*, 83.

GARNER, W. W., 1920a. A capillary transmission constant and methods of determining it experimentally. *Soil Sci.*, 10: 103–126.

GARNER, W. W., 1920b. The capillary potential and its relation to soil moisture contents. *Soil Sci.*, 10: 357–359.

Garner, W. W., 1933. Comparative responses of long-day plants to relative length of day and night. *Plant Physiol.*, 8: 347–356.
Garner, W. W. and Allard, H. A., 1920. Effect of the relative length of day and night and other factors of the environment on growth and reproduction in plants. *J. Agric. Res.*, 18: 553–606.
Garner, W. W. and Allard, H. A., 1923. Further studies in photoperiodism: the response of the plant to relative length of day and night. *J. Agric. Res.*, 23: 871–920.
Garner, W. W. and Allard, H. A., 1930. Photoperiod response of soybean in relation to temperature and other environmental factors. *J. Agric. Res.* 41: 719–735.
Garrett, W. N., Bond, T. E. and Kelly, C. F., 1960. Effect of air velocity on gains and physiological adjustment of Hereford steers in a high temperature environment. *J. Animal. Sci.*, 19: 60–66.
Garrett, W. N., Givens, R. L., Bond, T. E. and Hull, J. L., 1966. Observation on the need for shade in beef feed lots. *Proc. Western Sect. Am. Soc. Animal Sci.*, 17: 349–355.
Garza, R. T., Barnes, R. F., Mott, G. O. and Rhykerd, C. L., 1965. Influence of light intensity, temperature, and growing period on the growth, chemical composition and digestibility of culver and Tanverde alfalfa seedlings. *Agron. J.*, 57: 417–420.
Gates, D. M., 1964. Leaf temperature and transpiration. *Agron. J.*, 56: 273–277.
Gates, D. M., 1965. Heat, radiant and sensible. In: P. E. Waggoner (Editor), *Agricultural Meteorology*. Am. Meteorol. Soc. Boston, Mass.
Gates, D. M., Alderfer, R. and Taylor E., 1968. Leaf temperature of desert plants. *Science*, 159: 994–995.
Geiger, R., 1957, 1965. *The Climate near the Ground.* Harvard Univ. Press., Cambridge, Mass.
Gerald, C. J. and Namken, L. N., 1966. Influence of soil texture and rainfall on the response of cotton to moisture regime. *Agron. J.*, 58: 39–42.
Gerber, J. F. and Decker, W. L., 1961. Evapotranspiration and heat budget of a corn field. *Agron. J.*, 53: 259–261.
Gilman, D. L., Hibbs, J. R. and Laskin, P. L., 1965. *Weather and Climate Modification.* U.S. Weather Bur., US Dept. Commerce, Wash., D. C.
Gilmore, E. C. and Rogers, J. S., 1958. Heat units as a method of measuring maturity of corn. *Agron. J.*, 50: 611–615.
Gipson, J. R. and Joham, H. E., 1969. Influence of night temperature on growth and development of cotton. *Agrond. J.*, 61: 365–367.
Goode, J. E. and Ingram, J., 1971. The effect of irrigation on the growth cropping and nutrition of Cox's Orange Pippin apple trees. *J. Hort. Sci.*, 46: 195–208.
Graham, W. G. and King, K. M., 1961. Fraction of net radiation utilized in evapotranspiration from a corn crop. *Proc. Soil Sci. Soc. Am.*, 25: 158–160.
Green, D. E., 1961. Factor affecting soybean seed quality. *Agron. Abstr.*, p.72.
Grist, D. H., 1959. *Rice—Tropical Agriculture Series*, 3rd ed. Longmans, London.
Haddock, D. J., 1964. Soil warmth depends on air temperatures. *Texas Farming Citriculture*, Vol.7.
Hahn, L. and McQuigg, J. D., 1970. Evaluation of climatological records for rational planning of livestock shelters. *Agric. Meteorol.*, 7: 131–141.
Haines, F. M., 1952. The absorption of water by leaves in an atmosphere of high humidity. *J. Expt. Bot.*, 3: 95–98.
Halkias, N. A., Veihmeyer, F. J. and Hendrickson, A. H., 1955. Determining water needs for crops from climatic data. *Hilgardia*, 24: 207–233.
Hall, A. D., 1945. *The Soil—An Introduction to the Scientific Study of the Growth of Crops*, 5th ed. Murray, London.
Hamner, K. C., 1944. Photoperiodism in plants. *Ann. Rev. Biochem.*, 13: 575–590.
Hamner, K. C. and Bonner, J., 1938. Photoperiodism in relation to hormones as factors in floral initiation and development. *Bot. Gaz.*, 100:388–431.
Hancock, J. 1954. The direct influence of climate on milk production. *Dairy Sci. Abstr.*, 16(2): 89–102.
Hank, O. and Vásáhelyi, J., 1954. A talaj szinének hatása a gyapot fejlödesere (Effect of soil color on the development of cotton). *Idöjáras*, 58(3): 137–143.
Hanks, R. J., Gardner, H. R. and Florian, R. L., 1969. Plant growth–evapotranspiration relation for several crops in the central Great Plains. *Agron. J.*, 61: 30–34.
Hanna, W. F., 1924. Growth of corn and sunflowers in relation to climatic conditions. *Bot. Gaz.*, 78:200–214.
Haroon, M., Long, R. C. and Weybrew, J. A., 1972. Effect of day/night temperature on seedling establishment of (*Nicotiana tabacum* L.) in controlled environment. *Agron. J.*, 64: 491–493.

Harrison, G. A., 1963. Temperature adaptation as evidenced by growth of mice. *Federation Proc.*, 22: 691–697.

Harrold, L. L. and Dreibelbis, F. R., 1951. Agricultural hydrology as evaluated by monolith lysimeter. *US Dept. Agric. Tech. Bull.*, 1050.

Hart, J. S., 1957. Climatic and temperature induced changes in the energetics of homeotherms. *Rev. Can. Biol.*, 16: 133–174.

Hartt, C. E. 1963. Translocation as a factor in photosynthesis. *Naturwissenschaften*, 21: 1–2 (Sonderdruck).

Haude, W., 1952. Verdunstungsmengen und Evaporationskraft eines Klimas. *Ber. Dtsch. Wetterd., US Zone*, 42: 225–229.

Haun, J. R., 1976. Development of models for specific crop calendar events. *Int. J. Biometeorol.*, 20: 261–266.

Heinicke, D. R., 1969. Characteristics of McIntosh and Red Delicious apples as influenced by exposure to sunlight during growing season. *Proc. Am. Soc. Hort. Sci.*, 89: 10–13.

Heitman, H., Kelly, C. F., Bond, T. E. and Hahn, L., 1959. Modified summer environment and growing swine. *Proc. Western Soc. Amer. Soc. Animal Prod.*, 10: 1–4.

Heitman, H., Hahn, L., Bond, T. E. and Kelly, C. F., 1962. The effects of modified summer environment on swine behavior. *Animal Behavior*, 10: 15–19.

Hendricks, W. A. and Scholl, J. C., 1943. Techniques in measuring joint relationships: the joint effects of temperature and precipitation on corn yields. *North Carolina Agric. Expt. Sta., Tech. Bull.*, 74.

Henschel, A., 1964. Laboratory facilities for adaptation research: high and low temperature. In: D. B. Dill (Editor), *Handbook of Physiology*. Am. Physiol. Soc., Wash., D. C., pp.323–328.

Herath, W. and Ormrod, D. P., 1965. Some effects of water temperature on the growth development of rice seedlings. *Agron. J.*, 57: 373–376.

Hess, W. N. (Editor), 1974. *Weather and Climate Modification.* Wiley, New York, N.Y.

Hess, E. A. and Bailey, C. B. M., 1961. Comparative physiological effects of cold on farm and laboratory animal. *Animal Breed. Abstr.*, 29: 379–392.

Hesse, W., 1954. Der Einfluss meteorologischer Faktoren auf die Transpiration der Pfefferminze (*Mentha piperita* L.). *Angew. Meteorol.*, 2: 14–18.

Hielman, J. L., Kanemasu, E. T. Bagley, J. O. and Rasmussen, V. P., 1977. Evaluating soil moisture and yield of winter wheat in the Great Plains using Landsat. *Remote Sensing Environ.*, 6: 315–326.

Hofmann, G., 1958. Dew measurements by thermodynamical means. *Int. Union Geodes. Geophys., Int. Assoc. Sci. Hydrol., General Assembly, Toronto, Gentbrugge*, 2: 443–445.

Hogg, W. H., Hounam, C. E., Mallik, A. K. and Zadoks, J. C., 1969. Meteorological factors affecting the epidemiology of wheat rust. *WMO, Geneva, Tech. Note*, 99.

Holekamp, E. R., Hudspeth, E. R. and Ray, L. L., 1960. Soil temperature—a guide to timely cotton planting. *Texas Agric. Expt. Sta. Misc. Pub.*, 465.

Holmes, R. H. and Robertson, Geo. W., 1959a. Estimating irrigation water requirement from meteorological data. *Can. Dept. Agric., Ottawa, Publ.*, 1054.

Holmes, R. H. and Robertson, Geo. W., 1959b. A modulated soil moisture budget. *Mon. Weather Rev.*, 87: 101–106.

Holmes, R. H. and Robertson, Geo. W., 1963. Application of the relationship between actual and potential evapotranspiration in dry land agriculture. *Trans. Am. Soc. Agric. Eng.*, 6: 65–68.

Holt, R. F. and van Daren, C. A., 1961. Water utilization by field corn in west Minnesota. *Agron. J.*, 53: 43–45.

Holub, A., 1957. Development of chemical thermoregulation in piglets. *Nature*, 180 (4591): 858–859.

Holzman, B., 1943. The influence of stability on evaporation. *Ann. N.Y. Acad. Sci.*, 44: 13–18.

Hounam, C. E., 1971. *Report of the CAgM Working Group on Assessment of Drought.* Working Group, WMO, Manuscript.

House, C. C., 1977. A within year growth model approach to forecasting corn yields. *US Dept. Agric. Statistical Reporting Serv., Wash., D. C.*

Howell, R. W., 1956. Heat, drought and soybean. *Soybean Dig.*, 16(10): 14–17.

Howell, R. W., 1963. Physiology of the soybean. In: A. G. Norman (Editor), *The Soybean.* Academic Press, New York, N.Y.

Howell, R. W. and Cartter, J. L., 1953. Physiological factors affecting composition of soybeans. *Agron. J.*, 45: 526–528.

Howell, R. W. and Cartter, J. L., 1958. Physiological factors affecting composition of soybeans, II.

Response of oil and other constituents of soybeans to temperature under controlled condition. *Agron. J.*, 50: 664–667.

HUDSON, J. P., 1965. Evaporation from Lucerne under advective conditions in the Sudan, I. factors affecting water losses and their measurement. *Exp. Agric.*, 1: 23–32.

HUNTER, J. R. and ERICKSON, A. E., 1952. Relation of seed germination to soil moisture tension. *Agron. J.*, 44: 107–109.

HURD, R. G. and COOPER, A. J., 1970. The effect of early low temperature treatment on the yield of single-inflorescence tomatoes. *J. Hort. Sci.*, 45: 19–27.

HURST, G. W., 1964. Meteorological aspects of the migration to Britain of *Laphygma exigna* and certain other moths on specific occasions. *Agric. Meteorol.*, 1: 271–281.

HUTCHINSON, J. B., 1959. *The Application of Genetics to Cotton Improvement.* Cambridge Univ. Press, Cambridge.

INGRAM, D. L. and WHITTOW, G. C., 1962a. The effect of heating the hypothalamus on respiration in the Ox (Bos taurus). *J. Physiol.*, 163: 200–210.

INGRAM, D. L. and WHITTOW, G. C., 1962b. The effects of variations in respiratory activity and in the skin temperature of the ears on the temperature of the blood in the external jugular vein of the Ox (Bos taurus). *J. Physiol.*, 163: 211–221.

INOUE, E., 1963. The environment of plant surface. In: L. T. EVANS (Editor), *Environment Control of Plant Growth.* Academic Press, New York, N.Y.

INOUE, E., MIHARA, Y. and TSUBO, Y., 1965. Agrometeorological studies on rice growth in Japan. *Agric. Meteorol.*, 2: 85–107.

INTERNATIONAL RICE RESEARCH INSTITUTE, 1970. *Annual Report.* The IRRI, Los Banos, Laguna, Philippines.

JACKSON, J. E., 1967. Effects of shading on apple fruits. *East Malling Res. Sta. Maidstone Kent, Annual Rept.*

JACOBS, W. P., 1951. The growth of peanut plants at various diurnal and nocturnal temperature. *Science*, 114: 205–206.

JAWORSKI, J., 1961. Potential evaporation in areas sheltered with windbreaks and in open area. *Ekologia Polska*, Séria A, tome IX, No.10, pp.165–182.

JENSEN, M., 1954. *Shelter Effect Investigations into the Aerodynamics of Shelter and its Effects on Climate and Crops.* Danish Tech. Press, Copenhagen.

JENSEN, R. D., 1971. Effects of soil water tension on the emergence and growth of cotton seedlings. *Agron. J.*, 63: 766–769.

JOHNSON, F. A., 1954. The solar constant. *J. Meteorol.*, 11: 431–439.

JOHNSON, H. D., 1967. Climatic effects on physiology and productivity of cattle. In: R. H. SHAW (Editor), *Ground Level Climatology.* Am. Assoc. Advan. Sci., Wash., D. C., pp.189–206.

KAISER, H., 1960. Untersuchungen über die Auswirkungen von Windschutzstreifen auf das Bodenklima. (Investigations of the effects of shelter belts on the soil climate.) *Z. Acker. Pflanzenbau*, 111: 47–72.

KASPERBAUER, M. J., 1969. Photo- and thermo-control of pretransplant floral induction in burley tobacco. *Agron. J.*, 61: 898–905.

KASPERBAUER, M. J., 1970. Photo- and thermo-control of flowering in tobacco. *Agron. J.*, 62: 825–827.

KERSTEN, M. S., 1949. Thermal properties of soil. *Univ. Minn. Inst. Tech. Bull.*, 28.

KIESSELBACK, T. A., 1916. Transpiration as a factor in crop production. *Nebr. Agric. Expt. Sta. Res. Bull.*, 6.

KINCER, J. B. and MATTICE, W. A., 1928. Statistical correlation of weather influence on crop yields. *Mon. Weather Rev.*, 56(2): 53–57.

KLAGES, K. H. W., 1942. *Ecological Crop Geography.* MacMillan, New York, N.Y.

KUHN, P. M. and SUOMI, V. E., 1958. Airborne observation of albedo with a beam reflector. *J. Meteorol.*, 15: 172–174.

KUO, TSUNG-MIN and BOERSMA, L., 1971. Soil water suction and root temperature effects on nitrogen fixation in soybeans. *Agron. J.*, 63: 901–906.

LAING, D. R., 1965. *The Water Environment of Soybean.* Unpublished Ph. D. thesis. Iowa State Univ., Ames, Iowa.

LANDSBERG, H. E., 1968a. A comment on land utilization with reference to weather factors. *Agric. Meteorol.*, 5: 135–137.

LANDSBERG, H. E., 1968b. Climate, man and some world problems. *Scientia*, 102: 661–662.

LANGE, T. C., KINGSOLVER, C. H., MITCHELL, J. E. and CHERRY, E., 1958. Determination of the effects of different temperature on uredial infection with *Puccinia graminis var. tritici. Phytopathology*, 48: 658–660.

LeDuc, S. K., 1976. Yield–weather regression models for the Canadian Prairies. *CCEA, NOAA, Columbia, Miss., Tech. Note*, 76–2.

Lemon, E. R., 1960. Photosynthesis under field conditions, II. An aerodynamic method for determining the turbulence carbon dioxide exchange between the atmosphere and a corn field. *Agron. J.*, 52: 697–703.

Lemon, E. R., 1963. Energy and water balance of plant communities. In: L. T. Evans (Editor), *Environment Control of Plant Growth.* Academic Press, New York, N.Y.

Leonard, W. H. and Martin, J. H., 1949. *Cereals and Crops.* MacMillan, New York, N.Y.

Leontievsky, N. P., 1934. The plans of shelterbelt planting in raising the agricultural yield. US For. Serv. transl. 36. *J. Geophys.*, 4: 127–140.

Lewis, C. F. and Richmond, T. R., 1960. Genetic of flowering response in cotton, II. Inheritance of flowering response in *Gossypium bardanse* cross. *Genetics*, 45(1): 79–85.

Lieth, H., 1972. Modelling the primary productivity of the world. *Nature and Resources.* UNESCO, VIII, 2: 5–10.

Ligon, J. T. and Benoit, G. R., 1966. Morphological effects of moisture stress on Burley tobacco. *Agron. J.*, 58: 35–38.

Linacre, E. T. 1964. A note on a feature of leaf and air temperature. *Agric. Meteorol.*, 1: 66–72.

Lind, A. R., 1964. Physiological responses to heat. In: S. Licht (Editor), *Medical Climatology.* Waverly Press, Baltimore, Md., 8: 164–195.

Ljones, B., 1970. Ved Gravensteins Nordgrenser (At the northern limit of Gravensteins). *Frukt og Boer*, 1970, 119–132.

Lomas, J., 1971. *Meteorological Factors Favoring and Limiting the Economic Production of Rice and Wheat.* Commiss. Agric. Meteorol., 5th Session, WMO, Geneva.

Lomas, J. and Gat, Z., 1967. The effect of windborne salt on citrus production near the sea in Israel. *Agric. Meteorol.*, 4: 415–425.

Lourence, F. J. and Pruitt, W. O., 1971. Energy balance and water use of rice grown in the central valley of California. *Agron. J.*, 63: 827–832.

Lowry, R. L. and Johnson, A. F., 1942. Consumptive use of water for agriculture. *Trans. Am. Soc. Civil Eng.*, 107: 1243–1302.

Lowry, W. P. and Vehts, D. B., 1968. Accumulation of total incident radiation as a predictor of plant maturity date. *Ann. Mtg., Pacific Div., Am. Assoc. Advan. Sci., Logan, Utah*

Ludlam, F. H., 1958. The hail problem. *Nubila*, 1: 12.

Luxmoore, R. J., Millington, R. J. and Marcellos, H., 1971. Soybean canopy structure and some radiant energy relations. *Agron. J.*, 63: 111–114.

Lysenko, T. D., 1925. *The Theoretical Basis of Vernalization.* Selskosgys, Moscow.

MacFarlane, W. V., 1963. Endocrine functions in hot environments. *Rev. Res., UNESCO, Paris*, pp.153–222.

MacNeill, M. M., 1969. Conditions controlling growth of apple trees in pots under glass. *Annu. Rept. Malling Res. Sta.*, 1969A, 53: 107–109.

Mack, A. R. and Ferguson, W. S., 1968. A moisture stress index for wheat by means of a modulated soil moisture budget. *Can. J. Plant Sci.*, 48: 535–543.

Makkink, G. F., 1957. Ekzameno de la formula de Penman (Examining the Penman's formula). *Neth. J. Agric. Sci.*, 5: 290–305.

Marshall, J. K., 1967. The effects of shelter on the productivity of grass land and field crop. *Field Crop Abstr.*, 20: 1–14.

Matches, A. G., Mott, G. O. and Bula, R. J., 1962. Vegetative development of alfalfa seedlings under varying level of shading and potassium fertilization. *Agron. J.*, 54: 541–543.

Mather, J. R., 1954. The measurement of potential evapotranspiration. *Publications in Climatology, Laboratory of Climatology*, 7(1).

Matjakin, G. I., 1963. Effect of forest shelterbelt on microclimate (Polezashchitny Polosy). *Exp. All-Union Res. Inst. for the Improvement of Farmland by Forestation*, 6: 51–90. (US For. Serv. transl.)

Matjakin, G. I., 1952. *Forest Shelterbelts and Microclimate.* Geografgizd, Moscow.

Matsuo, T., 1955. *Rice Culture in Japan.* Yokendo, Tokyo.

Matsuo, T., 1957. *Rice Culture in Japan.* Ministry of Agriculture and Forestry, Tokyo.

Matsushima, S. and Tsunoda, K., 1958. Analysis of development factors determining yield and yield prediction in lowland rice. *Proc. Crop Sci. Soc. Jpn.*, 25(4): 432–434.

McCormick, W. C., Givens, R. L. and Southwell, B. L., 1963. Effects of shade on rate of growth and fattening of beef steer. *Ga. Agric. Exp. Sta. Tech. Bull., N.S.*, 27.

McDonald, M. A. and Bell, J. M., 1958. Effects of low fluctuating temperature on farm animals, IV. Influences of temperature on milk yield and milk composition. *Can. J. Animal Sci.*, 38: 160–170.

McDowell, R. E., 1967. Factors in reducing the adverse effects of climate on animal performance. In: R. H. Shaw (Editor), *Ground Level Climatology*. Am. Assoc. Advan. Sci., Wash., D. C., pp.277–291.

McGinnis, D. F. Jr., Pritchard, J. A. and Wiesnet, D. R., 1975. Snow depth and snow extent using VHRR data from the NOAA-2 satellite. *NESS, NOAA, Wash., D. C., NESS Tech. Mem.*, 63.

McGuiness, J. L. and Bordne, E. F., 1972. A comparison of lysimeter derived potential evapotranspiration with computed value. *Agric. Res. Serv., US Dept. Agric., Wash., D. C., Tech. Bull.*, 1452.

Michaels, P. J. and Scherer, V. R., 1977. *An Aggregated National Model for Wheat Yield in India.* Inst. for Environmental Studies, Univ. of Wisconsin, Madison, Wisc.

Miller, D. H., 1955. Snow cover and climate in the Sierra Nevada, California. *Univ. Calif. (Berkeley) Publ. Geogr.*, 11: 218.

Miller, D. H., 1959. Transmission of insolation through pine forest canopy, as it affects the melting of snow. *Mitt. Schweiz. Anst. Forstl. Versuchswes.*, 35: 57–59.

Miller, D. H., 1965. The heat and water budget of the earth's surface. *Advan. Geophys.*, 11: 176–277.

Modlibawska, I., 1961. Effect of soil moisture on frost resistance of apple blossom including some observations on ghost and parachute blossom. *J. Hort. Sci.*, 36(3) 186–196.

Molga, M., 1962. *Agricultural Meteorology, Part II. Outline of Agrometeorological Problems.* Warszawa.

Monteith, J. L., 1959. The reflection of short-wave radiation by vegetation. *Q. J. R. Meteorol. Soc.*, 85: 386–392.

Monteith, J. L., 1963. Dew facts and fallacies. In: A. J. Rutter and F. H. Whitehead (Editors), *The Water Relations of Plants.* Blackwell, London.

Monteith, J. L., 1965. Light distribution and photosynthesis in field crops. *Ann. Bot.*, 29: 17–37.

Moss, D. N., 1964. Optimum lighting of leaves. *Crop Sci.*, 4: 131–136.

Moss, D. N., 1965. Capture of radiant energy by plant. In: P. E. Waggoner (Editor), *Agricultural Meteorology.* pp.90–108.

Moss, D. N., Musgrave, R. B. and Lemon, E. R., 1961. Photosynthesis under field condition, III. Some effects of light, carbon dioxide, temperature and soil moisture on photosynthesis, respiration, and transpiration of corn. *Crop Sci.*, 1: 83–87.

Murata, Y., 1964. On the influence of solar radiation and air temperature upon the local differences in the productivity of paddy rice in Japan. *Proc. Crop Sci. Jpn.*, 33: 59–63.

Nestorova, E. I., 1939. *Vliyaniye temperaturnykh i svetovykh usloviy na chislo zeren v kolose pshenitsy dan.* Tom 24, Vyp. 8.

Newman, J. E., Cooper, W. C., Reuther, W., Cahoon, G. A. and Pennado, A., 1967. Orange fruit maturity and net heat accumulation. In: R. H. Shaw (Editor), *Ground Level Climatology.* Am. Assoc. Advan. Sci. Wash., D. C., pp.127–146.

Nittler, L. W. and Kenny, T. J., 1964. Induction of flowering in alfalfa birdsfoot trefoil and red clover as an aid in testing varietal purity. *Crop. Sci.*, 4: 187–190.

Noffsinger, T. L., 1961. Leaf and air temperature under Hawaii conditions. *Pacific Sci.*, 40: 304–306.

Nuttonson, M. Y., 1955. *Wheat-Climate Relationships and the Use of Phenology in Ascertaining the Thermal and Photo-Thermal Requirements of Wheat.* Am. Inst. Crop Ecol., Wash., D. C.

Oelke, E. A. and Mueller, K. E., 1969. Influences of water management and fertility on rice growth and yield. *Agron. J.*, 61: 227–230.

Olmsted, C. E., 1944. Growth and development in range grasses, III. Photoperiodic responses in twelve geographic strains of side oats grama. *Bot. Gaz.*, 106: 46–74.

Oury, B., 1965. Allowing for weather in crop production, model building. *J. Farm Econ.*, 47: 270–283.

Palmer, W. C., 1965. Meteorological drought. *Weather Bur., US Dept. Commerce, Wash., D. C., Res. Pap.*, 45.

Palmer, W. C., 1968. Keeping track of crop moisture conditions nationwide: The new crop moisture index. *Weatherwise*, 21: 156–161.

Parker, M. W. and Borthwick, H. A., 1943. Influence of temperature on photoperiodic reactions in leaf blades of *Biloxi* soybean. *Bot. Gaz.*, 104: 612–619.

Parker, M. W. and Borthwick, H. A., 1951. Photoperiodic responses of soybean varieties. *Soybean Dig.*, 11(11): 26–30.

Pasquill, F., 1951. Some further consideration of the measurement and indirect evaluation of natural evaporation. *Q. J. R. Meteorol. Soc.*, 76: 287–301.

Pearson, K., 1934. *Tables of the Incomplete Beta-Function.* Biometrika Office, Univ. College, London.

Pelton, W. L., 1961. The use of lysimetric methods to measure evapotranspiration. *Proc. Hydrol. Symp., 2nd, Toronto*, pp.106–134.

PELTON, W. L. and EARL, A. U., 1959. *The Influence of Field Shelterbelts on Wind Velocity and Evapotranspiration.* Can. Dept. Agric., Swift Current, Mimeogr. Annu. Rept.

PENDLETON, J. W. and HAMMOND, J. J., 1969. Relative photosynthetic potential for grain yield of various leaf canopy levels of corn. *Agron. J.*, 61: 911–913.

PENDLETON, J. W. and WEIBEL, R. O., 1965. Shading studies on winter wheat. *Agron. J.*, 57: 292–293.

PENDLETON, J. W., SMITH, G. E., WINTER, S. R. and JOHNSTON, T. J., 1968. Field investigations of the relationships of leaf angle in corn (*Zea mays* L.) to grain yield and apparent photosynthesis. *Agron. J.*, 60: 422–424.

PENMAN, H. L., 1948. Natural evaporation from open water, bare soil and grass. *Proc. R. Soc. Ser. A*, 193: 120–145.

PENMAN, H. L., 1949. The dependence of transpiration on weather and soil conditions. *J. Soil Sci.*, 1: 74–89.

PENMAN, H. L., 1952. The physical bases of irrigation control. *Int. Hort. Congr., 13th, London.*

PENMAN, H. L., 1956. Evaporation: an introductory survey. *Neth. J. Agric. Sci.*, 4: 9–29.

PENMAN, H. L., 1963. Vegetation and hydrology. *Common Wealth Bur. Soil, Harpenden, Tech. Comm.*, 53.

PENMAN, H. L. and LONG, I. F., 1960. Weather in wheat—an essay in micrometeorology. *Q. J. R. Meteorol. Soc.*, 86: 16–50.

PHATAK, S. C., WITTWER, S. H. and TEUBNER, F. G., 1966. Top and root temperature effect on tomato flowering. *Proc. Am. Soc. Hort. Sci.*, 88: 527–531.

PHILLIPS, R. E., 1968. Water diffusivity of germinating soybeans, corn, and cottonseed. *Agron. J.*, 60: 568–571.

PIERCE, L. T., 1958. Estimating seasonal and short-term fluctuations in evapotranspiration from meadow crops. *Am. Meteorol. Bull.*, 39.

PITTER, R. L., 1977. *A Model of Wheat and its Application to Food Management.* Preprint from: 13th Agric. Forest Meteorol. Conf., Purdue Univ., West Lafayette, Ind.

PLATT, C. M. R., 1972. Surface temperature measurement from satellite. *Nat. Phys. Sci.*, 235: 29–30.

POCHOP, L. O., CORNIA, R. L. and BECKER, C. F., 1975. Prediction of winter wheat yield from short-term weather factors. *Agron. J.*, 67: 4–7.

POMEROY, R. W., 1953. Studies on piglet mortality, I. Effect of low temperature and low plane of nutrition on the rectal temperature of the young pig. *J. Agric. Sci.*, 43: 182–191.

PRIESTLEY, C. H. B., 1955. Free and forced convection in the atmosphere near the ground. *Q. J. R. Meteorol. Soc.*, 81: 139–143.

PRIESTLEY, C. H. B., 1958. Sensible heat transfer from ground to air. Climatology and microclimatology. *Proc. Camberra Symp., UNESCO Arid Zone Res.*, 11: 106–108.

PRIESTLEY, C. H. B., 1966. The limitation of temperature by evaporation in hot climate. *Agric. Meteorol.*, 3: 241–246.

PRUITT, W. O., 1963. Application of several energy balance and aerodynamic evaporation equations under a wide range of stability. *Univ. Calif. Final Rept., USAEPG Contr. DA* 36-039-*SC-80334.*

PRUITT, W. O. and ANGUS, D. E., 1960. Large weighting lysimeter for measuring evapotranspiration. *Trans. Am. Soc. Agric. Eng.*, 3(2): 13–15.

RADKE, J. K. and BAUER, R. E., 1969. Growth of sugar beets as affected by root temperature, I. Greenhouse studies. *Agron. J.*, 61: 860–863.

RAHN, J. J. and BROWN, D. M., 1971. Estimating corn canopy extreme temperature from shelter value. *Agric. Meteorol.*, 8: 129–138.

RAMIAH, K. and RAO, M. B. V. N., 1953. *Rice Breeding and Genetics.* Indian Council Agric. Res., New Delhi.

RANNEY, C. D., HURSH, J. S. and NEWTON, O. H., 1971. Effects of bottom defoliation on microclimate and the reduction of boll rot of cotton. *Agron. J.*, 63: 259–263.

REIFSNYDER, W. E.,and LULL, H. W. 1965. Radiant energy in relation to forests. *US Dept. Agric. For. Serv., Tech. Bull.*, 1344.

REIFSNIJDER, W. E., FURNIVAL, G. M. and HOROWITZ, J. L., 1971. Spatial and temporal distribution of solar radiation beneath forest canopies. *Agric. Meteorol.*, 9: 21–37.

RENSE, W. A., 1961. Solar radiation in the extreme ultraviolet region of the spectrum and its effect on the earth's upper atmosphere. *Ann. N. Y. Acad. Sci.*, 95: 33–38.

RHOAD, A. O., 1938. Some observations on the response of pure bred *Bos Taurus* and *Bos Indicus* cattle and their crossbred types to certain conditions of the environment. *Am. Soc. Animal Prod. Proc.*, 31: 284–295.

RHOAD, A. O., 1941. Climate and livestock production. *US Dept. Agric., Wash., D. C., Yearb. Agric.*, pp.508–516.

RICHARD, L. D., 1928. The usefulness of capillary potential to soil moisture and plant investigations. *J. Agric. Res.*, 37: 719–742.

RIDER, N. E., 1954. Evaporation from an oat field. *Q. J. R. Meteorol. Soc.*, 80: 198–211.

RIEDE, W., 1938. The German soybean problem. *Herbage Rev.*, 6: 245–258.

RILEY, J. A., 1962. *Solar Radiation Variation Within a Cotton Plant Zone.* Weather Bur., Mid-South Weather Proj., Memphis, Tenn. (mimeogr.).

ROBERTSON, G. W., 1968. A biometeorological time scale for a cereal crop involving day and night temperatures and photoperiod. *Int. J. Biometeorol.*, 12: 191–223.

ROBINS, J. S. and DOMINGO, C. E., 1953. Some effects of severe soil moisture deficits at specific growth stages in corn. *Agron. J.*, 45: 618–621.

ROBINS, J. S. and RHOADES, H. F., 1958. Irrigation of field corn in the west. *US Dept. Agric. Leafl.*, 440.

ROHWER, C., 1931. Evaporation from free water surface. *US Dept. Agric. Tech. Bull.*, 271.

ROSENBERG, N. J., 1966. Influence of snow fence and corn windbreaks on microclimate and growth of irrigated sugar beets. *Agron. J.*, 58: 469–475.

ROSENBERG, N. J., 1967. The influence and implications of windbreaks on agriculture in dry region. In: R. H. SHAW (Editor), *Ground Level Climatology, AAAS Symp.*, pp.327–349.

ROSENBERG, N. J., HART, E. and BROWN, K. W., 1968. *Evapotranspiration Review of Research.* Univ. of Lincoln, Nebr.

RÜESCH, J. D., 1955. Der CO_2-gehalt bodennaher Luftschichten unter dem Einfluss des Windschutzes. (The CO_2-content of the layers near the ground influenced by shelterbelts.) *Z. Pflanzenernähr. Bodenk., Düng.*, 71(116) No.2: 113–132.

RUNGE, E. C. A. and BENCI, 1975. Modeling corn production—estimating production under variable soil and climatic conditions. Reported from: *Proc. Annu. Corn and Sorghum Res. Conf., 13th, 1975.*

RUNGE, E. and ODELL, R. T., 1958. The relation between precipitation, temperature and the yield of corn on the Agronomy South Farm, Urbana, Ill. *Agron. J.*, 50: 448–454.

RUNGE, E. and ODELL, R. T., 1960. The relation between precipitation, temperature and the yield of soybeans on the Agronomy South Farm, Urbana, Ill. *Agron. J.*, 52: 245–247.

RUTTER, A. J., KERSHAW, K. A., ROBINS, P. C. and MORTON, A. J., 1972. A predictive model of rainfall interception in forests, I. Derivation of the model from observation in a plantation of Corsican pine. *Agric. Meteorol.*, 9: 367–384.

SAKAMOTO, C. M., 1976. An index for estimating wheat yield in Australia. *CCEA, NOAA, Columbia,* Mo., *Tech. Note*, 76-3.

SAKAMOTO, C. M. and JENSEN, R. E., 1975. *Final Report Wheat–Climate Models for Argentina and Australia.* Environ. Stud. Serv. Center, Auburn, Alabama.

SAKAMOTO, C. M. and SHAW, R. H., 1967a. Apparent photosynthesis in field soybean community. *Agron. J.*, 59: 73–75.

SAKAMOTO, C. M. and SHAW, R. H., 1967b. Light distribution in field soybean canopies. *Agron. J.*, 59: 7–9.

SANDHU, S. S. and HODGES, H. F., 1971. Effects of photoperiod, light intensity, and temperature on vegetative growth, flowering, and seed production in *Cicer arietinum* L. *Agron. J.*, 63: 913–914.

SARGEANT, D. H. and TANNER, C. B., 1967. A simple psychrometric apparatus for Bowen ratio determinations. *J. Appl. Meteorol.*, 6: 414–418.

SCHOLANDER, P. E., WALTERS, V., HOCK, R. and IRVING, L., 1950. Body insulation of some arctic and tropic mammals and birds. *Biol. Bull.*, 99: 225–236.

SCHERER, V. R., 1977. *Models of National (Aggregated) Wheat Yield in the People's Republic of China with a Preliminary Investigation of Explained Yield Variability Due to Weather Factors.* Inst. Environ. Stud., Univ. Wisc., Madison, Wisc.

SCHLEUSENER, P. E., NEMETHY, J. J., SHULL H. H. and WILLIAMS G. E., 1961. Pasture irrigation requirements calculated from climatological data. *Trans. ASAE*, 4: 6–7; 11.

SCHNEIDER, S. R., WIESNET, D. R. and MCMILLAN, M. C., 1976. River basin snow mapping at the National Environmental Satellite Service. *NOAA Tech Memo. NESS 83, NOAA, Washington, D. C.*

SCHNELLE, F., 1963. *Frostschutz im Pflanzenbau (Frost protection in the culture plants), Band I. Die meteorologischen und biologischen Grundlagen der Frostschadensverhütung, Band II. Die Praxis der Frostschadensverhütung.* BLV Verlagsgesellschaft, München, Basel, Wien.

SCHWANKE, R. K., 1963. *The Interrelationships of Plant Population, Soil Moisture, and Soil Fertility in Determining Corn Yields on Colo Clay Loam.* Unpublished M.S. Thesis, Iowa State Univ., Ames, Iowa.

SCOFIELD, R. A. and OLIVER, V. J., 1977. A scheme for estimating convective rainfall from satellite imagery. *NESS, NOAA, Wash., D. C., Tech. Mem.*, 86.

SHAW, R. H., 1954. Leaf and air temperature under freezing conditions. *Plant Physiol.*, 29: 102–104.

SHAW, R. H., 1955. In: G. F. SPRAGUE (Editor), *Corn and corn improvement.* Acad. Press, New York, N.Y.

SHAW, R. H., 1962. *Prediction of Soil Moisture under Oat in Iowa, II.* Agron. Dept., Iowa State Univ., Ames, Iowa.

SHAW, R. H., 1963. Estimation of soil moisture under corn. *Iowa State Univ., Ames, Iowa, Res. Bull.*, 520.

SHAW, R. H., 1964. Prediction of soil moisture under meadow. *Agron. J.*, 56: 320–324.

SHAW, R. H., 1977. Use of a moisture-stress index for examining climate trends and corn yields in Iowa. *Iowa State J. Res.*, 51: 249–254.

SHAW, R. H.,and THOM, H. C. S., 1951. On the phenology of field corn, the vegetative period. *Agron. J.*, 43: 9–15.

SHAW, R. H. and WEBER, C. R., 1967. Effects of canopy arrangements on light interception and yield of soybeans. *Agron. J.*, 59: 155–159.

SHAW, R. H., RUKLES, J. R. and BARGER, G. L., 1958. Seasonal changes in soil moisture related to rainfall, soil types and crop growth. *Iowa Agric. Home Econ. Exp. Sta., Ames, Iowa, Res. Bull.*, 457.

SIMINOVITCH, D., BALL, W. L., DESJARDINGS, R. and GAMBLE, D. S., 1967. Use of protein-based foam to protect plants against frost. *Can. J. Plant Sci.*, 47: 11–17.

SIMONIS, W., 1947. Carbon Dioxide Assimilation unter Stuffproduktion trocken gezogener Pflanzen. *Planta (Berlin)*, 35: 188–224.

SIROTENKO, V. C., 1976. Parametric modelling of winter-crop yield time series. *Sov. Meteorol. Hydrol.*, 12: 86–88.

SLATYER, R. O., 1958. Availability of water to plants. In: *Climatology and Micrometeorology.* UNESCO.

SMITH, C. V., 1964. Animal housing and meteorology:

I. Ventilation and associated pattern of air flow. *Agric. Meteorol.*, 1: 30–41.

II. The rating and ventilation for animal house. *Agric. Meteorol.*, 1: 107–120.

III. A quantitative relationship between environment comfort and animal productivity. *Agric. Meteorol.*, 1: 249–270.

SMITH, J. W., 1904. Relation of precipitation to yield of corn. *US Dept. Agric., Yearbook.* Washington, D. C., pp.215–224.

SMITH, W. L. and HOWELL, H. B., 1971. Vertical distributions of atmospheric water vapor from satellite IR spectrometer measurement. *J. Appl. Meteorol.*, 10: 1026–1034.

SMITH, W. O. and BYERS, H. G., 1938. The thermal conductivity of dry soil of certain of the great soil group. *Soil Sci. Soc. Am. Proc.*, 3: 13–19.

SOMNERHOLDER, B. R., 1962. Design criteria for irrigation corn. *Agric. Eng.*, 43: 336–339; 348.

STAKMAN, E. C. and HARRAR, G., 1957. *Principles of Plant Pathology.* New York.

STALFELT, M. G., 1932. Der Einfluss des Windes auf die Kutifular und Stomatare Transpiration. *Svensk. Bot. Tidsskr.*, 26: 45–69.

STANHILL, G., 1961. A comparison of methods calculating potential evapotranspiration from climatic data. *Israel J. Agric. Res.*, 11: 159–171.

STANSEL, J. W., 1975. *Six Decades of Rice Research in Texas.* The Texas A & M Univ., College Station, Texas, pp.43–50.

STAPLE, W. J. and LEHANE, J. J., 1955. The influence of field shelter belt on wind velocity, evaporation, soil moisture and crop yield. *Can. J. Agric. Sci.*, 35: 440–453.

STEUBING, L., 1951. Der Einfluss von Heckenanlagen auf den Taufall (influence of hedges upon deposition of dew.) *Ber. Dtsch. Wetterdienstes, US-Zone*, 5 (32): 53–56.

STEVENSON, K. R. and SHAW, R. H., 1971. Effects of leaf orientation on leaf resistance to water vapor diffusion in soybean (*Glycine max L. Merr.*) leaves. *Agron. J.*, 63:327–329.

STEWART, R. E., PICKETT, E. E. and BRODY, S., 1951. Environmental physiology with special reference to domestic animal. *XVI Mo. Univ. Agr. Expt. Sta. Res. Bull.* 484.

STEYAERT et al., 1978. Atmospheric pressure and wheat yield modeling. Agric. Meteorol., 19: 23–34.

STOECKLER, J. H., 1962. Shelterbelt influence on Great Plains field environment and crop. *US Dept. Agr. Prod. Res. Rept.*, 62: 1–26.

STOUGHTON, R. H. and VINCE, D., 1954. Possible applications of photoperiodism in plants. *World Crops*, 6: 311–313.

SULAKVELIDZE, G. K., BIBILASHVILI, N. SH. and HAPCHEVCE, V. F., 1967. *Formation of Precipitation and Modification of Hail Processes.* Translation from Russian, 1965, by Israel Scientific Translation Program, Jerusalem.

SUOMI, V. E. and TANNER, C. B., 1958. Evapotranspiration estimates from heat-budget measurements over a field crop. *Trans. Am. Geophys. Union*, 39: 298–304.

SUTTON, O. G., 1953. *Micrometeorology*. McGraw-Hill, New York.
SWINBANK, W. C., 1951. The measurement of vertical transfer of heat and water vapor by eddies in the lower atmosphere with some results. *J. Meteorol.*, 8: 135–145.
TAERUM, R., 1964. Effect of moisture stress and climatic conditions on stomatal behavior and growth in Rome Beauty apple trees. *Proc. Am. Soc. Hort. Sci.*, 85: 20–32.
TANNER, C. B., 1957. Factors affecting evaporation from plants and soils. *J. Soil Water Cons.*, 12: 221–227.
TANNER, C. B., 1963. Plant temperatures. *Agron. J.*, 55: 210–211.
TANNER, C. B. and LEMON, E. R., 1962. Radiant energy utilized in evapotranspiration. *Agron. J.*, 54: 207–212.
TAYLOR, R. J. and DYER, A. J., 1958. An instrument for measuring evaporation from natural surface. *Nature*, 181: 408–409.
TAYLOR, R. J. and WEBB, E. K., 1955. A mechanical computer for micrometeorological analysis. *Commonwealth Scientific and Industrial Res. Organization, Australia, Tech. Pap.*, 6.
THARP, W. H., 1960. The cotton plant—How it grows and why its growth varies. *US Dept. Agric., Washington, D. C., Agric. Handbook*, 178: 1–17.
THOMAS, M. D. and HILL, G. R., 1937. The continuous measurements of photosynthesis, respiration and transpiration of alfalfa and wheat growing under field condition. *Plant Physiol.*, 12: 285–307.
THOMAS, M. D. and HILL, G. R., 1949. Photosynthesis under field condition. In: J. FRANCK and W. E. LOOMIS (Editors), *Photosynthesis in Plants.* Iowa State College Press, Ames, Iowa.
THOMPSON, L. M., 1962. Evaluation of weather factors in the production of wheat. *J. Soil Water Cons.*, 17: 149–156.
THOMPSON, L. M., 1963. *Weather and Technology in the Production of Corn and Soybean.* Center for Agr. and Econ. Development, Iowa State Univ., Ames, Iowa, CAED Rept., 17.
THOMPSON, L. M., 1969. Weather and technology in the production of wheat in the United States. *J. Soil Water Cons.*, 23: 219–224.
THORNTHWAITE, C. W., 1948. An approach toward a rational classification of climate. *Geograph. Rev.*, 38: 55–94.
THORNTHWAITE, C. W., 1953. The place of supplemental irrigation in post-war planning. *Publ. Climatol., Johns Hopkins Univ.*, 6(2) 11–29.
THORNTHWAITE, C. W. and HOLZMAN, B., 1939. The determination of evaporation from land and water surface. *Mon. Weather Rev.*, 67: 4–11.
THORNTHWAITE, C. W. and KASER, P., 1943. Wind-gradient observation. *Trans. Am. Geophys. Union*, 24: 166–182.
THORNTHWAITE, C. W. and MATHER, J. R., 1955. The water budget and its use in irrigation. *U.S. Dept. Agric., Washington, D. C., Agric. Yearbook*, pp.346–358.
TOMATO, S., 1956. Influence of windbreaks on evaporation. *12th Congr. Int. Union Forest Res. Organization.*
TOMLINSON, B. R., 1953. Comparison of two methods of estimating consumptive use of water. *Agric. Eng.*, 34: 459–464.
TULLER, S. E. and CHILTON, R., 1973. The role of dew in the seasonal moisture balance of a summer-day climate. *Agric. Meteorol.*, 11: 135–142.
TURC, L., 1961. Evaluation des besoins en eau d'irrigation; évapotranspiration potentielle. *Ann. Agron.*, 12: 13–49.
UENO, M. and SMITH, D., 1970. Influence of temperature in seedling growth and carbohydrate composition of three alfalfa cultivars. *Agron. J.*, 62: 764–767.
ULANOVA, E. S., 1975. *Agronometeorological Conditions and Winter Wheat.* Joint Publ. Res. Service, Arlington, Virginia.
VAN BAVEL, C. H. M., 1953. A drought criterion and its application in evaluating drought incidence and hazard. *Agron. J.*, 45: 167–171.
VAN BAVEL, C. H. M., 1959. Drought and water surplus in agricultural soils of the lower Mississippi Valley area. *US Dept. Agric. Tech. Bull.*, 1209.
VAN BAVEL, C. H. M. and EHRLER, W. L., 1968. Water loss from a sorghum field and stomatal control. *Agron. J.*, 60: 84–86.
VAN BAVEL, C. H. M. and MEYER, L. E., 1962. An automatic weighing lysimeter. *Agric. Eng.*, 43: 580–588.
VAN BAVEL, C. H. M. and WILSON, T. V., 1952. Evaporation estimates as critical for determining time of irrigation. *Agric. Eng.*, 33: 417–418.
VAN DER LINDE, J., 1962. Trees outside the forest. In: *Forest Influences. FAO Forestry Forest Production Studies, Rome*, 15: 141–208.

VAN EIMERN, J., 1964. Windbreaks and shelterbelts. *WMO, Geneva, WMO Tech. Note*, 59.

VAN EIMERN, J., FRANKEN, E. and HARRICS, H., 1964. *Ergebnisse von Windschutz Untersuchungen in Hamburg-Garstedt.* (Results of investigations on shelterbelts at Hamburg-Garstedt in 1952). Landwirtschaft-Angewandte Wissenschaft, Landw., Verlag Hiltrup/Westf.

VAN EVERDINGEN, E., 1926. Het Verband Tussen de Weersgesteldheid en de Aardappelziekte. *Tijdschr. Plantenziekten*, 32: 129–140.

VAN ROYEN, W., 1954. *The Agricultural Resources of the World.* Prentice-Hall, New York, N.Y., Vol. I.

VAN WIJK ,W. R., 1963. *Physics of the Plant Environment.* North-Holland, Amsterdam.

VAN WIJK, W. R. and HIDDING, A. P., 1955. Investigations on the change of the climate behind shelterbelt. *Landbouwkdg. Tijdschr. ('s-Gravenhage)*, 67: 707–712.

VEIHMEYER, F. J., 1927. Some factors affecting the irrigation requirements of deciduous orchards. *Hilgardia*, 2: 125–288.

VEIHMEYER, F. J. and HENRICKSON, A. H., 1955. Does transpiration decrease as the soil moisture decreases? *Trans. Am. Geophys. Union*, 36: 425–448.

VEIHMEYER, F. J., PRUITT, W. O. and MCMILLIN, W. D., 1960. Soil moisture as factor in evapotranspiration equation. *Ann. Meeting Am. Soc. Agric. Eng., 1960.*

VELLOSE, M. H., 1971. *Remote Sensing in the Economic Management of Coffee Production and Marketing.* National Aeronautics and Space Administration, Washington, D. C., Earth Resources Survey System, pp.195–200.

VENTSKEVICH, G. Z., 1961. *Agrometeorology.* National Science Foundation (Translated from Russian), Washington, D. C.

VERDIUM, J. and LOOMIS, W. E., 1944. Absorption of CO_2 by maize. *Plant Physiol.*, 19: 278–293.

VEZINA, P. E., 1961. Variations in total solar radiation in three Norway spruce plantations. *Forest Sci.*, 7: 257–264.

VEZINA, P. E., 1963. Solar radiation available below thinned and unthinned balsam fir canopies. *Can. Dept. For., For. Res. Branch Rept.*, 2133 (unpublished).

VON SCHWARZ, A. R., 1879. Vergleichende Versuche über die physikalischen Eigenschaften verschiedener Bodenarten. *Forsch. Geb. Agric.-Phys.*, 2: 164–169.

VON WILHELMI, TH., 1962. Globalstrahlung und Temperatur in einer Douglasiengruppe und die Einwirkungen auf das Radialwachstum. *Allgem. Forstwiss.*, 133: 38–43.

VOUGH, L. R. and MARTIN, G. C., 1971. Influence of soil moisture and ambient temperature on yield and quality of alfalfa forage. *Agron. J.*, 63: 40–42.

WAGGONER, P. E., 1969. Predicting the effect upon net photosynthesis of changes in leaf metabolism and physics. *Crop Sci.*, 9: 315–321.

WAGGONER, P. E. and SHAW, R. H., 1952. Bases for the prediction of corn yields. *Plant Physiol.*, 25: 225–244.

WAGGONER, P. E., MILLER, P. M. and DEROO, H. C., 1960. Plastic mulching principles and benefits. *Conn. Agric. Expt. Sta. Bull.*, 634.

WAGGONER, P. E., BEGG, J. E. and TURNER, N. C., 1969. Evaporation of dew. *Agric. Meteorol.*, 6: 227–230.

WAGGONER, P. E., HORSFALL, J. G. and LUKENS, R. J., 1972. EPIMAY—A simulator of southern corn leaf blight. *Conn. Agric. Expt. Sta. Bull.*, 729.

WAGNER, F., 1883. Untersuchungen über das relative Wärmeleitungsvermögen verschiedener Bodenarten. *Forsch. Geb. Agric.-Phys.*, 6: 1–51.

WALKER, J., 1951. Planning and planting field shelterbelts. *Can. Dept. Agric. Publ.*, 785.

WALLACE, H. A., 1920. Mathematical inquiry into the effect of weather on corn yield in the eight Corn Belt states. *Mon. Weather Rev.*, 48: 439–446.

WALLACE, H. A. and BRESSMAN, E. N., 1949. *Corn and Corn Growing.* Wiley, New York, N.Y.

WALLIN, J. R., 1967. Ground level climate in relation to forecasting plant diseases. In: R. H. SHAW (Editor), *Ground Level Climatology.* Amer. Assoc. Advan. Sci., Washington, D.C.

WANG, J. Y., 1963. A graphical solution on temperature-moisture responses to tomato yield. *Proc. Am. Soc. Hortic. Sci.*, 82: 429–445.

WANG, J. Y. and WANG, S. C., 1962. A simple graphical approach to Penman's method for evaporation estimates. *J. Appl. Meteorol.*, 1: 582–588.

WANJURA, D. F., BUXTON, D. R. and STAPLETON, H. N., 1970. Temperature model for predicting initial cotton emergence. *Agron. J.*, 62: 741–743.

WANJURA, D. F., HUDSPETH, E. B. Jr. and BILBRO, J. D. Jr., 1967. Temperature-emergence relations of cottonseed under natural diurnal fluctuation. *Agron. J.*, 59: 217–219.

WARK, D. Q., YAMAMOTO, G. and LIENESCH, J., 1962. Meteorological satellite. *US Weather Bur., Washington, D. C., Lab. Rept.*, 10.

WARREN, D. C., 1939. Effect of temperature on size of eggs from pullets in different latitudes. *J. Agric. Res.*, 59: 441–452.

WASSINK, E. C., 1953. Specification of radiant flux and radiant flux density in irradiation of plants with artificial light. *J. Hortic. Sci.*, 28: 177–184.

WATSON, D. J., 1947. Comparative physiological studies on the growth of field crops. I. Variation in net assimilation rate and leaf area between species and varieties and within and between years. *Ann. Bot. (N.S.)*, 11: 41–76.

Webb, E. K., 1960. On estimating evaporation with fluctuating Bowen ratio. *J. Geophys. Res.*, 65: 3415–3417.

WEICKMANN, H. K., 1953. Observational data on the formation of precipitation in cumulonimbus clouds. In: H. F. BYERS (Editor), *Thunderstorm Electricity*. Univ. of Chicago Press, Chicago, Ill., p.66–138.

WEICKMANN, H. K., 1964. The language of hail storms and hail stone. *Nubila*, 4: 7.

WENT, F. W., 1944. Plant growth under controlled conditions, II. Thermoperiodicity in growth and fruiting of the tomato. *Am. J. Bot.*, 31: 135–150.

WENT, F. W., 1945. Plant growth under controlled conditions, V. The relation between age, light, variety and thermoperiodicity of tomatoes. *Am. J. Bot.*, 32: 469–479.

WENT, F. W., 1950. The response of plant to climate. *Science*, 112: 489–494.

WENT, F. W., 1955. Fog, mist, dew and other sources of water. *US Dept. Agric., Washington, D. C. Agric. Yearbook*, pp.103–109.

WENT, F. W., 1956. The role of environment in plant growth. *Am. Sci.*, 44: 378–398.

WIEGAND, C. L. and NAMKEN, L. N., 1966. Influence of plant moisture stress, solar radiation and air temperature on cotton leaf temperature. *Agron. J.*, 58: 582–586.

WILLIAMS, G. D. V., 1973. Estimate of prairie provincial wheat yields based on precipitation and potential evapotranspiration. *Can. J. Plant Sci.*, 53: 17–30.

WILLIAMS, G. D. V., JOYAT, M. I. and MCCORMICK, P. A., 1975. Regression analyses of Canadian prairie crop-district cereal yields, 1961–1972 in relation to weather, soil and trend. *Can. J. Soil Sci.*, 55: 43–53.

WILLIAMS, G. E., 1954. *Calculating Irrigation Requirements Using Weather Data.* Unpubl. M.S. Thesis, Univ. of Nebraska, Lincoln, Nebr.

WILLIAMS, G. P., 1961. Evaporation from water, snow, and ice. *Proc. Hydrol. Symp., 2nd, Toronto*, pp. 31–47.

WILSIE, C. P., and SHAW, R. H., 1954. Crop adaptation and climate; In: A. G. NORMAN (Editor), *Advance in Agronomy*, 4: 199–252.

WILTSHIRE, S. P.. 1931. The correlation of weather conditions with outbreaks of potato blight. *Q. J. R. Meteorol. Soc.*, 57: 304–316.

WOLFE, T. K., 1927. A study of germination maturity and yield on corn. *Va. Sta. Tech. Bull.*

WOLLNY, E., 1878. Untersuchungen über den Einfluss der Bodenfarbe auf die Erwärmung des Bodens. *Forsch. Geb. Agric.-Phys.*, 1: 43–72.

WOODRUFF, J. M., MCCAIN, F. S. and HOVELAND, C. S., 1967. Effect of relative humidity, temperature, and light intensity during ball opening on cotton seed quality. *Agron. J.*, 59: 441–444.

WOODRUFF, N. P., READ, R. A. and CHEPIL, W. S., 1959. Influence of a field windbreak on summer wind movement and air temperature. *Agric. Exp. Sta., Kansas State Univ., Manhattan, Kansas, Tech. Bull.*, 100.

WORLD METEOROLOGICAL ORGANIZATION, 1963. Protection against frost damage. *WMO, Geneva, Tech. Note*, 51.

YAO, A. Y. M., 1969. The R-index for plant water requirement. *Agric. Meteorol.*, 6: 259–273.

YAO, A. Y. M., 1972. Agricultural drought and its probability in the USA. *Abstr. Bull. Am. Meteorol. Soc.*, 53: 1033.

YAO, A. Y. M., 1973. Evaluating climatic limitations for a specific agricultural enterprise. *Agric. Meteorol.*, 12: 65–73.

YAO, A. Y. M., and SHAW, R. H., 1956. A comparison of methods of estimating evaporation from small water surfaces in Iowa. *Final Rept. US Weather Bur. Contr. CWB-8808, Dept. Agron., Iowa State Univ., Ames, Iowa.*

YAO, A. Y. M. and SHAW, R. H., 1964. Effect of plant population and planting pattern of corn on the distribution of net radiation. *Agric. J.*, 56: 165–169.

ZINKE, P. J., 1967. Forest interception studies in the U.S. In: W. E. SOPPER and H. W. LULL (Editors) *Forest Hydrology*. Pergamon, Oxford, pp.137–161.

Chapter 3

City Climate*

H. E. LANDSBERG

Introduction

It has been known since ancient times that city climate differs from that of the surrounding country. Hippocrates, that shrewd observer, even commented on the differentiation of exposure of cities or their subdivisions and alleged effects on health. Among the earliest known effects of urbanization were the changes brought about in air composition. Especially in the cities of the cooler regions and during the days of emerging manufacturing the polluting consequences of combustion processes became well known. With the advent of coal as a universal fuel problems that have not been completely solved to this day began to concern the scientists (EVELYN, 1661).

During the dawn of scientific meteorology it became customary to record consistently the changes of the various elements then accessible to measurement. Regular series of thermometer observations, most of them at the centers of learning, were started. Some of these have been maintained since the late 17th and early 18th century. They are still of use in assessing changes brought about by growth of cities. It was quite natural that sooner or later differences between the city and the countryside were noted in the observations. We can credit HOWARD (1818) with the first climatography of a city and also with recognition of the essential features of city climate.

Since that time the climates of many cities have been analyzed and described on the basis of meteorological observations. We might mention here the early work of RENOU (1862) in Paris. In many instances the purpose was primarily a summary of these observations rather than a comparative climatology of the city and its environment. In some instances, even in modern times, parallel observations from representative sites inside and outside the city were missing.

A decisive advance in studying the climate of cities was made when SCHMIDT (1930) introduced the use of an instrumented motor vehicle for making traverses across a metropolitan area to obtain data for comparative purposes. This procedure, hinged on a number of fixed stations, has remained a useful technique. It has been supplemented in recent years by excursions into the third dimension through the instrumentation of towers, use of tethered balloon soundings, tracking of free-floating balloons or tetroons, and slow radiosonde ascents. In addition infra-red radiation observations from aircraft and satellites have been helpful.

A number of detailed surveys have been published and from these a number of generalizations can be derived which we propose to discuss below. An extensive review of the

* Part of this work was supported by the National Science Foundation under grant GA-13353.

literature can be found in KRATZER's (1956) excellent work. A useful annotated bibliography was given by BROOKS (1952). A series of papers on Japanese cities by SEKIGUTI et al. (1964) appeared in English abstract form in the *Tokyo Journal of Climatology*. CHANDLER (1969) has prepared a rather extensive list of citations to 1968. This was followed by two notable reviews of OKE (1974, 1980).

An attempt to distinguish how much of the changes ascribed to city climate were caused by human influence was made by LANDSBERG (1956), and later followed by a specific experiment in a new town (LANDSBERG, 1975). The ravages produced by air pollution have been discussed in numerous treatises. As it applied to city climate much has been brought together in a U.S. PUBLIC HEALTH SERVICE symposium (1962). A broad overview of the problems is contained in a comprehensive WMO Symposium (1970), and the inadvertent effects of cities on weather and climate have been assessed (LANDSBERG, 1974).

An exemplary climatography of a city has appeared for Vienna, Austria. It combines physical and biometeorological aspects of a large settlement (STEINHAUSER et al., 1955, 1957, 1959). The utility of such a detailed analysis for planning and urban renewal cannot be overestimated.

An urban model

An attempt has to be made to define what constitutes a city climate. The modification brought about by human settlements obviously exceeds those usually defined as microclimatic. The order of magnitude of the effect is very much akin to the changes produced by small mountains or lakes. We can as a latitude shift of several degrees justifiably range the city climate among the specific mesoclimates. There is, of course, a wide scale of differences that have to be included in this classification. The size of the city, its type of activity, its location in the macroclimatic zones will have a decisive influence on the degree of differentiation from the environment. Nature thus can readily accentuate or diminish the role of the man-made alterations.

The greatest difficulty lies in the fact that many cities have developed in their location because of some special topographic advantage to the early settlers. Natural harbors, river valleys, lake fronts, flat terrain in mountains, natural defenses and other features attracted settlements. They offered advantages of transportation, protection, or food and water supply. They undoubtedly, had even in their natural setting a climate measurably different from their surroundings. LINKE (1940) once voiced it this way: "In examining the available statistics I have always returned to the suspicion that not only the city *per se*, that is an accumulation of houses in a small space, has led to climatic departures but also the fact that most cities lie in valleys and depressions." Similar statements have been made since by other authors.

There can be no doubt that city influence has at times been overestimated. But even quite elementary reasoning shows that the changes introduced by settlement alone are quite profound. This goes, incidentally, also for other human interference with the natural surface. Locally at least deforestation or aforestation can bring about important micro- or mesoclimatic differences. Replacement of prairies by cultivated farmland, drainage of bogs, large-scale irrigation, are among other human interferences with the natural climate.

For the present we are still forced to deal with many of the city-induced changes in qualitative terms. But even though we are only partially able to formulate the changes in mathematical terms, there is no doubt that a fairly reliable picture exists on the orders of magnitude of climatic changes introduced by urbanization.

We can thus reasonably predict what changes should take place and compare prediction with observations. To enumerate a few:

(*1*) Replacement of natural surface by buildings, often tall and densely assembled, causes increased roughness. This should reduce surface wind speeds. The irregularity of the surface from street and parks to various roof heights should lead to increased turbulence.

(*2*) Replacement of natural soil by impermeable pavements and roofs combined with drainage systems should reduce evaporation and humidity and lead to faster run-off.

(*3*) Pavements and building materials have physical constants substantially different from natural soil. Generally, many have lower albedos and greater heat conductivity and capacity. This alters the radiation balance and has consequences for air temperature. In most instances, it leads to heat storage from the solar radiation received. Internal radiative exchange between walls and streets causes different reactions to the diurnal temperature cycle. Also, in effect a new primary radiative surface is created in densely built-up areas at roof level. This leads to considerable alteration of the lapse rates in the lowest layer, which, locally, is in turn profoundly influenced by alternate shading and exposure of surfaces to incoming radiation.

(*4*) Added by human activities is heat, which can be a substantial part of the local energy balance. This, together with the effects enumerated under (*1*) and (*3*) above leads to an increase in temperature above the surroundings. In turn, this will lead to convective rising of air and can cause cloudiness and promote precipitation. It will also promote a tendency for air flow into the built-up areas.

(*5*) Addition of foreign substances, such as water vapor, fumes, and gases from combustion and industrial processes has probably the most obvious effects. Because of large numbers of hygroscopic nuclei among these admixtures visibilities are reduced, radiation is intercepted, fogs are formed, and, together with the added turbulence under (*1*) and the convective effects mentioned under (*4*), precipitation is increased.

We shall see below if we can verify these statements. Unfortunately knowledge of three-dimensional air flow around cities is still very restricted. A great deal of information has been collected in the Los Angeles area, where air pollution is a formidable problem. But this is not a good prototype because nature has arranged it so that the city influence cannot be disentangled from mountain breezes, land and sea breezes, and the low-level dynamic inversion. In addition, there is a lot of sunshine causing photochemical effects in the contaminants. It is about as complex a setting as one could find and will not give much information on a basic urban model. Useful elements for such a model can be found in observations at Fort Wayne, Indiana (GRAHAM, 1968) and in the Metromex project (ACKERMAN, 1974a,b). Both of these studies were made in topographically relatively simple settings and contributed notably to our understanding. Sporadic experiments with tetroons in the New York City region were also very enlightening (ANGELL et al., 1973).

Thermal field of the city

Energy consumption on earth is continually rising. Most of it is derived from extensive use of fossil fuel in this era. Although world consumption is approaching 10^{20} calories per year this remains less than 1/50,000 of the solar radiation received and is thus negligible in the total energy balance of the earth. However, locally in concentrated settlements it becomes an appreciable factor. In U.S. cities of million inhabitant size in the northeast, it amounts to about 10–15% of the annual solar radiation income. SCHMIDT (1917) once estimated that in some of the largest central European cities with densely packed population it could reach 1/3 of the radiation received at the surface from the sun. This is what has been estimated to be the portion of man-contributed energy in the northeastern U.S. in the year 2000 (JASKE et al., 1970). The greatest effect of the enormous fuel consumption is as yet the resultant air pollution which will be discussed in greater detail in a later section.

However a major portion of the positive heat excess of cities is caused by the type of surface prevalant in urban areas. NISHIZAWA (1958) has tried to get an insight into the phenomena by heat balance measurements made in Tokyo. These were few in number and unfortunately lack simultaneous values for the same conditions in an undisturbed environment and hence can give only a glimpse into this complex problem. One of his examples shows on a midday in August an isolation of 0.90 ly/min, a net outgoing radiation flux of 0.15 ly and a radiation balance of 0.75 ly/min over a granite street with a surface temperature of 30°C and temperature of 28°C at 3 cm height. The heat transport into the ground was 0.50 ly and heat transport by eddy conductivity 0.25 ly to the air. At night the eddy conductivity over the street stayed appreciable, amounting early in the night to 0.1 ly/min and in the early morning before sunrise still 0.035 ly/min. No heat is lost by evaporation and much is stored in daytime and is released at night. Compare this to a grass surface which is moist from dew and guttation. It uses a considerable amount of heat in the morning for evaporation, conducts very little into the soil in daytime and cools rapidly in the early night hours. Nishizawa concludes that there are three distinct layers with different eddy conductivities in a city: (*1*) the ground layer to roof tops; (*2*) the roof top layer up to $1\frac{1}{2}$–2 times building height; and (*3*) the upper layer above 2 times building height.

A more recent study by MAISEL (1971) in the newly established town of Columbia, Maryland, gave a comparison between a parking lot in the urbanized area and an undisturbed field of clover and tall weeds nearby, during daytime. As an average of five days with clear skies the incoming global radiation Q_I was 1.12 ly/min and the net radiation $Q_N = 0.59$ ly/min, for the vegetated surface with $Q_I = 1.07$ ly/min and $Q_N =$ 0.59 ly/min. This shows that the lower albedo and higher emission ($\Theta = 315°K$) of the parking lot, is compensated by the higher albedo and lower emission of the vegetated plot ($\Theta = 300°K$). For five overcast days Maisel obtained the following daytime values for the parking lot: $Q_I = 0.55$ ly/min, $Q_N = 0.30$, $\Theta = 311°K$; and for the vegetated field: $Q_I = 0.52$ ly/min, $Q_N = 0.28$ ly/min, $\Theta = 294°K$.

These findings are supported by studies of GRASNICK (1961) who investigated the nocturnal cooling over various soil and surface types and its influence on air temperatures. The various soil constants such as conductivity, have a large influence on the amount of heat given off to the air after sunset. Forest soil and dry sand transfer very little heat into the air, humus more, and solid rocks, such as sandstone and granite, most. No data are given on soil with plant growth but by inference the transfer can be assumed to be small. The actual temperature change depends on the effective radiation and the exchange coefficient. For example, at shelter height of 2 m with an exchange coefficient of 1 g cm^{-1} sec^{-1} and a net outgoing radiation flux of 0.18 cal. cm^{-2} min^{-1} the temperature change in 8 h over sand and humus is 12° to 13°C, over sandstone or granit 7°–8°C. This author finds that at a height of 76 m the differences in nocturnal cooling over various soils with the usual low nightly exchange coefficients become negligible. The order of magnitude agrees quite well with observations of temperature stratification over cities by DUCKWORTH and SANDBERG (1954).

The different elements in the radiative and heat exchange balance during the light and night hours let us immediately deduct that the tempeıature contrasts of city and countıy are different between these two segments of the day. It also immediately points to the very important fact that much of the contrast is governed by the sky and wind conditions. Thus the numbeı of clear days and nights in a given macroclimate and the ventilation rate have a profound influence on the overall difference of temperatures between a city and its rural environs. This is, of course, not too surprising because these elements affect all types of microclimatic differentiation.

Many measurements of the temperature of various surfaces in bright sunshine are found in the literature. Most of them apply only to limited areas, such as a roof, a brick wall, a street surface, or grass plot. More enlightening for the scale of phenomena with which we are concerned here are integrated surface temperature measurements over larger, representative areas. This has only recently become possible by infrared radiation measurements from aircraft and satellites. LORENZ (1962) reported a few during daytime even though flight regulations prohibited surveys of a city proper from the air. His data are sufficiently indicative of the orders of magnitude that can be counted on. In cultivated growth, both in sunshine and shade he obtained surface temperatures not much more than ± 1°C from air temperatures. Forests were generally 1°–2°C cooler than air temperature. An airport lawn showed in sunshine +3°C, a concrete platform +8°C, hangar roofs and an asphalt street +17°C, an asphalt parking surface +20°C above air temperature. Even under overcast conditions did a small village show a surplus of 3°C above the surrounding pasture surfaces. All of these measurements refer to summer conditions.

In recent years these measurements have been supplemented by infrared aerial photography which gives very beautifully differentiated but qualitative micrometeorological patterns of the landscape (MATTSON, 1969). From a helicopter survey with an infrared thermometer MAISEL (1971) found, on a clear summer evening two hours after sunset, surface temperatures, as follows: business area with parking lot 28°C, small lake 26°C, grass and wooded area 22°C. Combining data obtained by LORENZ (1973) in Munich with infrared measurements made during the rapid growth period of the new town of Columbia, Maryland, 1968–1975 (LANDSBERG, 1975) gives a consistent picture. The more intensive the urban land use, the higher the mid-day temperatures of the surface on sunny days. The differences to rural areas, unaffected by urban construction is shown in Fig.1.

Occasionally there is opportunity to follow the changes in climate directly, as is the case when buildings encroach gradually on a station previously located in open country.

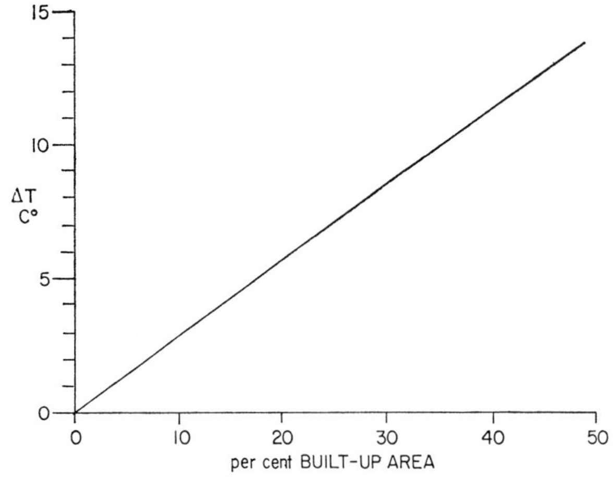

Fig.1. The influence of the degree of urbanization on the surface temperature as measured by infrared thermometry.

KONČEK (1962) found in Bratislava, by comparison with undisturbed stations that primarily the evening and night temperatures increase. In that particular case autumn was the season of greatest change which may be owing to the fact that clear, calm nights are frequent then.

In some cities the change can be noted in long-term records. A particularly noteworthy case is Paris. This has been placed on record by DETTWILLER (1970a). He showed that in the absence of an appreciable regional trend in the last 80 years the Paris temperature mean has risen about 1.1°C compared with the environs. This is shown in the regional isotherms (Fig.2). A glance at this figure makes it amply clear why the graphically descriptive term "heat island" has been applied to the urban temperature surplus. In

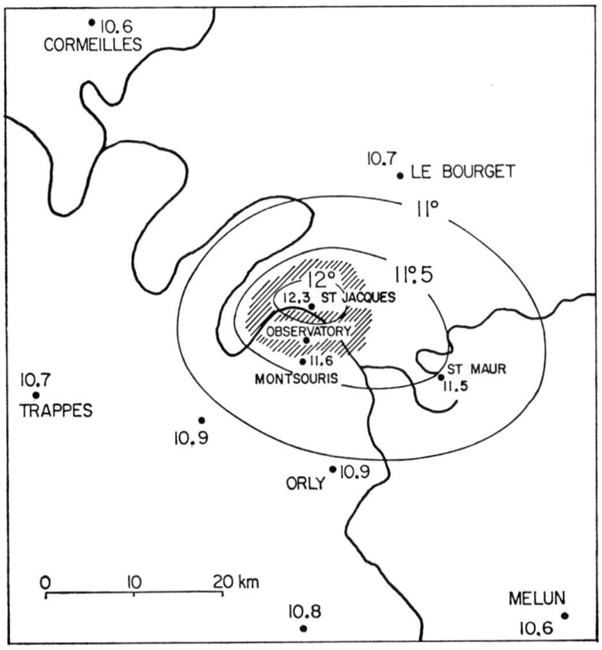

Fig.2. The urban heat island of Paris, France, shown by mean annual isotherms, in °C. The region is characterized by minimal orographic complexity. (After DETTWILLER, 1972.)

Paris this has not only affected the air temperature but also applies to the soil. Temperatures in a sub-basement at 27 m below the ground have also risen over 1°C. Measurements there since the 18th century had been very constant, as one would expect, until the beginning of the 20th century when the rise became notable.

The diurnal variation of temperature on a clear day indicates usually almost equally rapid rises in city and country after sunrise. At the time of maximum there is generally little difference. The reason for this is that convection and increased day-time wind speeds promote rapid equalization, although in very large cities even then a tendency for a temperature surplus remains. This is noted in densely built-up areas at street level, which is an environment shunned for regular meteorological stations, usually located in parks. Special surveys, however, reveal this surplus. The temperatures in city and country begin to diverge very rapidly after sunset. At that time the country cools quickly, the city feeds on its heat reservoir stored in the stony bulk of buildings and streets. The difference becomes largest about 3 to 5 h after sunset and gradually diminishes toward the early morning minimum.

A very typical set of differences between daily maxima and minima is shown in Fig.3. This shows histograms of frequency of various temperature differences in Lincoln, Nebraska, a place chosen because there are no local orographic difficulties that might cause complications. Also, this is not a very large metropolis. It had a little over 100,000 inhabitants at the time when these data were compiled. In the warm season the maximum temperatures are slightly higher on the airport, in a treeless environment with a superheated runway. The town station is on the park-like campus of the university of Nebraska. In winter the maxima show a very symmetrical distribution around the zero deviation class, obviously the typical picture of random variations. The minimum temperatures show a startling surplus of higher temperatures at night in the city in both the warm

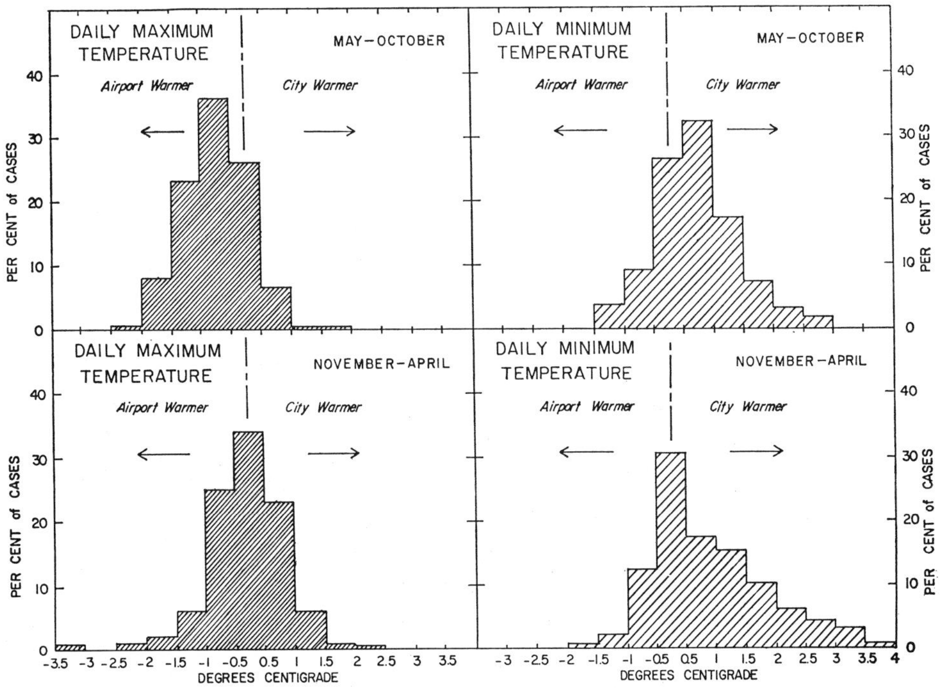

Fig.3. Frequency of differences of daily maximum and minimum temperatures between city and airport station at Lincoln, Nebraska, for the year 1953.

and cold season. In summer with more radiation nights and less wind the smaller differences prevail. In winter there are many cases of zero difference representing the cloudy, windy nights and many cases of up to 5°C difference when the weather is calm and clear. Air pollution is not a contributing factor in this case.

CHANDLER (1961a) has given a similar histogram for the much larger city of London in quite a different macroclimate. For the annual frequency of the difference city versus country he finds a modal and median value of 1°C for the maxima and a range from −2°C to +5°C. For the minima the median is at +2.2°C, the mode at +0.6°C but the range goes from −1.7°C to +9°C. The negative differences are ascribed to different cloud cover over city and country and frontal passages at different times, effects which might well be apart by a considerable time span in a sprawling metropolitan area such as London.

Much of our knowledge about city climates has been derived from traverses. Because of the fact that the nocturnal differences are so much larger the night-time phenomena have found more attention than those during light hours. The temperature differences are usually small during the light hours, even on bright days. In the areas of greatest concentration of buildings the values rarely exceed +1° to 2°C. Notable differentiation exists between park areas, business districts and suburban areas. An interesting feature is found in cities which are bisected by a river. In that case the air above the river, and on the banks adjacent to it, is measurably cooler and this leads to a dual heat center. It was found for Washington and Arlington on the two sides of the Potomac river (LANDSBERG, 1950). EMONDS (1954) showed a similar condition for Bonn and Beuel on opposite banks of the Rhein.

The effect of the city can be established by metropolitan networks of stations. These permit a better estimate of the average differences than occasional traverses. They also integrate over-all weather conditions. An example of long-term means based on regular observations of stations in the Washington, D.C. area is shown in Fig.4.

Clear nights, however, bring out the whole contrast, as they do in so many micro- and mesoclimatic patterns. In recent years the most intensive studies have referred again to that great metropolitan area: London (CHANDLER, 1960, 1961a, 1962a,b, 1965). He has demonstrated clearly that, even though the broad isotherm patterns generally show essentially concentric lines within the city area, the varying density of settlement, parks, and squares show a great amount of variation. He has also clearly brought out the role of the wind as an important factor in the size of the area with abnormally high nocturnal temperatures. The sharp contrasts between rural and urban temperatures in London develop with wind speeds of less than 6 m/sec and under clear skies usually run around 6°C. The gradients are not uniform but show a distinct crowding at the edge of the built-up area where a rapid rise of 4°C takes place. With winds in excess of 11 m/sec the city effect seems to vanish completely. This threshold value of wind speed is less in smaller cities. For example, in Reading (England) PARRY (1956) established it at about 7 m/sec. For Palo Alto (California) DUCKWORTH and SANDBERG (1954) arrived at a value of 3 m/sec, measured 8 m above ground in town which they estimate would correspond to about 6 m/sec for standard anemometer exposures. In Brno (Czechoslovakia) QUITT (1960) also found that the city excess was reduced to 0.5°C with winds above 11 m/sec but reached 2.5° to 3°C at lower wind values. Quitt's surveys were unique in one respect: he instrumented a street car and used it for traverses.

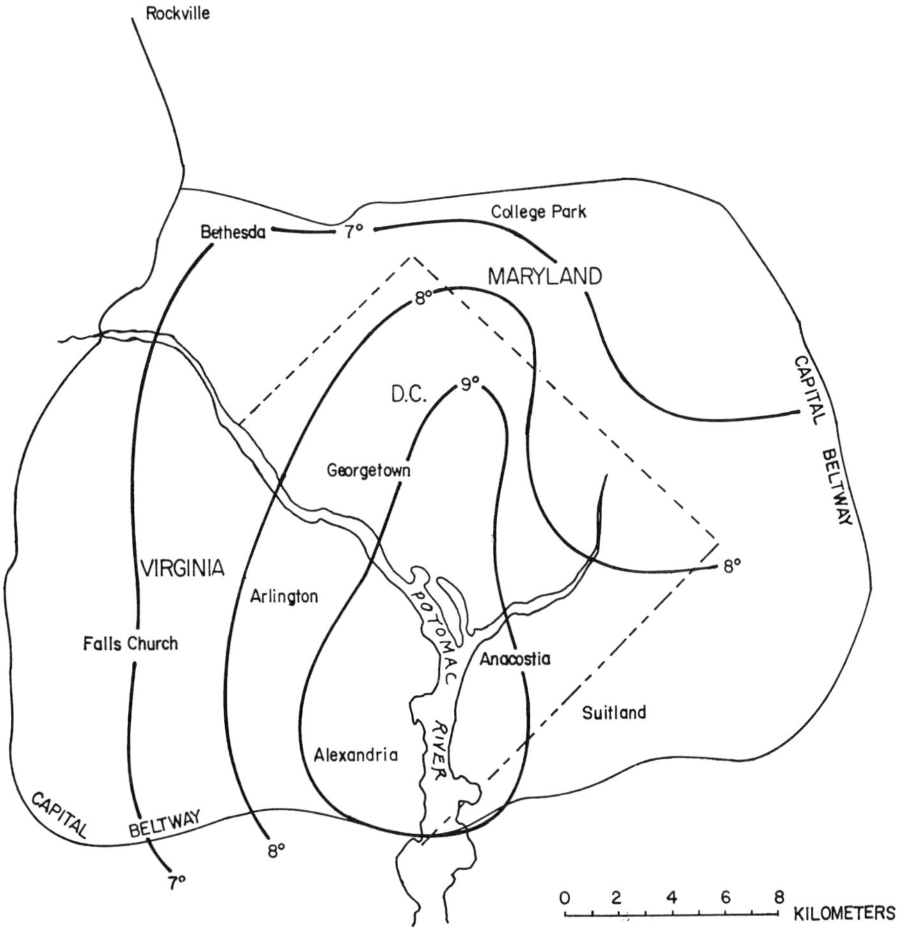

Fig.4. Mean annual value of daily minimum temperatures for the Washington, D.C., metropolitan area, based on 15 years of observations. (After WOOLLUM, 1964.)

Illustrative of the process is the gradual temperature rise in Tokyo of 0.032°C per year (FUKUI, 1970). Although the natural temperature rise in the area from 1936 to 1965 can be estimated at about 0.01°C per year. The annual values for Tokyo are shown in Fig.5. One may note the low value for 1945, although it was a cooler year, when the temperature drop was larger in Tokyo than in the general area, but Tokyo had been 90% razed by air raids in that year.

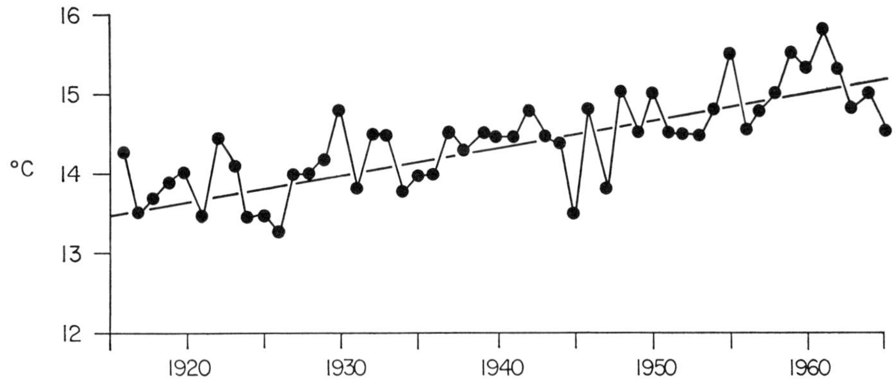

Fig.5. Tokyo annual air temperatures and trend line 1916–1965. (After FUKUI, 1970.)

SUNDBERG (1951) in Uppsala (Sweden) made a detailed investigation of the relative role of various meteorological elements on the urban-rural temperature difference. His results indicated clearly that cloudiness and wind are the dominant influences. In a general way the relations take the form:

$$\Delta T = \frac{a - bn}{v}$$

where ΔT is the temperature difference in degrees C, n cloudiness in 1/10 sky cover and v the wind speed in meters per second. The constants a and b are characteristics of the particular city: in Uppsala for night conditions they had the values $a = 4.6$ and $b = 0.28$. The factor a includes in essence the static city influences, such as albedo, heat conductivity, degree of building density.

The temperature differences that develop have, of course, a notable influence on all temperature-derived climatic elements. Most obvious is the effect on heating degree-days. In the United States a comparison of 18 pairs of city vs. airport observations showed an average 10% reduction of degree-days below 65°F (18°C) in the cities. A similar value was observed in Toronto between the city station and the Malton airport (SHENFELD and SLATER, 1960).

Similarly, the dates of the earliest freezing temperatures in autumn and the latest in spring are shifted to extend the freeze-free season in the city considerably. This is primarily a result of the much higher minimum temperatures in the city on calm, clear radiation nights. Fig.6 shows the average date of the last freezing temperature in the Washington, D.C. metropolitan area in spring. The difference between the urban center and the rural area is an astonishing three weeks. KRATZER (1956) reports for Munich (Germany) an increase of 61 days in the freeze-free season for the city. DAVITAYA (1958) found for Moscow a value of 30 days surplus of freeze-free days and an increase of the long-term mean of temperature sums above 10°C of 250°.

Only very few surveys of actual surface temperatures exist. But VILKNER (1961) reported for Greifswald (Germany) on a late May date a surface temperature of 3.7°C and 500 m away on a field −8.9°C. The temperature in the meteorological shelter in suburban surroundings on this calm, clear night after a polar outbreak was −0.6°C. From his surveys he derives the differences for various surfaces against shelter temperatures for calm, clear nights shown in Table I. It is not yet adequately established whether or not these temperatures measured by ordinary liquid-in-glass thermometers are equivalent to the infrared radiation temperatures referred to earlier.

MAISEL (1971) in Columbia, Maryland, showed also that very large surface temperature differences can develop in distances of less than 1 km in urbanized areas. Table II shows these foı a clear night at the time of minimum temperature and day with moderate radiation. The differences to air temperature are also shown. For the asphalt surfaces, differences to air temperature at 2 m of as much as 26°C were noted on particularly bright days.

These circumstances are among the few beneficial changes in climate produced by cities. They reduce fuel requirements in the cool climates; they permit earlier bloom of flowers and shrubs; they permit the survival of temperature-sensitive species, which otherwise would not survive in the particular macroclimatic zone. Another benefit in snowy climates is the reduced snowfall in cities compared with the rural surroundings. It might

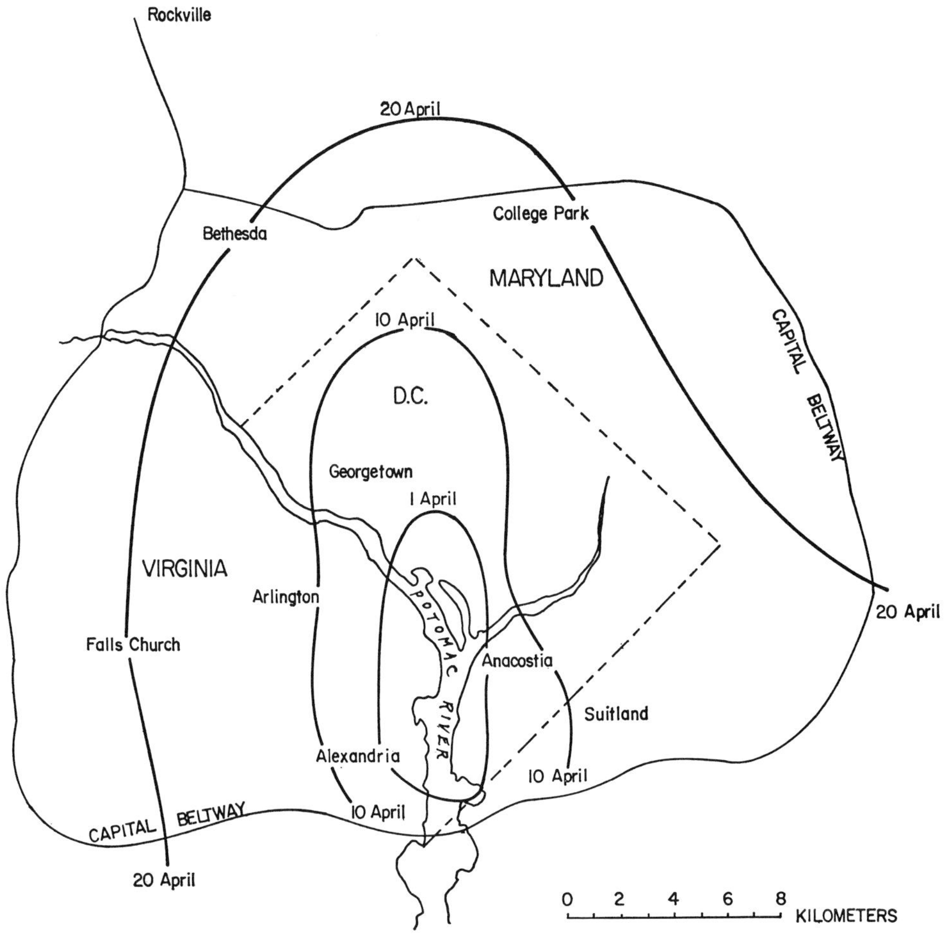

Fig.6. Mean date of last freezing temperature in spring in the Washington, D.C., metropolitan area, based on 15 years of observations. (After WOOLLUM, 1964.)

TABLE I

GROUND SURFACE TEMPERATURE DIFFERENCES IN URBAN AREAS ON CLEAR NIGHTS
(After VILKNER, 1961)

ΔT (°C)	environment
+2	center of town, densely built up, protected from outgoing radiation
0	shelter (2 m); town squares, parks, radiative protection
−2	suburb, unattached buildings, some radiative protection
−4	suburb, edge of town, no radiative protection
−6	dry lawns, sport fields
−8	large lawn surfaces (moist)
−10	pastures (moist), outside city

TABLE II

SURFACE TEMPERATURES IN °C AND DIFFERENCE TO AIR TEMPERATURE IN AN INCIPIENT URBAN AREA

Type surface:		grass	bare soil	asphalt	lake
Clear	surface T	2	6	12	12
night	ΔT air	−10	−7	−2	0
Day	surface T	30	35	41	26
	ΔT air	+3	+4	+14	−1

be appropriate to mention here the portioning of energy estimated by PERKINS (1962) needed for maintaining San Francisco's nocturnal "heat island". He found that the total required is about $7.5 \cdot 10^{12}$ cal./h. Of this about 1/3 is from combustion pıocesses and 2/3 from stored solar heat.

Within the mesoclimate of the city numerous microclimates develop. We do not refer to orographic influences that may be dominant in some localities. The emphasis is rather on the variations introduced by different styles and types of buildings and land use. Unfortunately little quantitative material is at hand. MAHRINGER (1963) made a useful contribution to this theme on courtyards in Vienna (Austria). One of them was 8 × 15 m, another only 2 × 4 m with building heights of 20 m. This is not untypical of the well-like areas found in old tenement districts. The larger of the two (*A*) gets sunshine on the ground in that latitude (48°) only from about 1 May to 1 August. The smaller yard (*B*) shows its deepest solar penetration on 21 June (summer solstice) on the south wall about 12 m above ground. Here are some values characterizing these exposures in mid-summer: daily temperature amplitudes in shelters: *A*: roof 10.2°C, ground 5.6°C; *B*: roof 12.1°C, ground 0.2°C. By large, the yards are warmer in winter than the undisturbed air, the smaller one considerably so; in summer they are cooler. With weak winds considerable inversions develop in these wells.

QUITT (1960) mentions that in Brno gradients of 1°C per 100 m horizontal distances are not uncommon in the city area and that the sun-exposed sides of streets are on the average 0.5°C warmer in all seasons than the shady ones.

LANDSBERG (1970) showed that a single square of buildings with a parking lot in the center will develop a nocturnal heat island. On a clear, calm summer evening air temperatures at 2 m were 1.2°C warmer than over a surrounding grass surface, while the simultaneous ground temperatures, measured by the infrared thermometer, were 7.5°C higher on the parking lot than on the grass. NORWINE (1973) also showed that an isolated shopping center created its own heat island.

Yet, our knowledge of the low-level vertical temperature distribution over cities is still limited. DUCKWORTH and SANDBERG (1954) obtained some low-level soundings over San Francisco and Palo Alto (California) to about 200 m in the evening during the formation of the nocturnal radiation inversion. In the larger city there existed often nearly isothermal conditions or only slightly increasing temperatures with elevation. In the surrounding countryside intense ground inversions develop but at about 25 m the temperatures actually became warmer than in the city ("cross-over effect", if plotted on an adiabatic chart) and remains so up to the highest points reached. In a smaller city this effect does not show but the inversion was much less than in the country and above 25–30 m the temperature differences became very small.

Temperature data collected for a year on a television tower in Louisville, Kentucky, by DEMARAIS (1961) at 18, 52, and 160 m above ground gave a picture of lapse rates for an urban area. The lowest level is approximately representative of the "active" (roof) surface of a city. No simultaneous local data from undisturbed environments are available but DeMarais compares his results with two series from rural sections, one in Idaho and the other in England. In these two locations night conditions in a similar shallow layer of air are characterized by intense inversions. In Idaho midday values usually show superadiabatic gradients. At the English location midday values in summer are also superadiabatic and in other seasons show a temperature decrease with height.

The averages for Louisville generally show temperature decreases with height throughout the day and year, with superadiabatic gradients from 10h00 to 16h00. In winter the heating effect of the city is noted even at night when decreases of temperature with height prevail. Only in the late night hours in spring and autumn are inversions in the 18 to 160 m layer present with some frequency (up to 40% of the time) but even then their presence is less frequent in the lowest layer, 18 to 50 m (usually below 20% in the bi-hourly observations). This confirms the fact that the nocturnal temperature stratification in cities is markedly disturbed in comparison with natural conditions. This is confirmed by a series of nocturnal temperature surveys by helicopter over Cincinnati by CLARKE and MCELROY (1970). They give some insights into the structure of the boundary layer over a sizeable city in rolling terrain. In a typical case, with a fairly well developed gradient wind of 12 m/sec, a surface wind of 3 m/sec, and a weak inversion upwind from the city in rural areas, there was a superadiabatic gradient to 65 m above the central part and an adiabatic to isothermal lapse rate to the interface of the urban boundary layer at 240 m.

A gradual consensus is developing about the influence of city size on the magnitude of the temperature difference city–country. One of the difficulties is the fact that the spectacular period of growth of the large cities in the moderate and higher latitudes of the Northern Hemisphere in the 20th century has been paralleled by a slow general warming of the climate until about 1950. To this are added the uncertainties of station changes and exposures. CHANDLER (1964) goes so far as to state that "heat-island intensities are by no means linear functions of the area of cities, or individually of their growth". He asserts that for London the heat-island intensities have been more dependent upon the changes in regional climate than the growth of the city.

This is not borne out everywhere. MITCHELL (1953) in a careful analysis of 77 cities in the United States found that the mean temperature changes were proportional to the changes of the square root of population figures. He confirmed this in later studies (MITCHELL, 1962), which showed a very close relation between city growth rate and summer temperature difference increases between rural and city stations. In winter there was still a positive correlation but evidently population can not be used as the only correlate.

The magnitude of the urban-rural temperature was first related by SUMMERS (1964) to meteorological parameters in the surface layers. His simple model expressed it as:

$$\Delta T_{(u-r)} = \frac{2xQ\,(\delta\Theta/\delta z)}{\bar{u}\,\varrho\,c_p}$$

where $\Delta T_{(u-r)}$ = urban-rural temperature difference; x = wind fetch from periphery to center of urban area radius of city area; Q = heat input of city; $\bar{u}$ = mean wind speed; ϱ = air density; c_p = specific heat of air at constant pressure; $\delta\Theta/\delta z$ = lapse rate of potential temperature in rural area.

The values of Q, $\bar{u}$, $\delta\Theta/\delta z$ are generally only approximately known. Hence checking of observations against the theory is difficult.

OKE (1973) developed from observations in a number of towns of varying size a relation for the maximum value of $\Delta T_{(u-r)}$:

$$\Delta T_{(u-r)\max} \sim \log \text{Pop.}$$

where "Pop." is the population number. The greatest differences develop in the evening about 2–3 h after sunset, when winds are calm ($\bar{u} \to 0$). The slope of the logarithmic population line is different for European than American towns, as shown in Fig.7. The reasons for the difference are not entirely clear but may be sought principally in the different construction and density of housing and the different mode of energy production and use. It is interesting to note that experiments in a town which grew in a few years from <1000 inhabitants to 25,000 people (Columbia, Maryland) yielded points falling exactly on Oke's American heat island line (LANDSBERG, 1975).

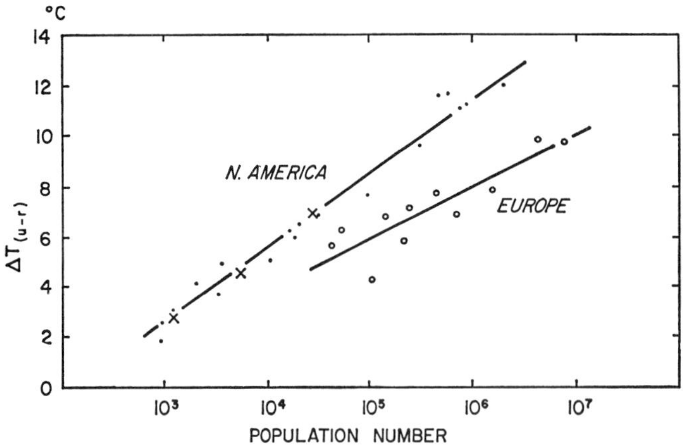

Fig.7. Maximum temperature difference urban–rural, $\Delta T_{(u-r)}$, as related to total population (on logarithmic scale). Note two crosses on line for American cities showing values for Columbia, Maryland, during early growth phase. (After OKE, 1973.)

The gradual growth of the rural-urban temperature-difference is documented in many cities. SCHERHAG (1963) showed it for Berlin. For a number of growing cities in the U.S., within simple orographic settings, one can approximate the numerical value of the *mean annual temperature difference* ΔT°C as roughly proportional to the city diameter d in kilometers:

$$\Delta T = 0.1\, d$$

Such a simple relation is obviously only a first approximation but has some foundation in the often highly similar pattern of city development; i.e., a densely built-up core or business district, a more open residential ring with lower building heights, finally a loose belt of suburban settlement. The temperature increase must be eventually self-limiting owing to the convective currents that will be induced.

There is evidence that heat production by industrial, domestic, and traffic activity has a measurable effect. MITCHELL (1962) found for four winter seasons in New Haven (Connecticut) that the mean daily temperature was on weekdays 0.6°C higher in the city than at the airport station. But on Sundays this difference was only 0.3°C. This difference is statistically significant at the 99% confidence level.

The wind field of the city

Even a single building can bring about a marked change in the wind flow. It will show decided lee-side effects, which may either lead to a notable speed reduction or even an increase, because of eddy formation. An instructive series of wind-tunnel tests has been conducted by JENSEN and FRANCK (1963). Such effects are not too easy to establish in nature but can be beautifully demonstrated for physical models of cities in wind tunnels (CERMAK, 1976). Canyon-like streets will often cause deflections of the wind direction by 90° from that prevailing in the country. According to DIRMHIRN and SAUBERER (1959) the channeling effect of streets reduces the wind speed close to the building fronts compared to the middle of the street. They found that if the wind speed in the center of the street is designated as 100%, the sidewalk on the windward side has 90%, and the protected sidewalk toward lee only 45%.

These authors also noted that the effect of trees is quite marked. Under trees in the parks of Vienna the speed was only 50% of that in open areas. They found that deciduous trees in leaf caused a 20–30% reduction. This is corroborated by observations at Nashville (Tennessee) analyzed by FREDERICK (1961). There the urban area wind speed is reduced to 30–60% compared to the airport, more in summer than in winter. After defoliation city wind speeds increase about 25%. In spring there is a gradual decrease as the leaves emerge. In assessing these observations it must be stressed that anemometer exposures were not those commonly used at meteorological stations, where every attempt is made to obtain an unobstructed wind field. However, correlations for 24-h wind movement between airport and city stations are generally high, usually between +0.80 and +0.90 so that once the speed differences between two observation points are known a fair estimate of wind movement in town can be based on the more permanent regular weather observing posts (FREDERICK, 1964).

But even when anemometers are located on buildings or in open areas the wind speed reduction by the city is unmistakable. This was first reported by KREMSER (1909) for a station which was located in 1883 at the outskirts of Berlin. In the next two decades an encroachment of apartment buildings took place. The anemometer was 32 m above ground but eventually the roof heights of the surrounding apartments were only 7 m below the anemometer level. This caused a mean wind speed reduction of 25%. In two rather open locations in New York City, LaGuardia Airport and Central Park Observatory, a reduction of 23% was observed for the latter location which is surrounded by dense settlement. The values for various seasons are shown in Table III.

Considering the fact that the anemometer at LaGuardia Field was 25 m above ground and that at Central Park 19 m one would expect *a priori* about 4% difference in the

TABLE III

SEASONAL AVERAGE WIND SPEED AT CENTRAL PARK OBSERVATORY AND LAGUARDIA AIRPORT, NEW YORK CITY (in m/sec)

Season	LaGuardia	Central Park	Difference
Spring	5.6	4.4	1.2
Summer	4.7	3.6	1.1
Autumn	5.0	3.8	1.2
Winter	6.3	5.0	1.3

two localities so that at least 19% of the wind reduction can be attributed to city influence. For peak gusts observed at the two localities the reduction is only 12%.
STEINHAUSER et al. (1959) report a comparative series of observations for Vienna (Austria) from a roof installation at the Technische Hochschule in the center of the built-up area and the Hohe Warte at the edge of the city; anemometer height 36 m. These authors give the ratio of the wind speed in the central area to that at the outlying station for two wind directions, various speed classes, and two seasons. Their results are reproduced in Table IV.

TABLE IV

WIND SPEED RATIOS, VIENNA CENTER AND SUBURB
(After STEINHAUSER et al., 1959)

	Speed class (km/h)	5	15	25	35
W-wind	Winter	0.5	0.6	0.7	0.77
	Summer	0.75	0.82	0.9	0.85
SE-wind	Winter	0.5	0.8	1.05	—
	Summer	1.0	1.25	1.20	—

The anemometer at Hohe Warte is influenced in summer by the leaf effect already referred to. Hence the reductions of speed in summer in the city are less for west winds than in winter. In case of the rarer southeast winds no doubt the different roughness with respect to the wind fetch has a decisive influence.
In Columbia, Maryland, during the initial growth phases of the town from 1000 in 1969 to 12,000 in 1974 comparisons of wind speeds with the Baltimore Airport 20 km away were made for observations every 3 h. Table V shows the change.

TABLE V

PROGRESSIVE CHANGES OF WIND SPEED AS PERCENTAGE OF SIMULTANEOUS OBSERVATIONS AT BALTIMORE AIRPORT
(% frequency in three classes)

Year	⩾70%	70–99%	<<<±100%β1969
1969	43	32	25
1970	46	29	25
1971	51	26	23
1972	51	28	21
1973	58	24	18
1974	65	24	14

* Percentage of Baltimore Airport wind speed.

Values of roughness coefficients for cities are still limited to a few cases but, obviously, this is a matter of wide variation even within one city according to the exposure of the anemometer and its position with respect to built-up areas. ARIEL and KLINCHNIKOVA (1960) have made some calculations for Kiev and Leningrad, based on observations from television towers located in the built-up area. Up to 180 m height the winds show adherence to the logarithmic distribution rule. The roughness coefficients, z_0, under thermal equilibrium conditions were 4.6 m for Kiev and 4.4 m for Leningrad. According to these authors the surface roughness has a greater influence on the wind speed than the thermal stratification. In Columbia, Maryland, we found values around 1 m. NICHOLAS (1974) has tried to approximate the roughness for various land uses in an urban area. The well known reversal of diurnal wind-speed variation (i.e., surface maximum in the afternoon, minimum in early morning) takes place between 50 and 80 m in Kiev and 100 and 150 m in Leningrad. PAPAI (1961) determined the exchange coefficient over Budapest. As has to be expected, it is larger in summer than in winter and increases up to 500 m elevation and then decreases. A cogent survey on the aerodynamic and turbulence problems of urban wind fields has been presented by MUNN (1970).

Much more difficult to establish than the obvious aerodynamic roughness influence on wind speed is the wind circulation pattern of cities. Under unstable thermal stratification in summer the central parts of cities are spots of general updrafts, a fact known to air travelers as "bumpiness" and occasionally exploited by glider pilots to gain altitude. The urban "plume" has been mapped for St. Louis. In day-time it bulges the mixed layer. This can be as much as 1–5 km in midday during summer. Downwind effects reach to 20 km or even more.

It has been noted that considerable differences in wind direction can be noted in metropolitan areas (SLADE, 1963). Low level wind vanes in the Washington, D.C. area showed only in a quarter of the cases coincidence of urban and suburban wind direction within 22.5°. However, tower data (95 m height) averaged over half-hour intervals show 78% coincidence of directions. With vertical instability the deviations are largest.

For Louisville, Kentucky there exists a fairly detailed flow analysis based on surface and television tower (160 m above ground) observations, made by POOLER (1963). This town is in an area where the Ohio river and terrain differences of 130 m within 8 km from urban area constitute probably only a very small topographic influence when thermal conditions are unstable. When the general pressure gradients are weak, as indicated by calms on the tower, the flow at the surface shows an inflow toward the warmer section of the urban area. This "country breeze"—an analogy to the sea breeze—is shown in Fig.8. This indicates a distinct convergence toward the city area. Such a

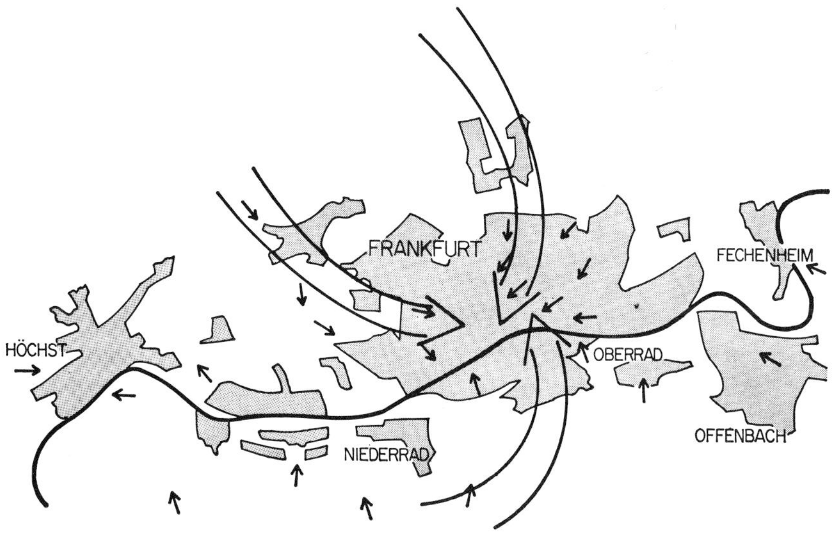

Fig.8. Country breeze observed in Frankfurt a.M. (After STUMMER, 1939.)

country breeze is also evident over parts of the area when the surface flow caused by the large-scale pressure gradient is blocked by the topography.

BERG (1947) estimated that a temperature difference of 5°C would lead to a breeze of 3 m/sec at the edge of town but it is very difficult to observe this flow. Actual observations in town generally show smaller values but the effects of channeling and induced turbulence make it difficult to obtain consistent values.

In Leicester (England) CHANDLER (1961b) noted pulsating air currents toward the city, which gradually decreased the city–country temperature contrast. In several cities observers had called attention to occasional front-like inflow of cold air from the country into the city during night hours. This effect was first reported by SCHMAUSS (1925).

In an unusual set of observations OKITA (1960) has been able to document the convergence of flow into the city during calm clear nights. He took advantage of the occasional simultaneous occurrence of fog and below-freezing temperatures at Asahikawa City (Hokkaido, Japan). Under those circumstances the hoarfrost deposits on branches reflected the wind direction that blew them against these obstacles. One set of his observations is shown in Fig.9.

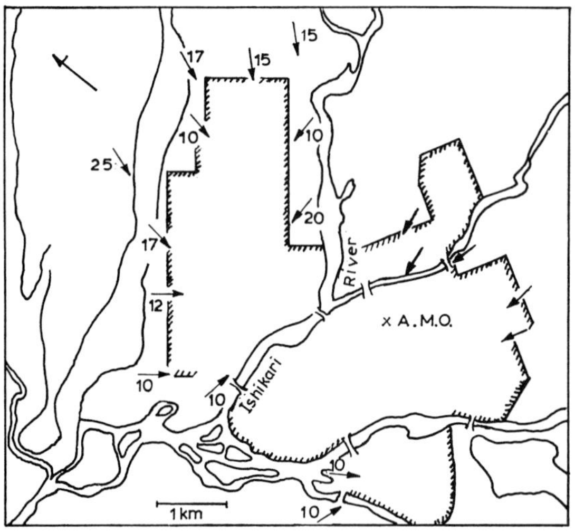

Fig.9. Example of convergence of air flow into city of Asahikawa (after OKITA, 1960), as obtained from hearfrost deposits. Number indicates length of deposit in rime. Hatching shows outline of city.

SCHMIDT and BOER (1963) have made a very detailed wind study around a refinery area in Holland. The results were to some extent affected by an adjacent cold waterway. In the center of largest heat production, an area of about 10^4 m^2 area, upward velocities of 15 cm/sec were observed. In the surroundings a downward stream of 25–30 cm/sec occurred. From regular wind observations some convergence of the surface wind toward the industrial area was noted.

These various observations and the earlier cited data on temperature fields, especially those of CLARKE and MCELROY (1970) permit us to construct a model of nocturnal circulation in an urban area, under weak general wind conditions. This is shown in Fig.10. A convecting rising current over the city, under the boundary layer interface is replaced by invasions of country air. There is evidence that this takes place in discrete pulsations rather than as a steady flow pattern. Frictional forces evidently play a large role in this city–country circulation.

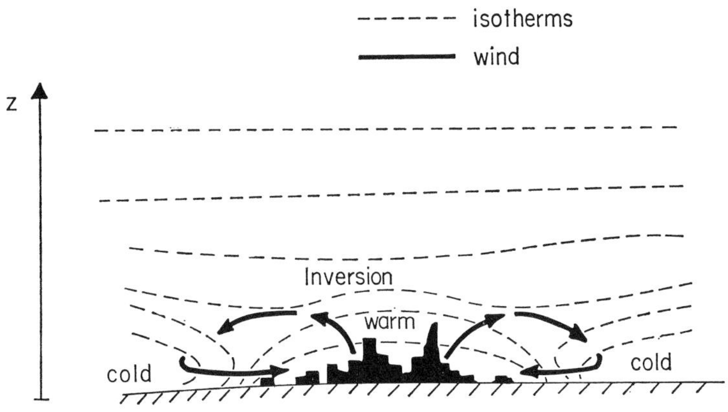

Fig.10. Idealized scheme of noctural city circulation in clear, calm weather. Diagram shows the urban heat island and the radiative ground inversions in the rural areas, causing a "country breeze" with an upper return current.

It is gratifying that the urban effect on the boundary layer can be quite satisfactorily modelled under various circumstances. In that respect solutions in this field are now as adequate as those for land and sea breeze (see FLOHN, 1969, Vol.2, Ch.4 of this series). A prototype of these numerical studies is a model by GUTMAN and TORRANCE (1975). It is two-dimensional, using the equations representing conservation of mass, momentum, energy, and water vapor. These are applied in a vertical profile from 1 m below the surface to 1,400 m into the atmosphere. A turbulent diffusion coefficient, equal for momentum, heat, and water vapor is used. Urban heat additions and roughness are incorporated. With reasonable values for these parameters and for areal land use, the finite difference solutions of the equations yield very realistic values for nocturnal temperature and wind fields by comparisons with observations in Montreal. The conclusion is that the temperature field is indeed controlled by the urban heat additions and the wind field by the increased roughness.

Atmospheric pollution in cities

The main difference of the atmospheric environment in cities and areas undisturbed by human interference is the radical alteration of atmospheric composition. Human activities, especially various combustion processes, bring a host of admixtures into the air. These are gaseous, liquid, and solid. They are often altered by photochemical reactions. Their accumulation or dissipation is wholly dependent upon the meteorological conditions.

Usually the contamination most commonly noted are the suspended solids. In recent years, around 1975, in the United States concentration in towns of less than 50,000 inhabitants averaged annually around 60 $\mu g/m^3$. In cities between 300,000 and 500,000 it was around 100 $\mu g/m^3$ and in metropolitan areas above 1 million inhabitants the average dust load was 120 $\mu g/m^3$. More significant is perhaps the number of days suspended particulates exceeded the limiting levels set for an "acceptable" environment, i.e. 260 $\mu g/m^3$. In Baltimore this was exceeded in 1974 on about 20 days, in Birmingham on 30 days, in Cleveland on 80 days, and in the steel town of Steubenville, Ohio on about

100 days. Most of this material falls out in the neighborhood of its origin. This fall-out has reached in many cities truly extraordinary proportions. A few values are listed in Table VI.

TABLE VI

AVERAGE MONTHLY DEPOSIT OF SOLID MATERIAL IN SELECTED CITIES ($tons/km^2$)

London, England	39	Pittsburgh, U.S.A.	20
Birmingham, England	32	Toronto, Canada	15
New York, U.S.A.	30	Los Angeles, U.S.A.	14
Praha, Czechoslovakia	27	Budapest, Hungary	14
Chicago, U.S.A.	26	Halle, Germany	13
Detroit, U.S.A.	25		

Much of the solid, heavy dust is a nuisance and more of a symptom of pollution rather than a health menace although dusts such as lead and asbestos can be dangerous. However, all suspensions do have a measurable effect on the radiation conditions, which will be dealt with later.

Usually the smaller dust particles of $\frac{1}{2}$ to 5 microns remain suspended and number concentrations of 10 to 25 per cm^3 in cities are not uncommon. At the outskirts only about 1 dust particle per cm^3 is found. Still smaller particles are often referred to as condensation nuclei. In the countryside about $10{,}000/cm^3$ are the average. In towns of less than 100,000 inhabitants the average concentration is $3\frac{1}{2}$ times higher, and in larger cities 15 times the amount in the country is found. Occasional concentrations of several millions condensation nuclei per cm^3 are found. Many of these particles are hygroscopic, some lead to droplet formation already at 70% relative humidity and fog formation takes place whenever the relative humidities exceed 95%. The optical effects of all these suspensions are profound.

A great deal of attention has been devoted to the gaseous contaminants. They include usually irritant and occasionally dangerous constituents of city air. A few ranges of concentrations are noted in Table VII.

TABLE VII

CONCENTRATIONS OF SOME AIR POLLUTANTS IN CITY ATMOSPHERES
(in parts per million by volume)

Carbon dioxide	300–1,000
Carbon monoxide	1–200
Sulfur dioxide	0.01–3
Oxides of nitrogen	0.01–1
Aldehydes	0.01–1
Oxidants, incl. ozone	0.00–0.8
Chlorides	0.00–0.3
Ammonia	0.00–0.21

Some of these, such as sulfur dioxide, have been blamed for the increased mortality caused during air pollution catastrophes. These have been associated with stagnant anticyclones accompanied by little wind motion and low-level inversions. Under these conditions the pollutants accumulate and can cause acute respiratory insults. The cases in the Belgian Meuse valley in 1930, in Donora (Pennsylvania) in 1948, and in London (England) in 1952 brought about many fatalities and have been extensively studied by public health authorities. But there can be no doubt that the subacute concentrations claim daily victims, if not in form of fatalities, but at least through eye irritation, bronchitis, emphysema, and asthma. Even cyclic hydrocarbons of the type suspected as carcinogenic substances, such as benzo(a)pyrene, have been found in small concentrations in many cities. The air pollution problem is one of the most crucial problems in 20th century urbanism.

The magnitude of the problem can be gauged by looking at a single phase only: motor vehicles powered by gasoline—driven internal combustion engines. Measurements made in the Los Angeles (California) area indicate that vehicular traffic produces daily about 1,000 tons of hydrocarbons, 300 tons of nitrogen oxides, 8,000 tons of carbon monixide. For the carbon monoxide this amounts to 6.5 million cubic meters of carbon monoxide or enough to cause a concentration of 30 parts per million in a layer 120 m deep over the metropolitan area of 1,760 km^2. It is generally assumed that an 8-h exposure to such a concentration would lead to adverse health effects.

One of the exhaust gases is nitrogen dioxide. It has obnoxious qualities of its own but its main nuisance feature is the fact that it gives off an oxygen atom under the influence of ultraviolet solar radiation which combines in a complex sequence of reactions with the oxygen molecules of the air to produce ozone, a highly toxic substance. Under clear skies and stagnant air conditions substantial concentrations can occur. In Table VIII some average concentrations of important pollutants are listed (U.S. Government sources) for a few typical cities in the 1970's.

TABLE VIII

AVERAGE CONCENTRATION OF IMPORTANT GASEOUS ADMIXTURES IN THE CITY AIR (parts per million by volume)

City	SO_2	CO	NO_2	HC	Ox
Baltimore	0.074	7	0.26	2.2	0.033
Cincinnati	0.030	4	0.23	2.8	0.028
Los Angeles	0.08	6	0.10	3.1	0.15

Ox = Oxidant HC = hydrocarbons

The absolute figures vary considerably from year to year because of weather conditions. Abatement measures also indicate slow reductions in the averages but the ranges of these contaminants vary widely: one order of magnitude less than the average on "clear" days, and one order of magnitude higher on days with stagnation gives an approximate picture. The seasonal variations are quite large. In most cities winter values are from 2 to 5 times higher than those measured in summer. But the most important, single variable is the wind speed. From data collected by NOACK (1963) at Halle (Germany) in a highly industrialized region on SO_2 concentrations we can derive Table IX.

This shows that the mean concentıation is approximately inversely proportional to the cube root of the wind speed. (It has to be remembered here that wind is only one factor influencing concentration and that mixing depth, not given by Noack, is also important, although these two factors are obviously correlated.) In areally less contaminated regions

TABLE IX

WINTER VALUES OF SO_2 AT HALLE
(After NOACK, 1963)

Wind speeds	0–1.9	2–3.9	4–5.9	6–7.9	8–9.9 (m/sec)
Mean concentration	0.55	0.39	0.30	0.26	0.25 (ppm)

the wind has a somewhat larger effect, but usually distance from polluting sources has the greatest influence.

It is interesting to note here that a very simple model is generally adequate to describe the concentration of pollutants χ of an area source after GIFFORD (1972) by:

$$\chi = \frac{c\,Q}{\bar{u}}$$

where Q is the area source strength and c a constant (e.g. for small particulates: $c \approx 225$), and $\bar{u}$ is mean wind speed.

A notable circumstance is that nearly all contaminants vary in the same way: small particles as well as the various gases. Cross-sections through cities show that on an average the mean concentration decreases about exponentially from the core of the polluted area toward the rural environs. In first approximation the change over a city area can be expressed in two equations:

$$n = 100\ \mathrm{e}^{-0.35\,d}$$
$$n' = 100\ \mathrm{e}^{-0.75\,d'}$$

Where n and n' are the concentrations, in per cent of the maximum concentration in the center of pollution, downwind and upwind, respectively; and d and d' are the distances downwind and upwind in kilometers. These relations apply to mean conditions only. In individual weather situations very wide variations are encountered.

A pertinent study on urban plumes of contaminants has been made for the Cologne region by BAND (1969). He shows that all elements, including pollutants, show a measurable effect downwind. This leeward distortion of various isopleths for area sources of pollution are not at all unlike the well-known concentration patterns from single smoke-stacks.

It has been known for many years that the concentration of air-borne microbes (bacteria, etc.) varies about in the same sense as the chemical pollutants. But it remains unexplored why this is the case. Whether they attach themselves to particles or whether their production is caused by or related to the masses of humans in cities is still a puzzle. In Vienna, for example, the annual average in the center is 800 microbes/m^3, at the edge of town only 200/m^3 (STEINHAUSER et al., 1959). After precipitation there are 40% less than in fine weather—a typical wash-out phenomenon that is also observed for dust and condensation nuclei. Lower temperatures brought about reduction in the number so that on the whole winter and summer are the seasons with lower values than spring and autumn in that particular environment.

The pollution of cities has also marked effects on certain types of vegetation. One of the peculiarities of town centers is the fact that lichens and other epiphytes are often completely missing (MÄGDEFRAU, 1960). This has often been referred to as the "lichen desert". Because their metabolism requires saturation for assimilation it has been claimed that the reduced relative humidity in cities is responsible. Present opinion, however, leans toward the explanation that air pollution is responsible. Some species are highly drought-resistant but most are sensitive to SO_2 and other pollutants. In several cities it has been possible to demonstrate that the area free of lichens is expanding with the growth of the city. For example, in Munich (Germany), which had 350,000 inhabitants in 1890, the "lichen desert" covered 8 km^2; in 1956, when the city had grown to 1 million people, the lichen-free area covered 58 km^2.

Radiation, illumination, visibility

Radiation conditions in cities are complicated by the variously oriented receiving surfaces, such as walls and roofs of houses and buildings. The direction of street and the density of dwellings is, of course, a determinant factor. The absorption of energy in the air, the spectral composition, and the partition of energy between direct and diffuse radiation is profoundly affected by air pollution.

The geometrical relations for insolation have been explored both by climatologists and architects. A number of nomographic devices have been designed which will permit calculation of sunrise and sunset times for various latitudes and solar declinations. These have then to be further interpreted for orientation of houses and streets and their height and width (see: LIST, 1951; OLGYAY, 1963). KAEMPFERT (1949, 1951) made a number of valuable contributions to this subject matter, stimulated by the conditions that are found in old European towns. He introduced the useful parameter "narrowness" which is the ratio of street width to house height. He gives an instructive example for a street in Trier (Germany) where the narrowness is 5 m/17.5 m = 3.5. Running from north-northwest to south-southwest, this street gets on the longest day (21 June) sunshine from 10h10 to 12h20 or 2 h and 10 min. In midwinter the sunrise and sunset value at street level are 10h10 to 10h40, or a total possible sunshine duration of 30 min. A street running in the same direction, at the same latitude (50°N) but with the narrowness value of 1, i.e. same street width as house height, the corresponding values on 21 June were: 08h20 to 15h10 (duration 6 h, 50 min); on 21 December 09h40 to 11h40 (duration 2 h). This, as stated, applied to street level. Windows on higher floors would receive more sunshine.

DIRMHIRN and SAUBERER (1959) performed similar calculations for Vienna (lat. 48°) and for narrowness 1 gave the values listed in Table X for selected days and street directions. These authors also give nomographs showing sunrise and sunset for sloping surfaces, such as roofs, permitting derivation of the possible insolation.

TABLE X

POSSIBLE SUNSHINE DURATION (h), AT STREET LEVEL FOR STREET WIDTH EQUAL TO HOUSE HEIGHT

Street direction	15 June	15 Mar./Sept.	15 Dec.
N–S	6:25	4:25	2:25
NE–SW, NW–SE	8:10	4:40	1:20
W–E	12:10	0.25	0

These examples will serve as illustrations what a wide variety of radiative microclimates are created in a city. Of course, the possible sunshine is only an upper limiting factor. Cloudiness will naturally control the actual amounts received.

Only a very few studies have been undertaken to measure the actual radiation on walls and slanting roofs. HAND (1946) measured at the Blue Hill Observatory near Boston (42°N) the actual radiation sums received on an unshaded surface facing south. The main purpose of his work was to obtain data on energy available for supplemental heating. His values, covering one heating season only, are shown in Table XI.

TABLE XI

ACTUAL RADIATIVE ENERGY (SUN + SKY) RECEIVED ON VERTICAL SURFACE FACING S, NEAR BOSTON (U.S.A.), HEATING SEASON

Month	Sep.	Oct.	Nov.	Dec.	Jan.	Feb.	Mar.	Apr.	1/2 May	Season
k cal./cm^2	8.5	9.7	6.5	8.5	9.1	9.6	9.6	8.0	3.5	73.0

This table shows two things: the relatively large energy received even in midwinter on a southern wall or window (in the Northern Hemisphere), and the effect of cloudiness, notably evident in the low November value.

Unfortunately, the natural radiation even on cloudless days is heavily attenuated, as a result of interception by polluting particles. This affects particularly the short wave lengths. As a result the ultraviolet radiation in cities is drastically reduced but other wave lengths are also affected. Measurements in England have yielded some quantitative values for the lowest 330 m (ROACH, 1961). These are shown in Table XII. Calculations from these measurements permit the inference that under extreme conditions temperatures of the polluted layer may rise as much as 5°C/day.

TABLE XII

COMPONENTS OF ATTENUATION IN LOWEST 330 M OF POLLUTED ATMOSPHERE (%)
(After ROACH, 1961)

	Visible	Infrared	All wave lengths
Absorption	5	35	20
Back-scattering	5	5	5
Forward-scattering	90	60	75

Most evident is the reduction of illumination in cities caused by suspended pollutants. STEINHAUSER et al. (1955) report for Vienna 10% less in summer and in winter 18% less compared to open country. On cloudless days these same authors (1959) were even able to show the shift of the dust calotte over Vienna with different wind directions. On a day with a weak north wind the center was displaced to the south, with a northwest wind to the southeast. In London the reduction is on an average 17% in summer and 50% in winter. GILDERSLEEVES (1962) reported on a case there on 1 December 1961 under a heavy overcast at 550 m when illumination because of the smoke accumulations under the cloud layer dropped in midday to 2 klx.

MORIK (1963) cites a situation in Budapest when nightlike darkness prevailed between 10 and 11 o'clock on a March day. Visibility dropped to 80 m although it was 2,000 m at the outlying airport. Particulates were 5.4 mg/m^3 and SO_2 was 4 mg/m^3.

FEDEROV (1958a,b) measured the illumination in the factory districts of Zaporozhe (Ukraine, U.S.S.R.) in the Dniepe valley. The reduction range, compared to the suburbs, was between 5 and 25%. During midday in May and June it was close to the lowest values and reached its maximum in December. The average annual illumination loss was 13%.

LINKE (1940) found that the turbidity factor, defined by him as the actual extinction compared to a pure Rayleigh atmosphere, in Frankfurt a.M. (Germany) reached average values of 9.

The attenuation of radiation by particles has been used by MCCORMICK and BAULCH (1962) to obtain turbidity values and estimates of particle numbers in city atmospheres. These authors used a Volz sun photometer which measures the illumination intensity for the wave length 500 Å. With measurements from a helicopter they found that in Cincinnati (Ohio, U.S.A.) the pollution calotte, in autumn and winter, is concentrated in the lowest 400 m during the morning hours. In the afternoon a layer of air up to 600 m thick is considerably cleaner. This is the logical consequence of higher daytime wind speeds and more convection than in the night and early morning hours.

Under light winds and anticyclonic conditions they find that extinction (B) at height z (in meters) follows a logarithmic law:

$$B_z = B_0 \, e^{-0.00346z}$$

where B_0 is the extinction at the surface. The particle mass m concentration in a layer z meters deep which was found to depend on the extinction, as follows:

mass concentration at z height $m(z) = 960 \, B_0 \, e^{-0.00346z} \mu g/m^3$
mass concentration at ground level $m(0) = 969 \, B_0 \, \mu g/m^3$

One of the important consequences of the extinction by polluting solids is the reduction in the hours of sunshine over cities. This effect is augmented by a general tendency for increased cloudiness over metropolitan areas.

SHELLARD (1959) has given us a very impressive example, again from that archetype of cities: London. His summary of bright sunshine for a 30-year interval appears in Table XIII.

TABLE XIII

BRIGHT SUNSHINE DURATION, LONDON, 1921–50
(After SHELLARD, 1959)

Environment	Average hours per day					Minutes lost per day				
	Mar.	June	Sep.	Dec.	Year	Mar.	June	Sep.	Dec.	Year
Central districts	3.0	6.8	4.2	0.7	3.6	68	28	40	52	44
Inner suburbs	3.5	6.9	4.6	1.1	4.0	34	19	14	28	23
Outer suburbs	3.7	7.0	4.7	1.2	4.1	22	10	10	23	16
Rural surroundings	4.1	7.2	4.9	1.6	4.3	—	—	—	—	—

Nothing could better demonstrate the gloomy change brought about by man's urbanization. The table also shows the usual annual variation with much smaller city–country differences in winter than in summer. The fact that the heating season coincides with high frequency of low-level inversions leads to the heavy smoke pall found over many of the localities in higher latitudes. Fortunately, the Clean Air Act has not only brought considerable improvement in suspended particulates in London, but a concomitant increase in sunshine.

The literature is full of examples of gradual increase, over the years, in the number of cases of low visibility in cities. In most places this parallels the growth and increased industrialisation. However, some of these time series are not entirely reliable because of changes in observational practices. The contrast between city and country is, however, in no doubt whatsoever. From Shellard's study cited above some parallel series of observations of the ten year interval 1947–1956 he obtained the following values for the London area from the center outward: Kingsway 940, Kew 633, London Airport 562, rural southeast England 494 h per year.

In the United States recent air traffic developments have led to the establishment of many new airports at considerable distance from urban influences while the older landing fields stayed in operation. Table XIV gives a good picture of the higher frequency of lower visibilities at the airport closer to the polluting influences of the large city. Three pairs of stations in different climatic zones show the effects of proximity to the center of pollution clearly.

TABLE XIV

VISIBILITY DIFFERENCES AT SIX PAIRED AIRPORTS IN THE VICINITY OF LARGE U.S. CITIES
(in % cases less than 10 km; from hourly observations, 1951–1960)

Airport	J.	F.	M.	A.	M.	J.	J.	A.	S.	O.	N.	D.	Yr.
New York, La Guardia	39	37	29	32	34	37	37	38	32	33	38	40	35
New York, Kennedy	33	29	23	26	27	33	37	34	24	27	31	32	30
Chicago, Midway	54	48	43	39	35	33	34	35	26	36	36	46	39
Chicago, O'Hare	39	36	28	20	22	20	21	25	20	26	24	32	26
Los Angeles	43	37	36	36	32	42	47	49	49	58	47	46	43
Burbank	26	25	24	29	30	46	43	46	41	47	30	27	34

In this set of date again the colder months show the greatest differences. It is reassuring to know that the optical effects of low-level particulate air pollution are rapidly reversible. This was shown in Pittsburgh when railroads changed from coal-fired steam engines to diesel locomotives with notable improvements in visibility. In London since enactment of "clean air" legislation particulates have decreased and both sunshine and visibility have increased.

Humidity, cloudiness, precipitation

On the whole, the relative humidity in cities, at the normal level of meteorological shelters, is somewhat lower than in the country. This can be attributed to the fact that the temperatures are higher and hence the same amount of water vapor in the air is farther from the saturation pressure in the city. In some instances this can lead to fairly large departures of simultaneous values in the city and the country. These occur simultaneously with the large temperature differences. CHANDLER (1962a) cites in one of his London traverses a difference of 26%. Another reason for a different humidity regime in the city and the country (in the layer near the ground where our meteorological shelters are) is the radically different evaporation regime. The natural soil and its plant cover evaporates water into the air, in moist climates almost continuously. Pavements and houses in cities let all rain water run off quickly and drainage systems carry it away rapidly. So the lowest layer gets very little water vapor from this source. On the other hand, fairly sizeable amounts of water vapor are released from combustion and industrial processes. These, however, are released from tall stacks or chimneys at high temperatures. This vapor tends to rise and benefits a layer considerably above the roof tops.

The conditions in the ground zone are again well reflected by two pairs of airport stations for which we reported visibilities in the previous chapter. The information is given in two partitions in Table XV. There percentage frequency of hourly values of relative humidity is given by steps. The second part of the table also indicates these frequencies for the 90 to 100% relative humidity interval for 6 h of the day. (The Los Angeles area was omitted because of the direct ocean influence which overwhelms the city effect.) Both in New York and Chicago the airport nearer the city has higher frequencies of lower steps of relative humidity.

TABLE XV

PERCENTAGE FREQUENCY OF RELATIVE HUMIDITIES, BY STEPS, AND DIURNAL VARIATION OF HIGH VALUES FOR THE WHOLE YEAR

Station	0–29	30–49	50–69	70–79	80–89	90–100%
New York, La Guardia	2	22	37	14	14	11
New York, Kennedy	1	16	33	16	16	18
Chicago, Midway	2	14	34	20	18	10
Chicago, O'Hare	1	11	32	22	20	15
Station	90–100% relat. humidity at:					
	00h	04h	08h	12h	16h	20h
New York, La Guardia	14	18	11	7	7	10
New York, Kennedy	24	30	17	10	10	18
Chicago, Midway	12	18	10	5	5	8
Chicago, O'Hare	21	27	13	7	7	11

The diurnal variation of frequency of high humidities shows that the outlying stations have higher values at all hours of the day. But the difference between the station under city influence and the country station is most pronounced in the night and early morning hours. It is not uncommon to have a fog in the country and a very low stratus at the same time on the airport nearest the city. Evidently the heat supplied by the city can bring about such effects. At the same time it is quite possible to have also fog in the most polluted areas of town. Obviously many, often opposing influences are at work here.

Cloudiness over cities is only a little less complex. Generally, the increased convection over the city brings about earlier condensation than over the countryside. Especially in summer there is a tendency for cumulus formation over the city area. The observations are generally unsatisfactory for obtaining the city–country contrast. Some of this effect can be noted from hourly cloud observations at the two New York City airports, shown for the midday hours of July from 10 years of simultaneous observations in Table XVI.

TABLE XVI

MIDDAY FREQUENCIES OF CLOUDINESS (TENTHS), BY STEPS, FOR THE MONTH OF JULY LA GUARDIA (L.G.) AND KENNEDY (K) AIRPORTS, NEW YORK, 1951–1960

Cloudiness:	0–3		4–7		8–10	
	L.G.	K.	L.G.	K.	L.G.	K.
11h	33	35	25	21	42	44
13h	24	31	35	27	41	42
15h	25	32	33	27	41	41
17h	31	31	26	24	43	45

This gives the percentage frequency for 3 different steps of cloudiness. At 11 o'clock the low cloudiness category is almost equal at the two facilities. At 13h00 there is a notable shift at the airport in the city area from the low cloudiness to the medium cloudiness class. This lasts through the early afternoon hours and by 17h00 the two distributions are again almost identical. The high cloudiness class shows nearly equal values. This shows that the city influence on cloudiness is most distinguishable on bright days.

Many longer series of observations in the larger metropolitan areas point toward long-term trends of decreasing number of clear days and an increase in the number of cloudy days (LANDSBERG, 1951).

The influence of cities on rainfall, too, is by no means easy to evaluate. Logical reasoning suggests an increase. This is predicated on the facts that the aerodynamic roughness of the city increases turbulence, the heating causes ascending currents, the air pollution multiplies the number of hygroscopic condensation nuclei and various processes add water vapor. The number of freezing nuclei also increases with the degree of air pollution. To demonstrate specifically the increase in precipitation is much more difficult.

The roughness effect seems to slow down frontal movements under certain circumstances. The first paper specifically devoted to this theme by BELGER (1940) showed for a small sample of four cases a retardation of 25% of the speed of migrating rain fronts over Berlin, Germany. The city appeared to deform the front and in two of his cases the maximal rainfall in the whole region was recorded over the city. With much better observational material LOOSE and BORNSTEIN (1977) come to very similar conclusions for frontal retardation by friction effects over New York City. In two cases this amounted to as much as 50% over upstream speeds. Interestingly enough these authors found that with a well-developed heat island fronts were retarded in the up-wind half of the city and accelerated in the down-wind half.

There are numerous individual reports and papers of very intense rainfalls over city areas. In many of them it is thoroughly documented that the surrounding countryside received considerably less (see, e.g., KRATZER, 1956; D.M.H., 1964; HULL, 1957; U.S. WEATHER BUREAU, 1964) than the city. It is quite likely that the city furnished the impulse for the release of convective instability, and there is some persuasive evidence of such cases for London by ATKINSON (1970, 1975).

He showed in individual cases as well as for a series of thunderstorms the influence of the city. There thundershowers yield 30% more precipitation than in the surrounding areas. Orographic conditions would have led one to expect the opposite, especially in hills to the northwest. HUFF and CHANGNON (1960) have tried to use the eleven raingauge network in the 26 km² area of the town of Urbana, Illinois, with a 10-year record. Although they find that the amounts are largest to the lee side of the town and that a gauge near the center seems to show heavier falls than the remainder of the record, comparison with a rural network of similar gauge density indicates to them that sampling variation may still be at work. A restudy by CHANGNON (1962), using an additional gauge in the area validates the 10-year rainfall pattern from two records over 30 years. The rural–urban rainfall contrast then turns out to be 762 mm:864 mm or about a 12% increase. Another study by CHANGNON (1961) establishes the urban effect for Chicago at about 7%, as shown in Table XVII.

TABLE XVII

RURAL–URBAN PRECIPITATION CONTRASTS (mm) IN CHICAGO, ILLINOIS, 1945–1956
(After CHANGNON, 1961)

Season	Urban average	Rural average	Difference
Winter	287	254	33
Summer	348	342	6
Transition	236	216	20
Year	871	812	59

HARNACK and LANDSBERG (1975) showed very conclusively that in summer the urban heat island can convectively produce isolated showers in and downwind of the city. These are not anticipated on the basis of the synoptic situation and are strictly an independent urban effect in cases of conditional instability. In Chicago the greatest differences show up in the winter season. In the summer with showers the difference is insignificant. The local effect of the land–lake contrast in the Chicago area is a factor of unknown influence.

The greater effect in winter has been noted in other areas too. A study made for Tulsa, Oklahoma (LANDSBERG, 1956, p.595) showed for that city, which has no disturbing topographic influences, a rainfall 11.5% higher than at the airport, from October through March. In the April–September interval the surplus was only 4.7%. This for a 14-year period is definitely within the "noise" of rainfall variability.

The winter increase is an effect already discussed by KRATZER (1937) in the first edition of his book. It is caused by small amounts of drizzle, rain or snow falling from low stratus clouds. It is generally confined to the urban area and may well be induced through a nucleating effect of polluting aerosols. The existence of an influence of air pollution on city rainfall has been suspected since a study by ASHWORTH (1929) in Rochdale, England. Not only was there a gradual increase of rainfall over a 30-year period but also a notable weekly fluctuation. For Sundays the average rainfall was 9.4 mm less than for the weekdays. This difference, which was not found at neighboring rural stations, was three times the size of the standard error.

The tendency for increased precipitation in industrial areas was confirmed by a study of WIEGEL (1938) for the Ruhr region in Western Germany. A 35-year interval there showed a 5 % increase in rainfall over the industrialized zones compared to the unaffected surroundings. The weekly variation, with a Sunday minimum, could be confirmed for some American stations during the cold season by FREDERICK (1970). An example of his results is shown for the station Schenectady, New York, in a highly industrialized district (Fig.11). A similar effect was observed in Paris, France (DETTWILLER, 1970b).

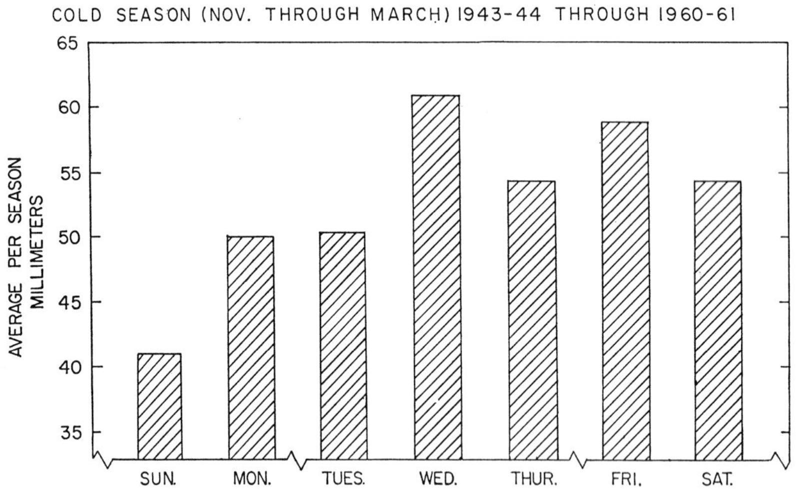

Fig.11. Weekly variation of precipitation at Schenectady, New York, during cold season, 18 years. (After FREDERICK, 1964.)

There is still some controversy on individual cases of pollution-induced precipitation increases (CHANGNON, 1970a,b; HOLZMAN and THOM, 1970), but the evidence makes such an influence at least possible. It is, of course, difficult to disentangle this influence from the effect of the convective lifting by the heat island. The lifting may, indeed be the more effective means for creating additional precipitation. However, it is hard to use it as an explanation for the rainfall increases claimed by HOBBS and RADKE (1970) in the rainfall increases claimed by HOBBS and RADKE (1970) in the state of Washington, where downwind from sources of nuclei, especially pulp mills, cloud masses have been noted 30 km downwind and up to 30 % more precipitation.

A very clear influence on summer precipitation has been established for St. Louis (HUFF and CHANGNON, 1973; HUFF and SCHICKEDANZ, 1974). Although it manifests itself downwind of the city as a 15 % increase over earlier years, the causes are again not easily disentangled. The heat plume is definitely involved but the effect of nucleation remains obscure. The industrial nuclei have been noted to spoil the effects of natural nuclei (BRAHAM and SPYERS-DURAN, 1974).

The fact that cities or large industrial installations, can, under suitable weather conditions, create their own precipitation is becoming better established by individual case studies. One of these was presented by VON KIENLE (1952) for Mannheim-Ludwigshafen, Germany. A ground fog and low stratus had formed over the twin cities at temperatures 4°C below freezing. On two days snow fell from these clouds. It left about 6 mm on the ground. This precipitation was entirely restricted to the city area. Another case was reported by CULKOWSKI (1962) for Oak Ridge, Tennessee. Snow was observed there downwind of the cooling towers of a gaseous diffusion plant. This plant uses about $75 \cdot 10^6$ liters of water per day. Most of it escapes as vapor. On December 22, 1960, a low stratus formed downwind from the towers. No clouds or only a few scattered cumulus were in the area. Temperatures were around 10°C below freezing. According to a personal communication from W. M. Culkowski there were about 2 mm of snow over an area of $6\frac{1}{2}$ km^2. This would amount to about 12% of the water vapor given off by the plant in the 4-h interval during which snow was observed.

Our knowledge of freezing nuclei in city areas is as yet incomplete. Some data suggest that, at least for very low temperatures, air pollutants can act as freezing nuclei (KLINE and BRIER, 1961). There are, however, substances in city air which by their surface and other characteristics can act as nucleants. This has been conclusively shown by SCHAEFER (1966, 1969) and HOGAN (1967).

The total influence of cities on snowfall in areas where this type of precipitation occurs is qualitatively established. It is, of course, a decrease. However, here again things are quite involved. Many cities in the Northern Hemisphere have been under the influence of the warming trend that was noted in the first half of the 20th century. This has resulted in a general decrease of the portion of annual precipitation falling as snow. For example, in Toronto POTTER (1961) reasoned that about half of the 11 cm annual decrease was caused by climatic change and half by city influence. In New York City where the combined influence of general climatic change and city growth is around 2°C, the snowfall has decreased by about 20%, i.e. from around 19 cm in the outgoing 19th century to 15 cm in the decades 1941–1960. In Chicago, for the 16-season interval 1928/29 to 1943/44 the downtown snowfall was 20 cm less than at either the airport or the less urban university station. However, wind influences on the catch cannot be quite ruled out because the city station was on top of a building.

It is unfortunate that snow statistics as to amount and depth on the ground are so unreliable. In cities subject to snowfall this is a meteorological factor of great practical importance because of its hindering effects on traffic and the costs of removal. FISCHER (1964a,b) has recently introduced a classification for the status of surfaces in cities that has considerable merit for assessing traffic hazards. He distinguished six categories: dry, moist, wet, slush, snow, ice. For Vienna, as an example, in ten winters he found street conditions on one of the main thoroughfares 25% of the cases dry, 41% moist and wet, 26% snow and slush, and 8% ice. This type of information is almost nowhere collected on a systematic basis.

Conclusion

Modelling of city influences on climate has made considerable progress. But many of the complex feedback mechanisms between undisturbed areas and those affected by human intervention remain to be explored.

For tropical cities there have been advances in documenting the heat island effect also. This has been done for Nairobi (NAKAMURA, 1967), Durban (PRESTON-WHYTE, 1970), and Mexico City (JAUREGUI, 1973). But there remain open questions. In particular, no detailed studies of urban influences on rainfall, if any, in the tropics have been made.

Another deficiency is the absence of comprehensive studies on the water balance in urban areas. We do not even have any emission inventories of artificial water vapor release.

What we do know is summarized, in relative terms, in Table XVIII.

TABLE XVIII

LOCAL CLIMATIC ALTERATIONS PRODUCED BY CITIES

Elements	Compared to rural environs
Contaminants:	
condensation nuclei	10–100 times more
particulates (dust)	10–50 times more
gaseous admixtures	5–25 times more
Radiation:	
total on horizontal surface	10–20% less
ultraviolet, low sun	30% less
ultraviolet, high sun	5% less
sunshine duration	5–15% less
Visibility (<10 km)	5–15% more
Cloudiness:	
couds	5–10% more
fog, winter	100% more
fog, summer	20–30% more
Precipitation:	
amounts	5–10% more
days with <5 mm	10% more
snowfall	5–10% less
Temperature:	
annual mean	0.5–1.0°C more
winter minima (average)	1–2°C more
heating degree days	10% less
Frost-free season	10% more
Surface relative humidity:	
annual mean	6% less
winter	2% less
summer	8% less
Wind speed:	
annual mean	20–30% less
extreme gusts	10–20% less
calms	5–20% more

References

ACKERMAN, B., 1974a. Wind fields over the St. Louis metropolitan area. *J. Air Pollut. Control Assoc.*, 24: 232–236.

ACKERMAN, B., 1974b. Wind fields over St. Louis in undisturbed weather. *Bull. Am. Meteorol. Soc.*, 55: 93–95.

ANGELL, J. K., HOECKER, W. H., DICKSON, C. R. and PACK, D. H., 1973. Urban influence on strong day time air flow as determined from Tetroon flights. *J. Appl. Meteorol.*, 12: 924–936.

ARIEL, N. Z. and KLINCHNIKOVA, L. A., 1960. Wind over a city (Veter v usloviiakh goroda). *Glav. Geofiz. Obs. Leningrad Trudy*, 94: 29–32. (U.S. Weather Bureau, translation by I. I. Donehoo, 1961.)

ASHWORTH, J. R., 1929. The influence of smoke and hot gases from factory chimneys on rainfall. *Q.J.R. Meteorol. Soc.*, 55: 341–350.

ATKINSON, B. W., 1970. A further examination of the urban maximum of thunder rainfall in London, 1951–60. *Trans. Pap. Inst. Brit. Geograph., Publ.*, 48: 97–119.

ATKINSON, B. W., 1975. Urban effects on precipitation: An investigation of London's influence on the severe storm in August 1975. *Dept. Geograph., Queen Mary Coll., Univ. London, Occ. Pap.*, 8, 31 pp.

BAND, G., 1969. Der Einfluss der Siedlung auf das Freilandklima. *Mitt. Inst. Geophys. Meteorol. Univ. Köln*, 9, 139 pp.

BELGER, W., 1940. *Der Groszstadteinfluss auf nichtstationäre Regenfronten und ein Beitrag zur Bildung lokaler Wärmegewitter.* Berlin, Dissertation, Orthen, Köln, 69 pp.

BERG, H., 1947. *Einführung in die Bioklimatologie.* Bouvier, Bonn, 131 pp.

BRAHAM, R. R. and SPYERS-DURAN, P., 1974. Ice nucleus measurements in an urban area. *J. Appl. Meteorol.*, 13: 940–945.

BROOKS, C. E. P., 1952. Selected annotated bibliography on urban climates. *Meteorol. Abstr. Bibliogr.*, 3: 734–773.

CERMAK, J. E., 1976. Aerodynamics of buildings. *Ann. Rev. Fluid Mech.*, 8: 75–106.

CHANDLER, T. J., 1960. Wind as a factor of urban temperature—a survey in North-East London. *Weather*, 15: 204–213.

CHANDLER, T. J., 1961a. The changing form of London's heat-island. *Geography*, 46: 295–307.

CHANDLER, T. J., 1961b. Surface breeze effects of Leicester's heat-island. *Midland Geograph.*, 15: 32–38.

CHANDLER, T. J., 1962a. London's urban climate. *Geograph. J.*, 128: 279–298.

CHANDLER, T. J., 1962b. Temperature and humidity traverse across London. *Weather*, 17: 235–242.

CHANDLER, T. J., 1964. City growth and urban climates. *Weather*, 19: 170–171.

CHANDLER, T. J., 1965. *The Climate of London.* Hutchinson, London, 292 pp.

CHANDLER, T. J., 1969. *Selected Bibliography on Urban Climatology.* W. M. O. Geneva, 289 pp.

CHANGNON, S. A., JR., 1961. Precipitation contrasts between the Chicago urban area and an offshore station in southern Lake Michigan. *Bull. Am. Meteorol. Soc.*, 42: 1–20.

CHANGNON, S. A., JR., 1962. A climatological evaluation of precipitation patterns over an urban area. *Symp. Air over Cities, U.S. Pub. Health Serv. SEC Tech. Rep.*, A62–5: 37–66.

CHANGNON, S. A., 1970a. Recent studies of urban effects on precipitation in the United States. *W. M. O., Tech. Note*, 108: 325–341.

CHANGNON, S. A., 1970b. Reply (to Holzman and Thom). *Bull. Am. Meteorol. Soc.*, 51: 337.

CLARKE, T. F. and MCELROY, J. L., 1970. Experimental studies of the nocturnal urban boundary layer. *W. M. O., Tech. Note*, 108: 108–112.

CULKOWSKI, W. M., 1962. An anomalous snow at Oak Ridge, Tennessee. *Mon. Weather Rev.*, 90: 194–196.

DAVITAYA, F. F., 1958. Principles and methods of agricultural evaluation of climates. In: F. F. DAVITAYA and M. SKULIK (Editors), *Voprosy Agrometeorologii.* USSR. Glavnoe Upravlenie Gidrometeorologicheskoi Sluzhby, Moskva, pp.62–70 (in Russian).

DEMARRAIS, G. A., 1961. Vertical temperature difference over an urban area. *Bull. Am. Meteorol. Soc.*, 42: 548–554.

DETTWILLER, J., 1970a. Deep Soil Temperature Trends and Urban Effects at Paris. *J. Appl. Meteorol.*, 9: 178–180.

DETTWILLER, J., 1970b. Incidence possible de l'activité industrielle sur les précipitations à Paris. *W. M. O., Tech. Note*, 108: 361–362.

DIRMHIRN, I. and SAUBERER, F., 1959. Das Strassenklima von Wien. In: F. STEINHAUSER, O. ECKEL and F. SAUBERER (Editors), *Klima und Bioklima von Wien, Wetter und Leben*, Sonderhefte, III, pp.122–135.

D.M.H., 1964. Just an odd thunderstorm. *Weather*, 19: 220–221.

DUCKWORTH, F. S. and SANDBERG, J. S., 1954. The effect of cities upon horizontal and vertical temperature gradient. *Bull. Am. Meteorol. Soc.*, 35: 198–207.

EMONDS, H., 1954. Das Bonner Stadtklima. *Arb. Rhein. Landeskd.*, 7, 64 pp.

EVELYN, J., 1661. *Fumifugium: or the Inconvenience of the Aer, and Smoke of London Dissipated.* Oxford (Reprinted, Manchester, National Smoke Abatement Society, 1933), 43 pp.

FEDOROV, M. M., 1958a. The effect of smoke on the illumination of a city. *Gigiena i Sanitarya*, 23: 14–18 (in Russian).

FEDOROV, M. M., 1958b. The distribution regarding size of dust and smoke particles in the air over an industrial city. *Dokl. Akad. Nauk, SSSR*, 118, No. 4 (in Russian).

FISCHER, P. L., 1964a. Der winterliche Strassenzustand 1962/63 in Wien, I. Schottengasse und Helferstorferstrasse. *Wetter Leben*, 15: 165–169.

FISCHER, P. L., 1964b. Eine vereinfachte Klassifikation der Strassenzustände. *Wetter Leben*, 15: 162–164.

FLOHN, H., 1969. Local wind systems. In: H. FLOHN (Editor), *World Survey of Climatology*, Vol.2, *General Climatology*. Elsevier, Amsterdam, pp.139–171.

FREDERICK, R. H., 1961. A study of the effect of tree leaves on wind movements. *Mon. Weather Rev.*, 89: 39–44.

FREDERICK, R. H., 1964. On the representativeness of surface wind observations using data from Nashville, Tennessee. *Int. J. Air Water Pollut.*, 8: 11–19.

FREDERICK, R. H., 1970. Preliminary results of a study of precipitation by day-of-the-week over the eastern United States. *Bull. Am. Meteorol. Soc.*, 51: 100.

FUKUI, E., 1970. The recent rise of temperature in Japan. In: *Japanese Progress in Climatology*, 1970. Laboratory of Climatology, Tokyo Univ. of Education, Tokyo, pp.46–65.

GIFFORD, F. A., 1972. Applications of a simple urban pollution model. *Conference on Urban Environment and Second Conference on Biometeorology*. Am. Meteorol. Soc., Philadelphia, pp.62–63.

GILDERSLEEVES, P. B., 1962. A contribution to the problem of darkness over London. *Meteorol. Mag.*, 91: 365–369.

GRAHAM, I. R., 1968. An analysis of turbulence statistics at Fort Wayne, Indiana. *J. Appl. Meteorol.*, 7: 90–93.

GRASNICK, K.-H., 1961. Der Einfluss der Bodenart auf die nächtliche Temperaturänderung in der bodennahen Luftschicht. *Z. Meteorol.*, 15: 27–36.

GUTMAN, D. P. and TORRANCE, K. E., 1975. Response of the Urban Boundary Layer to Heat Addition and Surface Roughness. *Boundary Layer Meteorol.*, 9: 217–233.

HAND, I., 1946. Solar energy received on a vertical surface facing south. *Bull. Am. Meteorol. Soc.*, 27: 416.

HARNACK, R. P. and LANDSBERG, H. E., 1975. Selected cases of convective precipitation caused by the metropolitan area of Washington, D.C. *J. Appl. Meteorol.*, 14, 1050–1060.

HOBBS, P. V. and RADKE, L. F., 1970. Cloud condensation nuclei from industrial sources and their apparent influence on precipitation. *J. Atm. Sci.*, 27: 81–89.

HOGAN, A. W., 1967. Ice nuclei from direct reaction of iodine vapors with leaded gasoline. *Science*, 158: 800.

HOLZMAN, B. G. and THOM, H. C. S., 1970. The La Porte precipitation anomaly. *Bull. Am. Meteorol. Soc.*, 51: 335–336.

HOWARD, L., 1818, 1820. *The Climate of London Deduced from Meteorological Observations Made in the Metropolis and at Various Places around it.* W. Phillips, London, 2 vols.

HUFF, F. A. and CHANGNON, S. A., 1960. Distribution of excessive rainfall amounts over an urban area. *J. Geophys. Res.*, 65: 3759–3765.

HUFF, F. A. and CHANGNON, S. A. Jr., 1973. Precipitation modification by major urban areas. *Bull. Am. Meteorol. Soc.*, 54, 1220–1232.

HUFF, F. A. and SCHICKEDANZ, P. T., 1974. Metromex: Rainfall analyses. *Bull. Am. Meteorol. Soc.*, 55: 90–92.

HULL, B. B., 1957. Once-in-a-hundred-year rainstorm, Washington D.C., 4 September 1939. *Weatherwise*, 10: 128–131.

JASKE, R. T., FLETCHER, J. F. and WISE, K. R., 1970. A national estimate of public and industrial heat rejection requirements in decades through the year 2000 A.D. *Am. Inst. Chem. Engrs., 67th Nat. Mtg. Atlanta, Ga., Paper No.* 37.

JAUREGUI, E., 1973. Urban climate of Mexico City. *Erdkunde*, 27: 298–307.

JENSEN, M. and FRANCK, N., 1963. *Model-Scale Tests in Turbulent Wind*, I. The Danish Technical Press, Copenhagen, 96 pp.

KAEMPFERT, W., 1949. Zur Frage der Besonnung enger Strassen. *Meteorol. Rundsch.*, 2: 222–227.

KAEMPFERT, W., 1951. Ein Phasendiagramm der Besonnung. *Meteorol. Rundsch.*, 4: 141–144.

KLINE, D. B., and BRIER, G. W., 1961. Some experiments on the measurements of natural ice nuclei. *Mon. Weather Rev.*, 89: 263–272.

KONČEK, M., 1962. Teplotné Pmery Bratíslavy Podlá Súčasnych Pozorovani Nilkol'kych Staníc. *Meteorol. Zpravy*, 15: 55–60.

KRATZER, A., 1937. *Das Stadtklima*. Vieweg, Braunschweig, 143 pp.

KRATZER, A., 1956. *Das Stadtklima* (2nd ed.). Vieweg, Braunschweig, 184 pp.

KREMSER, V., 1909. Ergebnisse vieljähriger Windregistrierungen in Berlin. *Meteorol. Z.*, 26: 259–265.

LANDSBERG, H. E., 1950. Comfortable living depends on microclimate. *Weatherwise*, 3: 7–10.

LANDSBERG, H. E., 1951. Some recent climatic changes in Washington, D.C. *Arch. Meteorol. (B)*, 3: 65–71.

LANDSBERG, H. E., 1956. The climate of towns. In: *Man's Role in Changing the Face of the Earth.* Univ. of Chicago Press, Chicago, pp.584–606.

LANDSBERG, H. E., 1962. City air—better or worse? *Symp. Air over Cities, U.S. Publ. Health Serv. SEC Tech. Rep.*, A62-5: 1–22.

LANDSBERG, H. E., 1970. Micrometeorological temperature differentiation through urbanization. *W. M. O., Tech. Note*, 108.

LANDSBERG, H. E., 1974. Inadvertent atmospheric modification through urbanization. In: W. H. HESS (Editor), *Weather and Climate Modification.* Wiley, New York, N.Y., pp.726–763.

LANDSBERG, H. E., 1975. Atmospheric changes in a growing community. *Inst. Fluid Dyn. Appl. Math., Univ. of Maryland, Tech. Note*, BN 823, 54 pp.

LINKE, F., 1940. Das Klima der Groszstadt. In: B. DE RUDDER and F. LINKE (Editors), *Biologie der Groszstadt.* Steinkopff, Dresden-Leipzig, pp.75–90.

LIST, R. J., 1951: Smithsonian meteorological tables. *Smiths. Misc. Coll.*, 114: 497–520.

LOOSE, T., and BORNSTEIN, R. D., 1977. Observations of mesoscale effects on frontal movement through an urban area. *Mon. Wheather Rev.*, 105: 563–571.

LORENZ, D., 1962. Messungen der Bodenoberflächentemperatur vom Hubschrauber aus. *Ber. Deut. Wetterd.*, 11 (82): 29 pp.

LORENZ, D., 1973. Meteorologische Probleme bei der Stadtplanung. *FBW-Blätter, Baupraxis*, 9/73, 57–62.

MÄGDEFRAU, K., 1960. Flechtenvegetation und Stadtklima. *Naturwiss. Rundsch.*, 13: 210–214.

MAHRINGER, W., 1963. Ein Beitrag zum Klima von Höfen im Wiener Stadtbereich. *Wetter Leben*, 15: 137–146.

MAISEL, T. N., 1971. *Microclimatic Differentiation by Urbanization.* M.S. Thesis, Univ. of Maryland (unpublished).

MATTSON, J. O., 1969. Infrared thermography: a new technique in microclimatic investigations. *Weather*, 24: 106–112.

MCCORMICK, R. A. and BAULCH, D. M., 1962. The variation with height of dust loading over a city as determined from the atmospheric turbidity. *J. Air Poll. Cont. Assoc.*, 12: 492–496.

MITCHELL, J. M. JR., 1953. On the causes of instrumentally observed secular temperature trends. *J. Meteorol.*, 10: 244–261.

MITCHELL, J. M. JR., 1962. The thermal climate of cities. *Symp. Air over Cities, U.S. Publ. Health Serv. SEC Tech. Rep.*, A62-5: 131–143.

MORIK, J., 1963. Probleme der Verhütung von Luftverunreinigungen in Ungarn. *Angew. Meteorol.*, 4: 274–279.

MUNN, R. E., 1970. Airflow in urban areas. *W. M. O., Tech. Note*, 108: 15–43.

NAKAMURA, K. 1967. City temperature of Nairobi. In: *Japanese Process in Climatology*, pp.61–65.

NICHOLAS, F. W., 1974. *Parameterization of the Urban Fabric: A Study of Surface Roughness With Application to Baltimore, Maryland.* Ph.D. Dissertation, Univ. of Maryland (unpublished).

NISHIZAWA, T., 1958. The influence of buildings on urban temperature. *Miscell. Rep. Res. Inst. Nat. Resour.*, 48: 40–47.

NOACK, R., 1963. Untersuchungen über Zusammenhänge zwischen Luftverunreinigung und meteorologischen Faktoren. *Angew. Meteorol.*, 4: 299–303.

NORWINE, J. R., 1973. Heat island properties of an enclosed multilevel shopping center. *Bull. Am. Meteorol. Soc.*, 54: 637–641.

OKE, T. R., 1973. City size and the urban heat island. *Atmosph. Environ.*, 7: 769–779.

OKE, T. R., 1974. Review of urban climatology, 1968–1973. *W.M.O., Geneva, Tech. Note*, 134: 132 pp.

OKE, T. R., 1980. Review of urban climatology, 1973–1976. *W.M.O., Geneva, Tech. Note.*

OKITA, T., 1960. Estimation of direction of air flow from observations of rime ice. *J. Meteorol. Soc. Japan*, 38: 207–209.

OLGYAY, V., 1963. *Design With Climate.* Princeton Univ. Press, 190 pp.

PAPAI, L., 1961. A Kicserélödesi együttható meghatarozása Budapest felett. *Idöjárás*, 65: 113–118.

PARRY, M., 1956. Local temperature variations in the Reading area. *Q. J. R. Meteorol. Soc.*, 32: 45–47.

PERKINS, W. A., 1962. Some effects of city structure on the transport of airborne material in urban areas. *Symp. Air over Cities, U.S. Publ. Health Serv. SEC Tech. Rep.*, A62-5: 197–207.

POOLER, F., 1963. Airflow over a city in terrain of moderate relief. *J. Appl. Meteorol.*, 2: 446–456.

POTTER, J. G., 1961. Changes in seasonal snowfall in cities. *Can. Geograph.*, 5: 37–42.

PRESTON-WHYTE, R. A., 1970. A spatial model of an urban heat island. *J. Appl. Meteorol.*, 9: 571–573.

QUITT, E., 1960. Die Erforschung der Temperaturverhältnisse von Brno und Umgebung. *Wetter Leben,* 12: 311–322.
RENOU, E., 1862. Différences de température entre Paris et Choisy-le-Roi. *Ann. Soc. Metéorol. France,* 10 (2): 103–109.
ROACH, W. T., 1961. Some aircraft observations of fluxes of solar radiation. *Q. J. R. Meteorol. Soc.,* 87: 346–363.
SCHAEFER, V. J., 1966. Ice nuclei from automobile exhaust and iodine vapor. *Science,* 154: 1555–1557.
SCHAEFER, V. J., 1969. The inadvertent modification of the atmosphere by air pollution. *Bull. Am. Meteorol. Soc.,* 50: 199–206.
SCHERHAG, R., 1963. Die grösste Kälteperiode seit 223 Jahren. *Naturwiss. Rundsch.,* 16: 169–174.
SCHMAUSS, A., 1925. Eine Miniaturpolarfront. *Meteorol. Z.,* 42: 196.
SCHMIDT, F. H. and BOER, J. H., 1963. Local circulation around an industrial area. *Ber. Dt. Wetterdienstes,* 12 (9): 28–31.
SCHMIDT, W., 1930. Kleinklimatische Aufnahmen durch Temperaturfahrten. *Meteorol. Z.,* 47: 92–106.
SEKIGUTI, T., FUKUI, E. and YOSHINO, M. (Editors), 1964. Collected papers and abstracts. *Tokyo J. Climatol.,* I (1): 91 pp.
SHELLARD, H. C., 1959. The frequency of fog in the London area compared with that in rural areas of East Anglia and Southeast England. *Meteorol. Mag.,* 88: 321–323.
SHENFELD, L. and SLATER, D. F. A., 1960. The climate of Toronto. *Meteorol. Br. Dep. Transp. Canada, CIR-3352, TEC-327,* 52 pp.
SLADE, D. H., 1963. *Comments on the Measurement of "Surface" Wind.* U.S. Weather Bur., manuscript, 21 pp.
STEINHAUSER, F., ECKEL, O. and SAUBERER, F. (Editors), 1955, 1957, 1959. *Klima und Bioklima von Wien,* Pts. 1–3. *Wetter Leben,* Sonderhefte, 120, 136, 136 pp., resp.
STUMMER, G., 1939. Kleinklimatische Untersuchungen in Frankfurt am Main und seinen Vororten. *Ber. Inst. Meteorol. Geophys., Univ. Frankfurt,* Nr.5.
SUMMERS, P. W., 1964. *An Urban Ventilation Model Applied to Montréal.* Ph. D. Dissertation, McGill University (unpublished).
SUNDBERG, Å., 1951. Climatological studies in Uppsala, with special regard to the temperature conditions in the urban area. *Geographica,* 22: 111 pp.
U.S. PUBLIC HEALTH SERVICE, 1962. *Symposium on Air over Cities.* SEC Tech. Rep. A62–5, Cincinnati, 290 pp.
U.S. WEATHER BUREAU, OFFICE OF HYDROLOGY, 1964. *The Intense Rainstorm of August 19–21, 1963 at Washington, D.C.* U.S. Weather Bur., manuscript, 49 pp.
VILKNER, H., 1961. Die Nachttemperaturen am Erdboden einer Stadt. *Z. Meteorol.,* 15: 141–147.
VON KIENLE, J., 1952. Ein Stadt-gebundener Schneefall in Mannheim. *Meteorol. Rundsch.,* 5: 132–133.
WIEGEL, H., 1938. Niederschlagsverhältnisse und Luftverunreinigung des Rheinisch-Westfälischen Industriegebiets und seiner Umgebung. *Veröff. Meteorol. Inst. Berl.,* 3 (3): 52 pp.
WOOLLUM, C. A., 1964. Notes from a study of the microclimatology of the Washington, D. C. area for the winter and spring seasons. *Weatherwise,* 17: 262–271.
WOOLLUM, C. A., 1970. Washington Metropolitan area precipitation and temperature patterns. *W.M.O., Geneva, Tech. Note,* 108.

Chapter 4

Technical Climatology

R. REIDAT

This survey is concerned with the interrelation between technology and climatology and with applications of climatological information to engineering projects. During the last decades meteorological and particularly climatological support for engineering objectives has been much in demand.

Goal and scope of technical climatology

The interrelation between technology and weather

Many production processes are influenced by the weather, but they, in turn, also play a role in weather modification.

Often a process of manufacture needs environmental conditions requiring exact threshold levels of temperature, humidity, and air quality in order to avoid product defects. The technical effort needed for creating this production environment depends on the climate of the locality of manufacture. Durability and serviceability of projects also depend on the climatic conditions of the locality where they are used. It is essential to know the climatic features of this locality in order to minimize the technical effort for protection of the products from the deteriorating influences of climate.

On the other hand, industry has increasingly shown influences on our climatic environment. This includes the amounts of dust and effluents caused by large concentrations of motor vehicles up to the noteworthy imbalance of the heat and moisture budget as well as the air pollution due to the growth of industry. Unfavorable conditions for the diffusion of the exhaust plumes cause the smog problem as, for example, in Los Angeles. Or they may even result in disastrous episodes as in the Meuse valley in Belgium in 1930 or the Donora valley in Pennsylvania, U.S.A., in 1948 (see section on air pollution, p.381).

Meteorological measurements of turbulence as well as the heat and moisture budget of the atmospheric boundary layer furnish the data necessary for calculating appropriate stack heights. The sites of new industrial installations should be selected according to climate conditions and the climatic impact studies should deal with spreading of effluents and with the potential weather modification which might take place after a plant has been put into operation. This would include heat and moisture changes, deflection of air currents, and, interference with visual range. A plant site poorly located from the climatic point of view may lead to a permanent deterioration of the environment or require supplementary expensive remedies, but is rarely completely cured. This is especially true for cement and fish-meal factories, chemical plants and smelters.

Engineering demands on climatologists

An effective climatological support for technical projects is based on knowledge of the influence which the weather has on technical manufacture and manufactured products. An operations analysis has to reveal the pertinent weather factors.

Laboratory tests can reveal influences of individual weather elements or their combined effects on materials and the operation of equipment. Thus properties of materials and the operating conditions of equipment are tested as functions of the ambient temperature and moisture conditions. This determines maximum permissible load, the necessary protection from weather for operational reliability.

These tests are run either in the individual industrial plants, or material-testing laboratories, sponsored by governments or industries. For example, the European railway systems have established a joint laboratory in the Vienna arsenal for carrying out climatic tests of major units. Two huge climate chambers enable one to test the behavior of locomotives and even subdivisions of entire trains under various climatic conditions as they occur on earth, by simulating wind, temperature, moisture and precipitation.

Laboratory tests are supplemented by field tests, e.g., materials are kept in the open at the sea side, in the high mountains or in industrial areas for long periods of time to measure resistance against the damaging influence of weather. In W. Germany, England, Canada, and South Africa standardized model buildings, made of different materials, are tested for long periods under a variety of weather loads.

The laboratory tests are advantageous because intensity and duration of individual weather elements or their combinations can be programmed. But field tests of materials and equipment are subject to weather conditions which must be observed and documented carefully.

Clearly, in this field close collaboration between meteorologists and engineers is called for. Where possible teams including various professions are jointly working on projects such as plant location, product processes, shipping problems, and design and construction tasks.

Many industries aim at mass production. Universal protection for all types of climate on earth would be prohibitively expensive. Hence, a classification for the use of materials and equipment for protection under a wide variety of temperature and moisture conditions will be helpful. Table I (BURCHARD, 1963) is an example of such a classification which may be useful for electrical, electronic and fine mechanical equipment. It contains for constant load the monthly means of temperature and relative humidity, for short-term load the extreme daily means of temperature and relative humidity and for critical load the absolute annual temperature extremes at a selected locality.

The number of classes necessary for engineering purposes depends on the desired scale and expense for weather protection. For example, the construction of electric motors and motor vehicles requires a less detailed classification than does the design of air conditioning installations, or the weather obstructions to the construction of buildings. After agreements on weather limits by meteorologist and user have been laid down by the engineers, special technoclimatic maps (e.g., Fig.22, p.373) can delineate climatic

TABLE I

CLASSIFICATION FOR USE OF PRODUCTS (EQUIPMENT, INSTALLATIONS, ETC.) TO BE PROTECTED FROM HEAT, COLD AND MOISTURE
(After BURCHARD, 1963)

No.	Usage		Temperature of black body, sunlit, max. (°C)	Environmental conditions (air temperature and rel. humidity): Heat load		Dryness load		Humidity load		Cold load		Temperature of black body, at night, clear, min. (°C)
				Critical (°C)	Constant (°C/%)	Short-term (°C/%)	Constant (°C/%)	Short-term (°C/%)	Constant (°C/%)	Critical (°C)	Constant (°C)	
1	Dry rooms under control and poss. air-conditioned			30	22/50	25/40	22/50	23/80	22/70	12	18	
2	Warm	Inner rooms		35	25/40	31/30	25/40	28/90	23/80	− 2	+ 5	
3	Moderate	Outer rooms		35	25/40	31/30	25/40	28/90	23/80	−25	−12	
4	Environment	In the open air	70									−40
5	Multipurpose	Outer rooms		40	31/30	35/20	31/30	30/95	28/90	−40	−25	
6	Usage	In the open air	80									−55
7		Warm environment:										
	Almost	outer rooms		45	35/20	38/10	35/20	32/98	30/95	− 8	+ 2	
8	Unlimited	All environments:										
	Usage	outer rooms		45	35/20	38/10	35/20	32/98	30/95	−55	−40	
9		All environments: in the open air	90									−70

zones useful for operations, packing and shipping. Technoclimatic diagrams for various localities can also give a quick orientation on potential limiting conditions for weather-affected technologies (see Fig.8, p.344).

Climatic aids for the engineers

Observations made at weather stations, usually for other purposes, must be adapted. For this purpose long meteorological records are generally desirable. Often these records do not directly furnish the answer desired by engineers and need to be interpreted for such purposes by experts.

Some elements, for example, radiation, exchange, and turbulence are observed at a few locations only because of the expense and complexity of the measuring equipment. The representativeness of these measurements is for the local scale only; e.g., the interpolation of radiation measurements must account for deviations of climatic elements such as cloudiness, visibility and temperature caused by topographic peculiarities (see GEIGER, 1969). For turbulence short periods of observations at the locality and investigation may give the necessary information.

The classical climatological averages are generally too restricted in informational value for engineering use. Often complete frequency analyses, probabilities of specified threshold values (THOM, 1966b), and durations of certain weather conditions are needed. They can often be modeled by a particular statistical function (KING, 1971).

Often technical needs may not require comprehensive frequency distributions. Many refer to absolute threshold values or to values related to given return periods or percentiles. Tables II and III give specific examples for Hamburg temperatures. More information can be gained from a graphic presentation of cumulative frequencies in form of a diagram (Fig.1) or of isopleths (Fig.2). These enable one to read either the threshold temperatures related to a specific frequency of occurrence or the frequencies of occurrence of specific temperature intervals.

TABLE II

DAILY MEAN AND EXTREME TEMPERATURES (°C) AT HAMBURG (1891–1930), RELATED TO SPECIFIC RETURN PERIODS OF FREQUENCY

Once in	Maximum	Minimum	Daily mean	Temperature
1 year	30.5°	−13.5°	24.0°	−10.8°
2 years	30.9°	−15.0°	24.8°	−11.7°
10 years	33.0°	−18.0°	27.0°	−15.5°
40 years	34.7°	−22.5°	28.0°	−17.8°

TABLE III

THRESHOLD VALUES OF HOURLY TEMPERATURES (°C) AT HAMBURG (1948–1964), RELATED TO SPECIFIC PERCENTILES OF PROBABILITY

Maximum = 34.9°	Median = 7.9°	Minimum = −22.9°
1% = 24.8°	(50%)	1% = − 9.9°
5% = 20.3°		5% = − 5.3°
10% = 17.9°		10% = − 2.2°
25% = 15.0°		15% = + 2.3°

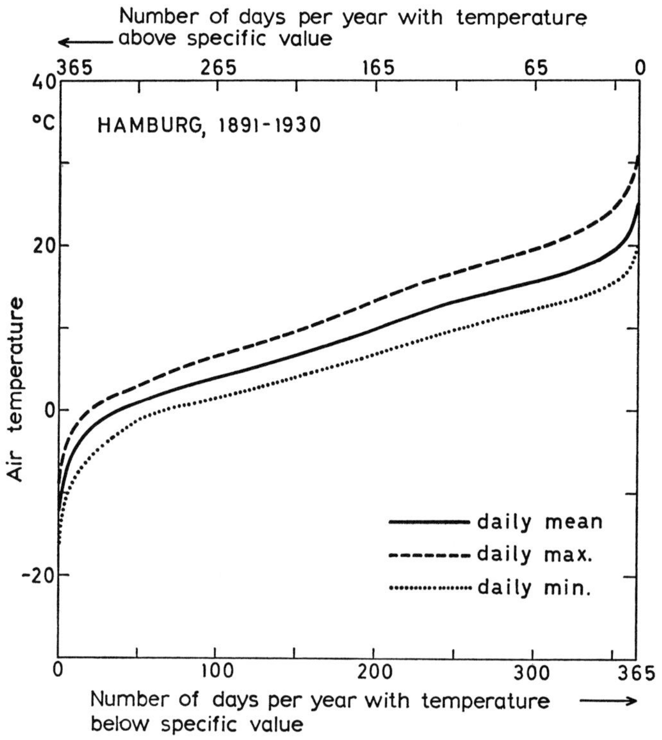

Fig.1. Cumulative frequency curves of air temperature at Hamburg (1891–1930).

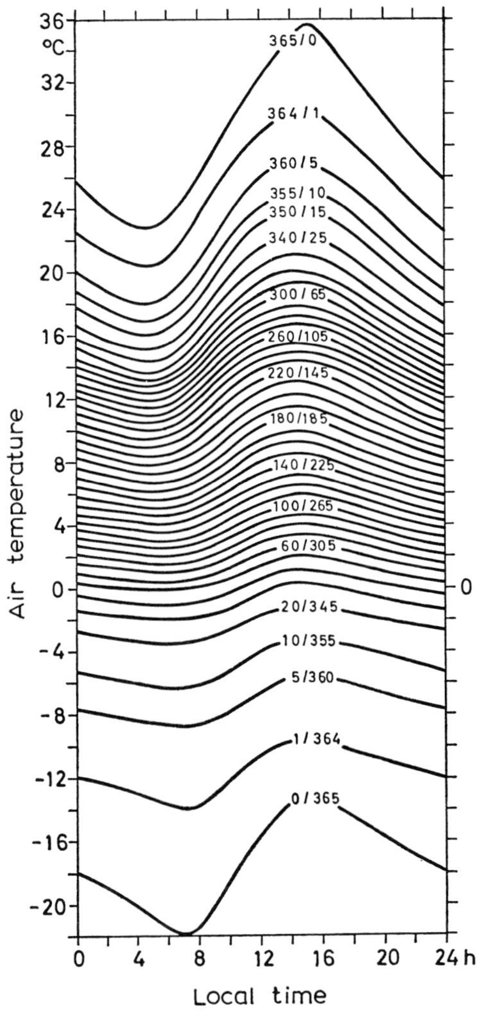

Fig.2. Cumulative frequency curves of hourly temperature values at Hamburg (1891–1930). The figures labeling the curves denote the numbers of days per year with temperature below/above a specific temperature value.

It is often advantageous to transform climatological frequency distributions in order to meet technical requirements. For instance, the frequency distribution of the daily mean of vapor pressure has been transformed into frequency distributions of relative humidity in relation to fixed air temperatures ranging from +30°C to −5°C and presented in graphic form in Fig.3. Thus, the engineer can read out how often at a given indoor air temperature he has to humidify or to dehumidify. He can also read (for mean daily vapor pressure) the amount of heating or cooling needed to keep outside air introduced for air conditioning at a specific moisture level.

In the case of cooling below the dew point water condenses. Fig.3 shows isolines of condensed water in grams of water per kilogram of air (g/kg). This is shown by dashed lines below the 100% relative humidity curve. These isolines show the frequency and the amount of water of condensation which can be expected for specific cooling temperatures.

Prototype tests of architectural structures permit evaluation of climatological data in terms of indoor conditions. Using again Hamburg data as an example, Fig.4 shows

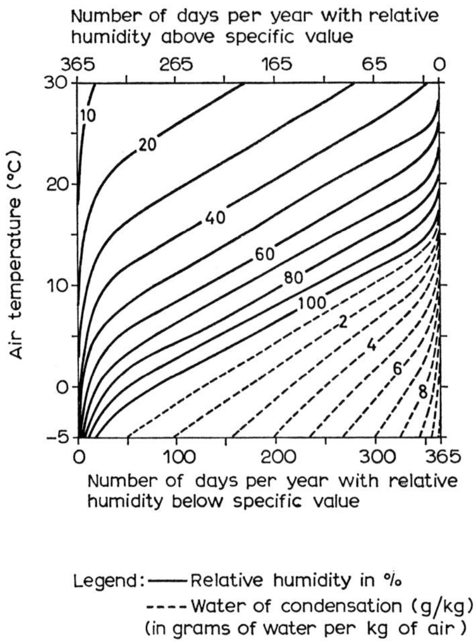

Fig.3. Cumulative frequency curves of relative humidity in relation to fixed air temperature values at Hamburg-Seewarte (1891–1930). (Derived from cumulative frequencies of daily means of vapor pressure.)

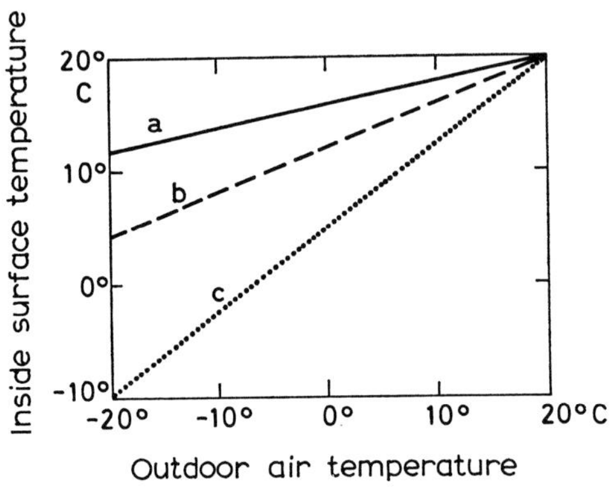

Fig.4. Functional relationship between inside surface temperatures in a room kept at a fixed indoor air temperature of 20°C and the outdoor air temperature. *a* = wall, $1^1/_2$ bricks in width; *b* = double window; *c* = single window.

expected temperatures on inside surfaces for a fixed inside air temperature of 20°C. The diagram clearly shows the difference of insulating qualities of single windows, double windows, and walls of $1\frac{1}{2}$-brick thickness. For the same locality Fig.5 portrays these indoor conditions as a cumulative frequency distribution with a scale showing the expected number of days per year given specific inside surface values.

Such isopleth frequency distributions can be treated easily as a two-dimensional problem. The isopleths delineate areas comprising values of the frequency of two parameters that correspond to a specific cumulative frequency value. An example is the so-called e-*t*-*U* diagram, developed by H. Burchard (Böer and Götschmann, 1963). Coordinates are temperature (*t*, linear scale on the abscissa) and vapor pressure (e, logarithmic scale on the ordinate). It also has a set of isolines of relative humidity (*U*). Fig.6 shows the cumulative frequency distributions of temperature and moisture in this e-*t*-*U* diagram,

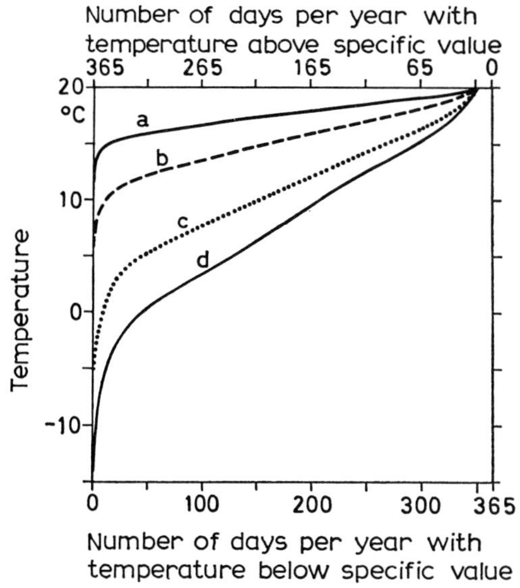

Fig.5. Cumulative temperature frequency curves for outdoor air as well as inside surfaces of windows and an outer wall $1^1/_2$ bricks in width, at a fixed indoor air temperature of 20°C at 21h00 local time at Hamburg (1891–1930). a = outer wall; b = double window; c = single window; d = outdoor air.

of the warmest and coldest months, for the stations Irkutsk, Potsdam, Cairo and Bombay (BÖER and GÖTSCHMANN, 1963). It can be seen that in July at Potsdam the hourly values of the temperature range between +5° and +37°C, the vapor pressure between 3 and 23 Torr and the relative humidity can become as low as 10%. 50% of all hours of this month experience temperatures between 10° and 20°C with vapor pressure values from 8 up to 15 Torr and the relative humidity surpassing 65%. However, in the same month at Cairo 50% of all hours experience temperatures between 20° and 35°C with vapor pressure values from 7 up to 16 Torr and the relative humidity ranging between 20% and 80%.

Air conditioning engineers prefer climatological frequency distributions to be presented in thermodynamic charts: the common type in the USA is the psychrometric chart (Fig.7), whereas in Europe a temperature–humidity diagram, that has been developed by MOLLIER (1923) and was called i–x diagram, is widely used. In this special nomograph for engineers Mollier presented the equation of the enthalpy of the air:

$$i_{(t+x)} = 0.24t + 4.6xt + 595x$$

where i = the enthalpy of the air in kcal. per kg of dry air; t = the air temperature in °C; x = the water vapor content of the atmosphere in g of vapor per kg of dry air (mixing ratio).

An example is given in Fig.8 which shows the temperature–humidity isopleths for the coldest month at Potsdam and the warmest month at Bombay. The diagram has as abscissa the water vapor content of the atmosphere (x, in g/kg) and as ordinate the air temperature (t, in °C). The slanting lines are the enthalpy values (i, in kcal.kg^{-1}) and the curved lines represent the relative humidity (φ) (BÖER and GÖTSCHMANN, 1963).

Such presentation of climatic data for engineers has developed into a specialized art in recent years. Other examples will follow under other topics of application discussed below.

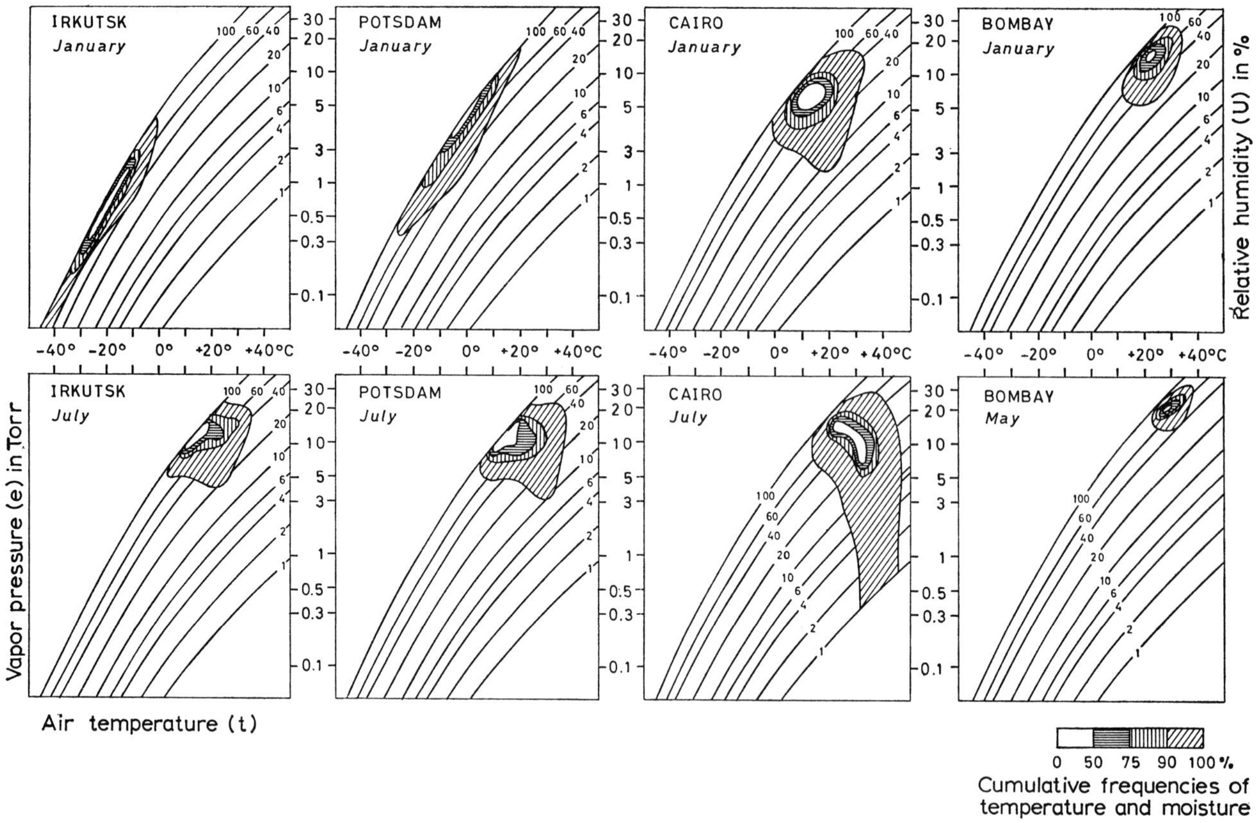

Fig.6. e–*t*–*U* diagram. Climatographs for the stations Irkutsk, Potsdam, Cairo and Bombay for the coldest and warmest month each.

Fields of application for technical climatology

Indoor climate

The oldest urbanization sprang up in areas of arable land with a dry and warm climate (e.g., Palestine–Jericho, Mesopotamia, Indus valley). The temperatures in those places enabled man to survive the winter season without using artificial sources of heat. In the warm season the thick walls cooled at night to warm up only slowly because of their high thermal capacity, thus keeping the house during day-time cooler than the surrounding air. After sunset the walls emit the heat which they accumulated during the day so that the indoor air cools less quickly than the surrounding air. Therefore, flat roofs become sleeping quarters. The development of heating devices enabled urbanization to spread poleward into the temperate latitudes. The improvement of indoor cooling today enables the design of healthy living and working spaces in hot-dry, as well as hot-humid, climatic zones.

The comfort standards of air conditioning vary somewhat according to different ethnic groups, country of origin and acclimatization to the native climate in the course of generations. Comfort is also related to the level of work to be performed in the air conditioned rooms (SCHINZEL, 1965). PEEL'S (1954) measurements taken in native houses in the Sierra Leone showed that, especially at night, the natives accepted as comfortable levels of air temperature and humidity considerably higher than those accepted by

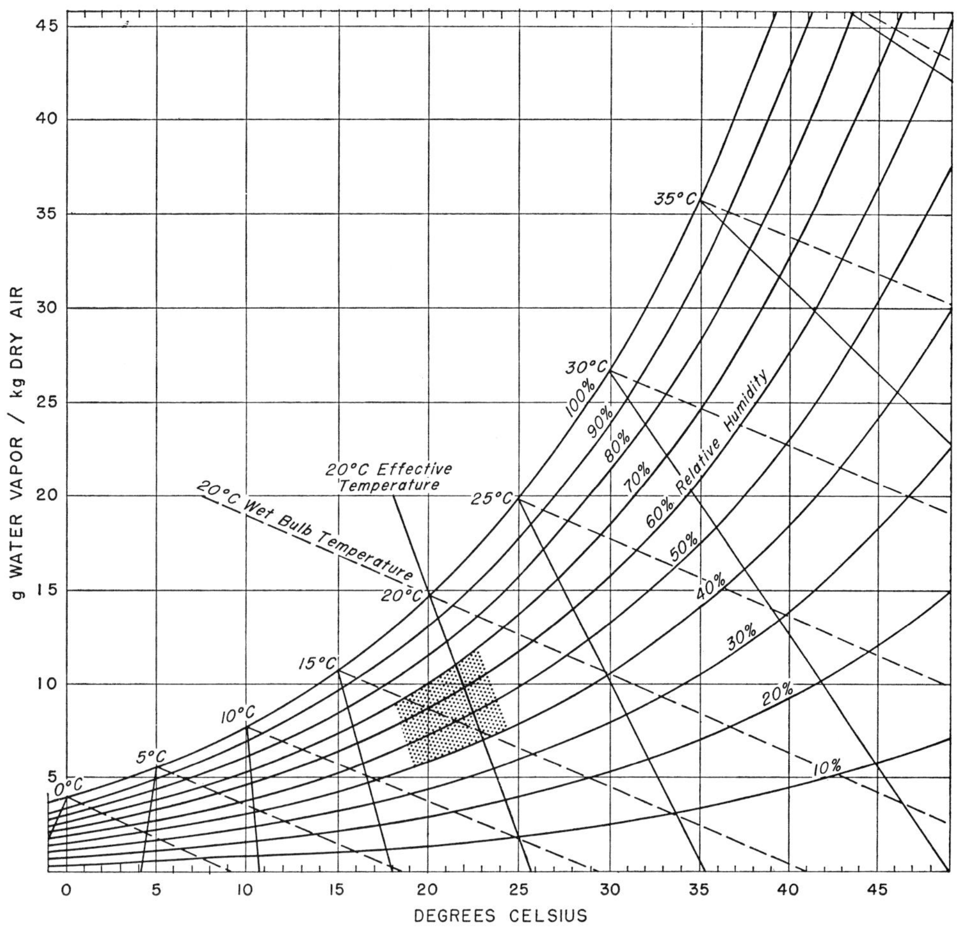

Fig.7. Thermodynamic comfort diagram (temperature vs. mixing ratio). Curved lines are relative humidities. Solid slant lines are effective temperatures; dashed straight lines are wet-bulb temperatures. Hatched area general comfort zone.

temperate-zone groups. The changes in the comfort standards of indoor climate in England have been investigated (BLACK, 1954). In 1936, Bedford found the comfort zone in Great Britain ranging between air temperatures of 15.6° and 20.2°C whereas in winter 1950/51, a poll of 5,240 male and 5,209 female office workers in London yielded the comfort zone ranging between air temperatures of 17.8° and 22.2°C (BLACK, 1954). In Germany, LEUSDEN and FREIMARK (1951) investigated the influence of air temperature, humidity and ventilation on the feeling of well-being. In the case of light physical exercise, they found that the comfort zone was between air temperatures of 18° and 24°C, relative humidities between 35% and 70% with an air movement of less than 20 cm per second. Whereas heavy workers felt comfortable at air temperatures between 17° and 23°C, relative humidities between 20% and 55% with an air movement between 20 and 30 cm per second (LEUSDEN and FREIMARK, 1951).

The feeling of comfort in an indoor climate is based upon the thermal balance of the body. 42% of the human heat production is released by radiation to the surroundings, 26% by heat conduction to the surrounding air, 18% by evaporation and 14% by exhaling (BRADTKE and LIESE, 1952). This heat release is influenced by the following meteorological factors: surrounding surface temperatures (radiation), surrounding air

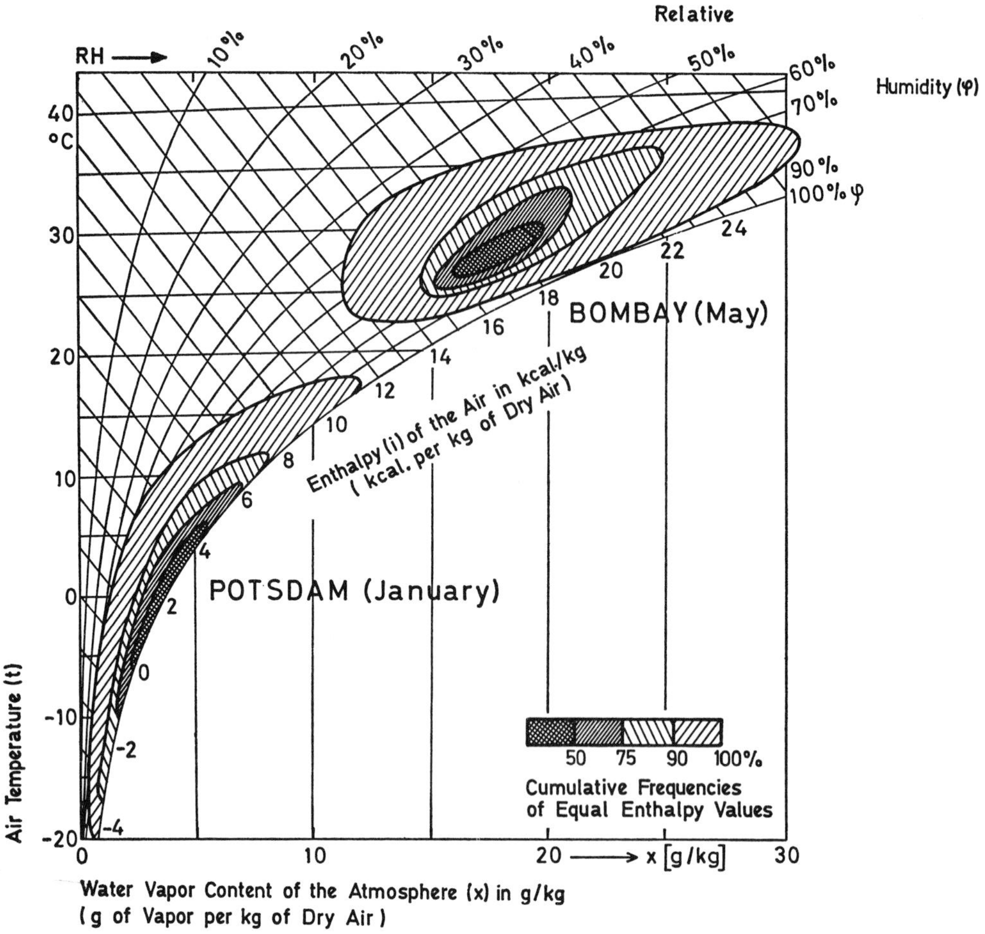

Fig.8. *i–x* diagram. Climatographs for the stations Bombay (May) and Potsdam (January).

temperature (heat loss due to convection, evaporation or exhaling), atmospheric humidity (evaporation, heat loss through the respiratory system) as well as air movement (heat loss caused by convection or evaporation). The interaction of these factors in forming the properties of indoor environmental conditions can be visualized by the use of a comfort chart developed by OLGYAY (1954) for the inhabitants of the temperate climatic zone of the USA (Fig.9). It has as ordinate the dry-bulb temperature (°C) and relative humidity (%) as abscissa. On this diagram the usually accepted comfort zone for indoor conditions is hatched for summer and cross-hatched for winter. Variations brought about by changes in humidity, air movement and radiation can also be seen from this diagram: in winter the comfort zone lies between air temperatures of 19° and 24°C and in summer between 21° and 27°C with a relative humidity of between 30% and 65%. However, in winter warm walls with temperatures up to 24°C can compensate an indoor air temperature as low as 17.5°C. Additional insolation of 400 $kJm^{-2}\ h^{-1}$ (= 10 ly n^{-1}) decrease the lower temperature threshold of the comfort zone by 1.5°C. An air temperature of 14°C is still accepted as comfortable if the usual clothing is doubled and 7°C is still accepted as comfortable by applying insulation clothing. In summer, the upper threshold of the comfort zone can be raised to 31°C by cooling the walls to 18.5°C, or to 33°C by ventilation up to 3 m per second, or to 43°C in case the

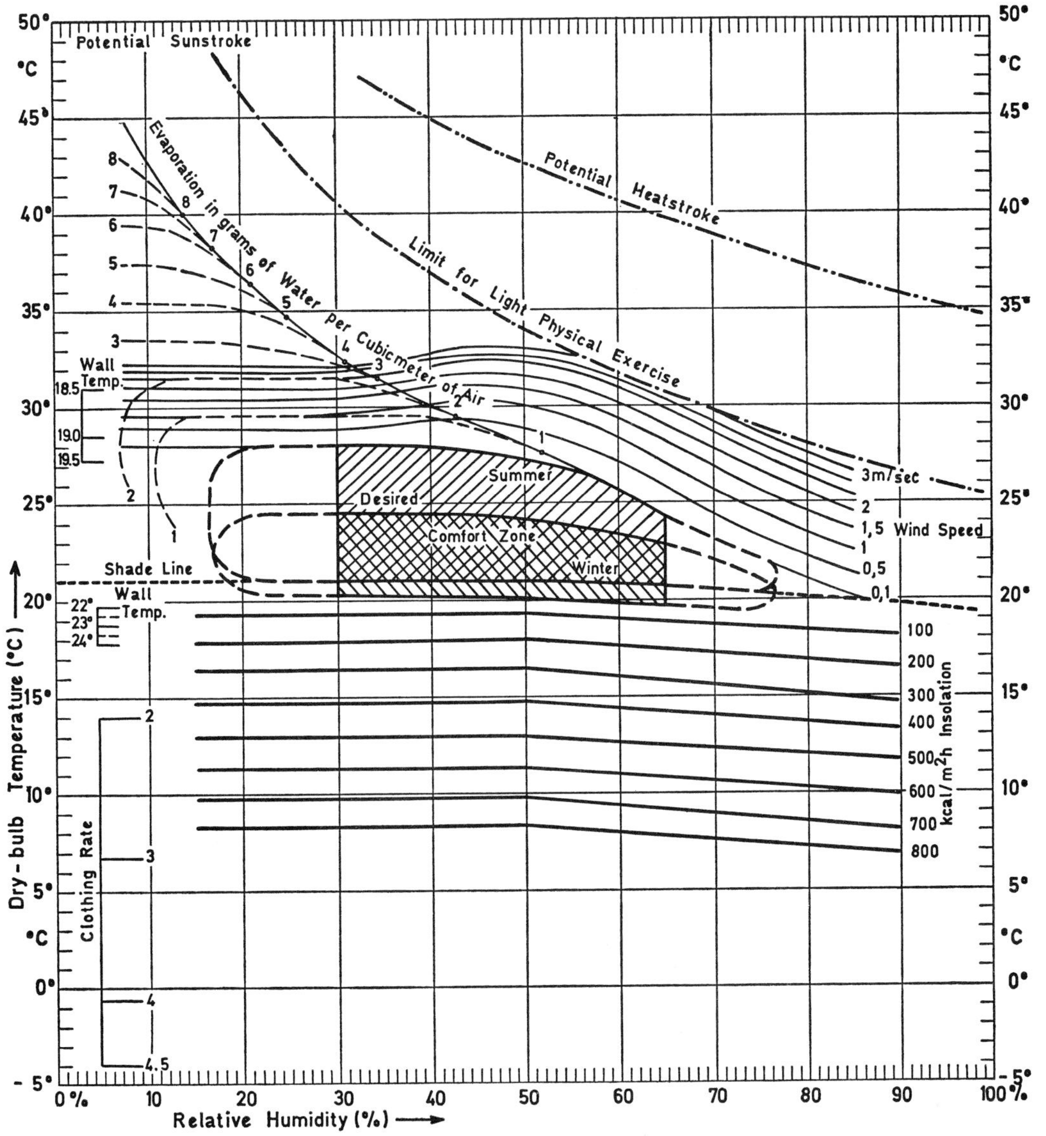

Fig.9. Temperature–humidity diagram for determining comfort zone in the temperate climatic zone of U.S.A. (after OLGYAY, 1954).

relative humidity is less than 10% by producing evaporative cooling through spraying 8 grams of water per cubic meter of air (OLGYAY, 1954).

Heating

As we have seen, controlling factors for indoor comfort in winter are the indoor air temperature and the surface temperatures of the surrounding walls. They must be kept within the range of the comfort zone under varying outdoor temperature conditions. The amount of artificial heat has to compensate for heat loss due to the conductivity of the walls and infiltration of air. Heating engineers use degree-days as a measure for the amount of heat to be applied. In 1937, A.D. Marston in the USA suggested a standard temperature of 65°F (18.3°C) for heating as well as for cooling (THOM, 1966b). In British usage, one heating degree-day is given for each degree that the daily mean outdoor temperature departs below the base of a normal indoor temperature of 65°F (18.3°C). In Russia, a heating day is a day with a daily mean outdoor temperature of less than +8°C; the outdoor temperature is related to an indoor temperature of +18°C (ANAPOLSKAJA and GANDIN, 1966); following the suggestions made by Hottinger and Raiss (RAISS, 1933), in German usage, a heating day (H_d) is defined as a day with a daily mean outdoor temperature $\leqslant 12°$. The sum of heating degree-days (G) is the sum of departures of the mean outdoor temperatures below the standard room temperatures (t_s) of 19°. Thus, the accumulated heating degree-days (G) are expressed with:

$$G = \sum_{1}^{i} (t_s - t_{hi})\ (°C)$$

Due to the rise of comfort demands and changes of the heating practice, the standard commission for heating engineers of W. Germany (VDI—Norm) concluded to lift the standard room temperature, t_s from 19°C to 20°C and the temperature of heating days, t_h, from $\leqslant 12°$ to $\leqslant 15°$ with the beginning of the heating season 1975/76. This change of basic values causes an increase of the number of heating days and the sum of degree-days. They rise with an average of 15–20% in Central Europe and more than 40% in the mediterranian peninsulae. Table IV shows the heating characteristics for a world-wide sample of locations, calculated for t_s 19°, $t_h \leqslant 12°$ and t_s 20°, $t_h \leqslant 15°$.
BREZINA and SCHMIDT (1939) investigated the heat loss through walls. It was almost proportional to the temperature difference between the temperatures of the wall surface (t_W) and the air (t_L) and the square root of wind speed ($\sqrt{v}$). In addition to that, the indoor climate is influenced by the infiltration of air around window and door frames and diffusion through walls. In the cold winter 1939/40, the fuel consumption of the Hamburg district heating plant was found to be influenced more by the cooling of the walls due to the wind speed than by the air infiltration due to the wind pressure (REIDAT, 1951). The relationship between the wind induced heat losses due to heat conductivity and due to air change has been used in Russia for determining an "effective heat loss temperature of a building". This figure corresponds to a temperature at which motionless air would cause the same heat loss as that caused by the actual wind speed. For a frequency of occurrence of 0.1% of all heating days it reaches a value of −152°C at Dickson on the Jenissei estuary, in central Siberia it falls in between −60°C and −70°C and in the Baltic area its value amounts to −30°C (GANDIN, 1965).

Cooling

Hot weather comfort conditions call for lowering of air temperatures and humidities but optimal designs will also try to achieve lowering the wall surface temperature, control of ventilation, evaporative heat regulation at low atmospheric humidities, and protection from direct sunshine.
Control of indoor humidity plays a minor role for summer air conditioning in temperate climates. For example, at Hamburg from June through September, air temperatures exceeding 23°C occur in only 157 h (5.2% of the total of hours of these months); only 44 of these hours (1.5%) have a relative humidity of more than 65%. Air temperatures exceeding 25°C occur in only 82 h (2.8%) in these months and only 5 of these hours (0.2%) have a relative humidity of more than 65%, on an average.

TABLE IV

HEATING SEASON

	t_s 19°; $t_h \leqslant$ 12°			t_s 20°; $t_h \leqslant$ 15°		
	H_d	D_d	t_h	H_d	D_d	t
Gothaab	365	7,650	−2.0	365	8,100	−2.2
Thorshaven	365	4,540	6.6	365	5,000	6.3
Bergen	271	3,900	4.6	365	4,710	7.1
Stockholm	278	4,480	2.9	313	5,040	3.9
Kopenhagen	258	4,130	2.8	327	4,590	6.0
London	229	2,890	6.4	285	3,080	6.0
Berlin	217	3,240	3.1	274	3,950	5.6
Paris	214	2,810	5.9	262	3,370	7.1
Marseille	152	1,600	8.5	202	2,050	9.9
Madrid	172	2,020	7.2	211	2,680	7.3
Rome	140	1,430	8.8	185	1,860	9.9
Athens	107	1,080	8.9	158	1,560	10.1
Wien	206	3,150	3.6	240	3,570	5.1
Bucharest	188	3,090	3.6	220	3,470	4.3
Archangelsk	272	6,160	−2.7	359	6,970	0.6
Leningrad	268	5,160	−0.3	303	5,640	1.4
Moscow	254	5,310	−1.9	294	5,680	0.7
Kiew	225	4,120	0.7	257	4,580	2.2
Odessa	204	3,350	2.6	230	3,680	4.0
Werchojansk	308	12,500	−21.5	354	12,670	−15.8
Irkutsk	279	7,090	−6.4	320	8,730	−1.3
Wladiwostok	241	5,320	−3.0	275	5,410	0.3
Tientsien	174	3,060	2.6	199	3,410	2.9
Hankau	130	1,630	6.5	159	1,970	4.1
Tokyo	157	1,870	7.1	195	2,390	7.8
Baghdad	65	590	10.0	99	860	11.3
Adelaide	59	470	11.0	145	1,060	12.3
Melbourne	113	990	10.2	192	1,610	11.6
Tunis	90	850	9.6	179	1,567	11.2
Johannesburg	77	660	10.4	151	1,213	12.0
S. Francisco	124	1,040	10.6	327	2,452	12.5
Chicago	202	3,360	2.5	235	3,860	3.6
New York	191	2,780	4.5	220	3,320	4.9
Washington	174	2,490	4.7	202	2,850	5.9
Santiago (Chile)	151	1,580	8.5	216	2,060	10.4
Punta Arenas	365	4,600	6.4	365	4,960	6.4
Montevideo	95	780	10.8	167	1,360	11.8
Buenos Aires	103	910	10.1	164	1,420	11.3

t_s = standard room temperature; t_h = mean outdoor temperature on heating days; H_d = heating days; D_d = heating degree-day, i.e., $t_s - t_h$; t = mean outdoor temperature of a heating day.

In low latitudes, however, air conditioning must include the control of both air temperature and air humidity (GIVONI, 1966). For full use of the natural climate there is a notable difference between the dry hot subtropical areas and the humid warm tropical regions. An example is given in Fig.10 which shows the diurnal variation of outdoor temperature, indoor temperature and relative humidity in July at Baghdad, a hot and dry place, and at Conakry, where it is humid and warm (AYOUB, 1966). Baghdad experiences a diurnal variation of air temperature of more than 20°C and the relative humidity varies between 20% during noontime and 50% at night. The walls there are

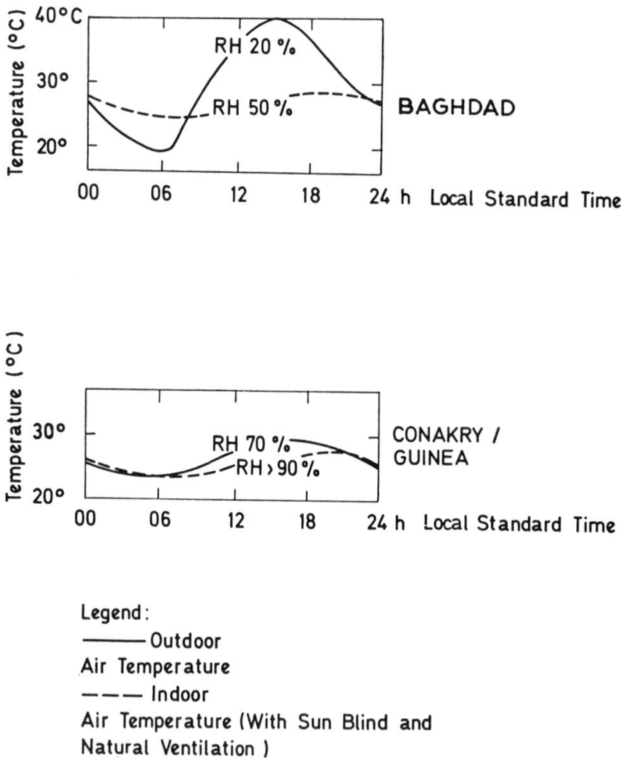

Fig.10. Diurnal variation of outdoor as well as indoor temperature and relative humidity in July.

thick and have a high thermal capacity. By shading all openings adequately from the sun and restricting day-time ventilation to minimal requirements the thick walls pre-cooled at night help to keep the indoor air from exceeding a daily maximum temperature of 29°C. At night the walls radiate the heat which they absorbed during the day. Artifical nocturnal ventilation helps to cool the buildings more quickly. Nevertheless, the indoor temperature even at night usually is about 5°C higher than that of the surrounding air. Thus, the physiological demand can be lowered by sleeping in the open. At Colomb-Béchar, at the fringe of the Sahara, experiments showed that an evaporative watei spray modified the indoor climate from 41°C temperature and 15% relative humidity to 27°C and 60% relative humidity (TIREL, 1961).

Conakry, which is representative for the humid tropical climate, experiences a diurnal temperature variation of no more than 5°C on the average, and the relative humidity undergoes variations only between 70% and 100%. Artificial ventilation during daytime lowers the indoor temperature by no more than 3°C but the high humidities still keep conditions uncomfortable. Acclimatization may help but the discomfort caused by high humidity can only be mitigated by air movement and dehumidification. The transition from an uncomfortable outdoor climate zone to the comfort zone of an air-conditioned room should be gradual to facilitate the adjustment of the thermoregulatory functions of the body, e.g., regulation of sweat rate. Abrupt transition may result in injuries to health (GIVONI, 1966).

For estimating the energy requirements for air cooling "of cooling degree hours" have been used by engineers in the temperate climate zones (in analogy with the heating degree-days). In Germany one cooling degree-hour is given for each degree that the temperature of each cooling hour (the temperature of which is ⩾ 23°C) departs above the base of 19°C (SPRENGER and KRUGER, 1950). This computation yields the mean values (1948–1964) for Hamburg given in Table V.

TABLE V

COOLING DEGREE HOURS FOR HAMBURG (1948–1964)

	May	June	July	Aug.	Sep.	Oct.	Year
Cooling hours	5	11	44	56	46	16	178
Cooling degree-hours	25	60	280	380	295	90	1,130

For economic reasons, heating and cooling plants cannot be dimensioned in such a way as to meet all extremes of environmental conditions. Generally, a confidence level of 0.05 is accepted. The dimensioning of heating plants is often based upon the mean annual temperature minimum according to DIN Norm 4701 (1950) (German industrial standards No.4701). This extreme, however, has been beaten in Germany during the coldest winters by as many as five to six consecutive days. For the design of cooling plants in the USA THOM (1960) suggests to stay within the confidence interval of 0.05 of the daily temperature maxima during the months June through September; this almost corresponds to common design criteria.

TABLE VI

SUMMER WEATHER DESIGN DATA FOR SELECTED CITIES IN THE U.S.A., 5% WITHIN THE SUMMER RISK (After THOM, 1960)

	Frequency of occurrence			Guide value (common use)
	1/20*	1/10	1/2	
Chicago, Ill.	35.8	35.3	34.1	35.0°C
Fort Worth, Texas	40.7	40.2	38.7	37.8°C
Miami, Florida	33.1	32.9	32.0	32.8°C
New Orleans, La.	36.3	36.0	35.1	35.0°C
New York, N.Y.	34.1	34.0	32.4	35.0°C
Philadelphia, Penn.	35.8	35.3	33.8	35.6°C
St. Louis, Mo.	39.2	38.4	36.7	35.0°C
Washington, D.C.	36.7	36.6	35.9	35.0°C

* 1/20 denotes once in 20 years.

The exact estimation of the potential cooling capacities with simultaneous humidity control requires enthalpy frequency considerations, i.e., multiparametric frequency distributions of air temperature and humidity as presented in the Figs.6 and 8 (BERLINER, 1957; FLUOR CORPORATION, 1958; REIDAT, 1970).

Application of climatology to building design

The microclimate of our buildings must be considered in two parts: They are adapted to offer shelter to escape unfavorable environmental conditions. Yet they are subject to these environmental conditions. For economic reasons, optimal use should be made of all favorable weather influence and avoidance of much of the unfavorable weather influence by economical engineering means. In the planning stage, the production costs and the operating costs have to be balanced. First and final expenses for the control of heat, cold and moisture which compiise the selection of building shape and materials may later result in considerable lowering of operating costs. Timing of construction activities according to meteorological conditions may lower production costs. Therefore, the architect needs to know about the climatic peculiarities of the locality as well as the modifications brought about by the influence of the prospective building. The architect needs to know the following: climatological means and extremes are by not-constants to be inserted into formulas without reservations. Frequencies and sequences enable the designer to estimate the critical meteorological values. Meteorological elements are closely interrelated and their influence on the house may be in different directions. The total weather load on a house is composed of the partial loads of the individual meteorological elements, the extremes of which usually do not occur simultaneously. When the temperature load reaches a maximum, the wind load usually is only small and the precipitation load is missing. In many localities, severe cold occurs with calms or with light winds; the greatest wind loads, however, are in the moderate latitudes of the Northern Hemisphere almost connected with gales from westerly direction.

Wind

Wind influences the ventilation of buildings. Inlets should be sited to avoid suction in case of strong winds, and chimneys should be protected from katabatic winds. Otherwise the fireplaces may experience downdrafts in the flue causing smoke to enter the house. Wind roses indicating the range of wind speeds from each direction are useful tools for building design. They permit an assessment of wind pressure on the structure. For a flat surface it comprises two components, the first being the dynamic pressure exerted on the windward side of the surface. This is equal to $\frac{1}{2}\,\varphi\,v^2$, where φ is the air density and v is the wind speed normal to the surface. Moving over and around the obstacle, the wind is deflected. Along the side walls and over the roof the wind speed is strengthened. The second component of wind pressure is the pressure decrease, or suction, produced on the entire lee side of the building (GARSTON, 1968). This is equal to $\frac{1}{2}\,c\,\varphi\,v^2$, where c is a structural constant varying from -0.3 for cylindrical objects to 1.0 for long plates. The wind pressure p is the sum of these two: $p = \frac{1}{2}(1+c)\,\varphi\,v^2$.

Measurements made at an office building of eighteen stories in London showed that the dynamic pressure exerted on the windward side was by far the major part of the wind pressure, while the load on the leeward side was only small. Suction mainly affected the windward edges of the side walls (NEWBERRY et al., 1968). The greatest suction effect has to be expected behind roofs inclined less than 15°. From an incline of 30° onwards, suction does not occur any more; steeper roofs are subject to dynamic pressure only.

A gap between two buildings does not only raise the wind speed, it also increases the suction effect on the border walls. Furthermore, openings through, or under, big square-shaped buildings considerably accelerate the wind in comparison with that felt at the windward side of the building. If these openings are panelled, there is a risk of surface destruction due to high suction loads.

Strong wind affects eddies between buildings which are different in height so that the wind speed near the ground may amount to more than twice the wind speed at the

height of the roof of the higher building. This may result in the serious interruption of traffic (SEXTON, 1970) and the breakage of plate-glass windows.
The roughness of the surface lowers the wind speed and increases the turbulence. Thus, the height of the gradient wind level is raised. On the average, this height is 270 m over open flat terrain, 390 m over suburbs and 420 m over cities (MUNN, 1970).
The increase of the mean hourly wind speed ($\bar{v}$) with height (Z) is described with the power law:

$$\frac{\bar{v}_Z}{\bar{v}_{10}} = \left(\frac{Z}{Z_{10}}\right)^m$$

where Z_{10} denotes the height of the anemometer level given as 10 m. The exponent m has been derived from various wind recordings taken at high masts. It has been found that it depends on the roughness of the surface and the distribution of atmospheric temperature in the vertical (HELLMANN, 1914; BORISENKŎ and SAVARINA, 1967; FRANKENBERGER, 1968). The measurements taken at Quickborn near Hamburg yielded values of m ranging between 0.1 and 0.9 and amounting to an average value of 0.17 in strong wind (FRANKENBERGER, 1968). In Russia (ORLENKO, 1970) the ratio of wind speed at the height Z to that at the anemometer height Z_{10} (10 m) has been measured in the case of neutral stability for various heights of surface irregularities; the results are listed in Table VII.

TABLE VII

RATIO v_Z/v_{10} FOR DIFFERENT Z AND z_0 (IS ROUGHNESS COEFFICIENT)
(After ORLENKO, 1970)

Height above ground, Z (m)	Roughness coefficients, z_0: even snowcover (20 cm): 0.05	ploughed land: 2	town: 100	city: 200 cm
10	1.0	1.0	1.0	1.0
50	1.16	1.25	1.67	1.94
100	1.22	1.36	2.08	2.58
150	1.27	1.46	2.46	3.15
200	1.32	1.54	2.76	3.65
300	1.39	1.69	3.37	4.32

Note: v_Z is wind speed at height Z: v_{10} is wind speed at the height of the anemometer (10 m).

A structure should be designed to stand the strongest gale, noted in the region. The absolute maximum wind speed amounting to 103 m/sec was measured on top of Mt. Washington, N.H., on 12 April 1934. Peak gust recordings of 55 m/sec occurred in Japan during the passage of a typhoon on 26 September 1959. Twenty-year anemometer recordings gave evidence of the occurrence of gusts up to 50 m/sec along the British northwest coast and on the peaks of the German highlands.
Such sporadic measurements are not very useful for building design. Hence, statistical methods for extreme wind speeds have been resorted to for return periods of 50 or 100 years. Statistical extreme value distributions have been derived by GUMBEL (1935) and FISHER and TIPPET (1926). Computed maximum wind speeds have been mapped for the USA, Great Britain and Russia (GARSTON, 1968; THOM, 1967; ORLENKO, 1970).

In the USA THOM (1967) has based his computations on the "extreme mile". This is the shortest time interval in which the wind travels an aggregate distance of one statute mile (1,609 m). For an average wind speed of 5.3 m/sec it amounts to 5.3 min and when the wind speed is raised to 20 m/sec, it drops to 1.3 min. In England the computation of the maximum wind speed has been based upon the strongest gust recordings of 1 to 2 sec duration. Therefore, the maps of maximum wind speed which have been plotted for several countries are not compatible.
A gust exerts an impact on the building resulting in a damped oscillation. The next gust may exert another damping effect if, e.g., the gust period is half the free oscillation period of the building. However, if the forced oscillation of the gust and the free oscillation of the building are in resonance, this may cause far greater stress on the building than the maximum wind speed alone. Taking this potential load into account requires a detailed analysis of the gust structure as it has been suggested by Davenport (DAVENPORT, 1961; VICKERS and DAVENPORT, 1968; KÖNIG and ZILCH, 1970).

The heat balance of buildings

The heat balance of buildings is controlled by insolation, heat radiation to the surroundings and heat exchange with the surrounding air.

Radiation

The effect of incoming solar radiation on the heating of a building depends on the solar elevation, cloud cover, turbidity of the air and absorptivity of the building.
The unit of radiation used by engineers is since 1978 by international convention the kilo Joule (kJ) per square meter (m^{-2}) and hour (h^{-1}). Until 1977 engineers made use of the unit kilo calorie (kcal.) per square meter per hour (kcal.$m^{-2}h^{-1}$). British engineers used the British Thermal Unit (BTU) per square foot (ft.$^{-2}$) and hour (h^{-1}), (BTU ft.$^{-2}h^{-1}$).
For transformation of these units is effective:

1 kJ $m^{-2}h^{-1}$ = 0.239 kcal.$m^{-2}h^{-1}$ = 0.647 BTU ft.$^{-2}h^{-1}$
1 kcal.$m^{-2}h^{-1}$ = 4.185 kJ $m^{-2}h^{-1}$ = 2.713 BTU ft.$^{-2}h^{-1}$
1 BTU ft.$^{-2}h^{-1}$ = 1.544 kJ $m^{-2}h^{-1}$ = 0.369 kcal.$m^{-2}h^{-1}$

Table VIII shows the effect of geographical latitude on incoming solar radiation. The computation was based on the measurements taken at 70 observation sites, valid for the 15th of June at 12 o'clock local time (PERL, 1935/36).
The solar radiant intensity received on a surface normal to the incident radiation on 15 June at 12 o'clock local time on the North Pole is 11% less than that received at 30° latitude. At summer solstice, the horizontal surface at the Pole receives only a third of what the horizontal surface at 30° latitude receives, whereas the south wall created at the Pole would receive the three fold amount of a south wall at 30° latitude.
These computations take into consideration the direct solar radiation only; the influence of diffuse radiation on the amount of radiant energy received by horizontal and normal surfaces is neglected. Measurements taken at the meteorological observatory at Potsdam during the years 1937–1941 revealed that the portion of diffuse radiation amounted to

TABLE VIII

SOLAR RADIANT INTENSITY AND DAILY TOTALS ON 15 JUNE AT 12 O'CLOCK LOCAL TIME IN VARIOUS LATITUDES (After PERL, 1935/36)

Geograph. lat. °N	Solar elev.	Radiant intensity ($kJ\ m^{-2}h^{-1}$) on normal surface	on horizontal surface	on south wall	daily total
30	75°	3,365	3,244	879	26,116
50	60°	3,139	2,716	1,582	24,610
75	30°	3,114	1,557	2,662	26,660
90	23.5°	2,988	1,168	2,762	27,915

31% on cloudless days in December and to 65% on the average of all days in December, whereas it amounted to 14% on cloudless days in June and to 42% on the average of all days in June (DIRMHIRN, 1964). VALKO (1967) determined the portion of scattered radiation from both the sky and the surface of the earth in the total incoming radiation for the 1st of July under cloudless sky; he found 74% for the north wall, 47% for the south wall and 37% for the west or east wall. Computations of the radiant energy received by the walls have been carried out by KAEMPFERT (1942) under consideration of turbidity, also by Dirmhirn and Sauberer (STEINHAUSER et al., 1957) under consideration of the diffuse sky radiation. There are considerable discrepancies between the data of various authors as seen in Table IX.

TABLE IX

DAILY TOTALS OF INCOMING RADIATION RECEIVED BY VERTICAL WALLS IN JUNE ($kJ\ m^{-2}\ d^{-1}$)

Author	Exposure of the wall: north	east–west	south
Perl	2,595	10,920	8,120
Kämpfert	2,679	12,140	8,700
Dirmhirn/Sauberer	3,767	12,850	10,800

Recordings of radiation incident upon vertical and inclined planes have been obtained at Hamburg, Oslo, Locarno, Pretoria, Kinshasa and various Russian observation sites (GRÄFE, 1956; DE COSTER et al., 1955; VALKO, 1966; BIRKELAND, 1966; ADAMENKO and HAGRULLIN, 1969). VALKO (1970) has made an attempt to use the results of the field measurements made at Locarno for establishing a numerical model of the interaction of turbidy, sky radiation and terrestrial radiation. The greatest problem is the calculation of terrestrial radiation in densely built-up cities; it can be obtained by control tests only. The absorption of solar radiation by buildings depends on the albedo of the surface of the building materials. The differences in albedo can be seen from Table X.

The surface temperatures of aluminum foil normal to the incident solar radiation have been measured at Aachen with normal ventilation and air temperatures ranging between 27° and 29°C; the uncovered aluminum foil yields surface temperatures between 44°C and 48°C; after being covered by black paper the temperatures were raised to 72° and 78°C (SCHROPP, 1931).

TABLE X

ALBEDO OF VARIOUS SURFACE TYPES
(After VICK, 1951)

Surface	*Albedo* (%)
Glossy aluminum foil	90
White glazed tile	82
Marble	58
Raw brick	46
Concrete	40
Weather-worn lead coating	25
Slate, asphalt shingles	11
Coat of paint	
Zinc varnish, aluminum paint	78
Lemon yellow paint	50
Green or bright red	45
Blue	30

The heat transmission through uninsulated roofs was measured in Southeast Asia (KÖNIGSBERGER and LYNN, 1965). With an air temperature of +30°C and the incoming radiation amounting to 3,733 kJ m^{-2} h^{-1}, the temperature right under the roof and under a wooden floor of the attic, of 1.3 cm thickness was measured. Table XI shows the results.

TABLE XI

TEMPERATURES BELOW ROOFS IN SOUTHEAST ASIA (°C)
(After KÖNIGSBERGER and LYNN, 1965)

Roofing material	Underside of: roof		Wooden attic floor	
	temp.	excess over air temp.	temp.	excess over air temp.
Red clay tiles	55.3	+25.3	40.9	+10.9
Corrugated iron				
new and shiny	41.8	+13.8	32.9	+ 2.9
rusted or painted red	66.3	+36.6	38.8	+ 8.8
Corrugated asbestos sheets				
new	43.3	+13.3	35.8	+ 5.8
old	58.2	+28.2	41.7	+11.7
Reinforced concrete (10 cm)	52.1	+22.1		
Corrugated aluminum sheets				
new sheets	41.8	+11.8	32.9	+ 2.9
dust covered	66.3	+36.3	38.8	+ 8.8

These figures demonstrate the controlling influence of building materials on the excess heat inside the houses. However, they also show the protection offered by a thin wooden ceiling underneath the ventilated roofing.

86% of the global radiation are transmitted through regular window glass into the room which results in heating inside. Measurements taken at Holzkirchen, in southern Germany (FRANK et al., 1970), towards 4 o'clock in the afternoon showed that a global radiation of 2,218 kJ m^{-2} h^{-1} yielded a heat transfer of 1,256 kJ m^{-2} h^{-1} through a double window. At the same time, the heat transfer through a wall made of light building material amounted to only 167 kJ m^{-2} h^{-1}, through a wall made of hollow bricks it was no more than 84 kJ m^{-2} h^{-1}.

In the case of window glass the maximum heat release into the room occurs simultaneously with the maximum incoming radiation. However, solid building materials delay the maximum heat release up to 8 h as a result of the slow process of heat transfer by conduction within and through the wall. Therefore, the windows must be shaded from the sun either by fixed or movable shades or by heat-absorbing window glass. According to VDI, 1970 (German industrial standards, 1970) the fractions, shown in Table XII, of incident global radiation are transmitted into the interior space.

TABLE XII

PROTECTION FROM THE SUN: PERCENTAGE FRACTION OF RADIATION INCIDENT UPON THE WINDOW AT THE CULMINATION OF THE SUN THAT IS TRANSMITTED INTO THE INTERIOR SPACE

Sun protection	Radiant transfer (%)
Window glass	86
Heat-absorbing glass	60
Reflecting glass	35–50
Venetian blinds with a slope less than 45°	13
Awnings	26
Blinds between double panes of glass, slats 45°	43
Blinds inside, slats 45°	60
Light curtains inside	43
Plastic foil inside	60

The table shows that to be effective heat insulation must be applied outside the window. If heat-absorbing glass is used, the window panes are heated up considerably, resulting in an additional heat release into the room caused by heat conduction and infrared radiation (FRANK et al., 1970).

Many designs are available using landscaping for shading. Many publications refer to the planting schemes used (OLGYAY, 1954; PENNINGTON et al., 1964; ANQUEZ et al., 1965; TONNE, 1966; KITTLER, 1970).

Temperature

The interior space is protected from outside heat, as well as cold, by the walls. The controlling factors are the heat loss of the wall due to heat conduction and heat storage. In structural engineering it is common practice to express the heat loss due to heat conduction as a thermal resistance D or $1/\lambda$, with λ denoting the rate of heat transfer per hour per unit temperature difference per square meter of a plane structural component of any thickness (kJ m^{-2} h^{-1} $deg.^{-1}$). The thermal resistance decreases with increasing weight of material. To give an example: the same thermal resistance applies to a wall of lightweight boards 5 cm in thickness, timber wood 7.5 cm in thickness, cellular concrete 25 cm in thickness, or hollow bricks and brickwork 38 cm in thickness.

The heat capacity S is the ratio of the heat absorbed (kJ) (or released) by one cubic meter

of material to the temperature rise (or fall) of 1°C (kJ m^{-3} $deg.^{-1}$). It depends on the specific heat of the building material.
The structural stability of a building is affected by the heat expansion α_t of the individual types of building material. It is the increase (decrease) in millimeters of the unit length (m) at the temperature of 0°C due to heating (cooling) by 1°C (mm m^{-1} $deg.^{-1}$).
Typical values of these three measures have been compiled for various types of material. A sample is given in Table XIII.

TABLE XIII

WEIGHT (kg m^{-3}), HEAT TRANSFER (λ, kJ m^{-2} h^{-1} $degr.^{-1}$), HEAT CAPACITY (S, kJ m^{-3} $degr.^{-1}$) AND HEAT EXPANSION (α_t, mm m^{-1} $degr.^{-1}$) FOR VARIOUS BUILDING MATERIALS
(After SAUTTER, 1948)

Materials	Weight	λ	S	α_t
Iron	7,800	201	3,766	0.012
Aluminum	2,700	732.4	2,486	0.024
Compact natural stone	2,800	10.5	2,461	0.011
Gravel concrete	2,200	4.6	1,933	0.011
Brickwork	1,800	3.1	1,582	0.005
Cellular concrete	1,200	2.3	1,055	0.010
Pine wood	400	0.4	753	0.005

The interaction between heat conductibility and heat storage is shown in Table XIV comparing 4 walls of equal heat conductibility, but with differing wall thickness.

TABLE XIV

HEAT STORAGE OF WALLS OF EQUAL HEAT CONDUCTIBILITY
(After SAUTTER, 1948)

Construction of walls	a	b	c	d
38 cm brickwork	38	210	8,516	2.5
25 cm cellular concrete	38	210	4,907	4.8
7.5 cm timber wood	38	210	1,708	11.2
5 cm lightweight board	38	210	829	25.3

a = Thickness (cm) of brickwork of equal heat conductivity; b = heat loss (kJ m^{-2} h^{-1}) per 35 degr. temperature difference (e.g., 20°C indoors and −15°C outdoors); c = heat storage (kJ m^{-3}) per 35 degr. temperature difference; and d = hourly loss of accumulated heat as percentage of the heat storage.

The loss of accumulated heat of the lightweight board is ten times, and that of the timber wood wall, more than four times the loss of accumulated heat of the brickwork. Thus, lightweight buildings of small heat capacity can be heated in wintertime more quickly, however, they cool just as quickly (barracks climate), so that permanent heating is required. Thick walls of great heat capacity are useful in hot, dry climates; their temperature is only slowly raised during the day, thus keeping the house much cooler than the surrounding air. However, at night they have a temperature which is higher than the outside air temperature, thus, thick-walled buildings are comfortable for sleeping quarters in these climates. Measurements which have been taken in the U.S.A. prove (Fig.11) that in the case of equal heat conductivity a wooden house yields a diurnal temperature range amounting to about 2/3 and a brick house yields one amounting to only 1/3 of the temperature range of the air outside. In comparison with the outside

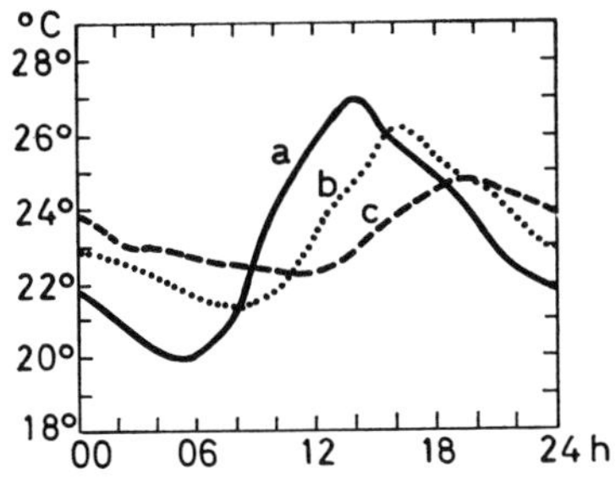

Fig.11. Diurnal temperature variation in July in the U.S.A.: a = open-air temperature; b = room temperature (wooden walls); c = room temperature (brick walls).

temperature variation, the daily temperature extremes inside a wooden house are delayed by about 2–3 h, in a brick house by a bout 6 h (OLGYAY, 1954).

The differences in the linear heat expansion of building materials may, especially in cases of abrupt temperature changes, result in tension in the composite structure; e.g., an abrupt drop of temperature due to a thundershower on a hot sunny day may cause the panelling of brick, ceramics or aluminum to come off concrete walls.

Considerable tension occurs in porous building materials when the water which is encased in the pores freezes and afterwards thaws again. Thus, these cycles of freezing and thawing may cause disruptive effects. The number of these cycles is equal to the difference of the number of frost days (days with minimum temperature below 0°C) minus the number of ice days (days on which the maximum air temperature in a thermometer shelter does not rise above 0°C). It is increased when on ice days the incoming solar radiation heats the wall surfaces to above 0°C.

Freezing of water in soils can cause frost heaving. This is caused by the 9% volume expansion in the transformation to ice. There are effects on drive ways, terraces, slabs and even, occasionally, on foundations (RUCKLI, 1950; ZIPPEL, 1954; GEIGER, 1961).

The frost penetration depth depends on the soil structure; this relationship can be seen in Table XV listing measurements taken in Finland.

TABLE XV

MEAN DEPTHS OF FROZEN GROUND DURING THE WINTER

(After KERÄNEN, 1951)

	Sand and gravel	Field	Clay	Moor
Northern Finland	126	100	90	88 cm
Southern Finland	72	47	50	42 cm

In Germany (ANIOL, 1952) the maximum depth of frozen ground during the cold winter 1928/29 was found to be 125 cm in sandy soil in Oldenburg (northern lowlands) and 80 cm in intermediate Devonian slaty soil at Trier (Moselle river).

The frost penetration characteristics have been studied during the winter 1939/40 at Giessen, Germany, by means of four lysimeters filled with four different types of soil; the results are presented in tabular (Table XVI) as well as graphic (Fig.12) form.

TABLE XVI

FROZEN GROUND AT GIESSEN, GERMANY, DURING THE WINTER 1939/40
(After KREUTZ, 1942)

Soil type	Depth of frozen ground (cm)	Frost penetration speed (cm day^{-1})	End of freezing	Delay against surface (days)
Basalt gravel	67	2.0	28 Feb.	+ 2
Sand	52	1.7	28 Feb.	+ 5
Loam	52	1.1	7 Mar.	+12
Humus	32	0.6	22 Mar.	+27

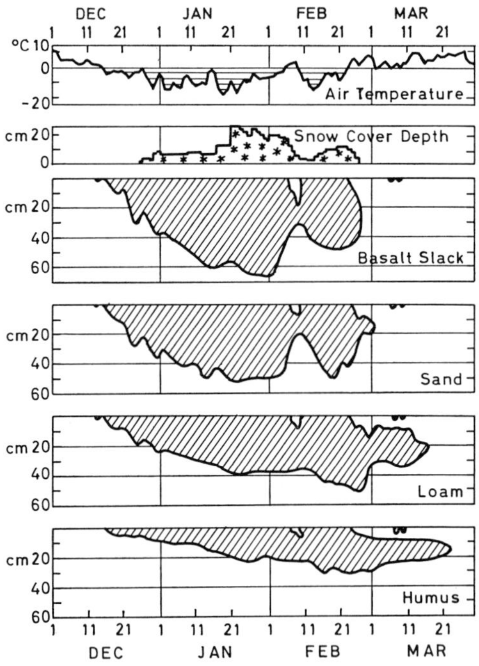

Fig.12. Frost penetration in four different types of soil at Giessen, W. Germany, in winter, 1939–40.

The difference of frost penetration and later melting in the various soil types is obvious from Fig.12.

In Russia, "continuous frost" denotes a spell of at least 30 days with below-freezing temperatures (ORLOWA, 1958). Within the last 40 years this 30 day-limit has been surpassed only once at Hamburg with 41 ice days in the winter 1946/47, however, in most of the territory of the U.S.S.R. a mean duration of more than 3 months of continuous frost can be expected. The duration increases to more than 7 months in Yakutsk and the remainder of northeastern Siberia. The summer heating fails to descend to the base of the layer of frozen ground: permafrost is present. Permafrost is a word coined by Muller (MULLER and SIEMON, 1945) to mean permanently frozen ground. In this condition a layer of soil or rock, at a variable depth below the surface of the earth, is at a temperature below freezing for at least 2 years. Fig.13 shows the vertical distribution of soil temperature in the permafrost region of Skovorodino in Siberia during the years 1928 through

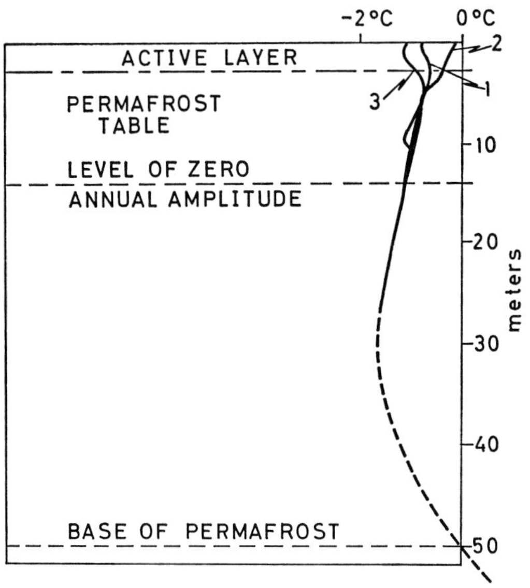

Fig.13. Temperature gradient in permafrost at Skovorodino, Siberia. *1* = mean temperature, 1928–1930; *2* = temperatures, 1930; *3* = temperatures for 1929.

1930. That part of the soil existing above permafrost which usually freezes in winter and thaws in summer, is called the active layer; as can be seen in the graph, its depth is no more than 1 m. Precipitation as well as melted snow and ice cannot percolate. Thus the active layer in summer is saturated with water and thus, muddy. The annual temperature range goes to a depth of 15 m. In some places in Siberia the base of permafrost can be as deep as 600 m, in Alaska 400 m and in Greenland 450 m.

Permafrost is estimated to cover an area of 35,000,000 km^2 on the globe; Table XVII shows its breakdown by continents. The regional distribution of permafrost is shown in Fig.14. It can be seen that in North America the continuous permafrost spreads to the Arctic Circle in Alaska only, whereas in Siberia it reaches southward up to 62°N and sporadic permafrost occurs as far south as 47°N in the upper Amur valley (DEGOES and NEEDLEMAN, 1960).

TABLE XVII

DISTRIBUTION OF PERMAFROST (millions of km^2)

Permafrost	Northern Hemisphere		Antarctica	World
	Eurasia	America		
Continuous	3.7	3.9	12.8	20.4
Discontinuous	3.7	3.7		7.3
Sporadic	3.8	3.5		7.3
Total	11.2	11.1	12.8	35.0

The exploitation of the mineral resources in the permafrost regions (petroleum, precious and heavy metals) is inhibited by the great technical difficulties which arise: buildings and industrial plants release heat which causes the underlying permafrost soil to thaw; the consequence is a loss of stability. Water supply is a problem too (KÖPPEN, 1931). The active layer, which is unfrozen in summer and often saturated with water, starts

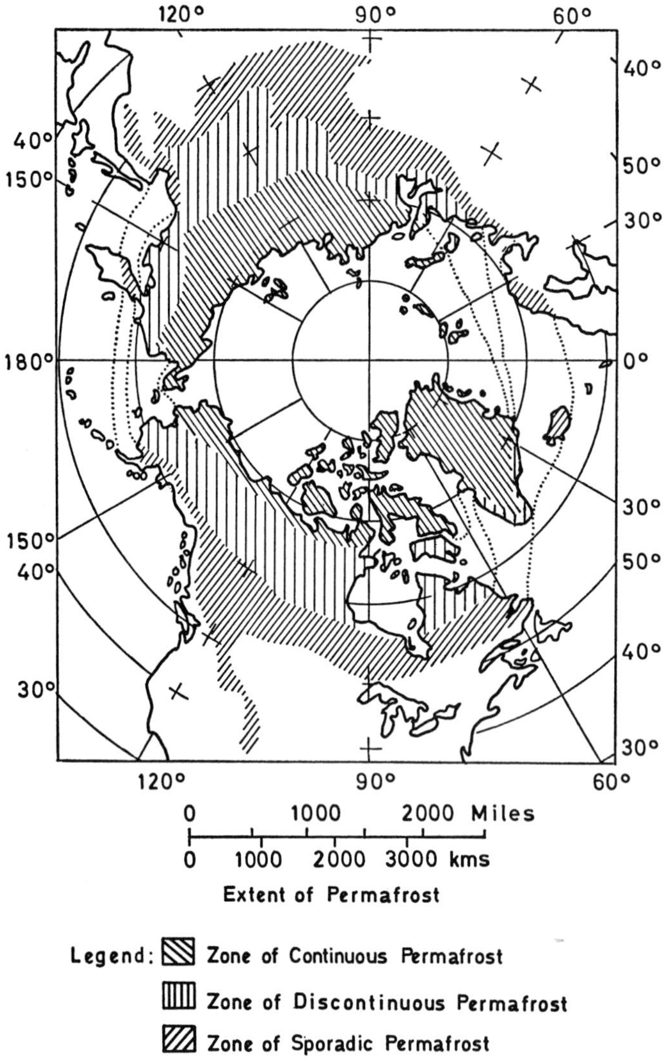

Fig.14. Map showing extent of permafrost.

sliding across the permafrost table underneath. Finally, there is the need of heat insulation which complicated the laying of pipelines from the oil fields on the north slope in Alaska to the ice-free seaports.

Moisture in building materials

Building materials are porous. A wall is said to be wetted when its pores are soaked with water. The soakage may be due to: (*1*) intake of water when the rain is driven against the wall (see next section); (*2*) capillary action which lifts water from the foundation ground; and (*3*) diffusion of water vapor through a building resulting in condensate water.

The diffusion of water vapor through a building is directed from the warm towards the cold side of the wall. If on its passage the diffusive current is cooled to its dewpoint, condensation occurs. The condensate water accumulated in the pores is moved by capillary action to the cold side of the wall where it might evaporate.

Due to their capillarity some walls never dry out completely; e.g., a permanent moisture of 2% by volume must be expected in brickwork and one of 7% by volume in concrete and clay. Table XVIII shows the extremes of moisture content in various types of building material in percentages by volume.

TABLE XVIII

MOISTURE CONTENT (%) OF BUILDING MATERIAL
(After SAUTTER, 1948)

	Brickwork	Concrete	Clay
Minimum moisture	0.3	3.0	4.2
Maximum moisture	4.0	17.0	14.5

Fig.15 gives an example for the soakage due to rain: After a two-day driving rain a third of the total depth of a wall consisting of hollow block bricks is soaked up with rain water. No sooner than two weeks later the original moisture condition of the wall has been restored.

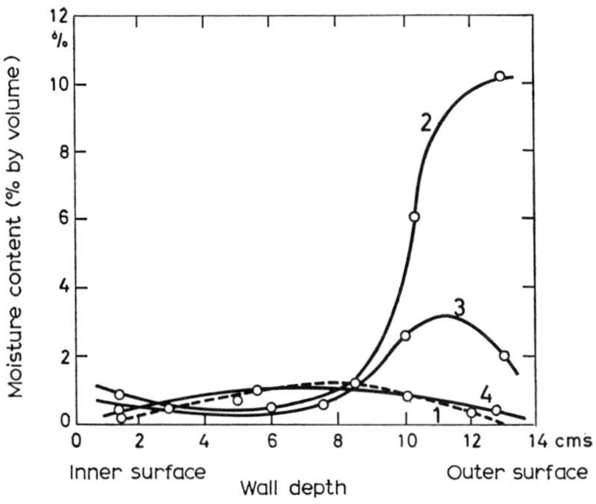

Fig.15. Moisture distribution within a wall after a two-day driving rain. (According to SCHÜLE et al., 1952). *1* = at begin of the rain period; *2* = immediately after end of the rain period; *3* = 7 days after end of the rain period; *4* = 13 days after end of the rain period.

The water vapor diffusion results in an accumulation of moisture in the core of the wall which is enhanced in the winter season because of the great temperature differences between the air indoors and outside. Vapor barriers inside the wall may prevent the soaking by rain, but this may result in an accumulation of moisture which after the end of the heating period seeps to the inner side of the wall. There it may give rise to the growth of fungi. Fig.16 demonstrates the accumulation of moisture in such a wall: The moisture supply and consequently the accumulation of moisture is much greater in kitchens (graph *a*) than in living rooms (graph *b*) (SCHÜLE et al., 1952).

Organic building materials are hygroscopic. By means of intake or release of moisture they tend to adjust to the given air temperature and relative humidity to approach moisture equilibrium, which, e.g., amounts to 3.5% by weight for spruce-wood in case of an air temperature of 20°C and a relative humidity of 15%. At the same temperature it rises to 9.2% for a relative humidity of 50% and to 20.0% for a relative humidity of

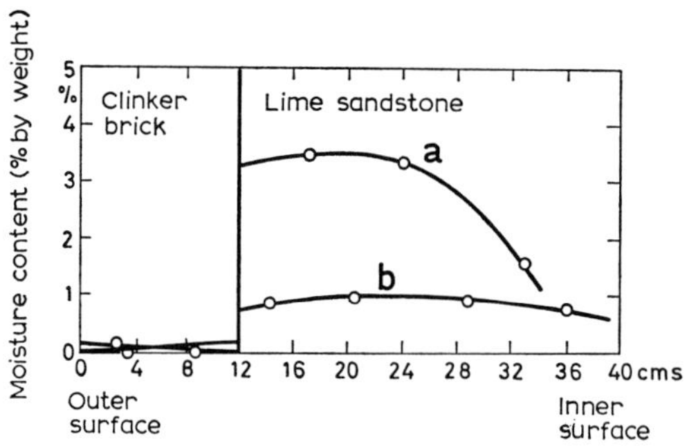

Fig.16. Horizontal moisture distribution in clinker brick faced lime sandstone outer walls. (According to SCHÜLE et al., 1952.)

90%. A change in moisture content results also in a change in the volume and length of the wood (KOLLMANN, 1951) (Table XIX).

The considerable differences in hygroscopic equilibrium in various types of climate are demonstrated by Fig.17 which shows the seasonal variation of the hygroscopic equilibrium of wood in Heligoland, Munich, Duala, Assuan and Ouagadougou.

TABLE XIX

CHANGES IN VOLUME AND LENGTH OF WOOD AS FUNCTIONS OF THE INCREASE OF RELATIVE HUMIDITY FROM 3 TO 20%
(After KOLLMANN, 1951)

	Percentage change in: volume	length
Beech-tree	10	5
Spruce	5	2.5

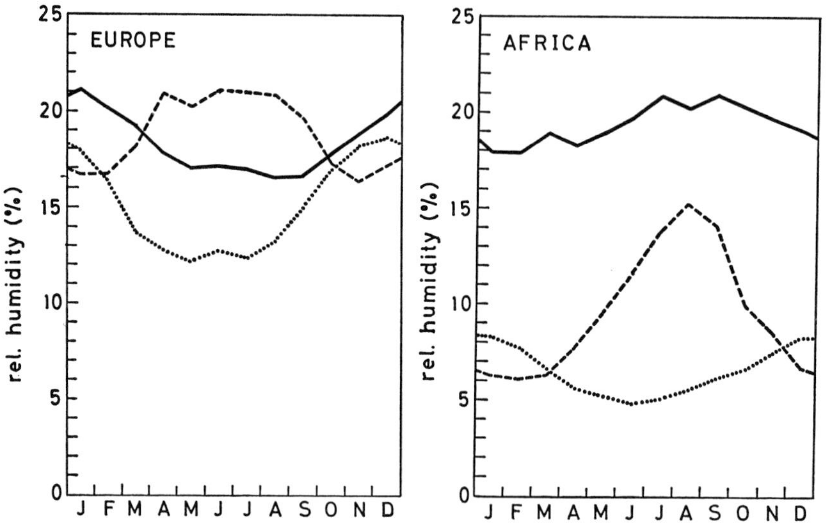

Fig.17. Seasonal variation of the hygroscopic equilibrium of wood. Europe: ———: Heligoland, 54°10′N 07°51′E, 40 m above MSL; . . . : Munich, 48°09′N 11°34′E, 522 m above MSL; — — —: Zugspitze, 47°25′N 10°59′E, 2962 m above MSL. Africa: ———: Duala, 04°03′N 09°41′E, 8 m above MSL; — — —: Ouagadougou, 12°22′N 01°31′W, 316 m above MSL; . . . : Assuan, 24°02′N 32°53′E, 111 m above MSL.

In the hot dry climate of Assuan and in the tropical humid climate of Duala the annual variation of the hygroscopic equilibrium of wood is less than 5%, whereas in the monsoon climate of Ouagadougou it rises up to 13%; off the German coast, in Heligoland, the difference between winter and summer amounts to 4% and is increased to 6% in the Alpine foreland, at Munich.

Rain

The meteorological services measure the amount of rainfall in rain gages with horizontal receiving areas. However, the angle of incidence of rain drops is changed by the wind: if the wind speed ⩾ 8 m/sec, the deflection is greater than 45°, even for large drops. Under these conditions a wall exposed to the wind will receive more precipitation than a horizontal surface of equal area. Rain windroses show the relationship between rainfall amount, wind direction and wind speed (Fig.18). For example, in Hamburg, rain accompanying southwest and west winds has the greatest rainfall amounts and the highest numbers of rainy hours. Rainfall coincident with a windspeed ⩾ 8 m/sec occurs

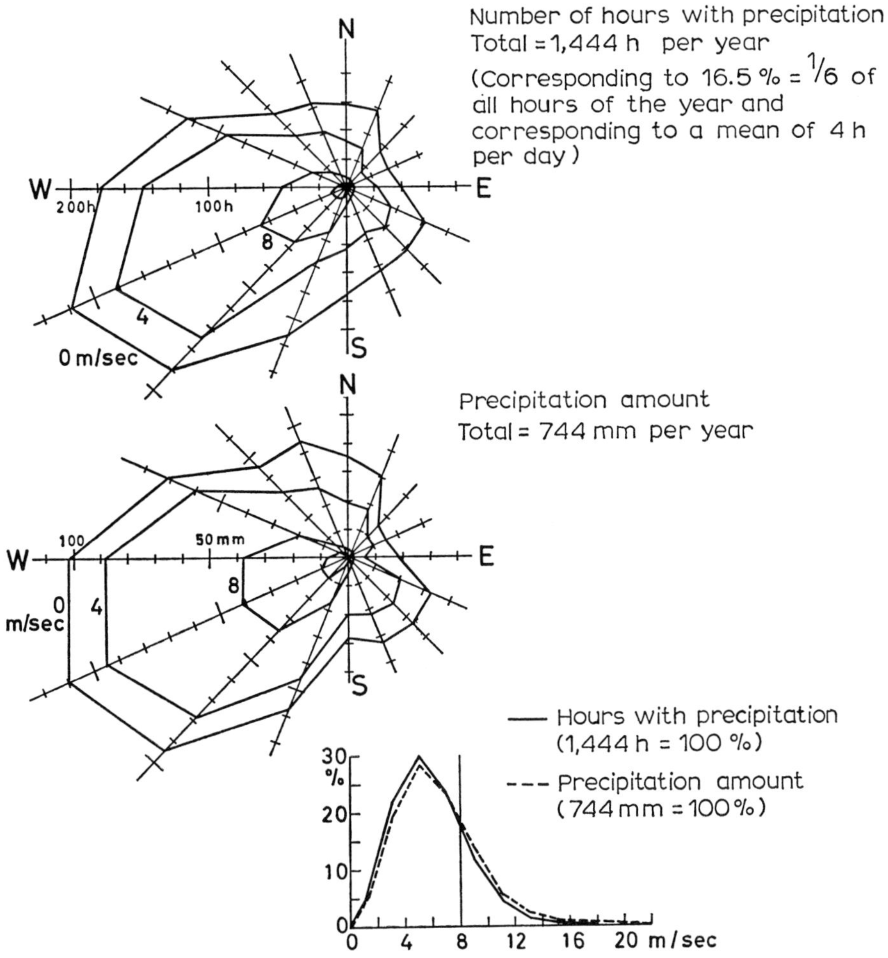

Fig.18. Precipitation windroses (mean values of hours with precipitation as well as precipitation amounts as functions of wind direction and wind speed) and correlation between precipitation and wind speed at Hamburg–St.Pauli (1959–1963).

most frequently with west-southwest wind and has never been observed with easterly wind. The coincidence of a wind speed $\geqslant$ 8 m/sec with rainfall reaches a percentage of 74% on top of the Brocken (in the Harz Mts. in central Germany, 1,100 m above MSL), 44% on the island of Heligoland, 25% at Hamburg, 12% at Berlin, 7% at Bamberg and 5% at Munich (REIDAT, 1963). Climatological information of this type will indicate where rain protection of walls is necessary.

Rainwater incident upon a wall is at first received by the pores of the wall and then moved towards the inner part of the wall by capillary action. After the capillaries are saturated with water, the rain forms a water film on the wall which flows downward bridging cracks and fissures between 0.1 mm and 5 mm in width. The coincidence of high wind speed and large rain drops may raise the speed and mass of the rain drops to such an extent that they enter and penetrate into broad joints. Eddies may form on the wall which force the water, instead of running off, to move upward even over sills, thus being pressed into lap joints—e.g., in case of slate or shingle sidings (TVEIT, 1970).

The assessment of the rain effect on walls should be based on the following meteorological criteria (BIRKELAND, 1966): (*1*) frequency of occurrence and duration of the incident rain as well as the rain intensity, especially in the case of heavy showers; (*2*) raindrop-size distribution for each type of rainfall; (*3*) pressure distribution on the surface of the wall during rainfall; (*4*) evaporation from an open water surface.

Raindrop spectra have been investigated by members of the Technical University of Karlsruhe, Germany, at 8 observation sites of different climate. It has been found that small drops with diameters up to 1 mm amount to a percentage of 98% in Iceland, however, to only 78% in the tropical rainy climate at Entebbe in Africa. The diameters of the largest drops observed in Iceland remained under 3 mm, but at tropical observation sites at Entebbe and Lwiro, and in summer at Karlsruhe, drop diameters of more than 4.5 mm have been recorded.

Excessive rainfall of an intensity of $\geqslant$ 6 mm/h has never been observed in Greenland. However, the frequency of occurrence was 3.8% at Karlsruhe, 8.4% at Barza, Italy, and 16% at Entebbe. These excessive rainfall rates occur usually at high air temperatures with high convective instability in the lower and middle troposphere (DIEM, 1968).

Rain gages with vertical receiving areas have been developed in Norway for the four main compass points of wind direction and in England for the four main and the four secondary compass points of wind direction. Measurements obtained with these devices have been correlated with simultaneous measurements of rainfall obtained with the conventional rain gages with horizontal receiving areas and recordings of wind speed (HOPPESTAD, 1955; KOORSGARD et al., 1964; LACY, 1965; THÖRNBLAD and RYD, 1970). Measurements have been made of maximum daily rainfall amounts on a vertical receiving area (Table XX).

TABLE XX

MAXIMUM 24 HOURLY RAINFALL AMOUNTS ON VERTICAL RECEIVING AREAS

	Rainfall (mm day^{-1})
Trondheim	40 (1954)
Copenhagen	36 (1964)
Glasgow	50 (1958)
Holzkirchen (Bavaria, Germany)	77 (1964)

Comparative measurements with rain gages built into walls showed strong influence on the impact of rain caused by deflection of the air current hitting the wall. The impact of rain at corners and edges of the building was considerably greater than in the center part of the wall (LACY, 1964).

Investigations at Garston (Great Britain) showed that the amount of rain received in a rain gage with a horizontal receiving area multiplied by the wind speed during a rainfall correlates well with the amount collected by a rain gage with vertical receiving area. Lacy suggests to characterize the rain exposure of a building by means of a nondirectional "driving rain index" which is the product of the mean annual precipitation multiplied by the mean annual wind speed (LACY and SHELLARD, 1962). This driving rain index has been mapped for Great Britain and Canada (BOYD, 1963; LACY, 1965).

Directional rain gages have been operated in Norway and Sweden. The data permit correlation of the mean precipitation amount versus wind direction and wind speed, and the mapping of the percentages of driving rain from the main points of wind direction (HOPPESTAD, 1955; THÖRNBLAD and RYD, 1970).

Water resource agencies need to know rainfall intensities for safety reasons. Fig.19 represents a compilation of the world record rainfalls from one minute to one year duration, plotted in logarithmic scale with the time as abscissa and the maximum rainfall in cm as ordinate (JENNINGS, 1941). The fitted curve of all these entries is a straight line for which FLETCHER (1950) empirically derived an envelope of world record values represented by:

$$R = 37.2 - \sqrt{D} \text{ (mm)}$$

with R denoting the rainfall (in mm) and D the duration in minutes.

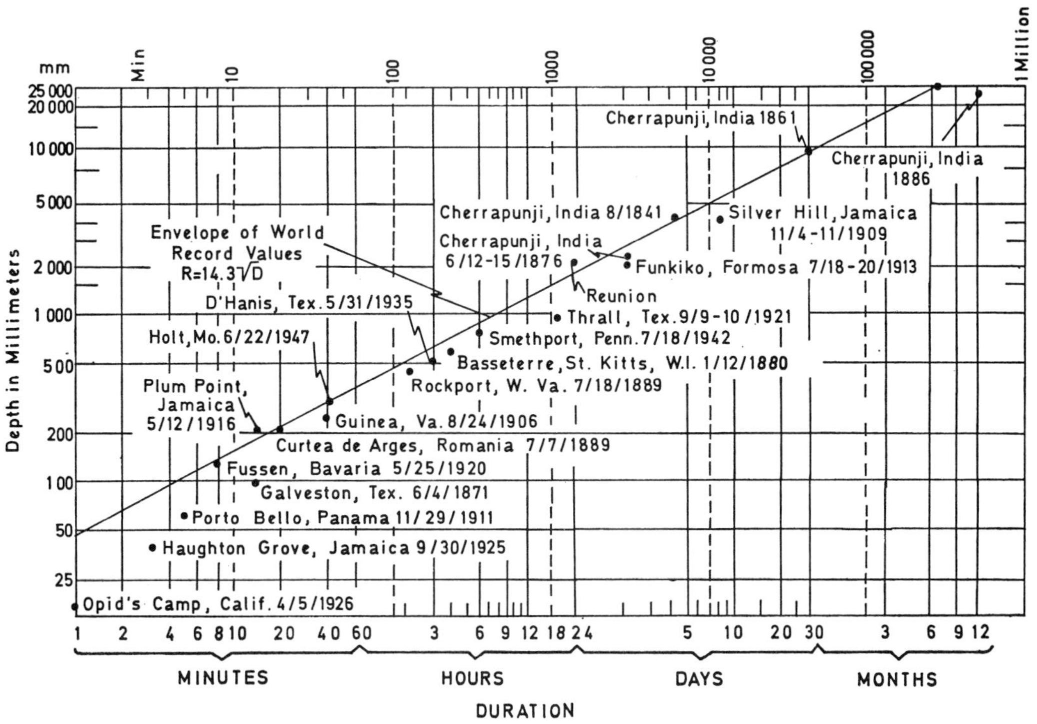

Fig.19. Maximum rainfall rates observed.

The drainage for carrying off the rainwater has to be based on the frequency of occurrence and the intensity of heavy, as well as, persistent rain. From the standpoint of water supply, the meteorological criteria are as follows.
Heavy rain: at least 10 min duration with an intensity:

$\geqslant 0.1\ l\ m^{-2}\ min^{-1}$ (i.e., 0.1 mm/min)

Persistent rain: at least 6 h duration with an intensity:

$\geqslant 1\ l\ m^{-2}\ h^{-1}$ (REINHOLD, 1937)

The rain record at Hamburg, evaluated following Reinhold's method are shown in Fig.20. The duration of rain in minutes is on the abscissa and the intensity of rain in $l\ m^{-2}\ min^{-1}$ or in $l\ sec^{-1}\ ha^{-1}$ (1 ha = 10,000 m^2) resp. as ordinate. The curved lines represent the frequencies (n) of occurrence for heavy rain, e.g., $n = 0.2$ means once in five years, $n = 1$ means once a year, and $n = 2$ means twice a year etc. Sewers in German cities are dimensioned for carrying off the heaviest rain, which, on the average, can be expected once in two years ($n = 0.5$), with a duration of at least 15 min. In order to carry off heavier rainfall in local areas which need to be protected from flooding (e.g., subway crossings or tunnels), it is necessary to build retaining reservoirs. The planning of their dimensioning also requires information on cumulative rainfall amounts of several consecutive days (CIMPA, 1968). A few comparative figures might be of interest: within the last century the record rainfall at Hamburg during one day amounted to 76 $l\ m^{-2}$, during three consecutive days to 95 $l\ m^{-2}$ and during five consecutive days to 135 $l\ m^{-2}$, whereas at Munich a total of 135 $l\ m^{-2}$ can be expected to fall during one day and in that area rainfall rates of more than 200 $l\ m^{-2}\ d^{-1}$ have been experienced (KERN, 1961). Extreme daily rainfall amounts have been recorded at Scotchi, U.S.S.R. with 186 $l\ m^{-2}$, at Peking with 251 $l\ m^{-2}$ and at Cherrapunji with 1,037 $l\ m^{-2}$ (SÜRING, 1939). The world

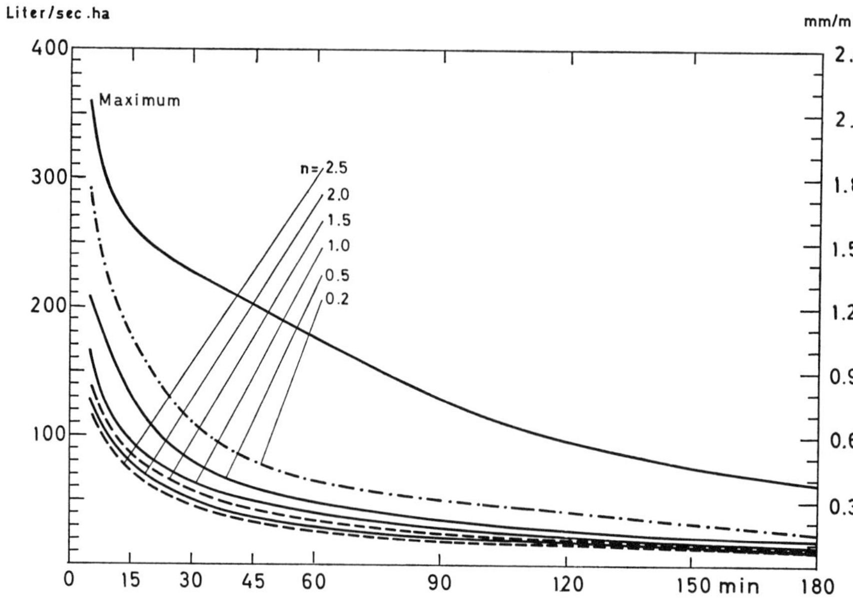

Fig.20. Frequencies of heavy rain at Hamburg–St.Pauli (1948–1967).

records of 24 hourly rainfall were due to typhoon rains with 1168 l m^{-2} on the Philippine Islands and 1880 l m^{-2} on the island of Réunion in the Indian Ocean, east of Madagascar.

Snow

Freshly fallen snow quickly packs resulting in an increase in density and water equivalent. Thus, a new snow cover of a depth of 100 cm and a volumetric weight of 60–80 kg m^{-3} turns into 28 cm deep packed snow of 200–300 kg m^{-3}, then into a 13 cm firn of 400–600 kg m^{-3} with the last step in the transformation into glacier ice of 900 kg m^{-3} equivalent to 7 cm of melt water.

The thermal conductivity of snow increases with increasing water equivalent of volumetric weight rising from 0.04 kcal. (0.167 kJ) m^{-1} h^{-1} per a volumetric weight of 100 kg m^{-3} to 0.20 kcal. (0.837 kJ) m^{-1} h^{-1} per 300 kg m^{-3} to 0.55 kcal. (2.304 kJ) m^{-1} h^{-1} per 500 kg m^{-3} and finally to 1.92 kcal. (8.035 kJ) m^{-1} h^{-1} for glacier ice.

Fig.21 compares the heat conductivity of snow of different density with that of various building materials. The heat conductivity of freshly fallen snow corresponds to that of granulated cork, the compacted snow to insulating board, dry firn to brickwork and wet firn to concrete (REIDAT, 1951).

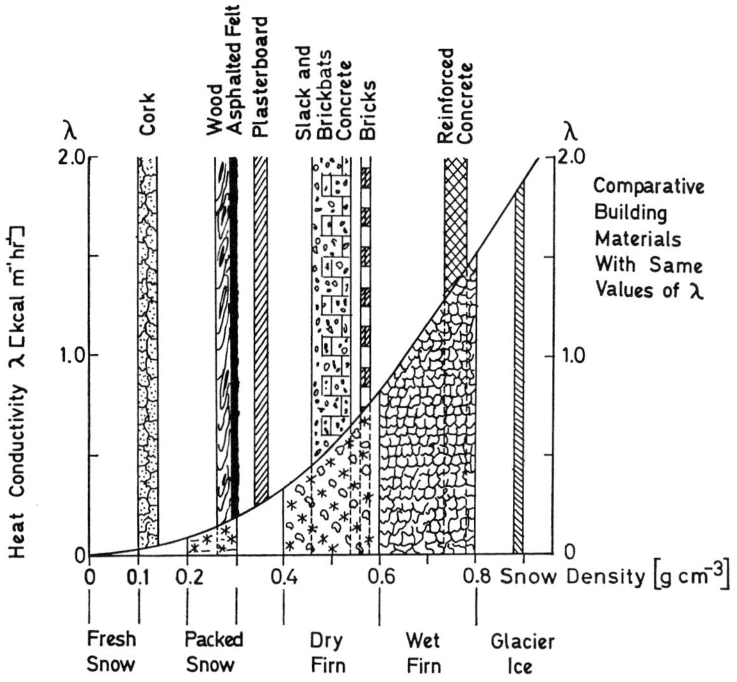

Fig.21. Heat conductivity of snow as function of density.

A dry snowcover prevents rapid and deep frost penetration in the soil. Thus, it is also a good insulating cover for roofs, but snow load requires roof constructions which can stand the heaviest expected snow loads. A variety of factors influence the snow load on roofs, such as wind drift, slope of the roof, the exposure of the building, heat transfer through the roof, run-off conditions for the melted snow and sliding for snow masses.

Usually, the snow load on a sloped roof is smaller than that on a horizontal surface. However, snowdrifts behind protrusions on the roof, such as chimneys, and snow slides from a higher to a lower roof may bring about extreme local snow loads.

Building codes usually define the snow load on roofs as a fraction of the snow load on a horizontal surface. The calculation is based upon the maximum snow depth during a long time interval, the return period being 10 or 30 years. This maximum snow depth is multiplied with the mean density of the snowcover, which in some countries is set equal to 0.2, in other countries, however, to 0.4 g cm^{-3}. Considering the inclination of the roof this basic snow load is then multiplied with a projective factor $\leqslant 1$. But the safety criteria for snow load on roofs are not uniform in various countries (SCHRIEVER and OSTANOW, 1967).

Snow loads can get highly dangerous when they absorb rainwater. Canadian building codes include the mean weight of the rain to be expected at the beginning of the melting period to the safe snow load (BOYD, 1961).

Empirical "safe" snow loads on roofs have been checked by control measurements in Canada, Russia and Sweden with roofs of different inclination and different exposure to the wind for obtaining the "true" values of snow load (ALLEN, 1959; PETER et al., 1963; TAESSLER, 1970). The Swedish measurements provided the following. The snow load on a flat roof is less than that on a plane in open terrain. A roof incline of 22° yields a snow load on the windward side by 48% greater, and a roof incline of 42° yields a snow load on the side by 9% less, than that of a roof incline of 4°. The ratio of the snow loads (kg m^{-2}) of the windward to the leeward side amounted to 100:90 for the roof of 4° incline, to 148:204 for the roof of 22° incline and to 91:163 for a roof of 42° incline. Thus, the Swedish building code obviously prescribes safety snow loads which are overestimated for flat roofs, however, underestimated for the lee sides of roofs of great inclination.

The heat flux from the house to the base of the snow on the roof, may start melting while air temperatures are still below freezing. The melt water, which flows downwards between the surface of the roof and the base of the snowcover, freezes upon reaching the cornice. This results not only in the formation of icicles, but also of an ice mound on the projecting part of the roof so that subsequent melt water is dammed up. In the case of tiled roofs this accumulated melt water can seep through the overlapping slits between the tiles into the attic, and cause damage (TAESSLER, 1970).

Fog deposit and glaze

Hoar frost deposit from fog and glaze from freezing rain may become a notable load for structures adding static load especially for masts, towers, antennas and overhead lines. The surface which is exposed to the wind is also considerably enlarged by the deposit resulting in an increase of wind pressure; this is especially true with overhead lines and tv-towers. The fog deposit on an exposed object depends on: (*1*) the content of water drops and ice crystals in the air, W (g m^{-3}); (*2*) the wind speed, v (m sec^{-1}); (*3*) the duration of the depositing, t (h); (*4*) the diameter or the surface of the exposed object, f (m or m^2 resp.).

There are two more additional quantities, c_1 and c_2: (*5*) a factor, c_1, which characterizes the adhesive strength of the ice deposit to the exposed object, e.g., amounting to 0.9 for aluminum; (*6*) a coefficient, c_2, which characterizes the collection efficiency depending on the diameter of the hydrometeors, δ, the wind speed, v, and the diameter of the exposed object, d. Hence, the deposit is expressed with:

$$M = 3{,}600\ c_1\ c_2\ W\, v\, t\, f\ [\mathrm{g\ m^{-1}\ h^{-1}}]$$

with:

$$c_2 = \frac{1}{311.5\cdot 10^5} \cdot \frac{\delta^2 v}{d} \qquad \text{(Diem, 1952)}$$

Due to the higher moisture of the air deposits formed at temperatures between 0° and −1° are about three times those formed at −11°. The growth of the deposit with increasing wind speed is illustrated by the fact that the deposit which is formed at a wind speed of 17 m $\mathrm{sec^{-1}}$ is about three times that formed at 5 m $\mathrm{sec^{-1}}$.

The risk of severe icing is greatest in higher mountain regions because of low ceilings in winter or orographic clouds. However, over flat terrain the freezing of raindrops, or the refreezing of largely melted snowflakes upon impact, may cause a dangerous accretion of glaze when falling into layers with below-freezing temperature near the ground. In Norway, the most severe icing is to be expected at heights between 800 to 1,000 m above MSL. The maximum ice load formed on a test cable at 1,000 m above MSL amounted to 60 kg ice per meter; this large quantity had formed within 30 h from 1 to 4 November 1953 (Råstad, 1955).

In February 1952, a series of measurements was taken on top of the Feldberg, in the Black Forest in Germany; ice formation occurred during more than half the hours of observation. The average ice load was 32 g $\mathrm{m^{-1}\ h^{-1}}$, the daily maximum was found to be 2,570 g $\mathrm{m^{-1}\ d^{-1}}$. A total of 4,220 g $\mathrm{m^{-1}}$ had formed between 1 and 6 February 1952 (Diem, 1952).

Methodical investigations into ice accretion have also been carried out in Russia (Rajewski, 1953; Muretow, 1957; Rudnewa, 1958, 1960). The least ice accretion, at a height of 2 m was found in the flat terrain of western Siberia with 20–50 g $\mathrm{m^{-1}}$; in west Russia, the flat terrain experiences a maximum ice accretion of 150 g $\mathrm{m^{-1}}$, however, the windward slopes of the hilly terrain, e.g., the Valdai Hills, have maximum loads of up to 500 g $\mathrm{m^{-1}}$; in the Carpathians ice accretion amounting to 25 kg $\mathrm{m^{-1}}$ must be anticipated. In the Ural the maximum ice load at a height of 2 m was found to be 2 kg $\mathrm{m^{-1}}$; on an overhead line, however, with a wind speed of 17 m/sec, it reached the amount of 16 kg $\mathrm{m^{-1}}$. The regional distribution of ice accretion, at a height of 2 m has been mapped (Rudnewa, 1960). Although the figures give hints for potential icing risk, the numerical values do not necessarily represent the ice load which might accrete on masts and cables.

A great hazard is the accretion of wet snow in connection with gale force winds and temperatures slightly below freezing. A typical case occurred on 19 March 1969, the ice load due to supercooled rain destroyed the radio transmitter at Emely Moor in England, which was 250 m in height (Page, 1969).

Weather and construction

Construction work is greatly influenced by the weather. Rain, wind and cold are a handicap. Heavy rainfall may flood the excavation. Gale force winds render the employment of cranes for hoisting building material impossible. Brickwork and poured concrete require a certain amount of time for reaching the stability desired; meanwhile they are greatly affected by gale force winds, wetness and freezing temperatures. Structures may be damaged if they dry up too quickly in summer or freeze in winter. In former times, construction work in central and northern Europe was interrupted in the cold season of the year.

Today, the engineering progress results in an extension of the building season, if possible covering the entire year. This requires protective measures; however, these must be cost-efficient with regarding to working and operating time. Therefore, the planning of these protective measures must be based upon the impediments posed by adverse weather anticipated during the construction period. An effective tool for this planning is the presentation of climatic data such as frequency and duration of temperatures below specific threshold values, precipitation and gale force winds (REIDAT, 1959). During the construction work medium-range weather forecasts and short-range warnings against critical changes of the weather for the worse are helpful (ZAWADIL, 1955). Obstructive weather influences result in a prolongation of the construction time and thus a rise in costs. As an objective measure for the obstructive weather influence on a building project to be expected during a specific period of time, a "building weather index" has been developed in Austria (HADER, 1960).

Climatic aids for construction engineering

With the help of laboratory and field tests the civil engineers try to determine the influence of weather on building materials and structure. These findings must be extrapolated to other building sites. This requires an insight into the weather conditions to be anticipated there, so that the architect can base his planning on local climate data.

A first attempt to produce special climatological information tailored for construction engineering was made by P. Siple, in cooperation with the periodical *Housing Beautiful*, in the U.S.A. to show the influence of the meteorological parameters on the building without any textual explanations: for each station standardized climate information is given on three charts in the form of analyses of temperature, sunshine, wind, precipitation, and moisture, by months (SIPLE, 1950). Similar information has been developed for Germany (REIDAT, 1959).[1]

The present climatological statistics for building design suffer from the inhomogeneity of data processing and sources. The Working Group of International Federation for Housing and Planning (I.F.H.P.) in collaboration with W.M.O. (World Meteorological Organization), aims at issuing terms of references for building design in order to initiate a world-wide uniform processing of climate data for architects.

Climate data for building design need to be applied with caution because they are valid only for the observation site. An experienced climatologist may be able to extrapolate such information, but for major buildings some on-site observations may be needed.

[1] A detailed processing of voluminous climatic observations of Vienna for the purpose of building design has been performed by STEINHAUSER et al. (1955, 1957, 1959). Data for Glasgow have been worked out by PLANT (1965). National climate data for building design have been published in Norway (JOHANNESSEN, 1956), in Russia (S.N.I.P., 1963) and for New Guinea (BALLANTYNE and SPENCER, 1964). For Sweden a "Climate Data Book" has been developed by TAESSLER (1970). In Austria, the attempt has been made to develop special maps for building design (BRUCKMAYER, 1958; HADER, 1970).

Climatic conditions during transport

Routes and means of transportation as well as merchandise being transported are affected by the weather.
Climatological outlooks enable the design of new roads in such a way as to avoid, or at least reduce, interruption of traffic by adverse weather. The design of the high bridge near Innsbruck, which was part of the highway to the Brenner Pass, had to be based on detailed studies of the wind flow in the valley underneath. Routing the highway from Linz to Vienna required thorough investigations into the local snow and icing conditions as well as wind flow along the edge of the Vienna Forest (FRIEDRICH, 1953).
Research in the field of microclimatology enables the designer to avoid weather hazards along the roads, such as radiational cooling or flooding, by means of appropriate lay-out and design of roads (Research Institute for Engineering Biology, 1938).
Traffic safety may require special construction to prevent wind and snow drift hazards. Meteorological services have been instituted to warn against hazards such as katabatic winds blowing on the east coast of the Adriatic, the bora avalanches and road icing in mountain valleys (LIEBSCH, 1955).
A contributing factor to road safety is the effect of various weather elements such as solar radiation, air temperature and humidity on road surface. Asphalt and concrete show different behavior. They show notable differences in temperature range under the same exposure. Asphalt is a poor conductor of heat compared with concrete; it absorbs a greater fraction of the global radiation. During maximum heating the temperature lapse rate in the concrete is $0.7° cm^{-1}$, in the asphalt, however, $2.5° cm^{-1}$; during the strongest cooling the lapse rate in the concrete is $-0.1° cm^{-1}$ and in the asphalt $-0.3° cm^{-1}$. Hence an asphalt surface heats more and quicker during daytime, but cools faster at night than a concrete surface. At below-freezing temperatures the icing on asphalt roads starts sooner and is more intense than on concrete roads. Past experience in Switzerland has shown that in deep valleys, which receive sunshine only late in the afternoon, concrete roads due to their better heat conductivity were free of ice in the transitional seasons—fall and spring—for a much longer period of time than asphalt roads. But in flat terrain at sunrise the asphalt absorbs enough heat for melting ice on its surface quicker than concrete surface (WEIL, 1959).
The meteorological conditions inside of buses and trucks and their effect on persons and merchandise transported, largely depend on the characteristic properties of the vehicles and on the weather conditions along the route. Air conditioning of highway and rail transport for passengers has capabilities for both heating and cooling the interior of the vehicles (DIRMHIRN, 1952). The railroad cars in the U.S.A. are kept at constant temperature by thermostatic regulation, the ventilation rate is 25–30%. There is no humidity control, because in summer as in winter, the moisture stays within the comfort limits unless the cars are very crowded (KÄLTETECHNIK, 1954).
For passenger ships bound to the tropics the air conditioning needs to include dehumidification to a relative humidity of 50%. The adequate ventilation of day-rooms and cabins is usually achieved without excessive drafts. The temperature differences between outside and inside air need to be controlled (TER LINDEN, 1940).
Bulk goods which are transported over large distances are affected by the change in air temperature and solar radiation for the entire length of the route. At Foggia, in southern

Italy, temperature measurements were made inside a tank car with a capacity of 30 m^3 filled with gasoline; the mean daily temperature was 26°C and the daily total of solar radiation 30.600 kJ. m^{-2} d^{-1}. The mean temperature of the gasoline was 34°C, with the excessive temperature of 52.6°C at the surface of the liquid (SCHÖN, 1957).

Horizontal temperature differences of 15°C to 20°C, at a distance of 1,000 km, are not uncommon in Europe. If the quantity of 30 m^3 of gasoline had been decanted in an Upper Rhine refinery, where the temperature of the liquid in the storage container was 10°C, the heating along the route of transportation to southern Italy, with a final temperature of 26°C, would have caused an increase in volume of 1 m^3. In order to avoid fluid losses, or the danger of explosion, the filling volumes must be adjusted to the change in volume which has to be anticipated by temperature variations en route.

Overseas shipment is particularly affected by the change in weather and climate. Exudation is very detrimental to the cargo. The cause of exudation can be either the cooling of the hull below the dewpoint of the cargo during the passage of cold ocean areas or condensation on supercooled cargo during the passage of warm humid ocean areas. Overheating may lead to spontaneous ignition of the cargo. Both the hazards of exudation and spontaneous ignition can only be avoided by appropriate ventilation. Research expeditions have been arranged for observing the environmental influence on the indoor climate of the cargo space and testing the effects of ventilation measures (HÖLLER et al., 1956).

In 1957 a research expedition with the motor ship *Cape Ortegal* was undertaken for studying the indoor climate of the cargo space. During the voyage from Antwerp to Rio de Janeiro (6–18 March) the surface water temperature rose from 6°C to 28°C at the Equator; simultaneously the air temperature inside of an unventilated cargo compartment below deck with engine at half a meter above the bottom was raised from 7°C to 20°C. The sudden activation of ventilation would have caused serious exudation with the risk of corrosion. On the way back (18 April–4 May) to Rotterdam, the surface water temperature dropped from 24°C to 9°C and the temperature fell from 22°C to 2°C, whereas the temperature inside of a millet and maize cargo in the hold, remained almost constantly 24°C at a depth of one meter. Strong ventilation toward the end of the voyage would have resulted in a great cooling of the surface of the cargo, thus increasing the temperature lapse rate in the upper layer of the cargo with condensation in the diffusing moist air from the cargo. The consequence would have been a soaking of the upper layer of the cargo. A coffee cargo stowed in space above the decks was heated to 38°C, caused by strong solar radiation at the Equator, but the inside air temperature 50 cm underneath the hatchway was 26°C; at night the cargo cooled down to 27°C (BULLIG, 1958).

The results of such research voyages led to design of special climatic maps for the sea-routes and the development of rules for ventilation. They show the potential diurnal drops of temperature in the month of May during the voyage to America and around Africa (HÖLLER et al., 1956). A meteorological outlook on the weather on the route and at the destination, and severe weather warnings, improve the safety of ship traffic and reduce the risk of deterioration of merchandise.

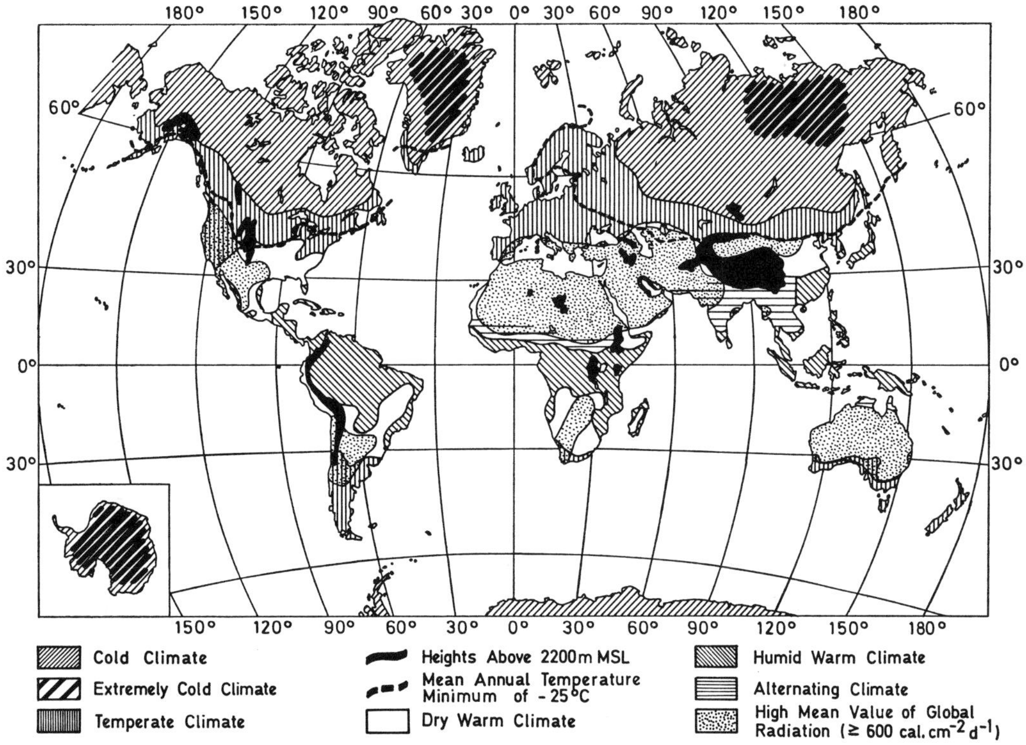

Fig.22. World climate map (proposed supplementation of German Industrial Tentative Standards No.50019; draft of Feb.1961) Van der Grintens circular projection. (BOËR et al., 1961.) Note the extreme climate stress along the coast of the Persian Gulf and the Gulf of Oman. Consider the possibility of extremely high air temperature occurring in some desert regions.

Climatic support for industry

The planning of industrial establishments has to take the influence of weather and climate on raw materials, semi-finished products and finished products into account. The manufacturing plant and the mode of operation must be adjusted to the climatic conditions for optimizing the manufacturing process. Quality defects due to the influence of adverse weather must be checked with a minimum of engineering. Only very rarely do natural climatic conditions meet the requirements for the manufacture and storage of products, indoors as well as outdoors. A good example is a Roquefort-cheese dairy in France. The production takes place in caverns under optimal conditions of constant temperature, humidity, and sufficient ventilation (FLEURI, 1951), the year around.

This and other optimal indoor climates for various processes are listed in Table XXI.

Technical products should be durable and perform well. The raw materials used in their manufacture must stand exposure to adverse weather. Corrosion reduces durability. Plastic surfaces are adversely affected by ultraviolet solar radiation. Metallic surfaces undergo deterioration by large temperature variations, high moisture content, and aerosols such as salt crystals, dust and industrial pollutants. The resistance of materials against adverse weather can be determined by laboratory and field tests. The results of a steel surface field test is shown in Table XXII.

Safety and reliability in operation are mainly affected by temperature and moisture. Low temperatures alter the elastic properties of materials. In severe cold iron can get as

TABLE XXI

OPTIMUM INDOOR CLIMATIC CONDITIONS FOR PRODUCTION AND STORAGE
(After GRUNDKE, 1956)

	Temp. (°C)	Rel. humidity (%)
Production:		
Cotton weaving mill	20–25	65–85
Wool weaving mill	20–25	50–80
Synthetics weaving mill	20–22	55–60
Viscose factory (Rayon spinning mill)	20	80–90
Chocolate works	18–20	40–50
Cheese-dairy (Roquefort)	4–8	about 90
Cheese-dairy (Camembert)	13–15	80–90
Paper offset printing	20–22	50–65
Film fabrication	20	60
Storage:		
Salted meat	−3–1	75–90
Preserved meat	10–12	50–70
Tobacco products	16–20	55–65
Paper	15–20	40–65
Leather	10–15	50–70
Films	12–18	50–65

TABLE XXII

CORROSION OF STEEL SURFACES (g m^{-2} $year^{-1}$)

Climate	Location	Corrosion
Dry and hot	Khartoum	0.05
Cold	Abisco (Sweden)	0.05
Humid and hot	Nigeria, Singapore	0.4
Temperate (countryside)	Wales	1.0
Temperate (coast)	Clashot, Redear	1.5
Little industrialized area	Motherwell	2.0
Highly industrialized area	Sheffield	5.0

brittle as glass. Lubricants become sticky. Electrical and electronic devices suffer in very moist air.

Special problems arise from cooling needs. Cooling agents are usually water or air; water cooling is more efficient than air cooling but not always available. In ordinary electric power plants the fraction of total heat production which is not transferred into electrical energy is 62–67% and in nuclear reactors 68% (DIECKAMP, 1971). The excess heat discharged into the rivers often upsets the biological equilibrium. In automobiles a limited amount of cooling liquid is cooled by air in radiators. This mode of cooling is insufficient for great industrial plants where the cooling rate is raised by evaporation in cooling towers.

The effectiveness of cooling towers depends mainly on the enthalpy of the air injected for the purpose of cooling. The enthalpy, and consequently the cooling effect, are equal in air having a temperature of 15°C and a relative humidity of 100%, or 20°C and 59%, or 25°C and 33%, or 30°C and 17%. The wet-bulb depression of the injected air limits the maximum cooling rate as listed in Table XXIII.

TABLE XXIII

WET-BULB TEMPERATURES (°C) AS FUNCTIONS OF AIR TEMPERATURE AND RELATIVE HUMIDITY

Dry-bulb air temp. (°C)	Relative humidity: 20%	40%	60%	80%
10	2.6°	4.6°	6.5°	8.3°
15	6.0°	8.5°	10.8°	13.0°
20	9.3°	12.4°	15.2°	17.7°
25	12.7°	16.2°	19.5°	22.4°
30	15.7°	20.2°	23.8°	27.1°

The evaporative loss of cooling water is proportional to the difference in the vapor tension of the air at the cooling tower's intake and outlet; in the temperate central European climate it amounts to about 2% of the cooling water supply (BISHOF, 1939; DIN, 1947, 1959; SPANGEMACHER, 1970).
Equipment for unrestricted global use requires such a high degree of protective measures against the deteriorating effects of radiation, temperature and moisture that in many climatic regions such a technical effort would not be cost-effective because it is unnecessary. The industries aim their products at large markets. Tailoring manufacturing of equipment for a specific climatic region is profitable only in case of sufficient demand. In consideration of the technical requirements, a joint board of German engineers and meteorologists developed the German industrial tentative standard No. 50 019 (DIN Vornorm 50 019) "Materials, unitized construction elements and equipment tests as well as climatological review on the types of outdoor climates". Special requirements apply to electronics, precision mechanics and optics, and improvement and supplementation is still needed. A tentative classification is shown in Table XXIV.

TABLE XXIV

MAIN TYPES OF CLIMATE FOR TECHNOLOGICAL PURPOSES

Meteorological parameters	Main types of climate			
	cold	temperate	dry and warm	humid and warm
Air temperature	Lowest monthly mean less than −15°C	Monthly means between −15°C and 25°C	Highest monthly mean above 25°C	At least one monthly mean above 20°C with the rel. humidity being above 80%
Air humidity	—	—	Rel. humidity below 80% if monthly temperature mean is above 20°C	
Air pressure				770 mbar

Later criteria were added for extremely cold climates with some monthly mean temperatures below −30°C, and for monsoon climates with alternating dry-warm and humid-warm seasons. The regional distribution of these main types of climate for technological requirements can been seen on the world climate map, Fig.22, which is supplemented in

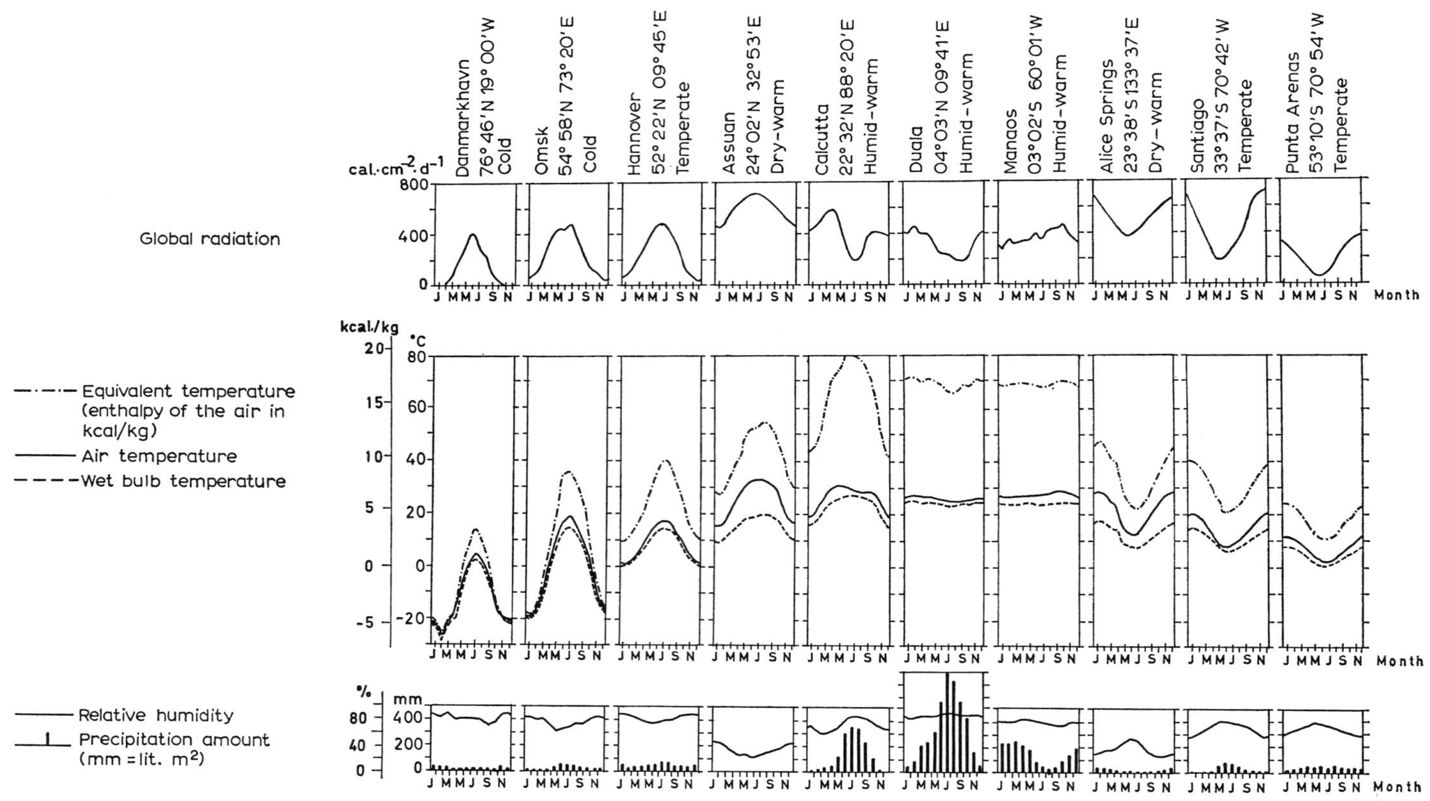

Fig.23. Examples of characteristic climatic features expressed with monthly means. (By courtesy of the German Standards Committee, extract from draft German Industrial Tentative Standards No.50019.) *Note*: The difference of the equivalent temperature minus the air temperature is a measure of the latent heat; the difference of the air temperature minus the wet bulb temperature is a measure of the possible evaporative cooling.

Fig.23 by some diagrams of typical climatic features. Information of this type can give a quick orientation to engineers, but specific industrial planning must be based upon detailed climatological analysis by an expert.

Air pollution

The combustion of solid and liquid fuel, and industrial effluents, lead to air pollution consisting of either solid residual combustion products (such as ashes or soot) or waste gases (CO_2, CO, SO_x, NO_x, F). The increase in industrialization since the end of World War II has led to a tremendous load on the dispersing properties of the air.

Dust

Dust particles have a radius of the order of magnitude of 10^{-3} cm. The smaller the diameter of the particle, the slower their settling velocity. These fine particles remain suspended for some time in the atmosphere and can drift considerable distances. In the early twenties of this century, dust counters konimeters and nuclei counters have been used for decades for instantaneous counts of aerosol particles. Modern methods of measuring solid pollutants use generally bulk filter samples aspirated over specific time intervals or photoelectric counters.

The contents of particulate matter in the air originally depend on the trajectories of air masses, but local emission from industrial processes, combustion of fuels and road traffic usually play the predominant role. The diurnal variation of particulate matter in the atmosphere depends on the diurnal variation of vertical temperature lapse rate, wind speed and man's activity cycle. At Hamburg, the maximum of emitted particulate matter is to be expected some hours after sunrise, the minimum shortly after midnight. Fall-out of dust is measured in open glass receptacles (Löbner, 1949) or by means of sensitized sampling plates (Diem, 1956).

Effenberger (1971) tested the common dustfall meters for the errors in measurement. They amounted to $\pm$ 1% for the open glass receptacles (Löbner) and to $\pm$ 1.7% for the sensitized sampling plates (Diem).

Simple measuring devices, exchanged every week or every fortnight, will give a gross estimate for local variations of dustfall in cities. Such measurements have been made in many urban areas (Berlin: Löbner, 1950; Cologne: Wiemer and Brambach, 1958; Hamburg: Gräfe and Schütze, 1966; a number of stations in Austria: Steinhauser, 1966). In the Hamburg area, measurements at 236 locations showed the following results. The mean daily dustfall over the entire Hamburg area during the years 1964–68 was 171 mg m^{-2} d^{-1}. Local "hot" spots of 400–500 mg m^{-2} d^{-1} were found in the harbor and in the industrial area; at the junction Stephansplatz the maximum was measured at 239 mg m^{-2} d^{-1} which corresponds to 139% of the overall mean. These results show the high production rate of particulate matter caused by traffic, in the form of unburned or partly burned fuel or of raised dust. Partial railway electrification reduced the dustfall on the platforms from 283% to 175% of the overall mean.

Waste gases

Stack effluents are subject to transportation and diffusion by the surrounding air. SUTTON (1947) developed a model of pollution dispersal from factory stacks on which most of the more modern developments are based. (More details on diffusion processes can be found in Volume 2 of this series: DEACON, 1969.)
Under some restrictive assumptions relations for the horizontal distance $x_{\max}$ where the effluent from a stack of height R reaches the surface of the earth and for the concentration of the pollution $\chi_{\max}$ at this point of contact have been developed:

$$x_{\max} = \left(\frac{R}{C_z}\right)^{\frac{2}{2-n}} \quad \text{and} \quad \chi_{\max} = \frac{2\,Q}{e\pi\bar{u}R^2}\left(\frac{C_x}{C_y}\right)$$

where C_x, C_y and C_z denote the virtual diffusion coefficients depending on the environmental lapse rate, and associated stabilities in the lower atmosphere; R the stack height or height of the source of pollution; n meteorological exponent derived by Sutton from the vertical wind profile which correlates well with the lapse rate; $\bar{u}$ mean wind speed at a height of 10 m/height; Q emission rate.
Sutton's equations imply that the horizontal distance at which the concentration of pollution reaches its maximum is almost proportional to the stack height in the range of small values of n and that this maximum concentration is directly proportional to the emission and inversely proportional to the square of the stack height.
DIEM and TRAPPENBERG (1953) showed that Sutton's equations can also be applied to the distribution of dustfall.
The exact evaluation of the virtual diffusion coefficients C_x, C_y and C_z meets with difficulties because their numerical values depend on vertical temperature lapse rate. From his measurements at Risö in Denmark JENSEN (1962) derived mean values ranging between 0.04 and 0.20 for C_y, which undergoes great diurnal and seasonal variations, and mean values ranging between 0.04 and 0.12 for C_z. Jensen also found that the numerical value of n very frequently falls between 0.2 and 0.6.
Modification of Sutton's equations have been developed in an ever-increasing literature. For details the reader is referred to one of many pertinent handbooks (STERN, 1976).
The rate of exchange of the stack effluents depends largely on the turbulent diffusion of the atmosphere, which PASQUILL (1962) described in terms of the meteorological parameters: wind speed, incoming solar radiation and cloud cover. He classified seven weather types or stability categories respectively: A = great instability; B = moderate instability; C = slight instability; D = neutral; E = slight instability; F = moderate stability; and G = great stability. The meteorological criteria for these stability categories are listed in Table XXV.
Determination of hourly frequencies of these categories from climatic observations gives a good rapid insight into the local dispersion characteristics and are helpful for planning. Many aspects of the climatological air pollution potential have been discussed by HOLZWORTH (1974).
FORTAK (1970, 1972) describes a multiple source diffusion model for the simulation and prediction of long-term (climatological) ground-level sulfur dioxide concentrations in urban areas. The model calculates fields of steady-state ground-level concentrations that

TABLE XXV

STABILITY CATEGORIES FOLLOWING PASQUILL (1962)

Wind speed u at 10 m (m/sec)	Day, insolation G (cal. m^{-2} min^{-1})			Night, cloud cover N		
	strong, $\geqslant$1,100	moderate, 1,100–500	slight, $\leqslant$500	8/8	$\geqslant$4/8	$\leqslant$3/8
<2	A*	A–B	B	D	G	G
2–3	A–B	B	C	D	E	F
3–5	B	B–C	C	D	D	E
5–6	C	C–D	D	D	D	D
$\geqslant$6	C	D	D	D	D	D

* For explanation see text.

correspond to a given spatial distribution of emission sources and to any possible combination of relevant meteorological diffusion parameters.

Assuming a uniform environmental lapse rate within the entire layer of potential conical distribution of the smoke plumes from stacks Fig.24 shows the different shapes of these plumes as functions of the environmental lapse rate.

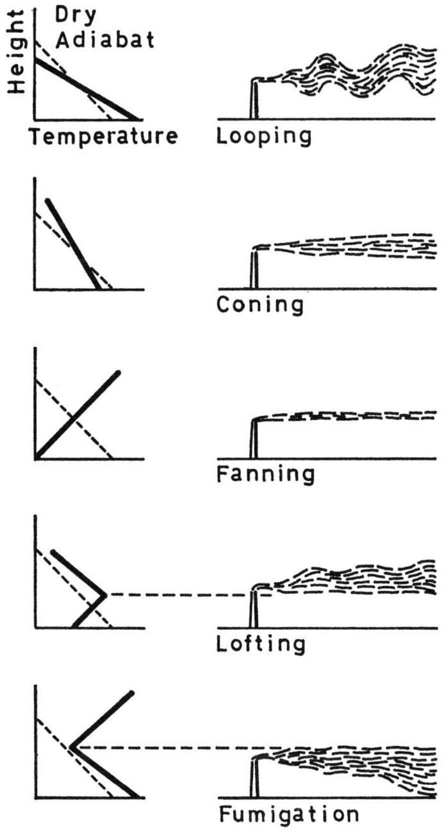

Fig.24. Characteristic forms of smoke plumes from chimneys (JENSEN, 1962).

In the case of superadiabatic lapse rates large eddies form, some of which may reach to the ground, thus resulting in highly intensified local air pollution (looping). If the environmental lapse rate is less than the dry-adiabatic lapse rate, conical distribution of the plume will occur (coning). In the case of temperature increase with height the plumes drift away in the height of thermal equilibrium without any noteworthy spreading (fanning). A surface inversion reaching to the top of the stack shields the ground from pollution; whereas in the layer where temperature decreases with height a conical plume can form (lofting). Inversions close above the chimney stack lead to an accumulation of the pollution in the boundary layer between ground and inversion (fumigation); this concentration may reach twice the intensity of that experienced in normal dispersion.

Registrations of air temperature at various heights of a stack on the Elbe estuary at Hamburg from mid-November 1962 until the end of May 1964 have been evaluated for the type of environmental lapse rate. In the layer up to 175 m above the ground 24% of all hours had an increase of temperature with height, 15% of all hours had absolute stability (the lapse rate being less than 0.6°C per 100 m), 30% of all hours had conditional instability (the temperature decrease with height falling between 0.6°C and 1.0°C per 100 m (i.e., less than the dry-adiabatic lapse rate but greater than the saturation-adiabatic lapse rate), and 31% of all hours had absolute instability (the lapse rate being greater than the dry-adiabatic lapse rate, i.e., greater than 1°C per 100 m).

The changes of environmental lapse rate followed a distinct diurnal variation. In the summer months June through August 1963, almost half of all hours between midnight and sunrise experienced an increase of temperature with height; only a quarter of all hours within this time interval had a temperature lapse rate greater than 0.6°C per 100 m. After sunrise the surface inversions are quickly destroyed and during noontime between 12 and 14 o'clock they do not occur (REIDAT, 1968).

The influence of these changes in environmental lapse rate on the distribution of airborne pollution is shown in the Fig.25. The formation and the removal of a surface inversion affect the plumes of three chimneys which are different in height. In case of a dry-adiabatic lapse rate (Fig.25a) all the three plumes are conical. Due to the formation of a surface inversion a stable surface layer is built up which expands to the height of the highest stack (b–d). The plumes of the two stacks underneath the top of the inversion

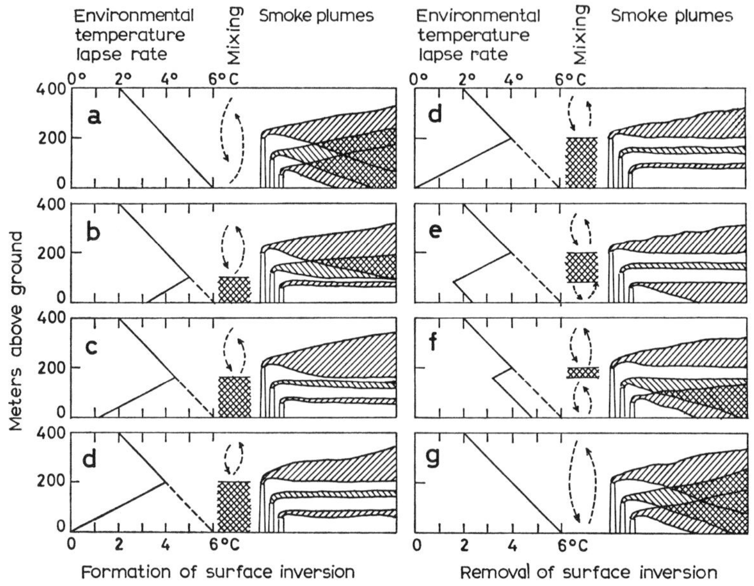

Fig.25. Response of industrial smoke plumes to the formation and removal of a surface inversion.

undergo fanning, the plume of the highest stack follows the type that lofts above the inversion.

The surface inversion is removed when the incoming solar radiation heats the ground so that a few centimeters of the lowest atmospheric layer near the ground are heated by convection. This gradually increasing convectional layer (Fig.25, e–f) underneath the gradually decreasing stable layer has a dry-adiabatic lapse rate. Thus, the plumes, starting with that of the lowest stack, are changed into the fumigating type resulting in a rapid increase of concentration of pollution near the ground, as often as another stack is affected. No sooner than after the surface inversion is entirely removed due to heating from the ground, all plumes exhibit conical shape again (Fig.25g).

It happens, especially in calm anticyclonic weather during winter, that the surface inversions are not fully removed during daytime. This is particularly the case in deep valleys. The meteorological peculiarities of valleys are of great importance for pollutant dispersal. The reader is referred for details to the discussion of GEIGER (1969).

In regimes of semipermanent or slow moving anticyclones meteorological conditions prevent rapid dispersions. The strong solar radiations in these regions will cause photochemical reactions among pollutants. The daughter products may be irritants and lead to reduced visibilities, a condition often designated as smog.

Tragic air pollution episodes with harmful concentrations of air pollutants have been experienced in 1930 in the Meuse valley in Belgium, 1948 in Donora, Pennsylvania, and 1952 in London.

References

ADAMENKO and HAGRULLIN, 1969. Results from observations of radiation incident upon exterior walls. *Tr. GGO*, 248: 69–73 (in Russian).

ALLEN, D. E., 1959. *Snow Loads on Buildings.* (Translation of 7 Russian papers.) National Research Council of Canada, Tech. Transl., No.830, Ottawa.

ANAPOLSKAJA, L. E. and GANDIN, L. S., 1966. Theoretical meteorology for building design. In: M. I. BUDYKO (Editor), *Modern Problems of Climatology.* Gidrometeoizdat, Leningrad (in Russian).

ANIOL, R., 1952. Die grössten Bodenfrosttiefen. *Naturwiss. Rdsch.*, 11: 493–494.

ANQUEZ, J., BOREL, J. C. and CROISET, M., 1965. La protection des baies vitrées. *Cahiers du Centre Scientifique et Technique du Bâtiment*, No.72, Cahier 608, 63 pp.

AYOUB, R., 1966. Natürliche Klimatisierung. *Z. Glasforum*, 1966: 1–7.

BALLANTYNE, E. R. and SPENCER, J. W., 1964. *Climatic Data for Building Design in the Territory of Papua and New Guinea.* CSIRO, Div. of Building Branch, TP 7.2—1, Melbourne, 37 pp.

BERLINER, P., 1957. *Die jahreszeitliche Häufigkeitsverteilung der Luftenthalpie in Deutschland.* Kältetech. 9: 138–142.

BIRKELAND, O., 1964. Climatological data important for the design of exterior walls. *Conseil Int. Bâtiment, Bull.*, 4: 9–13.

BIRKELAND, O., 1966. Climatological data for assessment of rain penetration. *Conseil Int. Bâtiment, Bull.*, 3: 29–32.

BISHOF, G. J., 1939. Cooling tower fundamentals and their application to cooling diesel and gas engines. *Am. Soc. Mech. Eng. (ASME) Trans.*, January 1939.

BLACK, F. W., 1954. Desirable temperatures in offices. *J. Inst. Heating Ventilating Eng. England*, 22 (319) Also Review in: *Gesundheitsingenieur*, 76: 281.

BLACK, R. F., 1954. Permafrost—a Review. *Bull. Geol. Soc. Am.*, 65: 839.

BÖER, W., 1964. *Technische Meteorologie.* Teubner, Leipzig, 232 pp.

BÖER, W., BURCHARD, H., HOFMANN, G., KERNER, G. and REIDAT, R., 1961. Third Draft for Supplement to DIN 50 019 (Draft Dated Febr. 1961). *Z. Metall.*, 15: 432–445.

BÖER, W. and FRITZSCHE, O., 1959. Über Grundsätze der Klimaeinteilung für technische Zwecke. *Elektrotech. Z.*, 80: 40–43.

Böer, W. and Götschmann, J., 1963. Zweidimensionale Häufigkeitsverteilungen der Lufttemperatur und der Luftfeuchte als Darstellungen der klimatischen Beanspruchungen technischer Erzeugnisse. *Z. Angew. Meteorol.*, 4: 194–198.

Borisenkŏ, M. M. and Savarina, M. V., 1967. Vertical profiles of wind velocities after measurements, taken on high meteorological towers. *Tr. GGO, Leningrad*, 210 (in Russian).

Boyd, W. D., 1961. Maximum snow depth and snow loads in Canada. *Natl. Res. Council, Div. Building Res., Res. Pap.*, 152, No.6312.

Boyd, W. D., 1963. Driving rain map of Canada. *Nat. Res. Council, Div. Building Res., Tech. Note*, 398.

Bradtke, F. and Liese, W., 1952. *Hilfsbuch für raum- und aussenklimatische Messungen.* Springer, Berlin, 108 pp.

Brezina, E. and Schmidt, W., 1939. *Das künstliche Klima in der Umgebung des Menschen.* Enke, Stuttgart, 212 pp.

Bruckmeyer, F. W., 1958. Bautechnische Klimakarten. In: *Wohnungsbauforschung in Österreich*, Monograph. 4, pp.117–130.

Bullig, J., 1958. *Bericht über die Ergebnisse der 2. laderaummeteorologischen Forschungsfahrt auf MS Cap Ortegal der H.S.D.G.* Deutscher Wetterdienst—Seewetteramt, Hamburg, 19 pp.

Burchard, H., 1963. Normung von Prüf- und Anwendungsklassen. *Elektrotech. Z.*, 15: 332–339.

Cimpa, F., 1968. Eine Tafel zur Bemessung von Hochwasserrückhaltebecken für grosse Bewirtschaftungsdauer und kleine Einzugsgebiete. *Deut. Gewässerkundl. Mitt.*, 12: 1–7.

Cole, A. E., 1960. Precipitation. In: *Handbook of Geophysics.* Geophysical Research Directorate of U.S. Air Force, New York, Chapter 6.1, 5 pp.

De Coster, M., Schüepp, W. and Van der Elst, N., 1955. Le rayonnement sur les plans verticaux à Léopoldville. *Mém. Acad. Roy. Sci. Coloniale, Sci. Tech., N. S.*, II: 1, 65 pp.

Davenport, A. G., 1961. The application of statistical concepts to the wind loading of structures. *Proc. Inst. Civ. Eng.*, 19, p.102.

Deacon, E. L., 1969. Physical processes near the surface of the earth. In: H. Flohn (Vol. Editor), *World Survey of Climatology*, Vol. 2. *General Climatology.* Elsevier, Amsterdam, pp.39–104.

Degoes, L. and Needleman, S. M., 1960. Permafrost. In: *Handbook of Geophysics.* Geophysical Research Directorate of U.S. Air Force, New York, Chapter XII.4, pp.54–59.

Dieckamp, H., 1971. The fast breeder reactor, a source of abundant power for the future. *EDS Trans., Ann. Geophys. Union*, 52: 756–762.

Diem, M., 1952. Eisansatz an Hochspannungsleitungen im Gebirge. *Wetter Leben*, 5: 19–20.

Diem, M., 1956. Messungen der Staubausbreitung aus den Schloten einer Industrieanlage am Niederrhein. *Mitt. Ver. Grosskesselbesitzer*, 42: 1–23.

Diem, M., 1968. Zur Struktur der Niederschläge III: Regen in der arktischen, gemässigten und tropischen Zone. *Arch. Meteorol. Geophys., Bioklimatol., Ser. B*, pp.347–390.

Diem, M. and Trappenberg, R., 1953. Staubniederschlag aus Rauchfahnen. *Mitt. Ver. Grosskesselbesitzer*, 23: 391–395.

DIN 1947 (Deutsche Industrienormen), 1958. *Leistungsversuche an Kühltürmen.* Beuth-Verlag, Berlin, 25 pp.

DIN 4701 (Deutsche Industrienormen), 1959. *Heizungen, Regeln für die Berechnung des Wärmebedarfes von Gebäuden.* Beuth-Verlag, Berlin, 32 pp.

DIN 4108 (Deutsche Industrienormen), 1960. *Wärmeschutz im Hochbau.* Beuth-Verlag, Berlin, 18 pp.

DIN 50 019 Vornorm (Deutsche Industrienormen), 1963. *Freiluftklimate, Klimaübersicht, Werkstoff-, Bauelemente- und Geräteprüfung.* Beuth-Verlag, Berlin, 12 pp.

Dirmhirn, I., 1952. Über das Klima in Wiener Strassenbahnwagen. *Wetter Leben*, 4: 158–162.

Dirmhirn, I., 1964. *Das Strahlungsfeld im Lebensraum.* Akademie-Verlag, Frankfurt, 426 pp.

Dreyfus, J., 1960. Le confort dans l'habitat en pays tropical. Eyrolle, Paris, pp.324–329.

Effenberger, E., 1971. Messfehler der gebräuchlichen Staubniederschlagsmessgeräte. *Staub—Reinh. Luft*, 31: 273–278.

Fisher, R. A. and Tippett, L. N. C., 1926. Limitating forms of the frequency distribution of the largest and smallest member of a sample. *Proc. Cambridge Phil. Soc.*, Vol.24, Part 2, p.180.

Fletcher, R. D., 1950. A relationship between maximum observed point and areal rainfall. *Trans. Am. Geophys. Union*, 31: 344–348.

Fleuri, H., 1951. Le frômage de Roquefort. *Actes de l'Institut d'Etude Economique Maritime et Commerciale de la ville de Sète*, 1951.

Fluor Corporation, 1958. *Evaluated Weather Data for Cooling Equipment.* Fluor Corporation, Los Angeles, Calif.

FORSCHUNGSSTELLE FÜR INGENIEURBIOLOGIE, 1942. Klimabeeinflussung durch Baumassnahmen. *Mitt. Forschungsstelle Ingenieurbiol., München*, 1942, pp.216–218.

FORTAK, H., 1970. Numerical simulation of temporal and spatial distributions of urban air pollution concentration. *Proc. Symp. Multiple Source Urban Diffusion Models.* U.S. Environmental Protection Agency, Research Triangle Park, N. C., pp.9.1–9.34.

FORTAK, H., 1972. *Anwendungsmöglichkeiten von mathematisch-meteorologischen Diffusionsmodellen zur Lösung von Fragen der Luftreinhaltung.* Herausgegeben vom Minister für Arbeit, Gesundheit und Soziales des Landes Nordrhein-Westfalen, 52 pp.

FRANK, W., GETTIS, K. and KÜNZEL, H., 1970. Sonneneinstrahlung—Fenster—Raumklima. *Ber. Bauforsch.*, 66, 66 pp.

FRANKENBERGER, E., 1968. Untersuchungen über Intensität, Häufigkeit und Struktur von Starkwinden über Quickborn in Holstein. *Meteorol. Rundsch.*, 21: 65–69.

FRIEDRICH, W., 1953. Das Autobahnprojekt und die Witterungsverhältnisse im Wiener Wald. *Wetter Leben*, 7: 203–205.

GANDIN, L. S., 1965. Computation of the heat loss of buildings as a function of various meteorological parameters. *Tr. GGO, Leningrad*, 178 (in Russian).

GARSTON, B. R. S., 1968. *Wind Loading on Buildings.* Building Res. Sta. Garston (England), Paper UDC 624.042, 14 pp.

GEIGER, R., 1961. *Das Klima der bodennahen Luftschicht.* Vieweg, Braunschweig, 646 pp. (English version: *Climate Near the Ground*, Harvard University Press, 1966, 628 pp.)

GEIGER, R., 1969. Topoclimates. In: H. FLOHN (Vol. Editor). *World Survey of Climatology*, Vol.2. *General Climatology.* Elsevier, Amsterdam, pp.105–138.

GEORGII, H. W. and HOFFMANN, L., 1966. Beurteilung von SO_2-Anreicherungen in Abhängigkeit von meteorologischen Einflussgrössen. *Staub*, 26: 511.

GIVONI, B., 1966. Review of hygienic requirements in buildings as related to climate. *Conseil Int. Bâtiment (CIB), Bull.*, 3: 10–28.

GRÄFE, K., 1956. Strahlungsempfang vertikaler ebener Flächen, Globalstrahlung in Hamburg. *Ber. Deut. Wetterdienstes*, 29, 15 pp.

GRÄFE, K. and SCHÜTZE, W., 1966. Staubniederschlagsuntersuchungen mit 230 Bergerhoff-Geräten. *Städtehygiene*, 8 (8): 170–177.

GREATHOUSE, G. A. and WESSEL, C. J., 1954. *Deterioration of Materials.* New York, N.Y.

GRUNDKE, G., 1956. Über die Bedeutung technoklimatischer Forschung. *Wiss. Z. Hochsch. Einzelhandel, Leipzig*, 1(3): 19–37.

GUMBEL, E. J., 1935. Les valeurs extrèmes des distributions statistiques. *Ann. Inst. Henri Poincaré*, 5: 115–158.

HADER, F., 1960. Indexzahlen der Witterungsbewertung bei Bauarbeiten. *Wetter Leben*, 12 (7–8): 143–147.

HADER, F., 1970. Schlagregen, Treibschnee und Starkniederschlag. *Z. Eternitwerke Vöcklabruck*, 29. 8 pp.

HELLMANN, G., 1914. Über die Bewegung der Luft in den untersten Schichten der Atmosphäre. *Ber. Akad. Wiss., Berlin*, I: 415–437 (1914); I: 174–197 (1917); I: 404–416 (1919).

HINZPETER, G., 1953. *Studie zum Strahlungsklima von Potsdam.* Akademieverlag, Berlin, 72 pp.

HÖLLER, E., BULLIG, J. et al., 1956. *Laderaummeteorologie.* Deutscher Wetterdienst, Seewetteramt, Hamburg, Einzelveröff., Nr. 9, 43 pp.

HOLZWORTH, C. G., 1974. Climatological aspects of the composition and pollution of the atmosphere. *W.M.O. Tech. Note*, 139: 43 pp.

HOPPESTAD, S., 1955. *Slagregen i Norge.* Norwegian Building Res. Inst., Rept., 13.

JAGANATHAN, P., ARLERY, R., TEN KATE, H. and SAVARINA, M. V., 1969. A note on climatological normals. *W.M.O. Tech. Note*, 84, 19 pp.

JENNINGS, A., 1941. The record rainfalls of the world. *Mon. Weather Rev.*, 69: 356.

JENSEN, K., 1962. Meteorological measurements at Risö 1958–61. *Danish At. Energy Comm., Res. Establishment Risö, Rep.*, 39 .

JOHANNESSEN, TH. V., 1956. Varmeutvekselingen i bygninger og klimat. *Nor. Byggefordkningsinst. Rap.*, 21, 258 pp.

KÄLTETECHNIK, 1954. Klimatisierung von Schienenfahrzeugen in den USA (Bericht einer Studienfahrt deutscher Techniker). *Z. Kältetech.*, 6, 1 p.

KAEMPFERT, W., 1942. Sonnenbestrahlung auf Ebene, Wand und Hang. *Abhand. Reichsamtes Wetterd.*, 9, 32 pp.

KERÄNEN, J., 1951. On frost formation in soil. *Fennica*, 73: 1–14.

KERN, H., 1961. Grosse Tagessummen des Niederschlags in Bayern. *Münchner Geogr. Hefte*, 21, 22 pp.

KING, J. R., 1971. *Probability Charts for Decision Making*. Industrial Press, New York, N.Y., 290 pp.
KITTLER, R., 1970. The Shading Effect of Trees with Regard to the Daylighting and Sun Control of Buildings. *Cons. Int. Bâtiment W 4, Symp. Weather Building, Vienna, 1970, Proc.*, pp.143–147.
KOENIG, G. and ZILCH, K., 1970. Ein Beitrag zur Berechnung von Bauwerken bei böigem Wind. *Mitt. Inst. Massivbau T.H. Darmstadt*, 15, 54 pp.
KÖNIGSBERGER, O. and LYNN, R., 1965. *Roofs in the Warm Tropics*. Architectural Association London, Paper No.1, 56 pp.
KÖPPEN, W., 1931. *Grundriss der Klimakunde*. De Gruiter, Berlin, 388 pp.
KOLLMANN, F., 1951. *Technologie des Holzes*. De Gruiter, Berlin.
KOORSGAARD, VAGN and MADSON LUND, Th., 1964. *A Note on Driving Rain Measurements*. Tech. University of Denmark, Copenhagen.
KRAPFENBAUER, R., 1969. Klimaforschung und rechnerische Schneebelastung. *Wohnungsbauforschung Österreich*, 7/8: 57–60.
KREUTZ, W., 1942. Das Eindringen des Frostes in Böden unter gleichen und verschiedenen Witterungsbedingungen während des sehr kalten Winters 1939/40. *Wiss. Abhand. Reichsamtes Wetterd.*, IX, 22 pp.
LACY, R. E., 1964. Driving Rain at Garston. *Cons. Int. Bâtiment, Bull.*, 4: 6–9.
LACY, R. E., 1965. Driving Rain Maps and the Onslaught of Rain on Buildings. *Building Res. Sta. Garston Watford, Building Current Pap., Res. Ser.* 54, 29 pp.
LACY, R. E. and SHELLARD, H. G., 1962. An index of driving rain. *Meteorol. Mag.*, 91: 177–184.
LEUSDEN, F. P. and FREIMARK, H., 1951. Darstellung der Behaglichkeit für den einfachen praktischen Gebrauch. *Gesundheitsingenieur*, 72: 271–273.
LIEBSCH, E., 1955. Die Bedeutung des amtlichen Wetterdienstes für den Betrieb der Österreichischen Bundesbahn. *Wetter Leben*, 7: 194–196.
LÖBNER, A., 1949. Zehn Jahre Regenwasseranalysen, ein Beitrag zur Ortsüblichkeit von Staubniederschlägen. *Gesundheitsingenieur*, 70: 196–200.
LÖBNER, A., 1950. Verstaubungskarten von Berlin 1949/50. *Raumforschung Raumordnung*, 10: 142–143.
LUGEON, J., 1963. *La Pollution Industrielle de l'Air dans un Pays Montagneux*. Sixth World Power Conf., Frankfurt, 19–26 June 1963, Sect. III, Paper 14, 16 pp.
MACLIN, W. C., 1962. The density and structure of ice formed by accretion. *Q. J. R. Meteorol. Soc.*, 88: 30–50.
MOLLIER, 1923. Thermodynamisches Arbeitspapier. *Z. Ver. Deut. Ing. (VDI)*.
MULLER and SIEMON, W., 1945. Permafrostly Frozen Ground and Related Engineering Problems. *U.S. Geol. Surv. Spec. Rep., Strategic Engineering Study*, 62, 121 pp.
MUNN, R. E., 1970. Airflow in urban areas. *W.M.O. Tech. Note*, 108: 15–39.
MURETOW, N. S., 1957. Organization for investigating into the ice accretion on wires. *Meteorol. Gidrol.*, 2 (in Russian).
NEIBURGER, M., 1959. Meteorological aspects of oxydation type air pollution. In: *The Atmosphere and the Sea in Motion. The Rossby Memorial Volume*. New York, pp.158–169.
NEWBERRY, C. W., EATON, K. J. and MAYNE, J. R., 1968. The nature of gust loading on tall buildings. *Building Res. Stat. Garston Watford, Current Pap.*, 66.
NETZMANN, E., 1960. Klimatauglichkeit. In: *Jahrbuch der Oberflächentechnik 1960*. pp.282–308.
OLGYAY, V., 1954. *Application of Climate Data to House Design*. Finance Agency Div. of Housing Research, Wash., D.C., 152 pp.
OLGYAY, V., 1963. *Design With Climate, Bioclimatological Approach to Architectural Regionalism*. Princeton Univ. Press, Princeton, N.J.
ORLENKO, L. R., 1970. Wind and Its Technical Aspects. *W.M.O., Tech. Note*, 109: 37–56.
ORLOWA, W. W., 1958. Permafrost in the U.S.S.R. *Tr. GGO*, 85: 32–49 (in Russian).
PAGE, J. K., 1969. Heavy glaze in Yorkshire, March 1969. *Weather*, 24: 486–495.
PASQUILL, F., 1962. *Atmospheric Diffusion*. Van Nostrand, London, Toronto, New York, Princeton, 292 pp.
PEEL, C., 1954. Thermal conditions in African dwellings in Sierra Leone. *J. Inst. Heating Ventilating Eng.*, 22: 125. (Review in: *Gesundheitsingenieur*, 76: 87).
PENNINGTON, SMITH, FABER and REED, 1964. *Experimental Analysis of Solar Heat Through Insulated Glasses With Indoor Shading*. Am. Soc. Heating, Refrigeration and Air Conditioning Engineers, Semi-annual Meeting, New Orleans, 1964.
PERL, G., 1935/36. Zur Kenntnis der wahren Sonnenstrahlung in verschiedenen geographischen Breiten, *Meteorol. Z.* 52, p.85; 53, p.467.
PETER, B. G. W., DOLGIECH, W. A. and SCHRIEVER, W. R., 1963. Variation of snow load on roofs. *Can. Nat. Res. Center (NRC), Res. Pap.*, 189.

PLANT, J. A., 1965. The climate of Glasgow. *Meteorol. Office Memor.*, 60.
RAISCH, E., 1930. Die Wärmeleitfähigkeit von Beton in Abhängigkeit von Raumgewicht und Feuchtegrad. *Gesundheitsingenieur*, Sonderheft, vol.53.
RAISS, W., 1933. Der Einfluss des Klimas auf den Heizwärmebedarf in Deutschland. *Gesundheitsingenieur*, 34: 397–403.
RAJEWSKI, A. N., 1953. Discussion of the frequency of glaze. *Meteorol. Gidrol.*, 1 (in Russian).
RÅSTAD, H., 1955. Erfahrungen über Vereisung von Fernleitungen in norwegischen Gebirgen. *Wetter Leben*, 7: 196–200.
REIDAT, R., 1951. Über die Zusammenhänge zwischen Heizwärmemengenabgabe und Witterung. *Gesundheitsingenieur*, 72: 247–252.
REIDAT, R., 1951. Bauwesen und Wetterdienst. *Deut. Holzbau*, 7: 98-106; 115–120.
REIDAT, R., 1959. *Wetterdaten für das Bauwesen—Hamburg.* Deut. Wetterdienst, Einzelveröff. des Seewetteramtes, Nr.23, 11 pp.
REIDAT, R., 1960. Erläuterungen zum Normenentwurf DIN 50 019: Freiluftklimate. *Deut. Industrienorm (DIN), Mitt.*, 39 (2): 62.
REIDAT, R., 1960. Klimadaten für Bauwesen und Technik—Temperatur. *Ber. Deut. Wetterdienstes, Offenbach*, 64, 81 pp.
REIDAT, R., 1963. Klimadaten für Bauwesen und Technik—Niederschlag. *Ber. Deut. Wetterdienstes, Offenbach*, 86, 62 pp.
REIDAT, R., 1968. Der tägliche Gang der Lufttemperatur bis in 200 m Höhe und sein Einfluss auf die Verbreitung von Rauchgasen, nach Registrierungen der Lufttemperatur an der Unterelbe bei Stade. *Z. Meteorol.*, 20: 310–317.
REIDAT, R., 1969. *Die Häufigkeit von Starkregen in Norddeutschland.* 1. Fortbildungskurs für Hydrologie am Leichtweiss-Institut der T.U. Braunschweig, Braunschweig, 1969, 13 pp.
REIDAT, R., 1969. Forming of ice on towers. *IASS (International Association for Shell Structures) Conf. Tower Shaped Structures, Proc., The Hague, 1969.* pp.105–114.
REIDAT, R., 1970. Climatological Data for Building Praxis. *W.M.O., Tech. Note*, 109: 199–210.
REIDAT, R., 1971. Meteorologische Unterlagen für die Klimatechnik. *Z. Wärme-, Kälte-, Sanitärtech.*, 1971, pp.293–302.
REIDAT, R., 1972/73. Klimabeitrag zum Hamburger Planungsatlas. *Veröff. Akad. Raumforsch. Landesplanung, Deutscher Planungsatlas.* Bd. VIII, Hamburg Lieferung 7. Jänecke-Verlag, Hannover, 1971.
REINHOLD, F., 1937. Einheitliche Richtlinien zum Auswerten von Schreibregenmessern. *Gesundheitsingenieur*, 60 (1): 22–26; 60 (2): 40–45; 60 (3): 55–61.
RUCKLI, R., 1950. *Der Frost im Baugrund.* Vienna, 1950.
RUDNEWA, A. W., 1958. *Maximum Amounts of Ice Accretion on Overhead Lines on the Territory of the Soviet Union.* Papers Issued by the Geophysical Main Observatory Voeikov, Leningrad, Part 85 (in Russian).
RUDNEWA, A. W., 1960. *The Amount of Ice Accretion on Overhead Lines as a Function of Terrain Configuration.* Papers Issued by the Geophysical Main Observatory Voeikov, Leningrad, Part 88 (in Russian).
SAUTTER, L., 1948. *Wärmeschutz und Feuchtigkeitsschutz im Hochbau.* Lippen, Berlin, 279 pp.
SCHINZEL, A., 1965. Hygienische Forderungen für das Wohnklima. *Wohnungsbauforsch. Österreich*, 10: 28–32.
SCHÖN, S. G., 1957. Maximale Temperaturen von Flüssigkeiten in liegenden zylindrischen Behältern bei starker Sonneneinstrahlung. *Mitt. Phys.-Tech. Bundesanst., Braunschweig*, 1957, pp.92–94.
SCHRIEVER, W. R. and OSTANOW, W. A., 1967. *Snow Loads.* Cons. Int. Bâtiment, Rep. No.9 on Methods of Load Calculation, Rotterdam, 1967.
SCHROPP, L., 1931. Die Temperatur technischer Oberflächen unter dem Einfluss der Sonneneinstrahlung und der nächtlichen Abkühlung. *Gesundheitsingenieur*, 54: 729.
SCHÜLE, W., CAMMERER, J. S. and GÖRLING, P., 1952. Feuchtigkeit in Aussenwänden. *Fortschr. Forsch. Bauwesen.* Reihe D, Heft 3.
SEXTON, D., 1970. Wind deflection on buildings. *W.M.O., Tech. Note*, 109: 57–61.
SIPLE, P., 1950. *Climatic Criteria for Building Construction.* Building Research Advisory Board Conference Report No.1, Washington, 1950.
S.N.I.P., 1963. Part II, Section A, Chapter 6: Application of Climatology and Geophysics to Building Design. Moscow, 1963 (in Russian).
SPANGEMACHER, K., 1970. *Kombination von Trocken- und Nasskühltürmen.* Haus der Technik, Vortragsveröff., 259, Vulkan-Verlag, Essen.
SPRENGER, E. and KRUGER, W., 1950. Kühlgradtage. *Gesundheitsingenieur*, 71: 117–118.

STEINHAUSER, F., 1966. Ergebnisse der Messungen der Staubablagerungen in Österreich. *Wetter Leben*, 18: 177–185.

STEINHAUSER, F., ECKEL, O. and SAUBERER, F., 1955, 1957, 1959. *Klima und Bioklima von Wien*. Teil I, Wien, 1955, 120 pp.; Teil II, Wien, 1957, 136 pp.; Teil III, Wien, 1959, 136 pp. Publ. Zentralanst. Meteorol., Geodyn., Wien.

STERN, A. C., (Editor), 1968. *Air Pollution*. Acad. Press, New York, N.Y., Vol. 1, 694 pp.; Vol.2, 684 pp.; Vol.3, 866 pp.

Studienreise Klimatisierung von Schienenfahrzeugen in den U.S.A., 1954. *Ber. Kältetech.*, 6: 245. (Review in: *Gesundheitsingenieur*, 76: 88.)

SÜRING, R., 1939. In: *Lehrbuch der Meteorologie*. Keller, Leipzig, 5.Aufl., Bd 1, pp.476–477.

SUTTON, O. G., 1947. The theoretical distribution of airborne pollution from factory chimneys. *Q. J. R. Meteorol. Soc.*, 73: 426–436.

TAESSLER, R., 1970. Problems caused by snow in the field of building sciences. *W.M.O., Tech. Note*, 109: 129–150.

TAESSLER, R., 1972. *Klimadata för Sverige*. Statens Inst. för Byggnads Forskening, Stockholm, 172 pp.

TER LINDEN, A. J., 1940. Klimaregelung auf Schiffen. *Gesundheitsingenieur*, 63: 377–381.

THOM, H. C. S., 1960. A rational method of determining summer weather design data. *Trans. Am. Soc. Heating, Refrigeration Air Conditioning Eng.*, 66: 248–263.

THOM, H. C. S., 1966a. Some methods of climatological analysis. *W.M.O., Tech. Note*, 81, 53 pp.

THOM, H. C. S., 1966b. Cooling degree-days and energy consumption. *Air Conditioning, Heating Ventilating*, 63: 53–54.

THOM, H. C. S., 1967. Towards a universal climatological extreme wind distribution. *Int. Res. Seminar: Wind Effects on Buildings and Structures, Nat. Res. Counc. Proc., Ottawa, 1967*. pp.669–683.

THÖRNBLAD, O. H. and RYD, H., 1970. Research in building climatology with special reference to the problem of snow and rain in connection with wind. *Cons. Int. Bâtiment, Symp. Weather Building, Vienna, 10–12 June 1970, Proc.*, pp.162–177.

TIREL, H., 1961. L'application des techniques du conditionnement de l'air en pays tropical. *Rev. Univers. Mines, Métall., Méc.*, 1961, pp.639–646.

TONNE, F., 1966. Sonnenschutz an Gebäuden. *Inst. Tageslichttech., Mitt.*, 11, 69 pp.

TVEIT, A., 1970. Moisture absorption, penetration and transfer in building structures. *W.M.O., Tech. Note*, 109: 151–158.

VALKO, P., 1966. Die Himmelsstrahlung und ihre Beziehung zu verschiedenen Parametern. *Arch. Meteorol., Geophys., Bioklimatol.*, 14: 336–359.

VALKO, P., 1967. Strahlungsmeteorologische Unterlagen für die Berechnung des Kühlbedarfes von Bauten. *Schweiz. Bl. Heiz. Lüft.*, 1967, 13 pp. (reprint).

VALKO, P., 1970. Radiation loads on buildings of different shape and orientation under various climatic conditions. *W.M.O., Tech. Note*, 109: 87–110.

VON HEIMENDAHL, F., 1963. Staub- und SO_2-Emissionen sowie Verbrennungsrückstände aus Dampfkraftanlagen der Bundesrepublik Deutschland. *Mitt. Ver. Grosskesselbebesitzer*, 87: 409–415.

VAN ZUILEN, D., 1939. Comfort formula. *Warmtetechniek (Netherlands)*, 10: 84–88; 91–94.

VDI (Verein Deutscher Ingenieure), 1970. Berechnung der Kühllast klimatisierter Räume. *VDI-Richtlinien*, 2078, 31 pp.

VICK, F., 1951. Montagefertige Häuser in Übersee. *Deut. Holzbau*, 7 (3/4): 33–49.

VICKERS, B. J. and DAVENPORT, A. G., 1968. A comparison of the theoretical and experimental determination of the response of elastic structures to turbulent flow. *Int. Res. Seminar: Wind Effects on Buildings and Structures, Nat. Res. Counc. Proc., Toronto, 1968*. Vol.1.

WEIL, 1959. Einige Ergebnisse aus neueren Versuchen für den Beton-Strassenbau. *Beton-Herstellung-Verwendung*, 1959, pp.3–12.

WIEMER and BRAMBACH, 1958. *Bericht über die Messungen des Flugstaubniederschlages im Kölner Raum*. Technischer Überwachungsverein, Köln, 76 pp.

WORKING GROUP OF THE COMMISSION FOR ATMOSPHERIC SCIENCES, 1972. Dispersion and forecasting of air pollution. *W.M.O., Tech. Note*, 121, 116 pp.

ZIMMERMANN, O., 1941. Die Gradtagszahlen Deutschlands und eines Teiles des europäischen Auslandes. *Gesundheitsingenieur*, 64: 375–384.

ZIPPEL, H., 1954. Schäden an Wohnungsneubauten infolge frostgefährdeter Böden. *Das Baugewerbe*, 19, 10 pp. (reprint).

ZUG, H., 1969. Maximale Schneelasten in der Schweiz. *Schweiz. Bauzeitung*, 86, 31.

ZAWADIL, R., 1955. Hitze und Kälte im Bauwesen. *Wetter Leben*, 7: 127–135.

Reference Index

ABBOTT, D. L., 250, 283
ABDELHAFEEZ, A. T. and VERKERK, K., 246, 283
ABE, S. and WADA, M., 232, 283
ACKERMAN, B., 301, 330
ADACHI, K. and INOUE, J., 233, 283
ADAR, R., *see* TENNENBAUM, J. et al.
ADOLPH, E. F., 115, 106, 177
ADOLPH, E. F. and MOLNAR, G. W., 103, 177
AICHINGER, F. and MUELLER, I. L., 27, 28, 177
ALBRECHT, E. and ALBRECHT, H., 62, 63, 177
ALBRECHT, H., *see* ALBRECHT, E. and ALBRECHT, H.
ALDERFER, R., *see* GATES, D. M. et al.
ALESSI, J. and POWER, J. F., 239, 283
ALEXANDER, G. and MCCANCE, I., 260, 283
ALISSOW, B. P., DROSDOW, O. A. and RUBINSTEIN, E. S., 222, 283
ALLARD, H. A., 202, 284
ALLARD, H. A., *see* GARNER, W. W. and ALLARD, H. A.
ALLEN, D. E., 368, 381
AMBROSETTI, FL., 125, 141, 177
AMELUNG, W., 70, 177
AMELUNG, W., JUNGMANN, H. and SCHULTZE, E.-G., 177
AMELUNG, W., *see* WIESNER, J. and AMELUNG, W.
ANALPOLSKAJA, L. E. and GANDIN, L. S., 346, 381
ANDERSON, B. C., *see* EDLEFSEN, N. E. and ANDERSON, B. C.
ANDERSON, M. C., 252, 253, 284
ANGELL, J. K., HOECKER, W. H., DICKSON, C. R. and PACK, D. H., 301, 330
ANGUS, D. E., *see* PRUITT, W. O. and ANGUS, D. E.
ANIOL, R., 357, 381
ANQUEZ, J., BOREL, J. C. and CROISET, M., 355, 381
ARIEL, N. Z. and KLINCHNIKOVA, L. A., 314, 331
ARLERY, R., *see* JAGANATHAN, P. et al.
ARMBURST, D. V., 242, 284
ASCHOFF, J., 177
ASCHOFF, J. and WEVER, R., 105, 177
ASHBURN, C. L., *see* DAIGGER, L. A. et al.
ASHWORTH, J. R., 327, 331
ASLYING, H. C., 269, 284
ATKINSON, B. W., 326, 331
ATLAS, D., 264, 284
ATLAS, D., CUMINGHAM, R. M., DONALDSON Jr., R. J., KANTOR, G. and NEWMAN, P., 277, 284
AXTHELM, L. S., *see* DAIGGER, L. A. et al.
AYOUB, R., 347, 381
AZMI, A. R. and ORMROD, D. P., 230, 284
BAGLEY, J. O., *see* HIELMAN, J. L. et al.
BAGLEY, W. T. and GOWEN, F. A., 270, 284
BAIER, W., 280, 284
BAIER, W. and ROBERTSON, G. W., 216, 284
BAILEY, C. B. M., *see* HESS, E. A. and BAILEY, C. B. M.
BAKER, D. G., 223, 284
BAKER, D. G., *see* BLAD, B. L. and BAKER, D. G.
BALL, W. L., *see* SIMINOVITCH, D. et al.
BALLANTYNE, E. R. and SPENCER, J. W., 370, 381
BALTRUSCH, M., 88, 89, 177
BAND, G., 320, 331
BARALT, G., *see* FUJITA, T. et al.
BARGER, G. L. and THOM, H. C. S., 267, 284
BARGER, G. L., *see* SHAW, R. H. et al.
BARMORE, M. A., *see* DAY, A. D. and BARMORE, M. A.
BARNES, R. F., *see* GARZA, R. T. et al.
BARTHOLIC, J. F., NAMKEN, L. N. and WIEGAND, C. L., 276, 284
BATES, C. C., 268, 284
BATTAN, L. J., 264, 284
BAUER, R. E., *see* RADKE, J. K. and BAUER, R. E.
BAULCH, D. M., *see* MCCORMICK, R. A. and BAULCH, D. M.
BAUMGARTNER, A., 91, 92, 177
BAUR, F. and PHILIPPS, H., 41, 177
BAVER, L. D., 215, 216, 284
BECKER, C. F., *see* POCHOP, L. O. et al.
BECKER, F., 165, 177
BEGG, J. E., *see* WAGGONER, P. E. et al.
BELGER, W., 326, 331
BELL, J. M., *see* MCDONALD, M. A. and BELL, J. M.
BENER, P., 16, 36, 37, 140, 177, 178
BENOIT, G. R., *see* LIGON, J. T. and BENOIT, G. R.
BERG, H., 3, 178, 316, 331
BERLINER, P., 349, 381
BERNHARDT, O., 35, 178
BERRY, I. L., SHANKLIN, M. D. and JOHNSON, H. D., 259, 284
BIBILASHVILI, N. SH., *see* SULAKVELIDZE, G. K. et al.
BIDER, M. and VERZÁR, F., 69, 70, 178
BIDLOT, R. and LEDENT, P., 155, 178
BILANSKI, W., DAVIS, J. R. and KIDDER, E. H., 262, 284
BILBRO Jr., J. D., *see* WANJURA, D. F. et al.
BIRKELAND, O., 353, 364, 381
BISHOF, G. J., 375, 381
BLACK, F. W., 343, 381
BLACK, R. F., 381
BLAD, B. L. and BAKER, D. G., 190, 284
BLANEY, H. F. and CRIDDLE, W. D., 223, 284

BLANEY, H. F. and MORIN, K. V., 223, 284
BLAXTER, K. L., 260, 284
BOER, J. H., *see* SCHMIDT, F. H. and BOER, J. H.
BÖER, W., 381
BÖER, W. and FRITZSCHE, O., 381
BÖER, W. and GÖTSCHMANN, J., 340, 341, 382
BÖER, W., BURCHARD, H., HOFMANN, G., KERNER, G. and REIDAT, R., 373, 381
BOERSMA, L., *see* KUO, T.-M. and BOERSMA, L.
BÖHNING, R. H. and BURNSIDE, C. A., 198, 284
BOHREN, C. F. and THORUD, D. B., 256, 284
BOND, T. E., 258, 284
BOND, T. E., HEITMAN, H. and KELLY, C. F., 259, 284
BOND, T. E., KELLY, C. F. and ITTNER, N. R., 259, 284
BOND, T. E., *see* GARRETT, W. N. et al.
BOND, T. E., *see* HEITMAN, H. et al.
BONNER, J., 206, 284
BONNER, J., *see* HAMNER, K. C. and BONNER, J.
BONSMA, J. C., 256, 284
BORDNE, E. F., *see* MCGUINESS, J. L. and BORDNE, E. F.
BOREL, J. C., *see* ANQUEZ, J. et al.
BORISENKŎ, M. M. and SAVARINA, M. V., 351, 382
BORNSTEIN, R. D., *see* LOOSE, T. and BORNSTEIN, R. D.
BORTHWICK, H. A. and PARKER, M. W., 243, 285
BORTHWICK, H. A., PARKER, M. W. and HENDRICKS, S. B., 202, 285
BORTHWICK, H. A., *see* PARKER, M. W. and BORTHWICK, H. A.
BOURKE, P. M. A., 227, 270, 285
BOUYOUCOS, G. J., 207, 285
BOWEN, E. G., 285
BOWEN, I. S., 220, 285
BOYD, W. D., 365, 368, 382
BRADTKE, F. and LIESE, W., 111, 118, 119, 120, 124, 178, 343, 382
BRAHAM, R. R. and SPYERS-DURAN, P., 328, 331
BRESSMAN, E. N., *see* WALLACE, H. A. and BRESSMAN, E. N.
BREZINA, E. and SCHMIDT, W., 346, 382
BRIER, G. W., *see* KLINE, D. B. and BRIER, G. W.
BRIGGS, L. J., 214, 285
BRIGGS, L. J. and MCLANE, J. W., 214, 285
BRIGGS, L. J. and SHANTZ, J. H., 224, 285
BRODY, S., 260, 285
BRODY, S., *see* STEWART, R. E. et al.
BROOKS, C. E. P., 300, 331
BROOKS, F. A., 191, 285
BROOKS, F. A., KELLY, C. F., RHOADES, D. C. and SCHULTZ, H. B., 262, 285
BROWN, D. M., 204, 205, 285
BROWN, D. M., *see* RAHN, J. J. and BROWN, D. M.
BROWN, K. W., 268, 270, 271, 285
BROWN, K. W. and ROSENBERG, N. J., 268, 270, 285
BROWN, K. W., *see* ROSENBERG, N. J. et al.
BROWN, P. L., 237, 285
BRUCKMEYER, F. W., 370, 382
BRUECK, K., *see* HENSEL, H. et al.
BRUENER, H., 58, 60, 61, 117, 121, 122, 127, 129, 153, 157, 158, 178
BUCKINGHAM, E., 215, 285
BUETTNER, K., 31, 32, 108, 109, 110, 111, 114, 126, 136, 178
BUETTNER, K., *see* CARLSON, L. D. and BUETTNER, K.
BUETTNER, K., *see* PFLEIDERER, H. and BUETTNER, K.
BUETTNER, K. J. K., 178
BULA, R. J. and MASSEGALE, M. A., 249, 285
BULA, R. J., RHYKERD, C. L. and LANGSTON, R. G., 250, 285
BULA, R. J., *see* MATCHES, A. G. et al.
BULLIG, J., 372, 382
BULLIG, J., *see* HÖLLER, E. et al., 372, 383
BULLRICH, K., 16, 54, 178
BURCHARD, H., 336, 337, 382
BURCHARD, H., *see* BÖER, W. et al.
BURMAN, R. D. and PARTRIDGE, J. R., 223, 285
BURNSIDE, C. A., *see* BÖHNING, R. H. and BURNSIDE, C. A.
BURRAGE, S. W., 226, 285
BURTON, A. C. and EDHOLM, D. G., 163, 178
BUSINGER, J. A., 221, 285
BUSS, S. and SHAW, R. H., 216, 285
BUXTON, D. R., *see* WANJURA, D. F. et al.
BYERS, H. G., *see* SMITH, W. O. and BYERS, H. G.

CAHOON, G. A., *see* NEWMAN, J. E. et al.
CALVERT, B. A., 246, 285
CAMBELL, R. B., *see* CHANG, J. H. et al.
CAMMERER, J. S., *see* SCHÜLE, W. et al.
CAMPBELL, C. A., PELTON, W. L. and NIELSON, K. F., 238, 285
CAPRIO, J. M., 205, 285
CARBORN, J. M., 268, 285
CARLSON, L. D., 260, 285
CARLSON, L. D. and BUETTNER, K., 155, 178
CARLSON, R. E., YARGER, D. N. and SHAW, R. H., 212, 213, 285
CARNUTH, W., *see* REITER, R. et al.
CARPENTER, L. G., 219, 286
CARSON, J. E., 210, 211, 286
CARTER, J. L., *see* HOWELL, R. W. and CARTER, J. L.
CASPERSON, G., 269, 286
CASTENS, G., 137, 178
CEHAK, K., *see* REUTER, H. and CEHAK, K.
CERMAK, J. E., 313, 331
CHAKO, O., *see* MANI, A. and CHAKO, O.
CHANDLER, R. F., 233, 286
CHANDLER, T. J., 300, 306, 311, 316, 325, 331
CHANG, J. H., 192, 194, 209, 222, 229
CHANG, J. H., CAMBELL, R. B. and ROBINSON, F. E., 225, 286
CHANGNON, S. A., 328, 331
CHANGNON, S. A., *see* HUFF, F. A. and CHANGNON, S. A.
CHANGNON Jr., S. A., 327, 331
CHAPMAN, A. L. and PETERSON, M. L., 232, 286
CHEN, T. H. and HSU, C. M., 246, 286

Chepil, W. S., see Woodruff, N. P. et al.
Cherry, E., see Lange, T. C. et al.
Chilton, R., see Tuller, S. E. and Chilton, R.
Christophersen, J., see Hensel, H. et al.
Cimpa, F., 366, 382
Clapstick, J. W. and Wood, T. B., 258, 286
Clarke, T. F. and McElroy, J. L., 311, 316, 331
Coblentz, W. W. and Stair, J., 30, 33, 178
Cole, A. E., 382
Colwell, R. N., 276, 286
Conrad, V., 172, 178
Cook, D., 280, 286
Cook, D., see Cooper, A. J. and Cook, D.
Cooper, A. J., 245, 246, 248, 286
Cooper, A. J. and Cook, D., 286
Cooper, A. J., see Hurd, R. G. and Cooper, A. J.
Cooper, C. S., see Dotzenko, A. D. et al.
Cooper, W. C., see Newman, J. E. et al.
Cordes, H., 64, 178
Cornia, R. L., see Pochop, L. O. et al.
Cotton, G., see Pueschel, R. F. et al.
Court, A., 130, 178
Courvoisier, P., 90, 178
Courvoisier, P. and Wierzejewski, H., 22, 178
Creasy, L. L., 286
Criddle, W. D., see Blaney, H. F. and Criddle, W. D.
Croiset, M., see Anquez, J. et al.
Crosier, W., 271, 286
Culkowski, W. M., 328, 331
Cumingham, R. M., see Atlas, D. et al.
Curry, M., 92, 178

Daigger, L. A., Axthelm, L. S. and Ashburn, C. L., 250, 286
Daingerfield, L., 241, 286
Dale, R. F., 279, 286
Dale, R. F. and Shaw, R. H., 225, 265, 280, 286
Dalrymple, P. C., 152, 153, 178
Dalton, J., 219, 227, 286
Dammann, W., 3, 141, 165, 178
Das, J. C., 279, 286
Das, J. C. and Ramachandran, G., 279, 286
Davenport, A. G., 352, 382
Davenport, A. G., see Vickers, B. J. and Davenport, A. G.
Davies, R., 72, 178
Davis, D. A., 286
Davis, J. R., see Bilanski, W. et al.
Davitaya, F. F., 308, 331
Day, A. D. and Barmore, M. A., 237, 286
Day, A. D. and Intalap, S., 237, 286
Day, R., see Forster, R. E. et al.
Deacon, E. L., 378, 382
Deacon, E. L. and Swinbank, W. C., 219, 286
Decker, W. L., 220, 223, 286
Decker, W. L., see Gerber, J. F. and Decker, W. L.
De Coster, M., Schüepp, W. and Van der Elst, N., 353, 382
Degoes, L. and Needleman, S. M., 359, 382
Delauche, J. C., 243, 287
DeMarrais, G. A., 310, 331
Denmead, O. T., 252, 287
Denmead, O. T. and Shaw, R. H., 195, 196, 218, 225, 240, 265, 287
De Quervain, A., 50, 178
DeRoo, H. C., see Waggoner, P. E. et al.
De Rudder, B., 3, 5, 71, 178
Desjardings, R., see Siminovitch, D. et al.
Dettwiller, J., 304, 328, 331
Dickson, C. R., see Angell, J. K. et al.
Dieckamp, H., 374, 382
Diem, M. and Trappenberg, R., 378, 382
Diem, M., 364, 369, 377, 382
Dieterichs, H., 148, 178
Dirmhirn, I., 16, 43, 45, 178, 353, 371, 382
Dirmhirn, I. and Sauberer, F., 313, 321, 331
D.M.H., 326, 331
Dobrenz, A. K., see Dotzenko, A. D. et al.
Dolgiech, W. A., see Peter, B. G. W. et al.
Domingo, C. E., see Robins, J. S. and Domingo, C. E.
Donaldson Jr., R. J., see Atlas, D. et al.
Dorno, C., 16, 49, 179
Dorno, C., see Thilenius, R. and Dorno, C.
Dotzenko, A. D., Cooper, C. S., Dobrenz, A. K., Laude, H. M., Massegale, M. A. and Fellner, K. C., 249, 287
Dreibelbis, F. R., see Harrold, L. L. and Dreibelbis, F. R.
Dreyfus, J., 382
Drosdow, O. A., see Alissow, B. P. et al.
Dubief, J., 134, 135, 179
Du Bois, E. F., 103, 179
Du Bois, E. F., see Hardy, J. D. and Du Bois, E. F.
Duckworth, F. S. and Sandberg, J. S., 303, 306, 310, 331
Dyer, A. J., Hicks, B. B. and King, K. M., 220, 287
Dyer, A. J. and Pruitt, W. O., 220, 287
Dyer, A. J., see Taylor, R. J. and Dyer, A. J.

Earl, A. U., see Pelton, W. L. and Earl, A. U.
Eaton, K. J., see Newberry, C. W. et al.
Eckel, O., see Steinhauser, F. et al.
Edgar, J. L. and Paneth, F. A., 93, 179
Edgerton, A. T., Mandl, R. M., Poe, G. A., Jenkins, J. E., Soltis, F. and Sakamoto, S., 277, 287
Edholm, D. G., see Burton, A. C. and Edholm, D. G.
Edlefsen, N. E. and Anderson, B. C., 215, 287
Effenberger, E., 377, 382
Egli, D. B., Pendleton, J. W. and Peters, D. B., 197, 198, 287
Ehrler, W. L., see Van Bavel, C. H. M. and Ehrler, W. L.
Eiden, R. and Eschelbach, G., 54, 73, 74, 179
Eik, K. and Hanway, J. J., 194, 287
Ellenberg, H., 84, 179
Ellis, H. T., see Pueschel, R. F. et al.
Emonds, H., 306, 331

ENOMOTO, N., 232, 287
ERICKSON, A. E., *see* HUNTER, J. R. and ERICKSON, A. E.
ESCHELBACH, G., *see* EIDEN, R. and ESCHELBACH, G.
EVELYN, J., 299, 331

FALLANSBEE, W. A., 278, 287
FEDOROV, M. M., 323, 331
FEBER, A., FORD, A. L. and MCCRORY, S. A., 270, 287
FELCH, R. E., 270, 272, 287
FELLNER, K. C., *see* DOTZENKO, A. D. et al.
FERGUSON, W. S., *see* MACK, A. R. and FERGUSON, W. S.
FERRIS, B. G., *see* FORSTER, R. E. et al.
FISCHER, P. L., 329, 332
FISHER, R. A., 278, 287
FISHER, R. A. and TIPPETT, L. N. C., 351, 382
FLACH, E., 4, 5, 6, 18, 21, 23, 25, 27, 28, 29, 50, 69, 125, 128, 139, 140, 142, 143, 145, 147, 159, 160, 175, 179
FLACH, E. and MOERIKOFER, W., 150, 151, 179
FLEISCH, A. and VON MURALT, A., 179
FLEISCHER, R., 179
FLEISCHER, R. and GRÄFE, K., 141, 179
FLETCHER, J. F., *see* JASKE, R. T. et al.
FLETCHER, R. D., 365, 382
FLEURI, H., 373, 382
FLOHN, H., 3, 62, 77, 79, 96, 177, 180, 317, 332
FLORIAN, R. L., *see* HANKS, R. J. et al.
FLOWERS, E. C., *see* PUESCHEL, R. F. et al.
Fluor Corporation, 349, 382
Food and Agriculture Organization, 287
FOITZIK, L. and HINZPETER, H., 51, 180
FORD, A. L., *see* FEBER, A. et al.
FORSCHUNGSSTELLE FÜR INGENIEURBIOLOGIE, 383
FORSTER, R. E., FERRIS, B. G. and DAY, R., 120, 180
FORTAK, H., 83, 180, 378, 383
FOSKETT, R. L., 279, 287
FOSTER, A. C. and TADMAN, E. C., 248, 287
FRANCK, N., *see* JENSEN, M. and FRANCK, N.
FRANK, W., GETTIS, K. and KÜNZEL, H., 354, 355, 383
FRANKEN, E., *see* VAN EIMERN, J. et al.
FRANKENBERGER, E., 351, 383
FREDERICK, R. H., 313, 328, 332
FREIMARK, H., *see* LEUSDEN, F. P. and FREIMARK, H.
FRIEDRICH, W., 371, 383
FRITZSCHE, O., *see* BÖER, W. and FRITZSCHE, O.
FRITSCHEN, L. J., 220, 226, 287
FUHRMAN, F. A., 131, 180
FUJITA, T., BARALT, G. and TSUCHIYA, K., 277, 287
FUKUI, E., 307, 332
FUKUI, E., *see* SEKIGUTI, T. et al.
FURNIVAL, G. M., *see* REIFSNYDER, W. E. et al.

GAASTRA, P., 197, 287
GAERTNER, W. and GOEPFERT, H., 109, 180
GAGGE, A. P., *see* WINSLOW, C. E. A. et al.
GALINDO, I. G. and MUHLIA, A., 53, 180
GAMBLE, D. S., *see* SIMINOVITCH, D. et al.
GANDIN, L. S., 346, 383
GANDIN, L. S., *see* ANAPOLSKAJA, L. E. and GANDIN, L. S.
GANGOPADHYAYA, M., HARBECK Jr., G. E., NORDENSON, T. J., OMAR, M. H. and URYVAEV, V. A., 225, 287
GARDNER, H. R., *see* HANKS, R. J. et al.
GARNER, W. W., 202, 215, 287, 288
GARNER, W. W. and ALLARD, H. A., 201, 243, 288
GARRETT, W. N., BOND, T. E. and KELLY, C. F., 260, 288
GARRETT, W. N., GIVENS, R. L., BOND, T. E. and HULL, J. L., 259, 260, 288
GARSTON, B. R. S., 350, 351, 383
GARZA, R. T., BARNES, R. F., MOTT, G. O. and RHYKERD, C. L., 250, 288
GAT, Z., *see* LOMAS, J. and GAT, Z.
GATES, D. M., 191, 193, 213, 288
GATES, D. M., ALDERFER, R. and TAYLOR, E., 211, 212, 288
GEIGER, R., 26, 92, 144, 175, 176, 180, 191, 192, 195, 252, 253, 254, 255, 256, 270, 288, 337, 357, 381, 383
GEIGER, R., *see* KÖPPEN, W. and GEIGER, R.
GEIGY DOCUMENTA, 60, 180
GEORGII, H. W., 77, 90, 180
GEORGII, H. W. and GRAVENHORST, G., 75, 180
GEORGII, H. W. and HOFFMANN, L., 383
GEORGII, H. W. and SCHAEFER, H. J., 83, 84, 180
GERALD, C. J. and NAMKEN, L. N., 242, 288
GERBER, J. F. and DECKER, W. L., 221, 288
GETTIS, K., *see* FRANK, W. et al.
GIAJA, J., 180
GIFFORD, F. A., 320, 332
GILAT, T., *see* TENNENBAUM, J. et al.
GILDERSLEEVES, P. B., 322, 332
GILGEN, A., *see* WANNER, H. U. and GILGEN, A.
GILMAN, D. L., HIBBS, J. R. and LASKIN, P. L., 263, 264, 288
GILMORE, E. C. and ROGERS, J. S., 206, 288
GIPSON, J. F. and JOHAM, H. E., 242, 288
GIVENS, R. L., *see* GARRETT, W. N. et al.
GIVENS, R. L., *see* MCCORMICK, W. C. et al.
GIVONI, B., 347, 348, 383
GLASER, E. M., 131, 180
GLASSER, M. and GREENBERG, L., 82, 83, 180
GOEPFERT, H., *see* GAERTNER, W. and GOEPFERT, H.
GOETZ, F. W. P., 180
GOODE, J. E. and INGRAM, J., 251, 288
GÖRLING, P., *see* SCHÜLE, W. et al.
GÖTSCHMANN, J., *see* BÖER, W. and GÖTSCHMANN, J.
GOWEN, F. A., *see* BAGLEY, W. T. and GOWEN, F. A.
GRÄFE, K., 353, 383
GRÄFE, K. and SCHÜTZE, W., 377, 383
GRÄFE, K., *see* FLEISCHER, R. and GRÄFE, K.
GRAHAM, I. R., 301, 332
GRAHAM, W. G. and KING, K. M., 221, 288
GRANDJEAN, E., 48, 49, 81, 82, 180
GRASNICK, K.-H., 302, 332

GRAVENHORST, G., *see* GEORGII, H. W. and GRAVENHORST, G.
GREATHOUSE, G. A. and WESSEL, C. J., 383
GREEN, D. E., 244, 288
GREENBERG, L., *see* GLASSER, M. and GREENBERG, L.
GRIST, D. H., 231, 288
GROSSE-BROCKHOFF, F., 103, 180
GRUNDKE, G., 374, 383
GUMBEL, E. J., 351, 383
GUTMAN, D. P. and TORRANCE, K. E., 317, 332

HADDOCK, D. J., 206, 288
HADER, F., 370, 383
HAHN, L. and MCQUIGG, J. D., 258, 259, 288
HAHN, L., *see* HEITMAN, H. et al.
HAINES, F. M., 227, 288
HALDANE, J. S., 154, 180
HALKIAS, N. A., VEIHMEYER, F. J., HENRICKSON, A. H., 223, 288
HALL, A. D., 235, 288
HALLIDAY, E. C., *see* KEMENY, E. and HALLIDAY, E. C.
HAMMOND, J. J., *see* PENDLETON, J. W. and HAMMOND, J. J.
HAMNER, K. C., 202, 288
HAMNER, K. C. and BONNER, J., 202, 288
HANCOCK, J., 260, 288
HAND, I., 322, 332
HANK, O. and VÁSÁHELYI, J., 260, 288
HANKS, R. J., GARDNER, H. R. and FLORIAN, R. L., 236, 237, 288
HANNA, W. F., 239, 288
HANWAY, J. J., *see* EIK, K. and HANWAY, J. J.
HAPCHEVCE, V. F., *see* SULAKVELIDZE, G. K. et al.
HARBECK Jr., G. E., *see* GANGOPADHYAYA, M. et al.
HARDY, J. D., 101, 102, 180
HARDY, J. D. and DU BOIS, E. F., 113, 180
HARNACK, R. P. and LANDSBERG, H. E., 327, 332
HAROON, M., LONG, R. C. and WEYBREW, J. A., 206, 288
HARRAR, G., *see* STAKMAN, E. C. and HARRAR, G.
HARRICS, H., *see* VAN EIMERN, J. et al.
HARRISON, G. A., 289
HARROLD, L. L. and DREIBELBIS, F. R., 240, 289
HART, E., *see* ROSENBERG, N. J. et al.
HART, J. S., 259, 289
HARTMANN, H. and VON MURALT, A., 62, 180
HARTT, C. E., 228, 289
HAUDE, W., 223, 289
HAUN, J. R., 279, 289
HAUSSER, K. W. and VAHLE, W., 30, 33, 180
HEBERDEN, W., 121, 180
HEINICKE, D. R., 251, 289
HEITMAN, H., HAHN, L., BOND, T. E. and KELLY, C. F., 260, 289
HEITMAN, H., KELLY, C. F., BOND, T. E. and HAHN, L., 259, 289
HEITMAN, H., *see* BOND, T. E. et al.
HELLMANN, G., 351, 383
HELLPACH, W., 49, 180
HENDRICKS, S. B., *see* BORTHWICK, H. A. et al.
HENDRICKS, W. A. and SCHOLL, J. C., 279, 289
HENRICKSON, A. H., *see* HALKIAS, N. A. et al.
HENRICKSON, A. H., *see* VEIHMEYER, F. J. and HENRICKSON, A. H.
HENSCHEL, A., 260, 289
HENSCHKE, U. and SCHULZE, R., 34, 180
HENSEL, H., 102, 113, 116, 117, 180
HENSEL, H., BRUECK, K. and RATHS, P., 104, 105, 106, 107, 180
HENSEL, H., PRECHT, H., CHRISTOPHERSEN, J. and LARCHER, W., 104, 105, 106, 107, 180
HERATH, W. and ORMROD, D. P., 232, 289
HERPERTZ, E., ISRAEL, H. and VERZÁR, F., 69, 181
HERRINGTON, L. P., *see* WINSLOW, C. E. A. et al.
HERRMANN, H., 141, 149, 181
HESS, E. A. and BAILEY, C. B. M., 260, 289
HESS, W. N., 264
HESSE, W., 227, 289
HIBBS, J. R., *see* GILMAN, D. L. et al.
HICKS, B. B., *see* DYER, A. J. et al.
HIDDING, A. P., *see* VAN WIJK, W. R. and HIDDING, A. P.
HIELMAN, J. L., KANEMASU, E. T., BAGLEY, J. O. and RASMUSSEN, V. P., 278, 289
HILDEBRANDT, G., 28, 181
HILL, G. R., *see* THOMAS, M. D. and HILL, G. R.
HILL, L., 121, 124, 180
HINZPETER, A., 11, 180
HINZPETER, G., 383
HINZPETER, H., *see* FOITZIK, L. and HINZPETER, H.
HITZLER, J. and LAUSCHER, F., 21, 22, 181
HOBBS, P. V. and RADKE, L. F., 328, 332
HOCK, R., *see* SCHOLANDER, P. E. et al.
HODGE, P. W. and LAUTAINEN, N., 54, 181
HODGES, H. F., *see* SANDHU, S. S. and HODGES, H. F.
HOECKER, W. H., *see* ANGELL, J. K. et al.
HOEGL, O., 181
HOESCHELE, K., 121, 181
HOFFMANN, L., *see* GEORGII, H. W. and HOFFMANN, L.
HOFMANN, G., 226, 289
HOFMANN, G., *see* BÖER, W. et al.
HOGAN, A. W., 329, 332
HOGAN, A. W., MOHNEN, V. A. and SCHAEFER, V. J., 69, 70, 181
HOGG, W. H., HOUNAM, C. E., MALLIK, A. K. and ZADOKS, J. C., 271, 289
HOINKES, H. C., 25, 53, 62, 153, 181
HOLEKAMP, E. R., HUDSPETH, E. R. and RAY, L. L., 241, 289
HÖLLER, E., BULLIG, J. et al., 372, 383
HOLLWICH, F., 14, 181
HOLMES, R. H. and ROBERTSON, G. W., 217, 265, 289
HOLT, R. F. and VAN DAREN, C. A., 240, 289
HOLUB, A., 260, 289
HOLZMAN, B., 289
HOLZMAN, B., *see* THORNTHWAITE, C. W. and HOLZMAN, B.
HOLZMAN, B. G. and THOM, H. C. S., 328, 332
HOLZWORTH, C. G., 378, 383

Hoppe, U., 47, 181
Hoppestad, S., 364, 365, 383
Horowitz, J. L., *see* Reifsnyder, W. E. et al.
Horsfall, J. G., *see* Waggoner, P. E. et al.
Houghton, F. C. and Yaglou, C. P., 181
Hounam, C. E., 264, 289
Hounam, C. E., *see* Hogg, W. H. et al.
House, C. C., 279
Hoveland, C. S., *see* Woodruff, J. M. et al.
Howard, L., 299, 332
Howell, H. B., *see* Smith, W. L. and Howell, H. B.
Howell, R. W., 243, 244, 289
Howell, R. W. and Carter, J. L., 244, 289
Hsu, C. M., *see* Chen, T. H. and Hsu, C. M.
Huber, B., 95, 181
Hudson, J. P., 222, 290
Hudspeth, E. R., *see* Holekamp, E. R. et al.
Hudspeth Jr., E. B., *see* Wanjura, D. F. et al.
Huff, F. A. and Changnon, S. A., 327, 328, 332
Huff, F. A. and Schickedanz, P. T., 328, 332
Hull, B. B., 326
Hull, J. L., *see* Garrett, W. N. et al.
Hunter, J. R. and Erickson, A. E., 244, 290
Huntington, E., 3, 181
Hurd, R. G. and Cooper, A. J., 245, 290
Hursh, J. S., *see* Ranney, C. D. et al.
Hurst, G. W., 228, 290
Hutchinson, J. B., 241, 290

Ingram, D. L. and Whittow, G. C., 260, 290
Ingram, J., *see* Goode, J. E. and Ingram, J.
Inoue, E., 229, 290
Inoue, E., Mihara, Y. and Tsubo, Y., 231, 232, 290
Inoue, J., *see* Asachi, K. and Inoue, J.
Intalap, S., *see* Day, A. D. and Intalap, S.
International Rice Research Institute, 231, 290
Irving, L., *see* Scholander, P. E. et al.
Israel, H., 78, 97, 98, 99, 181
Israel, H., *see* Herpertz, E. et al.
Ittner, N. R., *see* Bond, T. E. et al.

Jackson, J. E., 290
Jacobi, W., 65, 99, 181
Jacobs, M. B., 66, 181
Jacobs, W. P., 206, 290
Jaganathan, P., Arlery, R., Ten Kate, H. and Savarina, M. V., 383
Jansen, G., 181
Jaske, R. T., Fletcher, J. F. and Wise, K. R., 302, 332
Jauregui, E., 329, 332
Jauregui, E. and Sato, C., 131, 181
Jaworski, J., 269, 290
Jenkins, J. E., *see* Edgerton, A. T. et al.
Jennings, A., 365, 383
Jensen, K., 378, 379, 383
Jensen, M., 268, 269, 290
Jensen, M. and Franck, N., 313, 332
Jensen, R. D., 242, 290
Jensen, R. E., *see* Sakamoto, C. M. and Jensen, R. E.
Jessel, U., 75, 88, 94, 95, 181
Joham, H. E., *see* Gipson, J. F. and Joham, H. E.
Johannessen, Th. V., 370, 383
Johnson, A. F., *see* Lowry, R. L. and Johnson, A. F.
Johnson, F. A., 190, 290
Johnson, H. D., 257, 290
Johnson, H. D., *see* Berry, I. L. et al.
Johnston, T. J., *see* Pendleton, J. W. et al.
Jones, F. N., 181
Jones, G. E., *see* Pearson, J. E. and Jones, G. E.
Joyat, M. I., *see* Williams, G. D. V. et al.
Junge, Chr., 56, 68, 73, 74, 79, 80, 181, 182
Jungmann, H., 182
Jungmann, H., *see* Amelung, W. et al.
Jusatz, H. J., 182
Jusatz, H. J., *see* Rodenwaldt, E. and Jusatz, H. J.

Kaempfert, W., 321, 332, 353, 383
Kaiser, H., 269, 290
Kältetechnik, 371, 383
Kanemasu, E. T., *see* Hielman, J. L. et al.
Kantor, G., *see* Atlas, D. et al.
Kaser, P., *see* Thornthwaite, C. W. and Kaser, P.
Kasperbauer, M. J., 203, 290
Kawamura, R., 130, 182
Kayser, K., 88, 182
Keidel, W. D., 7, 182
Kelly, C. F., *see* Bond, T. E. et al.
Kelly, C. F., *see* Brooks, F. A. et al.
Kelly, C. F., *see* Garrett, W. N. et al.
Kelly, C. F., *see* Heitman, H. et al.
Kemeny, E. and Halliday, E. C., 89, 182
Kenny, T. J., *see* Nittler, L. W. and Kenny, T. J.
Keränen, J., 357, 383
Kern, H., 366, 383
Kerner, G., *see* Böer, W. et al.
Kershaw, K. A., *see* Rutter, A. J. et al.
Kersten, M. S., 208
Kidder, E. H., *see* Bilanski, W. et al.
Kiesselback, T. A., 224, 290
Kincer, J. B. and Mattice, W. A., 278, 290
King, E., 140, 182
King, J. R., 337, 384
King, K. M., *see* Dyer, A. J. et al.
King, K. M., *see* Graham, W. G. and King, K. M.
Kingsolver, C. H., *see* Lange, T. C. et al.
Kittler, R., 355, 384
Klages, K. H. W., 234, 290
Kleemann, A., 182
Klein, E., Seitz, E. D. and Meyer, H. A. E., 182
Klinchnikova, L. A., *see* Ariel, N. Z. and Klinchnikova, L. A.
Kline, D. B. and Brier, G. W., 329, 332
Kloepper, R., 164, 182
Knepple, R., 41, 140, 182
Knoch, K., 4, 164, 182
Knoch, K. and Schulze, A., 2, 182
Knoche, W., 137, 182
Koenig, G. and Zilch, K., 352, 384

KOLLMANN, F., 362, 384
KONČEK, M., 304, 332
KÖNIGSBERGER, O. and LYNN, R., 354, 384
KOORSGAARD, V. and LUND, M., 364, 384
KÖPPEN, W., 359, 384
KÖPPEN, W. and GEIGER, R., 3, 182
KRAMMER, M., 53, 182
KRAPFENBAUER, R., 384
KRATZER, A., 300, 308, 326, 327, 332
KREBS, A., 96, 98, 182
KREMSER, V., 313, 332
KREUTZ, W., 95, 182, 358, 384
KRUGER, W., *see* SPRENGER, E. et al.
KUHN, M., 25, 53, 182
KUHN, P. M. and SUOMI, V. E., 190, 290
KUHNKE, W., 81, 182
KÜNZEL, H., *see* FRANK, W. et al.
KUO, T.-M. and BOERSMA, L., 290

LACY, R. E., 364, 365, 384
LACY, R. E. and SHELLARD, H. G., 365, 384
LADELL, W. S., 116, 182
LAING, D. R., 244, 290
LANCASTER, A., 137, 182
LANDSBERG, H., 56, 68, 71, 90, 182
LANDSBERG, H. E., 3, 18, 19, 46, 127, 131, 149, 150, 155, 177, 182, 273, 275, 290, 300, 303, 306, 310, 312, 326, 327, 332, 333
LANDSBERG, H. E. and VAN MIEGHEM, M., 88, 182
LANDSBERG, H. E., *see* HARNACK, R. P. and LANDSBERG, H. E.
LANGE, T. C., KINGSOLVER, C. H., MITCHELL, J. E. and CHERRY, E., 271, 290
LANGSTON, R. G., *see* BULA, R. J. et al.
LARCHER, W., *see* HENSEL, H. et al.
LAUDE, H. M., *see* DOTZENKO, A. D. et al.
LAUSCHER, F., 173, 175, 182
LAUSCHER, F. and SCHWABL, W., 20, 183
LAUSCHER, F., *see* HITZLER, J. and LAUSCHER, F.
LAUTAINEN, N., *see* HODGE, P. W. and LAUTAINEN, N.
LEDENT, F., *see* BIDLOT, R. and LEDENT, F.
LEDUC, S. K., 279, 291
LEE, D. H. K., 3, 4, 131, 183
LEHANE, J. J., *see* STAPLE, W. J. and LEHANE, J. J.
LEHMANN, K., 124, 183
LEISTNER, W., 145, 149, 183
LEMON, E. R., 201, 228, 229, 291
LEMON, E. R., *see* MOSS, D. N. et al.
LEMON, E. R., *see* TANNER, C. B. and LEMON, E. R.
LEONARD, W. H. and MARTIN, J. H., 236, 291
LEONTIEVSKY, N. P., 270, 291
LEUSDEN, F. P. and FREIMARK, H., 343, 384
LEWIS, C. F. and RICHMOND, T. R., 241, 291
LIEBSCH, E., 371, 384
LIENESCH, J., *see* WARK, D. Q. et al.
LIESE, W., *see* BRADTKE, F. and LIESE, W.
LIETH, H., 282, 283, 291
LIGON, J. T. and BENOIT, G. R., 226, 291
LINACRE, E. T., 211, 212, 291
LIND, A. R., 260, 291
LINKE, F., 3, 51, 53, 136, 137, 183, 300, 323, 333
LIST, R. J., 321, 333
LJONES, B., 251, 291
LÖBNER, A., 377, 384
LOMAS, J., 235, 291
LOMAS, J. and GAT, Z., 228, 291
LONG, I. F., *see* PENMAN, H. L. and LONG, I. F.
LONG, R. C., *see* HAROON, M. et al.
LOOSE, T. and BORNSTEIN, R. D., 326, 333
LORENZ, D., 303, 333
LOTMAR, R., 183
LOURENCE, F. J. and PRUITT, W. O., 230, 291
LOWRY, R. L. and JOHNSON, A. F., 223, 291
LOWRY, W. P. and VEHTS, D. B., 205, 291
LUDLAM, F. H., 264, 291
LUEDI, W. and VARESCHI, V., 72, 183
LUFT, U. C., 63, 183
LUGEON, J., 384
LUKENS, R. J., *see* WAGGONER, P. E. et al.
LULL, H. W., *see* REIFSNYDER, W. E. and LULL, H. W.
LUND, M., *see* KOORSGAARD, V. and LUND, M.
LUXMOORE, R. J., MILLINGTON, R. J. and MARCELLOS, H., 196, 291
LYNN, R., *see* KÖNIGSBERGER, O. and LYNN, R.
LYSENKO, T. D., 204, 291

MACFARLANE, W. V., 131, 183, 260, 291
MACHTA, L., 96, 183
MACHTA, L., *see* PUESCHEL, R. F. et al.
MACK, A. R. and FERGUSON, W. S., 265, 291
MACLIN, W. C., 384
MACNEILL, M. M., 250, 291
MACPHERSON, R. K., 155, 183
MÄGDEFRAU, K., 320, 333
MAHRINGER, W., 310, 333
MAISEL, T. N., 302, 303, 308, 333
MAKKINK, G. F., 223, 291
MALLIK, A. K., *see* HOGG, W. H. et al.
MANDL, R. M., *see* EDGERTON, A. T. et al.
MANI, A. and CHAKO, O., 53, 183
MARCELLOS, H., *see* LUXMOORE, R. J. et al.
MARSHALL, J. K., 268, 270, 271, 291
MARTIN, G. C., *see* VOUGH, L. R. and MARTIN, G. C.
MARTIN, J. H., *see* LEONARD, W. H. and MARTIN, J. H.
MARTINI, E., 3, 183
MASSEGALE, M. A., *see* BULA, R. J. and MASSEGALE, M. A. *See also* DOTZENKO, A. D. et al.
MATCHES, A. G., MOTT, G. O. and BULA, R. J., 250, 291
MATHER, J. R., 223, 291
MATHER, J. R., *see* THORNTHWAITE, C. W. and MATHER, J. R.
MATJAKIN, G. I., 268, 269, 291
MATSUO, T., 230, 232, 291
MATSUSHIMA, S. and TSUNODA, K., 232, 291
MATTICE, W. A., *see* KINCER, J. B. and MATTICE, W. A.
MATTSON, J. O., 303, 333
MAYNE, J. R., *see* NEWBERRY, C. W. et al.

McCain, F. S., *see* Woodruff, J. M. et al.
McCance, I., *see* Alexander, G. and McCance, I.
McConnell, W. J. and Spiegelmann, M., 159, 183
McCormick, P. A., *see* Williams, G. D. V. et al.
McCormick, R. A. and Baulch, D. M., 323, 333
McCormick, W. C., Givens, R. L. and Southwell, B. L., 260, 291
McCrory, S. A., *see* Feber, A. et al.
McDonald, M. A. and Bell, J. M., 258, 292
McDowell, R. E., 256, 257, 292
McElroy, J. L., *see* Clarke, T. F. and McElroy, J. L.
McGinnis Jr., D. F., Pritchard, J. A. and Wiesnet, D. R., 278, 292
McGuiness, J. L. and Bordne, E. F., 223, 292
McLane, J. W., *see* Briggs, L. J. and McLane, J. W.
McMillan, M. C., *see* Schneider, S. R. et al.
McMillin, W. D., *see* Veihmeyer, F. J. et al.
McQuigg, J. J., *see* Hahn, L. and McQuigg, J. D.
Meyer, H. A. E. and Seitz, E. O., 31, 32, 33, 183
Meyer, H. A. E., *see* Klein, E. et al.
Meyer, L. E., *see* Van Bavel, C. H. M. and Meyer, L. E.
Michaels, P. J. and Scherer, V. R., 279, 292
Miescher, G., 38, 183
Mihara, Y., *see* Inoue, E. et al.
Miller, D. H., 253, 292
Miller, P. M., *see* Waggoner, P. E. et al.
Miller, W. E., *see* Yaglou, C. P. and Miller, W. E.
Millington, R. J., *see* Luxmoore, R. J. et al.
Missenard, A., 3, 183
Mitchell, J. E., *see* Lange, T. C. et al.
Mitchell Jr., J. M., 311, 312, 333
Modlibawska, I., 251, 292
Moeller, F., 183
Moeller, F. and Quenzel, H., 26, 183
Moerikofer, W., 183
Moerikofer, W., *see* Flach, E. and Moerikofer, W.
Mohnen, V. A., *see* Hogan, A. W. et al.
Mohring, D., 4, 148, 183
Molga, M., 255, 292
Molnar, G. W., *see* Adolph, E. F. and Molnar, G. W.
Monteith, J. L., 190, 194, 226, 292
Morik, J., 323, 333
Morin, K. V., *see* Blaney, H. F. and Morin, K. V.
Morton, A. J., *see* Rutter, A. J. et al.
Moss, D. N., 199, 201, 292
Moss, D. N., Musgrave, R. B. and Lemon, E. R., 197, 292
Mott, G. O., *see* Garza, R. T. et al.
Mott, G. O., *see* Matches, A. G. et al.
Mueller, E. A., 161, 162, 163, 183
Mueller, E. A. and Wenzel, H. G., 183
Mueller, I. L., *see* Aichinger, F. and Mueller, I. L.
Mueller, K. E., *see* Oelke, E. A. and Mueller, K. E.
Muhlia, A., *see* Galindo, I. G. and Muhlia, A.
Munn, R. E., 315, 333, 384
Murata, Y., 233, 292
Muretow, N. S., 369, 384
Musgrave, R. B., *see* Moss, D. N. et al.

Nakamura, K., 329, 333
Namken, L. N., *see* Bartholic, J. F. et al.
Namken, L. N., *see* Gerald, C. J. and Namken, L. N.
Namken, L. N., *see* Wiegand, C. L. and Namken, L. N.
Needleman, S. M., *see* Degoes, L. and Needleman, S. M.
Neiburger, M., 384
Nemethy, J. J., *see* Schleusener, P. E. et al.
Nestorova, E. I., 236, 292
Netzmann, E., 384
Neuroth, G., 119, 183
Neuroth, G., *see* Wezler, K. and Neuroth, G.
Neuwirth, R., 70, 87, 183
Newberry, C. W., Eaton, K. J. and Mayne, J. R., 350, 384
Newman, J. E., Cooper, W. C., Reuther, W., Cahoon, G. A. and Pennado, A., 205, 292
Newman, P., *see* Atlas, D. et al.
Newton, O. H., *see* Ranney, C. D. et al.
Nicholas, F. W., 315, 333
Nielson, K. F., *see* Campbell, C. A. et al.
Nishizawa, T., 302, 333
Nittler, L. W. and Kenny, T. J., 249, 292
Noack, R., 319, 320, 333
Noffsinger, T. L., 213, 292
Nordenson, T. J., *see* Gangopadhyaya, M. et al.
Norwine, J. R., 310, 333
Nuttonson, M. Y., 234, 235, 236, 292
Nyberg, A., 75, 183

Odell, R. T., *see* Runge, E. and Odell, R. T.
Oelke, E. A. and Mueller, K. E., 292
Oke, T. R., 300, 312, 333
Okita, T., 316, 333
Olgyay, V., 321, 333, 344, 345, 355, 357, 384
Oliver, V. J., *see* Scofield, R. A. and Oliver, V. J.
Olmsted, C. E., 202, 292
Olschowy, G., 91, 183
Omar, M. H., *see* Gangopadhyaya, M. et al.
Opitz, E., 59, 183
Orlenko, L. R., 351, 384
Orlowa, W. W., 358, 384
Ormrod, D. P., *see* Azmi, A. R. and Ormrod, D. P.
Ormrod, D. P., *see* Herath, W. and Ormrod, D. P.
Ostanow, W. A., *see* Schriever, W. R. and Ostanow, W. A.
Oury, B., 280, 292

Pack, D. H., *see* Angell, J. K. et al.
Page, J. K., 369, 384
Palmer, W. C., 264, 265, 266, 292
Paneth, F. A., *see* Edgar, J. L. and Paneth, F. A.
Papai, L., 315, 333
Parker, M. W. and Borthwick, H. A., 243, 244, 292

PARKER, M. W., *see* BORTHWICK, H. A. and PARKER, M. W.
PARKER, M. W., *see* BORTHWICK, H. A. et al.
PARRY, M., 306, 333
PARTRIDGE, J. R., *see* BURMAN, R. D. and PARTRIDGE, J. R.
PASKIN, P. L., *see* GILMAN, D. L. et al.
PASQUILL, F., 219, 292, 378, 379, 384
PASSEL, C. F., *see* SIPLE, P. A. and PASSEL, C. F.
PEARSON, J. E. and JONES, G. E., 96, 98, 183
PEARSON, K., 267, 292
PEEL, C., 342, 384
PELTON, W. L., 224, 292
PELTON, W. L. and EARL, A. U., 270, 293
PELTON, W. L., *see* CAMPBELL, C. A. et al.
PENDLETON, J. W. and HAMMOND, J. J., 197, 293
PENDLETON, J. W. and WEIBEL, R. O., 238, 293
PENDLETON, J. W., SMITH, G. E., WINTER, S. R. and JOHNSTON, T. J., 195, 293
PENDLETON, J. W., *see* EGLI, D. B. et al.
PENMAN, H. L., 217, 219, 221, 293
PENMAN, H. L. and LONG I. F., 229, 293
PENNADO, A., *see* NEWMAN, J. E. et al.
PERKINS, W. A., 310, 333
PERL, G., 352, 353, 384
PETER, B. G. W., DOLGIECH, W. A. and SCHRIEVER, W. R., 368, 384
PETERS, D. B., *see* EGLI, D. B. et al.
PETERSON, J. C., *see* PUESCHEL, R. F. et al.
PETERSON, M. L., *see* CHAPMAN, A. L. and PETERSON, M. L.
PFLEIDERER, H., 32, 166, 183, 184
PFLEIDERER, H. and BUETTNER, K., 113, 121, 184
PFOTZER, G., 184
PHATAK, S. C., WITTWER, S. H. and TEUBNER, F. G., 246, 293
PHILIPPS, H., *see* BAUR, F. and PHILIPPS, H.
PHILLIPS, R. E., 243, 293
PICKETT, E. E., *see* STEWART, R. E. et al.
PIERCE, L. T., 218, 293
PITTER, R. L., 293
PLANT, J. A., 370, 385
PLATT, C. M. R., 277, 293
PLEHN, I., *see* SCHMIDT-KESSEN, W. and PLEHN, I.
POCHOP, L. O., CORNIA, R. L. and BLECKER, C. F., 279, 293
POE, G. A., *see* EDGERTON, A. T. et al.
POMEROY, R. W., 260, 293
POOLER, F., 315, 333
PORTMANN, A., 5, 184
POTTER, J. G., 329, 333
POWER, J. F., *see* ALESSI, J. and POWER, J. F.
PRECHT, H., *see* HENSEL, H. et al.
PRESTON-WHYTE, R. A., 329, 333
PRIESTLEY, C. H. B., 211, 219, 293
PRITCHARD, J. A., *see* MCGINNIS Jr., D. F. et al.
PROETT, C. H., 137, 154, 184
PRUITT, W. O., 219, 293
PRUITT, W. O. and ANGUS, D. E., 224, 293
PRUITT, W. O., *see* DYER, A. J. and PRUITT, W. O.
PRUITT, W. O., *see* LOURENCE, F. J. and PRUITT, W. O.
PRUITT, W. O., *see* VEIHMEYER, F. J. et al.
PUESCHEL, R. F., ELLIS, H. T., MACHTA, L., COTTON, G., FLOWERS, E. C. and PETERSON, J. C., 52, 53, 184

QUENZEL, H., 54, 184, *see also* MOELLER, F. and QUENZEL, H.
QUITT, E., 306, 310, 334

RADKE, J. K. and BAUER, R. E., 207, 293
RADKE, L. F., *see* HOBBS, P. V. and RADKE, L. F.
RAHN, J. J. and BROWN, D. M., 213, 293
RAISCH, E., 385
RAISS, W., 346, 385
RAJEWSKI, A. N., 369, 385
RAMACHANDRAN, G., *see* DAS, J. C. and RAMACHANDRAN, G.
RAMIAH, K. and RAO, M. B. V. N., 231, 293
RANKE, O. F., 12, 184
RANNEY, C. D., HURSH, J. S. and NEWTON, O. H., 242, 293
RAO, M. B. V. N., *see* RAMIAH, K. and RAO, M. B. V. N.
RASMUSSEN, V. P., *see* HIELMAN, J. L. et al.
RÅSTAD, H., 369, 385
RATHS, P., *see* HENSEL, H. et al.
RAY, L. L., *see* HOLEKAMP, E. R. et al.
READ, R. A., *see* WOODRUFF, N. P. et al.
READER, J. D. and WHYTE, H. M., 112, 184
REGER, E., *see* SIEDENTOPF, H. and REGER, E.
REIDAT, R., 346, 349, 364, 367, 370, 380, 385
REIDAT, R., *see* BÖER, W. et al.
REIFSNYDER, W. E. and LULL, H. W., 253, 293
REIFSNYDER, W. E., FURNIVAL, G. M. and HOROWITZ, J. L., 256, 293
REINDELL, H., *see* WEIDEMANN, H. et al.
REINHOLD, F., 385
REITER, R., 184
REITER, R., CARNUTH, W. and SLADKOVIC, R., 184
RENOU, E., 299, 334
RENSE, W. A., 190, 293
REUTER, H. and CEHAK, K., 77, 184
REUTHER, W., *see* NEWMAN, J. E. et al.
RHOAD, A. O., 203, 257, 258, 260
RHOADES, D. C., *see* BROOKS, F. A. et al.
RHOADES, H. F., *see* ROBINS, J. S. and RHOADES, H. F.
RHYKERD, C. L., *see* BULA, R. J. et al.
RHYKERD, C. L., *see* GARZA, R. T. et al.
RICHARD, L. D., 215, 294
RICHMOND, T. R., *see* LEWIS, C. F. and RICHMOND, T. R.
RIDER, N. E., 229, 294
RIEDE, W., 243, 294
RILEY, J. A., 242, 294
ROACH, W. T., 322, 334
ROBERTSON, D. F., 38, 184
ROBERTSON, G. W., 281, 294

ROBERTSON, G. W., *see* BAIER, W. and ROBERTSON, G. W.
ROBERTSON, G. W., *see* HOLMES, R. H. and ROBERTSON, G. W.
ROBINS, J. S. and DOMINGO, C. E., 240, 294
ROBINS, J. S. and RHOADES, H. F., 240, 294
ROBINS, P. C., *see* RUTTER, A. J. et al.
ROBINSON, F. E., *see* CHANG, J. H. et al.
ROBITZSCH, M., 136, 137, 145, 184
RODENWALDT, E., 2, 184
RODENWALDT, E. and JUSATZ, H. J., 3, 184
ROGERS, J. S., *see* GILMORE, E. C. and ROGERS, J. S.
ROHWER, C., 219, 294
ROLLIER, A., 35, 184
ROSENBERG, N. J., 268, 269, 270, 294
ROSENBERG, N. J., HART, E. and BROWN, K. W., 219, 294
ROSENBERG, N. J., *see* BROWN, K. W. and ROSENBERG, N. J.
ROSKAMM, H., *see* WEIDEMANN, H. et al.
RUBINSTEIN, E. S., *see* ALISSOW, B. P. et al.
RUBNER, M., 137, 184
RUCKLI, R., 357, 385
RUDNEWA, A. W., 369, 385
RÜESCH, J. D., 270, 294
RUGE, H., 137, 184
RUKLES, J. R., *see* SHAW, R. H. et al.
RUNGE, E. and ODELL, R. T., 278, 279, 294
RUNGE, E. C. A., 279, 294
RUTTER, A. J., KERSHAW, K. A., ROBINS, P. C. and MORTON, A. J., 256, 294
RUTTNER, F., *see* SAUBERER, F. and RUTTNER, F.
RYD, H., *see* THÖRNBLAD, O. H. and RYD, H.

SAKAMOTO, C. M., 279, 294
SAKAMOTO, C. M. and JENSEN, R. E., 279, 294
SAKAMOTO, C. M. and SHAW, R. H., 196, 198, 294
SAKAMOTO, S., *see* EDGERTON, A. T. et al.
SAMEK, L., *see* WEIDEMANN, H. et al.
SANDBERG, J. S., *see* DUCKWORTH, F. S. and SANDBERG, J. S.
SANDHU, S. S. and HODGES, H. F., 203, 294
SARGEANT, D. H. and TANNER, C. B., 220, 294
SARGENT II, F., 131, 184
SATO, C., *see* JAUREGUI, E. and SATO, C.
SAUBERER, F., 41, 184
SAUBERER, F. and RUTTNER, F., 54, 55, 184
SAUBERER, F., *see* DIRMHIRN, I. and SAUBERER, F.
SAUBERER, F., *see* STEINHAUSER, F. et al.
SAUTTER, L., 356, 361, 385
SAVARINA, M. V., *see* BORISENKŎ, M. M. and SAVARINA, M. V.
SAVARINA, M. V., *see* JAGANATHAN, P. et al.
SCHAEFER, H. J., *see* GEORGII, H. W. and SCHAEFER, H. J.
SCHAEFER, V. J., 329, 334
SCHAEFER, V. J., *see* HOGAN, A. W. et al.
SCHARLAU, K., 3, 137, 184
SCHERER, V. R., 279, 294
SCHERER, V. R., *see* MICHAELS, P. J. and SCHERER, V. R.
SCHERHAG, R., 312, 334
SCHICKEDANZ, P. T., *see* HUFF, F. A. and SCHICKEDANZ, P. T.
SCHINZEL, A., 342, 385
SCHLEUSENER, P. E., NEMETHY, J. J., SHULL, H. H. and WILLIAMS, G. E., 223, 294
SCHMAUSS, A., 316, 334
SCHMIDT, F. H. and BOER, J. H., 316, 334
SCHMIDT, W., 299, 334
SCHMIDT, W., *see* BREZINA, E. and SCHMIDT, W.
SCHMIDT-KESSEN, W., 38, 165, 184
SCHMIDT-KESSEN, W. and PLEHN, I., 62, 184
SCHNEIDER, S. R., WIESNET, D. R. and MCMILLAN, M. C., 278, 294
SCHNELLE, F., 263, 294
SCHNELLE, K.-W., 165, 184
SCHOENBEIN, C. F., 92, 184
SCHOENHOLZER, G., 64, 184
SCHOLANDER, P. E., WALTERS, V., HOCK, R. and IRVING, L., 259, 294
SCHOLL, J. C., *see* HENDRICKS, W. A. and SCHOLL, J. C.
SCHÖN, S. G., 372, 385
SCHRAM, K. and THAMS, J. C., 45, 185
SCHRIEVER, W. R. and OSTANOW, W. A., 368, 385
SCHRIEVER, W. R., *see* PETER, B. G. W. et al.
SCHROEER, E., 93, 185
SCHROPP, L., 353, 385
SCHÜEPP, W., 51, 185
SCHÜEPP, W., *see* DE COSTER, M. et al.
SCHÜLE, W., CAMMERER, J. S. and GÖRLING, P., 361, 362, 385
SCHULTZ, H. B., *see* BROOKS, F. A. et al.
SCHULTZE, E.-G., 75, 185
SCHULTZE, E.-G., *see* AMELUNG, W. et al.
SCHULZE, A., *see* KNOCH, K. and SCHULZE, A.
SCHULZE, R., 9, 10, 14, 15, 39, 43, 47, 185
SCHULZE, R., *see* HENSCHKE, U. and SCHULZE, R.
SCHÜTZE, W., *see* GRÄFE, K. and SCHÜTZE, W.
SCHWABL, W., *see* LAUSCHER, F. and SCHWABL, W.
SCHWANKE, R. K., 226, 294
SCHWEIZERISCHE VEREINIGUNG DER KLIMAKURORTE, 185
SCOFIELD, R. A. and OLIVER, V. J., 278, 294
SEITZ, E. D., *see* KLEIN, E. et al.
SEITZ, E. O., *see* MEYER, H. A. E. and SEITZ, E. O.
SEKIGUTI, T., FUKUI, E. and YOSHINO, M., 300, 334
SEXTON, D., 351, 385
SHANKLIN, M. D., *see* BERRY, I. L. et al.
SHANTZ, J. H., *see* BRIGGS, L. J. and SHANTZ, J. H.
SHAW, J. H., *see* SLOAN, R. et al.
SHAW, R. H., 213, 216, 239, 279, 295
SHAW, R. H. and THOM, H. C. S., 240, 295
SHAW, R. H. and WEBER, C. R., 195, 295
SHAW, R. H., RUKLES, J. R. and BARGER, G. L., 240, 295
SHAW, R. H., *see* BUSS, S. and SHAW, R. H.

Shaw, R. H., *see* Carlson, R. E. et al.
Shaw, R. H., *see* Dale, R. F. and Shaw, R. H.
Shaw, R. H., *see* Sakamoto, C. M. and Shaw, R. H.
Shaw, R. H., *see* Stevenson, K. R. and Shaw, R. H.
Shaw, R. H., *see* Yao, A. Y. M. and Shaw, R. H.
Shaw, R. H., *see* Waggoner, P. E. and Shaw, R. H.
Shaw, R. H., *see* Wilsie, C. P. and Shaw, R. H.
Shellard, H. C., 323, 334
Shellard, H. G., *see* Lacy, R. E. and Shellard, H. G.
Shenfeld, L. and Slater, D. F. A., 308, 334
Shull, H. H., *see* Schleusener, P. E. et al.
Siedentopf, H. and Reger, E., 14, 185
Siminovitch, D., Ball, W. L., Desjardings, R. and Gamble, D. S., 263, 295
Simonis, W., 225, 295
Siple, P., 370, 385
Siple, P. A. and Passel, C. F., 130, 185
Sirotenko, V. C., 279, 295
Slade, D. H., 315, 334
Sladkovic, R., *see* Reiter, R. et al.
Slater, D. F. A., *see* Shenfeld, L. and Slater, D. F. A.
Slatyer, R. O., 227, 295
Sloan, R., Shaw, J. H. and Williams, D., 40, 185
Smith, C. V., 260, 295
Smith, D., *see* Ueno, M. and Smith, D.
Smith, G. E., *see* Pendleton, J. W. et al.
Smith, J. W., 278, 295
Smith, W. L. and Howell, H. B., 295
Smith, W. O. and Byers, H. G., 208, 295
S.N.I.P., 385
Sohar, E., *see* Tennenbaum, J. et al.
Soltis, F., *see* Edgerton, A. T. et al.
Somnerholder, B. R., 223, 295
Southwell, B. L., *see* McCormick, W. C. et al.
Spangemacher, K., 375, 385
Spencer, J. W., *see* Ballantyne, E. R. and Spencer, J. W.
Spiegelmann, M., *see* McConnell, W. J. and Spiegelmann, M.
Sprenger, E. and Kruger, W., 349, 385
Stabhill, G., 223, 295
Stair, J., *see* Coblentz, W. W. and Stair, J.
Stakman, E. C. and Harrar, G., 270, 295
Stalfelt, M. G., 227, 295
Stansel, J. W., 233, 295
Staple, W. J. and Lehane, J. J., 269, 270, 295
Stapleton, H. N., *see* Wanjura, D. F. et al.
Steinhauser, F., 83, 87, 88, 90, 94, 99, 100, 173, 174, 185, 377, 386
Steinhauser, F., Eckel, O. and Sauberer, F., 300, 314, 320, 322, 334, 353, 370, 386
Stern, A. C., 378, 386
Steubing, L., 270, 295
Stevens, S. S., 13, 28, 90, 185
Stevenson, K. R. and Shaw, R. H., 195, 295
Stewart, R. E., Pickett, E. E. and Brody, S., 259, 295
Stix, E., 72, 185
Stoeckler, J. H., 269, 295
Stone, R. G., 185
Stoughton, R. H. and Vince, D., 203, 295
Stummer, G., 315, 334
Suessenberger, E., 93, 185
Sulakvelidze, G. K., Bibilashvili, N. Sh. and Hapchevce, V. F., 264, 295
Summers, P. W., 311, 334
Sundberg, Å, 308, 334
Suomi, V. E. and Tanner, C. B., 220, 295
Suomi, V. E., *see* Kuhn, P. M. and Suomi, V. E.
Süring, R., 366, 386
Sutton, O. G., 219, 296, 378, 386
Swinbank, W. C., 219, 296
Swinbank, W. C., *see* Deacon, E. L. and Swinbank, W. C.
Szava-Kovats, J., 134, 185

Tadman, E. C., *see* Foster, A. C. and Tadman, E. C.
Taerum, R., 251, 296
Taessler, R., 368, 370, 386
Tanner, C. B., 213, 221, 296
Tanner, C. B. and Lemon, E. R., 221, 296
Tanner, C. B., *see* Sargeant, D. H. and Tanner, C. B.
Tanner, C. B., *see* Suomi, V. E. and Tanner, C. B.
Taylor, E., *see* Gates, D. M. et al.
Taylor, R. J. and Dyer, A. J., 220, 296
Taylor, R. J. and Webb, E. K., 220, 296
Ten Kate, H., *see* Jaganathan, P. et al.
Tennenbaum, J., Sohar, E., Adar, R. and Gilat, T., 130, 185
Terjung, W. H., 131, 132, 133, 185
Ter Linden, A. J., 371, 386
Teub, B., 95, 185
Teubner, F. G., *see* Phatak, S. C. et al.
Thams, J. C., 16, 17, 125, 141, 185
Thams, J. C. and Wierzejewski, H., 24, 185
Thams, J. C., *see* Schram, K. and Thams, J. C.
Tharp, W. H., 206, 241, 296
Thauer, R., 111, 117, 185
Thauer, R. and Zoellner, G., 185
Thauer, R., *see* Wezler, K. and Thauer, R.
Thilenius, R. and Dorno, C., 121, 124, 185
Thom, E. C., 130, 185
Thom, H. C. S., 337, 345, 349, 351, 352, 386
Thom, H. C. S., *see* Barger, G. L. and Thom, H. C. S.
Thom, H. C. S., *see* Holzman, B. G. and Thom, H. C. S.
Thom, H. C. S., *see* Shaw, R. H. and Thom, H. C. S.
Thomas, M. D. and Hill, G. R., 197, 198, 200, 296
Thomson, L. M., 279, 296
Thörnblad, O. H. and Ryd, H., 364, 365, 386
Thornthwaite, C. W., 216, 223, 296
Thornthwaite, C. W. and Holzman, B., 219, 296
Thornthwaite, C. W. and Kaser, P., 296
Thornthwaite, C. W. and Mather, J. R., 296
Thorud, D. B., *see* Bohren, C. F. and Thorud, D. B.
Tippett, L. N. C., *see* Fisher, R. A. and Tippett, L. N. C.

Tirel, H., 348, 386
Tomato, S., 269, 296
Tomlinson, B. R., 223, 296
Tonne, F., 355, 386
Toperczer, M., 21, 185
Torrance, K. E., *see* Gutman, D. P. and Torrance, K. E.
Trappenberg, R., *see* Diem, M. and Trappenberg, R.
Trendelenburg, W., 7, 8, 12, 185
Troll, C., 137, 177, 186
Tromp, S. W., 4, 186
Tronnier, H., 38, 186
Tsubo, Y., *see* Inoue, E. et al.
Tsuchiya, K., *see* Fujita, T. et al.
Tsunoda, K., *see* Matsushima, S. and Tsunoda, K.
Tuller, S. E. and Chilton, R., 226, 296
Turc, L., 223, 296
Turner, D. B., 83, 186
Turner, N. C., *see* Waggoner, P. E. et al.
Tveit, A., 364, 386

Ueno, M. and Smith, D., 249, 296
Ulanova, E. S., 234, 235, 296
UNESCO, 131, 186
UNESCO-FAO, 131, 186
Unz, F., 54, 186
Urbach, F., 38, 186
Uryvaev, V. A., *see* Gangopadhyaya, M. et al.
U.S. Public Health Service, 300, 334
U.S. Weather Bureau, Office of Hydrology, 326, 334

Vahle, W., *see* Hausser, K. W. and Vahle, W.
Valko, P., 52, 53, 186, 353, 386
Van Bavel, C. H. M., 265, 296
Van Bavel, C. H. M. and Ehrler, W. L., 213, 296
Van Bavel, C. H. M. and Meyer, L. E., 224, 296
Van Bavel, C. H. M. and Wilson, T. V., 221, 223, 296
Van Daren, C. A., *see* Holt, R. F. and Van Daren, C. A.
Van der Elst, N., *see* De Coster, M. et al.
Van der Linde, J., 268, 296
Van Eimern, J., 268, 269, 297
Van Eimern, J., Franken, E. and Harrics, H., 269, 297
Van Everdingen, E., 271, 297
Van Royen, W., 231, 297
Van Wijk, W. R., 191, 297
Van Wijk, W. R. and Hidding, A. P., 269, 297
Van Zuilen, D., 386
Vareschi, V., *see* Luedi, W. and Vareschi, V.
Vásáhelyi, J., *see* Hank, O. and Vásáhelyi, J.
V.D.I., 355, 386
Vehts, D. B., *see* Lowry, W. P. and Vehts, D. B.
Veihmeyer, F. J., 224, 297
Veihmeyer, F. J. and Henrickson, A. H., 218, 297
Veihmeyer, F. J., Pruitt, W. O. and McMillin, W. D., 218, 297
Veihmeyer, F. J., *see* Halkias, N. A. et al.
Vellose, M. H., 276, 297
Ventskevich, G. Z., 200, 235, 241, 244, 250, 261, 297
Verdium, J. and Loomis, W. E., 198, 297
Verkerk, K., *see* Abdelhafeez, A. T. and Verkerk, K.
Vernekar, A. D., 149, 186
Vernon, H. M., 154, 186
Vernon, H. M. and Warner, C. G., 154, 186
Verzár. F., *see* Bider, M. and Verzár, F.
Verzár, F., *see* Herpertz, E. et al.
Vesina, P. E., 253, 297
Vick, F., 354, 386
Vickers, B. J. and Davenport, A. G., 352, 386
Vilkner, H., 308, 309, 334
Vince, D., *see* Stoughton, R. G. and Vince, D.
Vincent, I., 121, 186
Von Betzold, 127, 137, 186
Von Heimendahl, F., 386
Von Kienle, J., 328, 334
Von Muralt, A., 61, 62, 186
Von Muralt, A., *see* Fleisch, A. and Von Muralt, A.
Von Muralt, A., *see* Hartmann, H. and Von Muralt, A.
Von Schwarz, A. R., 208, 297
Von Wilhelmi, Th., 253, 297
Vough, L. R. and Martin, G. C., 250, 297

Wachsmuth, W., 165, 184, 186
Wada, M., *see* Abe, S. and Wada, M.
Waggoner, P. E., 256, 288, 297
Waggoner, P. E., Begg, J. E. and Turner, N. C., 226, 297
Waggoner, P. E., Horsfall, J. G. and Lukens, R. J., 273, 297
Waggoner, P. E., Miller, P. M. and DeRoo, H., 261, 297
Waggoner, P. E. and Shaw, R. H., 211, 297
Wagner, F., 208, 297
Walker, J., 268, 297
Wallace, H. A., 278, 297
Wallace, H. A. and Bressman, E. N., 238, 239, 297
Wallin, J. R., 227, 297
Walters, V., *see* Scholander, P. E. et al.
Wang, J. Y., 247, 297
Wang, J. Y. and Wang, S. C., 221, 297
Wang, S. C., *see* Wang, J. Y. and Wang, S. C.
Wanjura, D. F., Buxton, D. R. and Stapleton, H. N., 207, 297
Wanjura, D. F., Hudspeth Jr., E. B. and Bilbro Jr., J. D., 241, 297
Wanner, H. U. and Gilgen, A., 94, 186
Wark, D. Q., Yamamoto, G. and Lienesch, J., 277, 297
Warner, C. G., *see* Vernon, H. M. and Warner, C. G.
Warren, D. C., 258, 298
Wassink, E. C., 193, 298
Watson, D. J., 194, 298

WEBB, C. G., 130, 186
WEBB, E. K., 219, 298
see also TAYLOR, R. J. and WEBB, E. K.
WEBER, C. R., *see* SHAW, R. H. and WEBER, C. R.
WEIBEL, R. O., *see* PENDLETON, J. W. and WEIBEL, R. O.
WEICKMANN, H. K., 264, 298
WEIDEMANN, H., ROSKAMM, H., SAMEK, L. and REINDELL, H., 186
WEIDEMANN, H., *see* REINDELL, H. et al.
WENT, F. W., 206, 227, 298
WENZEL, H. G., 186
WENZEL, H. G., *see* MUELLER, E. A. and WENZEL, H. G.
WESSEL, C. J., *see* GREATHOUSE, G. A. and WESSEL, C. J.
WEVER, R., *see* ASCHOFF, J. and WEVER, R.
WEYBREW, J. A., *see* HAROON, M. et al.
WEZLER, K., 101, 186
WEZLER, K. and NEUROTH, G., 113, 186
WEZLER, K. and THAUER, R., 158, 186
WHITEHEAD, F. H., *see* RUTTER, A. J. and WHITEHEAD, F. H.
WHITTOW, G. C., *see* INGRAM, D. L. and WHITTOW, G. C.
WHYTE, H. M., *see* READER, J. D. and WHYTE, H. M.
WIEGAND, C. L. and NAMKEN, L. N., 298
WIEGAND, C. L., *see* BARTHOLIC, J. F. et al.
WIEGEL, H., 328, 334
WIERZEJEWSKI, H., 126, 186
WIERZEJEWSKI, H., *see* COURVOISIER, P. and WIERZEJEWSKI, H.;
see also THAMS, J. C. and WIERZEJEWSKI, H.
WIESNER, J. and AMELUNG, W., 186
WIESNET, D. R., *see* MCGINNIS Jr. et al.;
see also SCHNEIDER, S. R. et al.
WILLIAMS, D., *see* SLOAN, R. et al.
WILLIAMS, G. D. V., 279, 298
WILLIAMS, G. D. W., JOYAT, M. I. and MCCORMICK, P. A., 279, 298
WILLIAMS, G. E., 223, 298
see also SCHLEUSENER, P. E. et al.
WILLIAMS, G. P., 219, 298
WILSIE, C. P. and SHAW, R. H., 202, 298
WILSON, O., 130, 187
WILSON, T. V., *see* VAN BAVEL, C. H. M. and WILSON, T. V.
WILTSHIRE, S. P., 271, 298
WINSLOW, C. E. A., HERRINGTON, L. P. and GAGGE, A. P., 187
WINTER, S. R., *see* PENDLETON, J. W. et al.
WISE, K. R., *see* JASKE, R. T. et al.
WITTWER, S. H., *see* PHATAK, S. C. et al.
WOERNER, H., 16, 187
WOLFE, T. K., 298
WOLLNY, E., 209, 298
WOLTERECK, H., 184
WOOD, T. B., *see* CLAPSTICK, J. W. and WOOD, T. B.
WOODRUFF, J. M., MCCAIN, F. S. and HOVELAND, C. S., 242, 298
WOODRUFF, N. P., READ, R. A. and CHEPIL, W. S., 269, 298
WOOLLUM, C. A., 307, 309
WORKING GROUP OF THE COMMISSION FOR ATMOSPHERIC SCIENCES, 386
WORLD METEOROLOGICAL ORGANIZATION, 175, 187, 262, 298, 300, 334

YAGLOU, C. P., 154, 156, 187
see also HOUGHTON, F. C. and YAGLOU, C. P.
YAGLOU, C. P. and MILLER, W. E., 154, 187
YAMAMOTO, G., *see* WARK, D. Q. et al.
YAO, A. Y. M., 265, 267, 268, 273, 274, 280, 298
YAO, A. Y. M. and SHAW, R. H., 195, 223, 298
YARGER, D. N., *see* CARLSON, R. E. et al.
YOSHIMURA, H., *see* ITOH, S. et al.
YOSHINO, M., *see* SEKIGUTI, T. et al.

ZADOKS, J. C., *see* HOGG, W. H. et al.
ZAWADIL, R., 370, 386
ZENKER, H., 140, 187
ZILCH, K., *see* KOENIG, G. and ZILCH, K.
ZIMMERMANN, O., 386
ZINKE, P. J., 254, 298
ZIPPEL, H., 357, 386
ZOELLNER, G., *see* THAUER, R. and ZOELLNER, G.
ZUG, H., 386

Geographical Index

Aachen, 353
Adriatic coast, 371
Africa, 132, 228, 230, 273, 364, 372
Alaska, 49, 53, 260, 359, 360
Algeria, 134
Alps, 28, 67, 125, 153, 171, 172
—, Eastern, 15, 21, 22
—, Swiss, 20, 26, 51, 175
America, 248, 359, 372 (see also U.S.A.)
Ames, Io., 280
Amu Darya Oasis, 222
Amur Valley, 359
Antarctic, 19, 53, 127, 130, 150
—, Highlands, 62
Antwerp, 372
Arabia, 150
Arctic, 19, 99, 127
— Circle, 359
Argentina, 54, 233, 238, 248
Argone, Ill., 210
Arizona, 220, 224, 248
Arlington, 306
Asahikawa City, 316
Ashford, 226
Asia, 228, 230, 233, 243, 248
— Minor, 18
—, Southeast, 354
Assuan, 362, 363
Athens, 18
Atlantic, 54, 69, 75, 148, 149
— European Coast, 148
— coastal areas, 134
Australia, 38, 149, 233, 238, 240
Austria, 100, 164, 300, 310, 314, 377

Bad Gastein, 100
Baghdad, 347
Bali, 52
Balkans, 248
Baltimore, 317
— airport, 314
Bamberg, 364
Barza, 364
Basel, 22, 23, 24, 25, 26, 53, 125, 126, 140, 141, 142, 143, 171, 172
Belgium, 335, 381
Belmar, 150, 153
Bergen, 23, 24, 153
Berlin, 51, 68, 69, 313, 326, 364, 377
—, Rummelsburg, 16, 17
Beuel, 306
Birmingham, Ala, 317
Black Forest, 24, 70, 369
Bolivian Plateau, 54
Bombay, 341
Bonn, 306
Boston, Mass., 322
Bratislava, 304
Brazil, 98, 230, 238, 240, 243
Brenner Pass, 371
Brno, 306, 310
Brocken, 364
Budapest, 315, 322

Cairo, 341
California, 227, 231, 248, 258, 259, 262, 263, 306, 310, 319
Camp Century, 151
Carpathian Mountains, 369
Canada, 233, 238, 245, 248, 272, 279, 336, 365, 368
Cape Verde Islands, 54
Central America, 148
Central Europe, 80
Cherrapunji, 366
Chicago, 325, 327, 329
Chile, 248
China, 25, 230, 233, 238, 240, 241, 245
Cincinnati, Ohio, 311, 323
Cleveland, Ohio, 317
Coachella Valley, Calif., 263
Cologne, 320, 377
Colomb-Bécher, 348
Colorado, 264
Columbia, Md., 302, 303, 312, 314
Conakry, 347
Connecticut, 312
Coshocton, Ohio, 224
Czechoslovakia, 306

Davis, Calif., 224, 231, 259
Davos, 6, 20, 21, 22, 23, 24, 25, 26, 27, 36, 37, 49, 51, 72, 125, 141, 142, 150, 153, 171, 172
Delhi, 53
Denmark, 269, 270, 378
Dickson, 346
Dnieper Valley, 323
Dodoma, 273, 274
Donora, Penn., 319, 335, 381

Dresden, 142
Duala, 362, 363
Dublin, 23

Eastern Alps, 15, 21, 22 (see also Alps)
Elbe Estuary, 380
El Centro, Calif., 258
Emely Moor, 369
England, 75, 243, 271, 306, 310, 317, 319, 322, 324, 327, 336, 343, 352, 369
Entebbe, 364
Europe, 22, 67, 69, 94, 140, 164, 168, 171, 243, 341, 346, 370, 372

Feldburg, 369
Finland, 23, 24, 357
Foggia, 371
France, 94, 243, 328, 373
Frankfort am Main, 83, 84, 315, 323
Freiburg im Breisgau, 23, 24, 70
Fort Wayne, Ind., 301

Garston, 365
Gelsenkirchen, 88
Germany, 24, 75, 164, 169, 192, 226, 243, 269, 308, 319, 321, 323, 325, 326, 328, 336, 346, 348, 349, 354, 357, 364, 369
—, North Sea Coast, 95, 149
Giessen, 357, 358
Giza, 18, 20, 21
Great Britain, 228, 273, 351, 365
Great Lakes, 250
Great Plains, 233, 236, 279
Greenland, 50, 75, 150, 151, 359, 364
Griefswald, 308

Halle, 319
Hamburg, 43, 140, 338, 339, 346, 349, 351, 353, 358, 363, 364, 366, 377, 380
Harbin, 245
Harry Mountains, 364
Hawaii, 52, 95, 225
Heidelberg, 99
Heligoland, 362, 363, 364
Helsinki, 23, 24, 79
Hohenpeissenburg, 23, 93
Hochenschwand, 23, 24
Hokkaido, 230, 316
Hokuriku, 230
Holland, 316
Holzkirchen, 354

Iceland, 364
Idaho, 310
Illinois, 278, 279, 327
India, 53, 98, 150, 230, 233, 238, 240, 243, 248, 279
Indian Ocean, 367
Indonesia, 53
Indus Valley, 342
Iowa, 238, 267
Ireland, 23, 94
Irkutsk, 341
Israel, 223, 228
Italy, 2, 233, 238, 364, 371

Japan, 230, 231, 232, 233, 243, 250, 316
Jenissei Estuary, 346
Jericho, 342
Jerusalem, 18
Jungfraujoch, 51, 69, 70
Kamannaja Steppe, 269
Kansas, 278, 279
Kanto, 230
Karlsruhe, 142, 364
Katmai, 49, 53
Kentucky, 280, 310, 315
Kerala, 98
Kiev, 314, 315
Kinshasa, 343
Kongwa, 274
Korea, 230, 243, 250
Krakatoa, 53

Leicester, 316
Leningrad, 314, 315
Lhasa, 245
Lincoln, Nebr., 305
Linz, 371
London, 72, 81, 84, 311, 319, 322, 323, 324, 325, 326, 342, 350, 381
Locarno, 22, 23, 45, 51, 141, 142, 353
Los Angeles, 301, 319, 325, 335
Louisville, Kentucky, 310, 311, 315
Lunzer Lake, 55
Lwiro, 18, 364

Madagascar, 367
Madison, Wisc., 202
Mannheim-Ludwigshafen, 88, 89, 328
Mauna Loa, 52, 95, 96
Mawson, 17, 18
Mediterranean, 18, 248
Menzenschwand, 70
Mesopotamia, 342
Meuse Valley, 319, 335, 381
Mexico, 131, 245, 248
— City, 53, 64, 329
Minnesota, 228
Mississippi, Valley, 250
Missouri, 220, 228, 278
Monazite, 98
Montana, 206
Montreal, 317
Moscow, 308
Moselle River, 357
Mount Agung, 52, 53
Mount Washington, N. H., 351

Munich, 303, 308, 321, 362, 363, 364, 366

Nairobi, 329
Nanga-Parbat, 62
Nashville, Tenn., 313
Nebraska, 223, 250, 279, 305
New Haven, Conn., 312
New Jersey, 228
New Mexico, 222
New York, 82, 83, 325, 326, 328
— City, 301, 313, 329
Norfolk, Va., 273
North Africa, 18, 99
North America, 69, 150, 222
North China Plain, 240
North Dakota, 278, 279
Northwest Europe, 75
North Sea, 148, 149
North Pole, 352
Norway, 23, 251, 364, 365, 369

Oak Ridge, Tenn., 278, 317, 323, 328
Ohio, 224, 317, 323
Oklahoma, 327
Oldenburg, 357
O'Neill, Nebr., 218, 221
Ontario, 204
Oslo, 353
Ougadougou, 362, 363

Pacific Ocean, 69, 150
Pakistan, 230, 240
Palestine, 342
Palo Alto, Calif., 306, 310
Paris, 299, 304, 305, 328
Philadelphia, 38
Philippine Islands, 230, 243, 367
Peking, 336
Pennsylvania, 319, 381
Peru, 240, 248
Pittsburgh, Penn., 324
Poona, 53
Potomac River, 306
Potsdam, 16, 17, 18, 148, 169, 170, 341, 352
Pretoria, 89, 90, 353

Quickborn, 351

Reading, 306
Red Sea, 134
Réunion, 367
Rhine River, 306
— Plain, 24, 70
Rio de Janeiro, 372
Riso, 378
Rochdale, 327
Roi Baudouin Coast, 53
Romania, 238
Rothamstead, 226
Rotterdam, 372
Russia, 250, 251, 346, 351, 358, 366, 369 (see also U.S.S.R.)
Ruhr, 328

Sahara, 18, 20, 54, 67, 134, 135, 348
Saint Louis, Miss., 328
Saint Paul, Minn., 190
Salzburg, 173
San Francisco, Calif., 310
San Salvador, 148
Scandinavia, 75, 94
Schenectady, N.Y., 328
Scotchi, 366
Siberia, 346, 358, 359, 369
Sierra Leone, 342
Skovorodino, 358
Somalia, 134
Sonnblick, 173
South Africa, 89, 134, 149, 336
South America, 238
South Dakota, 279
Southeast Asia, 150, 354 (see also Asia)
South Pole, 25, 53, 130, 153
Steubenville, Ohio, 317
Stockholm, 20, 21, 79
Sudan, 134
Sung Liao Plain, 243
Sweden, 308, 365, 368
Switzerland, 53, 70, 164, 371
Swiss Alps, 20, 26, 51 (see also Alps)

Tanzania, 273, 274
Tempe, Ariz., 220, 224
Tennesee, 312, 328
Tessin, 52, 53
Texas, 223
Thailand, 243
Tibet, 62
Tibetan Plateau, 245
Tifton, 273
Tokyo, 302, 307
Toronto, 308, 329
Trier, 321, 357
Tulsa, Okla., 327
Tuto West, 151

Ukraine, 323
Upper Bavaria, 93
Uppsala, 308
Ural, 369
Urbana, Ill., 195, 197, 278, 327
U.S.A., 80, 223, 230, 233, 240, 242, 243, 248, 257, 262, 268, 270, 271, 272, 273, 274, 278, 279, 280, 302, 308, 311, 317, 323, 324, 335, 341, 345, 351, 352, 356, 370, 371
—, Corn Belt, 226, 238, 240, 278, 279
U.S.S.R., 233, 234, 238, 240, 241, 243, 248, 269, 270, 279, 323 (see also Russia)

Valdai Hills, 369
Vienna, 88, 90, 93, 94, 300, 310, 313, 314, 320, 321, 322, 329, 336, 371
— Forest, 371

Wadi Halfa, 18
Washington, D.C., 79, 201, 226, 306, 308, 315, 328
Weissfluhjoch, 22, 23, 27, 51, 171, 172
West Germany, 23, 72, 88, 89, 165
Westerland on Sylt, 95
Wyoming, 223

Yakutsk, 358
Yangtze Valley, 240
Yugoslavia, 238

Zaporozhe, 323

Subject Index

Acclimatization, 61, 114, 115, 116, 157
Aerocolloid, 67
Aerosol, 11, 54, 66, 67, 171, 175
—, climatology, 169
—, continental, 74
—, electrical properties, 78
—, growth, 56
—, maritime, 74, 75
—, size spectrum, 73, 74
—, tropospheric, 73
Agriculture, 189
—, climate, 229
—, land utilization, 273–275
—, remote sensing, 276–278
Air, 6, 55
— circulation, urban, 317
— composition, 57, 58, 80
— concentration, 77
— conditioning, 346–349
— hygiene, 49, 66, 71
— masses, 56, 168–170
— movement, effect on farm animals, 259
— pollution, *see* Pollution, atmospheric
— residence time, 77
Aitken nuclei, *see* Condensation nuclei
Albedo, 23, 26, 27, 190
—, various surfaces, 354
Alfalfa climate, 249
Altitude, physiological effects, 64
Animal climate, 256–260
— breeding and housing, 260
Apple climate, 250
Atmospheric composition, *see* Air composition
Atmospheric stability, 380
Attic temperatures, 354

Balneology, 5
Bellani apparatus, 22
Bioclimatology, 1
—, definition, 3
—, evaluation, 6
—, history, 2
—, human, 1, 5
—, taxonomy, 2
Biometeorology, definition, 4
Brightness, 11, 12
Building design, 349
— ventilation, 350

Canopy temperature, 213
Carbon dioxide, atmospheric, 57, 89, 95, 96
— —, photosynthesis, 197, 198
Carbon monoxide, 84–86
Cargo climate, 372
City climate, 299–330
Climatic aids for construction engineering, 370
— — for transport, 371
Climate,
— alteration, urban, 330
— characteristics, 376
— classification, 275
— types for technology, 376
Climatic conditions, optimal for construction and storage, 374
Climatic elements,
— —, vertical distribution, 176
Climatograph, 344
Climatology, 3
—, agricultural, 189
—, solar, 14–25
—, synoptic, 3
—, technical, 335–381
—, ultraviolet radiation, 35–37
Clothing,
—, clo unit, 162
—, insulation, 163
—, permeability, 163
—, radiative response, 44–45
Cloudiness, urban, 326
Cloud modification, 263, 264
Cold protection of products, 337
Colour sensation, 12
Comfort zone, 117, 119, 154
— diagrams, 343–345
Condensation nuclei, 56, 67, 68–70, 73, 87, 90, 93
Cooling, comfort, 346–349
— degree hours, 249
— power, 119, 124–127, 141–143, 148, 150, 151, 168, 169
Corn climate, 238
Corrosion, 374
Cotton, 241
Crop,
— yield model, 278–281
— moisture index, 266
— productivity, 282

Degree day, growing, 204

— — heating, 345
Design temperature, summer, 349
Dew, 226
— in forests, 255
Drought,
— indices, 265
— modification, 265
— probability, 267
Dust, 66
— counts, 68
— damage, 80
— fall, 377
—, Sahara, 67

Effective temperature, *see* Temperature, effective
Electromagnetic spectrum, 7
Emission, noxious, 81
Energy flux
see also Radiation flux, 29
Enthalpy, 341
Environment, 164
Equivalent temperature, *see* Temperature equivalent
Erythema, 32, 33
— threshold, 34
Evaporation, skin, 110
—, cooling towers, 375
Evapotranspiration, 217
—, effect on wheat, 237
—, estimation, 219–224
—, potential, 221, 224
—, wind effect on, 227

Fog deposit, 368
Forest climate, 251–256
— —, mathematical models, 256
Frigorigraph, 121
Frigorimeter, 124
Frostbite, 130
— heaving, 357
— penetration, 358
— weather modification, 261–263
Frozen ground, depth, 357

Glaze, 369
Growing degree days, *see* Degree days, growing
Growing season, 204

Hail suppression, 264
Hay fever plants, 71
Haze, 144
Health resort climate, 4, 164, 165
Heat balance, buildings, 352
— —, human, 108
— —, urban, 302
— budget, human, 111
— cramps, 116
— exchange, 192
— island, see Urban heat island,
— modification, 260, 261
— production, body, 105, 106
— stroke, 116
— tolerance, 105, 106
— transfer through walls, 356
— transport, 107
Heating degree days, 345
— season, 347
Hemoglobin, 63
Humidity, 119, 121, 134, 135, 144, 145, 168, 171–173, 174
see also Vapor pressure
—, air,
—, effect on cotton, 242
— —, on farm animals, 259
— —, on plants, 227
— —, on wood, 362
— —, in forests, 254
—, relative, threshold values, 340
—, urban, 325
Hygiene, 5
—, air, 49, 66, 71
—, indoors, 126
Hypoxia, 59, 60, 62

Illumination, 10, 14–16, 27
— at forest edge, 20
Indoor climate, 342–349
Infrared, *see* Radiation, infrared,
Insects, wind transport of, 228
Inversion, 81, 380
Ions,
—, atmospheric, 78

Katathermometer, 121, 124

Leaf area, 194
— arrangement, 194
— temperature, 210–213
Light, 6, 7, 11, 13, 27, 30
see also Radiation, visible; Illumination
— distribution within canopies, 194–196
— intensity for plants, 198
— — for alfalfa, 250
— — for apple, 251
— — for cotton, 242
— — for rice, 233
— — for soybean, 244, 245
— — for tomato, 248
— — for wheat, 237, 238
— interception by canopies, 196
— physiological effects, 47, 48
— psychological effects, 48, 49
— reaction time, 28
— scattering, 54
— stimuli, 29
Lysimetry, 224

Measuring devices for cooling, 122
Medicine, 4, 5

—, preventive, 4
—, sports, 64
Meteorological hazards for plants, 260–270
Meteorotropic reaction, 2
Microbes, urban areas, 320
Moisture in building materials, 360
— index, crop, 266
— modification, 263
— need of apple, 251
— — of corn, 240
— — of cotton, 242
— — of soybean, 244
— — of tomato, 247, 248
— — of wheat, 236, 237

Nitrogen, atmospheric, 57
Nuclei, *see* Condensation nuclei

Oxygen, arterial saturation, 62, 63
—, atmospheric, 57
—, height dependence, 58, 59
Ozone, low level, 57, 92, 93, 94
—, stratospheric, 31, 38
—, urban, 319

Particles, size of, 66
Pathology, climatic, 5
Performance, 158
Permafrost, 359, 360
Photoperiodism, 201–203
— for alfalfa, 249
— for corn, 238, 239
— for cotton, 241
— for soybean, 243
— for tomato, 245
— for wheat, 234
Photosynthesis, effect of CO_2, 199
—, effect of light intensity, 199
— and water need, 201
Plant disease,
— — forecasting, 272
— —, weather effect, 270
— —, wind transport, 228
— growth, water requirements, 225
— —, irrigation requirements, 226
Physiology, climatic, 5
—, recreation, 165
—, respiration, 64–66
—, sensory, 13, 131
Pollen, 71–72
—, wind transport, 228
Pollution, atmospheric, 66, 377
— in cities, 317–321
— dispersal, 378
— episodes, 381
— mortality, 81, 82
— and maximal ground concentration, 378
— and urban radiation attenuation, 322, 323
Precipitation, 214
—, forest, 254
—, urban, 327, 328
Psychology, 5, 131
—, recreation, 165
Psychometric diagram, 155

Radiation, atmospheric, 29, 40
— —, cloudiness dependence, 41
— balance, 42–44, 141, 191
— —, urban, 302, 321
— bands, 193
—, circumglobal, 21, 23–26
— distribution in forests, 252, 253
— effect,
— — on buildings, 352
— — on animals, 258, 259
— energy, 9
— — on vertical surface, 322
— exchange, 91, 109
—, extraterrestrial solar, 9, 193
— fluxes, 29, 30, 44, 49
—, global, 10, 15, 17–19, 21, 45
—, ground reflection, 23, 26, 27, 29
—, heat, 140
—, infrared, 7, 9, 10, 30, 39, 40
— —, urban, 303, 304
—, latitudinal, 10
— modification, 260, 261
—, net, 191, 196
— penetration through windows, 355
— protection, 46
—, sky, 14, 16, 17, 37
—, solar, 7, 9, 14, 29
— spectrum, 9, 190
— terms, 8
—, thermal, 191
—, total, South Pole, 152
—, ultraviolet, 7, 9, 10, 30, 33–38
— —, dosage, 37
— —, carcinogenesis, 38
— units, 8
—, vertical walls, 353
—, visible, 9
Radioactive, atmospheric, 78, 96, 98, 99
— fall out, 100
Radon, 97, 98, 100, 101
Ragweed, 72
Rain,
—, frequency of heavy, 366
—, maximum amounts,
— — in 24 h, 364
— —, various intervals, 365
— windrose, 363
Remote sensing in agricultural climatology, 276–278
Respiration, 65
— rate, plants, 200
Rice,
— acreage, 231
— climate, 230–233

Roughness, 351
— length and plants, 229
— —, urban, 314

Salt particles, 74, 75
Sensation, 13, 14
—, climatic, 127
—, comfort, 138, 147
— scales, 128
—, thermal, 101
Skin, 30
— cancer, 38
— energy loss, 39
— humidity, 112
— layers, 31
— pigmentation, 34
—, reflection spectrum, 31, 32
— temperature, 112, 113, 120
Smog, 86, 381
Smoke, 66, 89
— plumes, 379, 380
Snow in forests, 255
—, heat conductivity of, 367
— load, 368
Soil moisture, 214–216
— temperature, 206–210
—, thermal capacity, 208
—, thermal conductivity, 208
—, thermal diffusivity, 208
—, windbreak effects, 260
Solar thermal unit, 205
Soybean climate, 243
Spectrum, *see* Radiation spectrum
Stability categories, 379
Stress, 105
— heat, 142
— on cargo, 372, 373
— protection,
— —, cold, 161
— —, heat, 161
Strontium-90, 100
Sulphate, 75
Sulphur dioxide, 82–84, 86–89
—, urban, 318
Sulphur fall-out, 76
Sultriness, 117, 135–137, 139–147
Sunshine duration, 22, 139, 140, 145
— protection, 355
—, urban, 321, 323
Sweating, 114

Tanning, 34
Temperature, 6, 101
—, attic, 354
—, body, 103–105
—, cardinal, 203
—, core, 107
—, summer design, 349
—, diurnal variation in rooms, 357
— — —, in urban areas, 305, 306
—, effective, 123, 133, 135, 155
— equivalent, 90, 136–138, 146–148, 156–159, 167
— for farm animals, 257, 258
— in forests, 253, 254
—, forehead, 118
— frequency distribution, 338
—, indoor–outdoor relation, 340
—, rectal, 113, 155
— response in alfalfa, 249
— — in corn, 239, 240
— — in cotton, 241, 242
— — in rice, 231, 232
— — in soybean, 243, 244
— — in tomato, 245–247
— — in wheat, 234–236
— sensation, 117
— in urban areas, 305, 306
—, nocturnal urban, 311
— urban–rural differences, 312
—, relation with vapor pressure, 342
—, wet bulb, 375
—, wet bulb globe (WBGT), 154
—, windbreak effect, 269
Therapy, climatic, 5
Thermoperiodicity, 206
Thermoregulation, 102
Tomato climate, 245
—, thermal response, 247
Topoclimatology, 275
Trace gases, atmospheric, 79
Transmission, atmospheric, 52, 53
Turbidity, atmospheric, 49–51
Turbulence, 201

Ultraviolet, *see* Radiation, ultraviolet
Urban circulation, 317
— climate alteration, 330
— — model, 300, 301
— cloudiness, 326
— freezing temperatures, 308, 309
— heat island, 302, 304
— humidity, 325
— plume, 315
— pollution, 317–321
— precipitation, 327
— radiation attenuation, 322, 323
— roughness, 314
— temperature, 307
— — lapse rates, 310
— —, nocturnal, 311
— — trends, 307
— visibility, 324
— wind field, 313–317

Vapor, diffusion, 195
— pressure, 134, 144
Vernalization, 204
Visibility, urban, 324

Vision,
—, daylight, 7
—, twilight, 7
Vitamin D, 34

Walls, moisture penetration, 361, 362
Waste gases, 378
Water vapor, atmospheric, 57, 150, 168
Wheat climate, 233–238
Weather effects and construction, 369, 370
— — and manufacturing, 335
— — and plant pathogens, 271
— forecasting (frost), 263
— modification, 260–270
— sensitivity, 1
Wind chill, 131
— damage to plants, 228
— effect on evapotranspiration, 227
— — on cotton, 242
— — on rice, 233
— — on forests, 252, 256
— gusts, 351
— speed,
— — effects on pollutants, 320
— — extremes, 351
— — profile (vertical), 92, 228, 351
— —, urban, 313
— transport of plant disease, 228
Windbreaks and crop yield, 270
— — and soil moisture, 269
— — and temperature, 269
— — and wind speed, 268